CHAMAEDOREA PALMS

The Species and Their Cultivation

Text and Photographs
by
DONALD R. HODEL

University of California

THE INTERNATIONAL PALM SOCIETY

Allen Press, Lawrence, Kansas

ISBN: 0-935868-56-9
Library of Congress No. 92-070479

Printed by Allen Press, Inc., Lawrence, KS 66044
Dust Jacket Design: Marianne A. and Donald R. Hodel
Book Design: Marianne A. and Donald R. Hodel

Legend for Front of Dust Jacket

Clockwise from upper left, *Chamaedorea tuerckheimii,* garden of P. Sullivan, Ventura, California; crown of *Chamaedorea tenerrima,* Alta Verapaz, Guatemala; *Chamaedorea sullivaniorum,* El Valle, Panama; *Chamaedorea amabilis,* El Valle, Panama; *Chamaedorea microspadix* in full mature fruit, Waimea Falls Park & Arboretum, Honolulu, Hawaii.

Legend for Back of Dust Jacket

Clockwise from upper left, *Chamaedorea pinnatifrons,* Fairchild Tropical Garden, Miami, Florida; *Chamaedorea metallica* on limestone rocks, Veracruz, Mexico; *Chamaedorea quezalteca,* University of San Carlos Botanical Garden, Guatemala; *Chamaedorea ernesti-augustii,* Toledo, Belize.

Legend for End Papers (front and back)

Plantation of *Chamaedorea tepejilote* cultivated for *pacaya,* in ever-present fog and mist, Coban, Alta Verapaz, Guatemala, November, 1989.

Affectionately dedicated to my wife Marianne; my children Robert and Christina; and the people of Latin America.

CONTENTS

PREFACE

The impetus for this book came from palm collectors in southern California and one individual in particular, Richard W. Palmer of Whittier. Dick, as he is known by most fellow palm collectors, has been a long-time member of the International Palm Society and maintained a large collection of palms since the 1960s. He has a fondness for *Chamaedorea* and in the 1970s purchased a large portion of the collection of these palms that the late David Barry, Jr. had assembled. Included were both staminate and pistillate plants of most of the commonly cultivated species of the time as well as many rarer ones.

In 1983 Dick approached me about the identity of a handsome, multi-stemmed *Chamaedorea* that he had grown since 1976. Alan Bredeson of Lemon Grove, California, had collected the seeds in El Salvador. Dick and I consulted the rather patchy accounts of *Chamaedorea* in the various floristic treatments but we could not conclusively arrive at a name for this handsome plant with any degree of confidence. In 1984, still in pursuit of a name for his palm, Dick asked if I could obtain a copy of Max Burret's (1933a) monograph of *Chamaedorea,* published in a relatively little-known German journal. I was successful in obtaining a copy for Dick that, except for the Latin descriptions of a few new species, was entirely in German. Dick translated the entire monograph into English and we thought we would be able to sort out the chamaedoreas. However, we found that the monograph was not much help since the only species with descriptions were the ones Burret named in the article and these were fairly obscure taxa. Also, there was no key to the species or illustrations. It became obvious that we needed an up-to-date account of the genus in cultivation with descriptions, keys, and illustrations.

About a year later, Pauleen Sullivan of Ventura, California, another long-time member of the International Palm Society and fancier of chamaedoreas, expressed her interest in the genus. She also related her frustration at being unable to identify her chamaedoreas because of the lack of a reference work. I volunteered to assemble a book on *Chamaedorea* in cultivation and to include all the elements necessary to make it a tool that could be used by all interested in this remarkable genus, whether research scientist or lay person and hobbyist.

I contacted Bill Gunther of Del Mar, California, another long-time supporter of the International Palm Society, and he lent his encouragement and support to the project. Bill and Pauleen gathered funds and later the International Palm Society at its 1986 Biennial Meeting in San Diego awarded a gift to the University of California to support my work on the project. Never did I envision the project being as comprehensive as that presented here. But over the nearly four years during which I worked on the chamaedoreas, I was fortunate to assemble an amazing wealth of information, much of it heretofore unavailable or unknown. The treatment far surpasses what I had originally intended but I feel the information will be useful to many and should be available.

The book represents those four years of work including travels to view chamaedoreas in gardens and/or in the wild in Hawaii, California, Florida, Panama, Costa Rica, Guatemala, Belize, Peru, and Mexico and to study herbarium specimens at herbaria throughout the United States and in Europe and Latin America. The results are reflective of a truly cooperative effort and the many individuals and institutions that opened their gardens, collections, or minds to me are recognized in the acknowledgments. The book is for the enjoyment and pleasure of all who feel a special affinity for palms, and especially that most remarkable genus *Chamaedorea.* I hope you all enjoy reading it as much as I have enjoyed writing it.

How to Use This Book

If one is interested only in identifying a chamaedorea, there are several possible approaches. If staminate flowers are available, then go directly to the key to the subgenera in Chapter 5. Once the subgenus is determined, use the key within that subgenus to determine the species. If the plant is cultivated and without flowers, go to the vegetative key to the cultivated species in Appendix E. If one knows the name of a specimen and simply desires to verify it or learn more about it, go to the index of names in Appendix F to find the page number on which that species is treated. If simply browsing, read on, and enjoy!

DONALD R. HODEL

ACKNOWLEDGMENTS

Several individuals and institutions have contributed financially to my research. They are Richard Palmer, Pauleen Sullivan, Bill Gunther, Inge Hoffmann, Carol and Lou Hooper, Audrey and Philip Keeler, Lynn Muir, Chris Hill, Loran Whitelock, the Southern California Chapter of the I.P.S., the International Palm Society and its Seed Bank, and the World Wildlife Fund.

I have had the pleasure of working with several botanists who have offered helpful suggestions and insight. They are Juan Jose Castillo Mont, Gregory de Nevers, Robin Chazdon, Birgitte Bergmann, John Dransfield, Juan Ibarra-Manriquez, Gloria Galeano, Henrik Balslev, Francis Kahn, and Andrew Henderson. Natalie Uhl and Michael Grayum were especially helpful and acted as sounding boards for my ideas about *Chamaedorea.*

Many Palm Society members have cooperated by opening their gardens and minds to me. In California these include Jim Degen, Jim Benzie, Richard Palmer, Pauleen Sullivan, Paul Mahalik, John Tallman, Rae Anderson, Lois Rossten, Lou Hooper, Frank Ketchum, Loran Whitelock, Jim Wright, Allan Bredeson, Ed Moore, Barry Osborne, Dan Sekella, and Richard Douglas. In Florida they are Nancy Hammond, Bill Shannon, Russ Hubbs, DeArmand Hull, Louise Futch, Teddie Buhler, Jeff Shimonski, Paul Drummond, Bob Verna, Carol Graff, and Frank Hausman. In Texas Horace Hobbs introduced me to members in the Houston area and offered his valuable insight on chamaedoreas. I corresponded with several overseas members including Ray Lawler, Richard Budd, Paul Anderson, Norma and Ian Edwards, Robert Tucker, Maria Walford-Huggins, Arden Dearden, and Stan Walkley in Australia and Dick Endt in New Zealand. In Costa Rica the late Robert Wilson made me feel right at home in his Las Cruces Tropical Botanical Garden (now Jardin Botanico Robert y Catherine Wilson).

Botanical gardens and other institutions and their personnel have cooperated freely and provided much useful information. In California these are Ron Harris, Mary Schaffler, Myron Kimnach, and Debra and James Folsom of the Huntington Botanical Gardens; Joan DeFato and John Provine from the Los Angeles State & County Arboretum; John Copeland of the Virginia Robinson Gardens; Chuck Coburn of the San Diego Zoo; Steve Timbrook and Chris Enright of Lotusland; and Sandra Eberhard of the Ag-Sci Library at the University of California, Riverside. In Hawaii they include Paul Weissich and Heidi Bornhorst formerly of the Honolulu Botanic Gardens; Ray Baker from the Lyon Arboretum; and Bill Garnett of the Waimea Arboretum. Others include Chuck Hubbuch of Fairchild Tropical Garden in Florida; Dennis Johnson of the World Wildlife Fund; in Costa Rica the Organization for Tropical Studies at La Selva and Luis Diego Gomez at the Jardin Botanico Robert y Catherine Wilson; and in Panama the Smithsonian Tropical Research Institute.

Several herbaria have cooperated by making loans available and/or providing other information. These include the Bailey Hortorium (BH), Missouri Botanical Garden (MO), Rancho Santa Ana Botanical Garden (RSA), Smithsonian Institution (US), National Arboretum (NA), Munich (M), Copenhagen (C), Brussels (BR), Kew (K), British Museum (BM), Stockholm (S), Gray Herbarium (GH), National University of Mexico (MEXU), University of Michigan (MICH), Field Museum (F), University of Goettingen (GOET), Florence (FI), National University of Panama (PMA), Madrid (MA), National Museum of Costa Rica (CR), Leiden (L), Utrecht (U), California Academy of Sciences (CAS), University of Wisconsin (WIS), Huntington Botanical Gardens (HNT), Natural History Museum, Paris (P), University of Vermont (VT), Fairchild Tropical Garden (FTG), University of South Florida (USF), Smithsonian Tropical Research Institute (SCZ), University of San Carlos, Guatemala (AGUAT), University of California, Riverside (UCR), and the National University of Honduras (UNAH).

Nurserymen and growers have freely given information about culture and production of chamaedoreas. These include Tamo Kitagawa of Kohala Nursery, Hawaii; Luciano Guerra of Guerras, Inc. and Hugo Dominguez and Gilbert Longoria of Longoria Seeds, Inc., all of Texas; Louis Thomas of Teakettle Farms, Belize; Jerry Hunter of Rancho Soledad Nurseries and Jack Ingwersen of Ingwersen Nurseries, both of California; and Robert Bernecker, Jr. of Bernecker's Nursery, Florida.

Several people accepted the unenviable task of reviewing the manuscript. They include Michael Grayum, Natalie Uhl, Barry Osborne, Tim Broschat, Richard Palmer, Jim Benzie, Jim Degen, John Dransfield, Pauleen Sullivan, and Marianne Hodel.

My immediate supervisors, formerly Nicelma King and currently George Rendell, have been supportive of my work as has Allyn Smith, regional director of Cooperative Extension at the University of California, Riverside.

I would be remiss if I did not mention my wife Marianne and children Robert and Christina who accompanied and supported me for many long and arduous hours often under difficult conditions in the field in Central America. They tolerated with grace and understanding my extended absences in the field, preoccupations, and seemingly endless hours at home writing and putting together this book. My mother Hazel Hodel accompanied us on trips to Belize, Costa Rica, and Panama and added immensely to our excursions. Finally, I recognize my father Robert Hodel who was an enjoyable and tireless companion on several memorable field trips in Mexico and Central America.

To all I extend my sincere gratitude and appreciation for the role each played in making this book what it is.

Donald R. Hodel
Los Angeles
December 15, 1991

PATRONS

The following people and institutions have generously contributed toward the publication of this book.

Elwood Bear
Mr. & Mrs. Byron Besse
Bud Blachly
Keith Boyer
Mrs. Theodore C. Buhler
Miriam R. Canon
Gertrude W. Cole
J. Burns Creighton, Jr.
Arden Dearden
Severn C. Doughty
Mr. & Mrs. Ian Edwards
John C. Elliott
Joe Ellis
Jay D. Fair
Mr. & Mrs. Alan Ferguson
Leonard Harrison
Hawaii Island (U.S.A.) Chapter of I.P.S.
Mr. & Mrs. Gunther Herman
Mr. & Mrs. Donald R. Hodel
Mr. & Mrs. Matthew A. Hodel
Mr. & Mrs. Robert J. Hodel
Walter H. Hodge
Mr. & Mrs. Louis Hooper
Houston, Texas (U.S.A.) Chapter of I.P.S.
Mr. & Mrs. Masahi Itano
Mr. & Mrs. Robert Judd, Jr.
Mr. & Mrs. Philip Keeler
Louisiana (U.S.A.) Chapter of I.P.S.
Geritt R. Ludwig
Margaretta T. McGehee
Martha M. McGovern
Jim Mintken
Mr. & Mrs. Lynn Muir
Masayuki Nako
New South Wales, Australia Chapter of I.P.S.
Palm Beach, Florida (U.S.A.) Chapter of I.P.S.
Palm & Cycad Society of MacKay, Australia
Palm & Cycad of New Zealand, I.P.S.
Palm & Cycad Society of Western Australia
Richard W. Palmer
Richard H. Phillips
John B. Reid
Mr. & Mrs. Bruce Robinson
James G. Sharpe
Mr. Soppeland
Southern California (U.S.A.) Chapter of I.P.S.
Maxwell L. Stewart
Pauleen Sullivan
John D. Tallman
Mr. & Mrs. Paul Weissich
Lloyd A. Wright
Mark Wuschke

CHAPTER 1

INTRODUCTION

Few other groups of palms can rival *Chamaedorea* for variety in foliage, size, and general habit. Highly ornamental with their neat, green, bamboo-like stems, they are among the most popular of palms and are used extensively for moist, shady areas in tropical and subtropical gardens where they fulfill a wide array of landscape uses. Relatively tough and durable and well suited to low light, they also make excellent house plants. Several, especially the parlor palm and the bamboo palm, are grown in vast quantities in North America, Europe, Australia, and the Orient for this use. In fact, the parlor palm, *Chamaedorea elegans,* is the most widely grown indoor palm.

That chamaedoreas are so widely grown in so many diverse areas attests to their popularity and relative ease of culture. They are especially popular in California and Florida. Just a small sampling of other areas where they have been noted as growing outdoors includes Hawaii (Baker 1980), Texas (Smith 1967), Louisiana (Doughty 1988), Bahamas (Pancoast 1985), France (Marnier-Lapostolle 1961), Papua New Guinea (Essig 1973), Spain (Smith 1975), Italy (Roster 1913, Ruffo 1920), India (Blatter 1926, Basu 1978), Indonesia (Sastrapradga and Davis 1983), and Australia (McGregor 1987). Of course, they are grown indoors in these areas and just about everywhere else, too.

Popularity

Chamaedoreas have several attributes that give them their popularity. Their tremendous diversity is exemplified in the vast array of species with large or small and pinnate or bifid leaves. Such great diversity is also reflected in stems that may be solitary or clustered, slender or somewhat stout, and relatively long or short, or sometimes even lacking. Plants can be dwarf, and flowering and fruiting when no more than 30 cm tall, or moderately large, with stems to 15 meters long. In fact, just about every conceivable combination of leaf and stem is represented somewhere in the genus.

Their great ornamental value is due to the diversity mentioned above and several other factors. Being relatively small palms, they are much more manageable and better suited to average residential landscapes. With few exceptions, chamaedoreas are easy to grow and are not particularly susceptible to pests or diseases. They are amazingly cold-hardy; in fact, most species will tolerate 0°C (32°F) without sustaining damage and a few species will not show damage as low as −5°C (23°F). Their tolerance of cool temperatures is one of the main factors in the popularity of chamaedoreas in southern California, especially in the coastal area from Santa Barbara to San Diego. They perform well there and, in fact, there are more species of *Chamaedorea* grown in southern California than in any other region.

Tropical areas with a more favorable climate like Hawaii, Florida, or Australia can grow various species of large genera such as *Ptychosperma, Pinnanga, Geonoma,* and others, yet these tropical palms are difficult to grow in southern California. Instead, collectors there rely on and exploit chamaedoreas, using the small, shade-loving palms for tropical accent. With its diversity, tolerance of cool temperatures, and relative ease of growth, *Chamaedorea* has not failed them.

Layout of the Book

The book is presented in much the same manner as most other reference works about groups of plants. The objective is to present the maximum amount of information about *Chamaedorea* in a form that is readily accessible and readable by all, whatever their level of interest or expertise.

A taxonomic or descriptive account and discussion of all the known species of *Chamaedorea* compose the main body of the book. Although the emphasis is on the cultivated species, those not considered cultivated are included also since they may soon be in collections and gardens.

Chapters addressing the history and taxonomy, distribution and ecology, and general morphology of *Chamaedorea* follow the introduction and precede the main body. The main body includes a key to the subgenera and the treatment of the species within each subgenus.

Within each subgenus, there is a key to the species that relies on both vegetative and floral characters and the descriptive account of each species. Each species is presented in alphabetical order within its appropriate subgenus. Chapters on hybrids, economic uses, conservation, and culture follow the main body. Completing the book are a glossary, the literature cited, and several appendices, including additional author citations, selected specimens examined, a list of the author's collections, a geographical listing of species by country, a vegetative key to the cultivated species, and an alphabetized index of names for easy accessing of the treatment of each species.

Format for Each Species

Each species is presented by the accepted, valid scientific name. Following each name is the authority or, in other words, the name of the person who formally described or named the species or who gave it its most recent, accepted nomenclatural status and the journal, page, and date of such action. The type specimen is given and where it was collected, along with the meaning of the specific epithet or species name. If I have examined a type, an exclamation mark (!) follows the acronym of the herbarium where the specimen is located.

Following the accepted, valid name but preceding the actual description there are, in many cases, additional scientific names. These scientific names, whether valid or not, are considered to be synonyms of the accepted name. Synonyms that are valid names are listed first and arranged chronologically and following each, as with the accepted names, are the authority, journal, page, and date published and, if applicable, the listing of the type specimen and explanation of the specific epithet. Horticultural names and names of no botanical standing are listed after the valid synonyms.

Ordinarily, additional citations of authors who discussed, listed, illustrated, or otherwise treated the species would follow the accepted names and synonyms. To streamline the treatment of each species, these additional citations are listed in Appendix A. Since names have often been improperly applied in the botanical and horticultural literature over the years, I attempted to ascertain and verify the correct application of a name in these additional citations by examining any description or illustration. If I felt a name was improperly used, this is so noted in Appendix A. Common or vernacular names, if any, are listed after the accepted name(s) and synonym(s).

Next is the actual description which is broken down into separate sections including habit, stem, leaves, inflorescences (both staminate and pistillate), flowers (both staminate and pistillate), and fruits. Each section of the description has the key organs in bold type to facilitate accessing the needed information. Following is a short note on the distribution of the species including the country in uppercase and, where appropriate, the department, state, or province in lowercase. Type of habitat and altitudinal ranges, where known, conclude the section on distribution. A discussion including pertinent or interesting information and, if cultivated, any important cultural tips, conclude the treatment of each species.

Definition of Cultivated

As used in this book, "cultivated" means that the species is currently well established somewhere in a collection or garden, usually in California, Hawaii, or Florida. "Well established" means that the plant is growing outdoors and has flowered. Unfortunately, I was unable to visit gardens in Australia, the Orient, and Europe to verify species cultivated there so I must rely on letters and photographs submitted by cooperating palm collectors and current published reports and illustrations. Most species cultivated in areas I was unable to visit are also grown in the United States. Europe, in particular, has had a long history of horticulture with respect to tropical plants, *Chamaedorea* included. In the 1800s and early 1900s, stovehouses in Germany, Austria, France, Belgium, and England contained many species of *Chamaedorea.* Some species were even grown outdoors along the Mediterranean coast of France and in Italy. Several are still commonly cultivated but many have been lost over the years, have disappeared from the annals of horticulture, and are presumed no longer cultivated in Europe or elsewhere.

A species occurring only as a single plant in one collection is considered not cultivated. Species that occur in cultivation only in our research collection in Los Angeles are considered not cultivated.

Nature of the Keys

The keys rely on vegetative characters wherever possible so that material, even in the sterile state, can be identified properly. However, in some cases vegetative characters are insufficient for proper identification and one must rely on comparisons of inflorescences, flowers, or fruits.

The Descriptions

The descriptions are from several sources including the original description and other verified accounts in the literature, the type and other herbarium specimens, and, in many cases, living plants from the wild or cultivation. Descriptions attempt to recognize the variability within a species, not only in the wild but between wild and cultivated specimens. Wherever possible, a range of dimensions, numbers, and shapes is given for various plant organs since they often vary.

One using the descriptions must be aware that occasionally individual plants will have organs whose size, shape, or number fall outside of the given ranges.

What Is a Species?

What seems to give people the most trouble in conceptualizing a species is recognizing and accepting natural variation within a group of closely related individuals. Often we tend to focus on one particular aspect of an individual plant and associate that character with the name for the species as we have learned it. As is often the case, though, we have learned an incorrect name. When we encounter another plant in which the particular character differs even slightly, we are immediately inclined to view the second individual as a different species. If we had the luxury of observing this particular character in several, as many as 10, 20, 50 or 100, individuals, we would probably find a range of variation within that particular character that tended to tie the extremes together. Picture yourself in a group of people and notice the individuals around you. They are all fairly different: short, tall, thin, wide, light, dark, an incredible amount of variation yet all belong to the same species, *Homo sapiens.* It is appropriate to recognize variation and consider it when we develop our concept of a species. Also, we should recognize that some people have a broader concept of a species than others.

Although we like to adhere to the most recent taxonomic and nomenclatural status of a species, one is certainly not bound to subscribe to a particular treatment and is free, rather, to follow as one's experience dictates. For the purposes of this book, I have attempted to strike a middle road that I hope will be acceptable to those with an interest in *Chamaedorea.* However, some taxonomists may find my species concepts are too broad, while others may say they are not broad enough. My treatment or any other, for that matter, is not the end but a beginning upon which something else can be built in the future. I have attempted to bring together all the currently available information on *Chamaedorea* so that its species can be recognized and understood. Future work will lead to better treatments of this group.

Nomenclature

One of the aspects of chamaedoreas and of all plants, for that matter, that is most perplexing to collectors is nomenclature or the scientific names of plants. How plants are named and why their names change occasionally seem imposing and frequently are a source of frustration resulting in resistance to use of the correct name. A short discussion of the nature of plant names may serve to alleviate what some view as the most negative aspect of botany and horticulture.

Rules agreed upon by international groups of botanists and horticulturists and more or less observed throughout the world govern the application of scientific names of plants. The botanists and horticulturists meet periodically and publish the latest revisions of the rules in the *International Code of Botanical Nomenclature* and the *International Code of Nomenclature for Cultivated Plants.*

Binomial System

Biologists, including botanists, use the binomial (literally two-name) system of naming living things first developed and implemented by the Swedish botanist, Karolus Linnaeus, in 1753. The binomial system is the basis for the nomenclatural codes mentioned above. The basic unit of this system is the **species** (plural, **species**). A group of closely related species composes a **genus** (plural, **genera**), the category next in rank. In the binomial system names of species consist of two words, the first indicating the genus, e.g. *Chamaedorea,* and the second the **specific epithet** or species within that genus, e.g. *fragrans.* The two names together form the binomial *Chamaedorea fragrans* that is understood and accepted by botanists and horticulturists around the world, whatever their native language. Genus and species names are written in Latin but are often derived from Greek words. The first letter of the genus is in uppercase and both genus and species are italicized or underlined in text. In the interest of space, the genus name is often abbreviated by its first letter but only when it is very clear which genus is being discussed.

The species name can be just about any Latinized name as long as it conforms to the rather elaborate rules of nomenclature. Imaginative botanists have come up with some very novel names. It is appropriate to devise names that signify something distinctive about the plant to aid in its identification, such as *serpens* for plants with prostrate or crawling stems or *nana* for small or dwarf plants. The name of the botanist (or its abbreviation) who named and described the species follows the binomial in the formal taxonomic presentation. The name is referred to as the authority and is helpful in tracing the species name through botanical history. Thus, *Chamaedorea glaucifolia* H. A. Wendl. shows that Hermann Wendland described and named this species.

Often the person who names and describes a species wants to honor the person who collected or studied that species. In such cases the suffix "ii" may be added to the end of the name. Thus, Max Burret described and named *C. lehmannii* Burret to honor the collector of the type, F. Lehmann.

If a plant is named after a person who did not collect the type or work on it and the author wished only to honor that person, than the suffix "iana" may be added to the name. Thus, Wendland named and described *C. martiana* H. A. Wendl. simply to honor the botanist Carl Martius even though Martius had nothing to do with his namesake. The rules are not hard and fast, however, and sometimes the suffix "e" is added to a name whether that person was the collector of the type or was simply being honored. Also, the rules are sometimes stretched for phonetic considerations.

Readers will note that some species names have more than one authority after them. Names with two authorities commonly occur when a species was first described in one genus and after later evaluation was transferred to another. The name of the original authority is enclosed in parentheses and placed before that of the person making the change. Thus, *Chamaedorea fragrans* (Ruiz & Pav.) Mart. shows that the Spanish botanists Hipolito Ruiz and Jose Pavon first described and named the species in one genus, in this case *Nunnezharia,* but later Martius transferred it to *Chamaedorea.*

The type or first species for *Chamaedorea* was actually *Borassus pinnatifrons,* named by N. Jacquin in 1797. However, Linnaeus had established *Borassus* in 1753 for a group of Old World fan palms. In 1859 the Danish botanist A. Oersted recognized this inconsistency and transferred *B. pinnatifrons* to *Chamaedorea.* Thus, we have *Chamaedorea pinnatifrons* (Jacq.) Oerst., the specific epithet retained since it was not already used for another species in *Chamaedorea.*

Occasionally, the word "ex" may link two authorities after a species name. If one botanist proposed a species name and another published it, "ex" is used to signify this, the latter authority officially describing it. Thus, *C. amabilis* H. A. Wendl. ex Dammer means that Wendland suggested the name but Udo Dammer formally published it with a description.

Sometimes the authority will name and describe the species in someone else's work. In these cases the word "in" links the two authorities. For example, *C. sartorii* Liebm. in Mart. means that Liebmann named and described the species but did so in Martius' work. Names of authorities linked by "in" can be shortened to the first authority while names linked by "ex" can be shortened to the latter authority. Thus, *C. sartorii* Liebm. and *C. amabilis* Dammer are acceptable.

Occasionally, the term "Hort." precedes the authority for a species indicating that horticulturists used the plant name or a nursery firm published the name in its sales catalogue. The authority following the term "Hort." is the earliest reference to the name in the literature. Although these horticultural names are usually Latinized and appear as good scientific names, they have not been published according to the rules of the *International Code of Botanical Nomenclature* and thus are names of no botanical standing. Generally, to be considered as a valid botanical name, several criteria must be met. The name must be Latinized, a type specimen designated, the article published in a journal or book commonly available or circulated in the botanical community and, after 1935, a description or diagnosis in Latin provided. O. F. Cook published several generic names for *Chamaedorea* but since they appear without Latin descriptions and postdate 1935, they have no botanical standing. Although they are invalid, I have included Cook's names as well as horticultural names in the formal synonymy for appropriate species because they are likely to be encountered in the literature. I designated Cook's names as having no botanical standing but horticultural names are generally accepted as such so I have not done so for these.

Why Do Names Change?

To many palm fanciers, one of the most frustrating aspects of scientific plant names is that they can change. The reasons for name changes vary. Occasionally, a species is cultivated under an erroneous name, a normal occurence in a genus as large and complex as *Chamaedorea.* Once a plant appears in cultivation and receives a name, it is a difficult and sometimes lengthy process to have the correct name accepted even when the original name can be shown to be incorrect. For example, in many southern California gardens and collections there is a single-stemmed, pinnate-leaved *Chamaedorea* which goes by the name of *C. concolor.* In habit, this palm is very similar to the common *C. oblongata.* It has leathery pinnae, staminate flowers with the petals spreading open, and black fruits. It is no doubt *C. oblongata.* Unfortunately, it will continue to be called *C. concolor* despite information to the contrary. The true *C. concolor* (= *C. pinnatifrons*) has thin pinnae, staminate flowers with the petals not spreading open, and red-orange fruits and is uncommon in cultivation.

More frequently, names change because our concepts of species or their relationships change or names do not conform to the rules and regulations of the code of nomenclature as illustrated below.

Concept of Priority in Names

Sometimes a species has been given more than one name, a situation not allowed according to the rules of the codes of nomenclature. The rules and regulations specify that a species can have only one name and that name (epithet) cannot be used for any other species in the same genus. The earliest validly published name for a species is the one that must be used. In 1853 Wendland described and named *C. martiana,* unaware that it was the same species as *C. cataractarum* that Martius had described and named only four years earlier. Although Wendland's plant was introduced to cultivation in Europe and there became an accepted name, we are required to use Martius' earlier name. Thus, *C. martiana* is rejected and becomes a **synonym** of *C. cataractarum.*

Occasionally, the same name has been given to more than one species in *Chamaedorea.* When this happens, the last described species must be given a new name. Thus, the first species to receive a name has priority to it and no other species in the genus can use the same name.

The reader might ask why *Chamaedorea* (Willdenow 1806) is the accepted name for this group of palms if *Morenia* was published 12 years earlier (Ruiz and Pavon 1794). If the same rule applies to genera that applies to species, the earlier name must be used. However, there are certain instances when, for the sake of clarity, special action of international botanical congresses exempts a later name from the rules and accepts it as the legitimate one. The accepted, later name is called a **conserved name.** Moore (1979) proposed that *Chamaedorea* be conserved over *Morenia* on the grounds that the latter genus contained only a few species restricted to northern South America whose names had not been used to any significant extent outside lists of plants native to the region. On the other hand, *Chamaedorea* contained about 100 species from Mexico to Bolivia with about 35 of them in cultivation. Moore argued successfully that the stability of nomenclature would be best served if *Chamaedorea* were conserved over *Morenia.*

Type Specimens and Determining Names

I am often asked how I decided what is the proper name for a specimen of *Chamaedorea.* Generally, I compared the specimen with descriptions and illustrations in the literature, which can be difficult since many species are poorly treated in or absent from contemporary or recent literature. One must go further and examine the original description of the species in question. However, many of the early original descriptions are very meager and lack sufficient information to diagnose a species adequately. The ultimate test is to examine the type, that specimen designated by the author of a species as the one upon which the species is based. That one must always go back to the type specimen cannot be overemphasized.

There are different classes of types. The holotype is the one specimen or other element (e.g. drawing) that the author used or designated as the nomenclatural type. As long as a holotype exists, it automatically fixes the application of the name. An isotype is any duplicate of a holotype; it is always a specimen. A syntype is any one of two or more specimens that the author cited when no holotype was designated or any one of two or more specimens designated as types. A lectotype is a specimen or other element selected from the original material to serve as the nomenclatural type when the holotype is missing or when the author designated two or more specimens as types or did not designate a holotype. Lectotypes must be chosen from among isotypes or syntypes. If the holotype is missing and it is impossible to select a lectotype because there are no isotypes or syntypes, then a neotype is selected.

A neotype is a specimen or other element selected to serve as the nomenclatural type when all of the material on which the name of the taxon was based is missing.

It is often difficult to identify types, especially of early names. Specimens may simply not have been identified as types and, as is the case with many of the early names in *Chamaedorea,* authors did not designate types in their articles. Adding to the confusion is the fact that many of the early collectors did not number their collections. Thus, for example, Liebmann named and described *C. sartorii* from the Barranca de San Francisco near Mirador, Veracruz, Mexico, without designating a type or numbering his collections. However, by examining itineraries and label data including location and date of collection one is often able to deduce or confirm with a fair degree of accuracy which of Liebmann's specimens of *C. sartorii* qualify as types.

An abitrary numbering of Liebmann's collections along with those of fellow countrymen Oersted further complicates citing of types of these two collectors. McVaugh (1987) reported that most of Liebmann's American collections were combined with Oersted's at Copenhagen into one systematic series in 1909. Copenhagen assigned new serial numbers to the combined collections and prepared new labels for all specimens, including duplicates; no other numbers except these assigned numbers appear on the labels. The numbers are meaningless in themselves, though, since they were assigned after the plants were arranged in systematic order. Also, no distinction was made between Liebmann's collections and those of Oersted. As a consequence, several of Liebmann's holotypes have more than one number each since individual sheets were often assigned different numbers. Thus, the holotype of *C. sartorii* has numbers *10801, 10802,* and *10803.* A few isotypes apparently distributed to other herbaria prior to the arbitrary numberization at Copenhagen are without numbers and further confuse the situation!

Wherever possible, I have examined the type specimens of species of *Chamaedorea* for positive confirmation of my findings. Type specimens of *Chamaedorea* are available in the United States, Europe, and Latin America. Many herbaria lent types or I traveled to examine types that could not be sent. In cases where the type was unavailable or had been destroyed, I was able to examine duplicates or photographs of types. In a few cases where neither of these alternatives was available, I consulted the original description along with other pertinent information and made a sometimes difficult decision as to the proper application of the name.

Observing Living Plants

Observing living plants, especially in the wild, is also very helpful in determining what constitutes a species and its correct name. One gains a different perspective of variation within a genus and species, especially if one's concept was based only on literature accounts, dried herbarium specimens, or even cultivated plants. Material in cultivation may not reflect the full range of variation within a species since it may have been introduced from only one or a few collections.

I not only traveled to various countries to observe *Chamaedorea* in the wild but returned with live material to grow in our research collection in the greenhouses in Los Angeles. By so doing, we were able to obtain flowers, heretofore unknown, for many named and unnamed taxa. Observing, collecting, and studying flowers of cultivated plants from documented, wild localities has added immeasurably to my knowledge of this group.

CHAPTER 2

HISTORY

Chamaedorea had a rather inauspicious beginning near the end of the 18th century when Spanish botanists Ruiz and Pavon (1794) named a palm from South America, *Morenia fragrans,* that would eventually be placed in *Chamaedorea.* A few years later, Ruiz and Pavon (1798) named three more species from South America, *Martinezia lanceolata, M. linearis,* and *Nunnezharia fragrans,* that would also end up in *Chamaedorea.* However, *Chamaedorea* still did not exist when Jacquin (1797) named *Borassus pinnatifrons* from Venezuela, a species that would remain hidden in relative obscurity for over 60 years until Oersted (1859) transferred it to *Chamaedorea.* By 1805, botanists had named five species of palms, all from northern South America, that would eventually be placed in *Chamaedorea,* but the genus had yet to be erected.

Finally, German botanist C. Willdenow formally published *Chamaedorea* in *Species Plantarum* (1806). Despite this, *Chamaedorea* contained only one species for nearly 20 years. Then German botanist Carl Martius named *C. pauciflora* (1823) and transferred *Nunnezharia fragrans* to *Chamaedorea* in the second volume of his epic *Historia Naturalis Palmarum* (1824b). Several years later and in another journal, Martius (1830) named three more species, all from Mexico and presently cultivated, *C. elatior, C. elegans,* and *C. schiedeana.* Continuing with volume three of *Historia Naturalis Palmarum,* Martius (1837) added two more species from Mexico, *C. oblongata* and *C. concolor.*

Twelve years later Martius (1849) and Liebmann named and described ten more species of *Chamaedorea* in volume three of Martius' work. Danish botanist Frederik Liebmann had collected all ten species in Mexico. Among them are some of the more commonly cultivated species such as *C. cataractarum, C. pochutlensis, C. sartorii,* and *C. tepejilote.* The Danish government had sent Liebmann to Mexico in 1840 to make collections of live plants and seeds for the botanical garden at Copenhagen. He explored Veracruz with W. Karwinski and later visted Oaxaca where he made numerous and important collections of *Chamaedorea* in the Chinantla region near present-day Valle Nacional. Three years later he returned to Copenhagen with his collections which numbered about 40,000 botanical specimens. Martius (1849) erected three sections within *Chamaedorea, Euchamaedorea, Collinia* and *Stachyophorbe.* The latter two are names that Liebmann suggested and which would figure prominently in *Chamaedorea* nomenclature in the future.

Although his contributions to the genus essentially ceased by 1850, Martius (1852) did name one more *Chamaedorea.* German botanists continued at the forefront of *Chamaedorea* study, however, as J. Klotzsch (1852) named *C. casperiana* and erected a new subgenus, *Stephanostachys,* for it. He also elevated section *Stachyophorbe* to a generic rank and to it transferred *C. cataractarum* and named *S. deckeriana.* At about this time, a young cohort of Klotzsch's, Hermann Wendland, first made his indelible mark in *Chamaedorea.*

Wendland, Director of the Royal Botanical Gardens at Herrenhausen near Hannover from 1879 until his death in 1903, had an illustrious career spanning well over half a century. He became a special figure in European botany and horticulture and is, perhaps, best known for his pioneering work with the large, intractable tropical monocots such as palms, Araceae, and Cyclanthaceae (Grayum and de Nevers 1988). During travels to Guatemala, El Salvador, and Costa Rica in 1856–1857, Wendland made important collections of these including herbarium specimens and live plants. A respected taxonomist and skilled gardener, he had a special fondness for *Chamaedorea* and returned to Europe with material for the greenhouses at Herrenhausen.

Surprisingly, other authors described and named most of Wendland's collections; for example, his associate Udo Dammer worked with the chamaedoreas after Wendland's death. Conversely, the new chamaedoreas that Wendland described and named were mostly from cultivated plants collected by others. As one of Europe's most prominent botanists and horticulturists, he carried on a lively seed and plant exchange not only with European botanical institutions but also with collectors working the American tropics like Hartweg, Ortgies, and Warscewicz (Grayum and de Nevers 1988). Wendland assembled an excellent living collection of *Chamaedorea* at Herrenhausen from these sources. He named several species from his plants at Herrenhausen and additional ones from other European gardens. Most of Wendland's species are cultivated today; however, a few, including *C. paradoxa,* seem to have disappeared from the annals of horticulture. The names are early, though, and difficult to ignore.

Wendland first became involved in the genus when he named three species (1852a, 1852b, 1852c) including two commonly cultivated ones, *C. ernesti-augustii* and *C. geonomiformis.* He followed quickly with five more new ones (1853b, 1853c) and then added seven a year later (1854) in his well known *Index Palmarum,* the latter including *C. arenbergiana, C. glaucifolia,* and *C. klotzschiana.* In *Index Palmarum* he arranged *Chamaedorea* by subgenera, adding his own new subgenus *Eleutheropetalum* to the previously recognized subgenera or sections *Collinia, Chamaedorea,* and *Stephanostachys.* Wendland continued to increase the size of *Chamaedorea* over the next 25 years by adding seven more new species (1859, 1862, 1880), among them *C. tenella* and *C. alternans.* Apparently, Wendland was working toward a much-needed monograph of *Chamaedorea* (Dammer 1904b) when he died. His friend and cohort in later years, Udo Dammer, intimated that he would finish Wendland's work and publish it but never did. There is some evidence that Wendland may have become frustrated and tired with *Chamaedorea* near the end of his career, perhaps from the difficulty of trying to identify the numerous specimens in cultivation. Dammer (1904b) reported that whenever he would show a plant to Wendland for identification, Wendland would respond, "Let it go; the chamaedoreas are like the dogs."

Wendland was not the only botanist of his time giving attention to *Chamaedorea.* Although not as well known in *Chamaedorea,* Danish botanist A. Oersted (1859) named five species of *Chamaedorea,* all from collections he made during his travels in Costa Rica about the middle of the 19th century. Rather than viewing *Chamaedorea* as one large genus, Oersted split it into several genera based mainly on previously named subgenera and sections such as *Collinia, Stachyophorbe, Eleutheropetalum,* and *Stephanostachys.* He even erected new genera, *Spathoscaphe* and *Dasystachys,* based on previously named species.

The latter half of the 19th century is often considered the golden age of European horticulture. Tropical plants were the rage and established nursery firms and horticultural houses had collectors roaming the tropics, supplying a steady stream of exotic foliage and flowering plant material for European stovehouses. The period lasted into the early 1900s and many of the finest private and public collections of tropical plants were developed or attained their zenith during the time. As one would expect, chamaedoreas were among those eagerly-sought-after tropical plants.

Easy to collect and cultivate, they were favorites of such famous collectors and plantsmen as Galeotti, Wendland, Hartweg, Ortgies, Zahn, Warscewicz, and above all Linden, the celebrated horticulturist of Ghent, Belgium, who introduced more chamaedoreas than anyone. The most important collections were in the great conservatories of botanical gardens at Kew, Paris, Berlin, Hannover, Munich, and Vienna. Some of the more famous private gardens of the time, including those of noted personages whose names often appear as epithets for palms, such as Borsig, Decker, Arenberg, van Houtte, and Loddiges, were well known for their collections of chamaedoreas. In fact, several species were described and named from plants cultivated in some of these collections.

Chamaedoreas were very popular during that period for much the same reasons as they are today and were quite common in the trade. The great nursery firms of the time that dealt in chamaedoreas included J. Linden of Brussels and Ghent, Belgium; Sanders & Sons of St. Albans, England; James Veitch & Sons of Chelsea, England; Ambroise Verschaffelt of Ghent, Belgium; and Vilmorin-Andrieux et Cie of Paris, France. Several species first appeared in catalogs of these firms. The horticultural literature of the time also popularized chamaedoreas. There were numerous journals and trade magazines featuring articles on chamaedoreas, many of them well illustrated, that served to promote these palms to gardeners and collectors. *Curtis' Botanical Magazine* was the most famous of these and has beautiful color plates of several of the more commonly cultivated species of the period including *C. stolonifera, C. tepejilote, C. ernesti-augustii, C. arenbergiana, C. glaucifolia, C. elegans,* and *C. geonomiformis* (see Plates 1, 2 and 3).

Europe lost one of its greatest plantsmen and a champion of chamaedoreas when Wendland died in 1903. Dammer continued Wendland's work and named nine new species of *Chamaedorea* shortly thereafter (1904a, 1904b, 1905a, 1915), several for

Plate 1. Historic Perspectives of Chamaedorea (courtesy of Curtis' Botanical Magazine). A. *C. tepejilote* (Hooker, 1873); **B.** *C. geonomiformis* (Hooker, 1874); **C.** *C. stolonifera* (Hooker, 1892); **D.** *C. arenbergiana* (Hooker, 1885).

A
B
C
D
Bot. Mag.
7265
Chamaedorea (Collinia) stolonifera H. Wendl.

which Wendland had been growing and preparing descriptions.

In the United States chamaedoreas were also popular in the latter part of the 19th century although not to the same extent as in Europe. They appeared in the literature and trade magazines by 1900. Riedel (1957) reported that F. Franceschi in southern California offered several species in his nursery catalogs between 1898 and 1908 including *C. ernesti-augustii, C. sartorii, C. oblongata, C. geonomiformis,* and various forms of *C. elatior.* Santa Barbara nurseryman Kinton Stevens had *C. elegans* for sale as early as 1893. *C. oblongata* was in the southern California nursery trade by 1900 and *C. glaucifolia* by 1912. Although information is somewhat sketchy, chamaedoreas appeared in the Florida nursery trade at about the same time as in California, if not earlier. The species were much the same, indicating a common source of seeds.

At the turn of the century, collectors for the Bureau of Plant Introduction of the United States Department of Agriculture began to send seeds of various chamaedoreas from the wild. In 1904 O. F. Cook sent seeds of several species from Guatemala. In 1934 C. A. Purpus sent seeds from Mexico including *C. elatior, C. elegans, C. sartorii, C. tepejilote, C. geonomiformis,* and *C. oblongata.* European growers and seedsmen also sent seeds of chamaedoreas to the U.S.D.A.

By the 1920s millionaire oilman E. Doheny in Los Angeles had assembled one of the most comprehensive collections of chamaedoreas in the United States. Doheny had large greenhouses and conservatories in which he maintained his collections of tropical plants including his favorites, palms and cycads. He had collectors searching for material for him in the jungles of Mexico and virtually no expense was spared for their efforts. Doheny is responsible for introducing *C. pochutlensis* from western Mexico to cultivation. Some of these original plants still exist in southern California at the Huntington Botanical Gardens and elsewhere. L. H. Bailey made note of Doheny's collection during one of his numerous trips to California.

William Hertrich developed another excellent collection of chamaedoreas, and easily the finest palm collection in the western United States, for Henry Huntington at the latter's estate in San Marino, California. Hertrich procured many chamaedoreas, mainly through U.S.D.A. introductions and by exchanging with other collectors in the area such as Doheny. Today, Huntington's former estate, known as the Huntington Library, Art Gallery & Botanical Gardens, still has a fine collection of chamaedoreas.

The calamitous events of that era, World War I, the Great Depression, and World War II, dimmed but did not extinguish interest in *Chamaedorea.* Frenchman A. Guillaumin (1923b) published a concise and informative report on the cultivated chamaedoreas and Cook (1937, 1939, 1943a, 1943b, 1947a, 1947b) extolled the virtues of the group for use as indoor plants. Cook's accounts are well illustrated and, in many ways, quite admirable. Unfortunately, his concept of taxonomic boundaries and ignorance of the rules of botanical nomenclature detracted from his otherwise informative work. The numerous generic names he proposed have no botanical standing and some, like *Neanthe bella,* have added nothing but confusion to the nomenclature of *Chamaedorea.*

The name of one individual does stand out during this period, if not for his contribution to our understanding of *Chamaedorea* then for the sheer number of species he named for the genus. Over a nine-year span beginning in 1930, German botanist Max Burret named 27 new species of *Chamaedorea* (1930, 1932, 1933b, 1934a, 1934b, 1936b, 1938), 19 of them in his monograph of the genus (1933a). Most are obscure species known only from the type collection although a few, including *C. microspadix* and *C. seifrizii,* are well established in cultivation. Burret also worked with *Morenia,* which he maintained as a genus separate from *Chamaedorea.* Burret named four species of *Morenia* (1936a, 1940), three of them in his 1936 monograph.

Burret's monographs of *Chamaedorea* and *Morenia* added only slightly to our knowledge of either genus. Other than the new species he named, the monographs consisted simply of lists without descriptions of the previously named species. They were useful, though, if simply for no other reason than for pulling together all the known names. With *Chamaedorea* he did arrange the genus by previously recognized sections although he maintained *Collinia, Eleutheropetalum,* and *Dasystachys* as separate genera. Perhaps Burret's greatest hindrance was that he did not collect in the field; all his work was based on dried specimens sent him by collectors.

→

Plate 2. Historic Perspectives of Chamaedorea (courtesy of Curtis' Botanical Magazine). A. *C. pumila* (as *C. nana*; Prain, 1916); **B.** *C. ernesti-augustii* (Hooker, 1855a); **C.** *C. linearis* (as *Morenia corallina*; Prain, 1913); **D.** *C. tenella* (as *Nunnezharia tenella*; Hooker, 1881).

A
Bot. Mag.
8652
Chamaedorea nana
B
C
Bot. Mag.
Morenia corallina, Karst.
D
Bot. Mag.

The well known horticulturist and botanist at Cornell University, L. H. Bailey, was a prolific student of palms in the 1930s and 1940s. He wrote the Palmae for the *Flora of Panama* (Bailey, 1943b) and in support of this project named eleven species of *Chamaedorea* (1943a) including *C. allenii* and *C. woodsoniana.* Like Burret, Bailey had not had the opportunity to collect *Chamaedorea* in the field and several of his species have been reduced to synonyms.

Based at the Field Museum in Chicago, Paul Standley and Julian Steyermark (1947) named 13 new species of *Chamaedorea* from their work on the Guatemalan flora. They spent much time in the field collecting and frequently had the opportunity to observe *Chamaedorea* in the wild. They were fortunate in working in Guatemala, an area especially rich in a great number of diverse chamaedoreas and not well collected. Most of the 13 species they named, while interesting, are not well known and only two, *C. brachypoda* and *C. quezalteca,* are cultivated.

Harold E. Moore, Jr., who followed Bailey at Cornell, made the last great contribution to our knowledge of *Chamaedorea.* Moore named several new species, among them the well known and distinctive *C. metallica.* Moore's greatest contributions, however, are largely unrecognized. He essentially prepared the treatment of *Chamaedorea* for Standley and Steyermark's (1958) *Flora of Guatemala.* More importantly, he was working toward a treatment of the subgenera of *Chamaedorea* and eventually a monograph of the genus when he died in 1980.

Although O. F. Cook extolled the virtues of *C. elegans* as an indoor plant in the 1930s, Vosters (1975) reported that a Florida nurseryman, Roger Nason of Delray Beach, is credited with developing the species as a house plant. Nason was producing plants commercially in 1945 from an initial shipment of 25 pounds (12 kg) of seeds from Mexico. Also about this time another Florida nurseryman, Leo Bernecker, began producing *C. seifrizii* on a commercial basis for the indoor trade. Bernecker became perhaps the most well known grower of *C. seifrizii* and that reputation has not declined today. Since seeds of *C. seifrizii* were readily produced in southern Florida and demand was still relatively low, growers did not have to look elsewhere for their propagative needs. However, as *C. elegans* gained popularity as an indoor plant and did not produce seeds in southern Florida, growers began to look elsewhere for seeds of this species.

Two individuals in southern Texas, Gilbert Longoria and Luciano Guerra, who began importing seeds of palms, including chamaedoreas, and other plants from Mexico, met the demand in the late 1940s and early 1950s. Initially working together and later separately, their names became synonymous with *C. elegans* as they supplied seeds of this popular plant to the burgeoning nursery industry in southern Florida. They were virtually the only suppliers of seeds of this species and deserve much of the credit for the development of *C. elegans* as one of the most popular potted plants in the world. Although they also supplied seeds of *C. seifrizii* and did play a role in its development as a staple of the indoor trade, their impact is much less with this species since growers in southern Florida produced most of their own seeds. Today *C. elegans* and *C. seifrizii* are major components of a billion-dollar, international foliage industry centered in southern Florida (see the chapter on economic uses) and are recognized around the world as among the best and most popular of plants for indoor use.

The late David Barry, Jr. of west Los Angeles did much to promote the use of chamaedoreas in the 1960s and 70s. The long-time Palm Society member carried on correspondence with palm fanciers and traveled to various European gardens where he was always looking to procure new chamaedoreas. Similarly, he had obtained seeds of several species from Mexico and maintained a long and productive correspondence with H. E. Moore, Jr. From these contacts and sources Barry developed an extensive listing of chamaedoreas for sale at his nursery. He was well known for having staminate and pistillate plants of a species so that the buyer could purchase both sexes and produce seeds.

In the 1960s a Florida nurseryman, Robert Wilson, moved to southeastern Costa Rica near the Panamanian border to develop a tea plantation. Although his tea-growing venture was not profitable, he parlayed his love and knowledge of plants into the development of one of the most significant collections of tropical ornamental plants in the world. Over a 20-year period he amassed at his residence, called Las Cruces Tropical Botanical Garden, an amazing treasure of palms, heliconias, gingers, aroids, ferns, bromeliads, orchids and other plants

---

→

Plate 3. Historic Perspectives of Chamaedorea (courtesy of Curtis' Botanical Magazine). *C. linearis* (as *Morenia fragrans*; Hooker, 1865).

Bot. Mag.
5492
Morenia fragrans.

from around the tropics. Wilson was especially fond of chamaedoreas and is responsible for introducing *C. brachyclada, C. minima,* and *C. stenocarpa* to cultivation. In his later years he made an arrangement for the Organization for Tropical Studies to administer his garden after his death. In 1987, just before his death, the garden was renamed Jardin Botanico Robert y Catherine Wilson in his and his wife's honor.

Also in the 1960s and 1970s nursery- and seedsman Horace Anderson of Leucadia, California, offered plants and later seeds of several chamaedoreas. His nursery north of San Diego was always popular with growers and collectors since he had species which at the time were quite rare. Also, one always delighted at hearing tales of his adventures in Mexico since he personally collected most of his material. Anderson is, perhaps, best known for rediscovering *C. metallica* and bringing it into cultivation. He had an affinity for students and young collectors and one always returned home from his nursery with a carload of choice palms and cycads purchased at very reasonable prices with a few gifts thrown in as an added bonus.

CHAPTER 3

DISTRIBUTION AND ECOLOGY

Moore (1973c) pointed out that studies of the biology of palms are in their infancy and although palms are conspicuous components of the intricate and complex tropical forest ecosystems, relatively little is known of their soil, water, light, and temperature relationships. Also, relatively little is known about the interrelationships of palms and their pollinators, predators, agents of dispersal, and other plants. However, an examination of the distribution and especially observations of the types of forests in which chamaedoreas occur can tell us much about their growth and cultural requirements.

Two Centers of Distribution

Chamaedorea includes about 100 species of dioecious, understory palms restricted to neotropical rain forests and cloud forests on the Atlantic and Pacific slopes from western and eastern Mexico through Central America to northwestern Ecuador and the Amazonian portions of Columbia, western Brazil, eastern Ecuador, eastern Peru, and northern Bolivia. A great number of species is concentrated in the mountainous area of southern Mexico and adjacent Guatemala. A secondary concentration occurs in the mountains of Costa Rica and Panama. Although I have reduced many of the named species from northern South America to synonymy and the region now appears to harbor few *Chamaedorea,* the situation may be more apparent than real. I feel that new species will emerge with further collecting and when palm taxonomists give more attention to *Chamaedorea* in this area.

There are a few species that are widely distributed, nearly throughout the entire range of the genus. The ubiquitous and highly variable *C. pinnatifrons* and *C. tepejilote* occur on both slopes of Mexico and Guatemala and extend throughout Central America into Colombia with the former extending as far south as Ecuador and Bolivia. Stout, single-stemmed *C. woodsoniana* occurs in cloud forests from Veracruz, Mexico, to the Panamanian/Colombian border.

Guatemala and Mexico

Guatemala and Mexico, with about 40 species each although some are common to both, have had more chamaedoreas named than any other countries. Mexico is large and diverse and in some areas has not been well collected. It may yield additional species not now recorded there or new to science. The east coast or Atlantic slope of Mexico is particularly rich in *Chamaedorea. C. radicalis,* recognized for its cold-hardiness and showy red fruits, occurs to the north in the state of Tamaulipas above the Tropic of Cancer. Just to the south, in San Luis Potosi and Hidalgo, one encounters the multiple-stemmed *C. microspadix,* also noted for its red fruits and tolerance of cold.

Further to the south in Veracruz, Oaxaca, Chiapas, and Tabasco is the heart of Mexican chamaedorea country. Here one finds the Mexican endemics *C. alternans, C. cataractarum, C. glaucifolia, C. hooperiana,* and *C. klotzschiana.* Other endemics from the area include *C. oreophila, C. queroana, C. schiedeana,* and the bifid-leaved *C. metallica, C. rigida,* and *C. stolonifera.* Although some such as *C. oreophila* are wide-ranging and encountered throughout this part of Mexico, others such as *C. hooperiana, C. rigida,* and *C. stolonifera,* appear to be narrowly distributed and restricted to a few areas. In fact, only a few scattered colonies of *C. stolonifera* remain in Chiapas.

Other species which occur in the area but also extend further south into Guatemala and, in some cases, into Belize, Honduras, and/or Nicaragua include *C. neurochlamys, C. oblongata, C. sartorii,* and the bifid-leaved *C. ernesti-augustii.* The diminutive, bifid-leaved *C. tenella* is known only from Veracruz and Chiapas in Mexico and southeastern Costa Rica.

The west coast of Mexico has its share of chamaedoreas. The wide-ranging *C. pochutlensis* occurs from the states of Durango and Nayarit in the north to Oaxaca in the south. In cloud forests on the Pacific slope of Oaxaca one finds *C. rhizomatosa, C. foveata,* and *C. whitelockiana,* the latter also extending into Guatemala.

For its size Guatemala is easily the richest area in *Chamaedorea.* The region of Coban on the Atlantic slope of Guatemala is famous for plants, especially chamaedoreas. Here one finds the endemics *C. lehmannii, C. verapazensis,* and the remarkable *C. tenerrima* with its pair of broadly flared terminal pinnae and sharply reflexed basal pinnae. Although common in cultivation, *C. brachypoda* is very rare now in Izabal on the Atlantic coast due to deforestation though it also occurs in adjacent Honduras. *C. nationsiana* with its unusual coblike infructescences, the attractive, bifid-leaved *C. castillo-montii* and *C. geonomiformis,* and the dwarf, pinnate-leaved *C. stenocarpa,* are very localized in Izabal, Guatemala. However, *C. geonomiformis* extends into Belize and Honduras and *C. stenocarpa* also occurs in Costa Rica and Panama where it is more widespread.

Although the Coban region is the most well known in Guatemala for *Chamaedorea,* the genus is equally abundant on the Pacific slope in the vast mountain cloud forests densely covering the volcanoes of Tajumulco, Santa Maria, and Zunil. Here are the endemics *C. skutchii, C. pachecoana, C. volcanensis,* and many of the species shared with Mexico, Honduras, and El Salvador.

Guatemala and Mexico share several species including *C. carchensis, C. elegans, C. liebmannii, C. parvisecta,* and the handsome, bifid-leaved *C. simplex,* all on the Atlantic slope. There is also one collection of *C. simplex* from Colombia. On the Pacific slope one finds *C. fractiflexa, C. vulgata,* and the stunning, bifid-leaved *C. stricta,* the latter only recently discovered in Costa Rica and Panama. The striking but elusive *C. tuerckheimii* is rare in forests in Veracruz, Mexico and Alta Verapax, Guatemala. The highly variable, vinelike *C. elatior* is common throughout the Atlantic slope of Mexico and both slopes of Guatemala. The bifid-leaved, multiple-stemmed *C. nubium* occurs in cloud forests on both slopes of Mexico and Guatemala and also extends into El Salvador and Honduras. The slender-stemmed *C. rojasiana* is recorded from the Pacific slope of Guatemala and the Pacific and Atlantic slopes of Chiapas, Mexico.

Chamaedorea arenbergiana and *C. quezalteca* occur on the Pacific and Atlantic slopes of Guatemala. The former, noted for its coblike infructescence of densely packed, black fruits similar to those of *C. nationsiana,* ranges from Mexico to Honduras and, perhaps, beyond. *C. quezalteca* is a handsome species with large, clustering stems that occurs from Chiapas, Mexico, into Honduras and El Salvador.

The region of northern Guatemala, Belize, and the Yucatan Peninsula of Mexico supports several interesting species of *Chamaedorea,* all predominantly found on limestone rocks. The highly variable, tightly cespitose *C. seifrizii* is scattered throughout the area, either in dense forest or open woodland while the loosely cespitose *C. graminifolia* is more uncommon. There are also a few collections of *C. graminifolia* from Costa Rica. *C. adscendens* with thick, velvety-green leaves is also rare in Alta Verapaz and Peten in Guatemala and in southern Belize.

South of Guatemala in El Salvador, Honduras, and Nicaragua chamaedoreas are not as common and this is probably due to several factors. Among them are the lack of high and large mountainous areas, deforestation, and paucity of collecting, the latter being especially so in Nicaragua. The most common species include *C. arenbergiana* and *C. oblongata,* which range into Mexico; *C. costaricana,* which ranges from Honduras to Panama; the widespread *C. woodsoniana*; and the omnipresent *C. pinnatifrons* and *C. tepejilote.*

Costa Rica and Panama

In the mountainous regions on both the Atlantic and Pacific slopes of Costa Rica and Panama, chamaedoreas once again become more common. They are especially abundant on the wet Atlantic slopes at about 800–1,500 meters elevation, which in many situations, is at or near the Continental Divide. Such places as the rich forest of the upper reaches of the Rio Sarapiqui and Braulio Carrillo National Park in Costa Rica and Fortuna and El Valle in Panama support six or more species of *Chamaedorea* each. Not to be overlooked are the dense rainforests of the Pacific slope in southeastern Costa Rica in the Golfo Dulce, Osa Peninsula, and adjacent mountainous areas.

The Costa Rican endemics include *C. chazdoniae, C. crucensis, C. macrospadix, C. parvifolia, C. pedunculata, C. undulatifolia, C. zamorae,* and the dwarf, bifid-leaved *C. minima* and *C. pumila. C. selvae* inhabits the Atlantic lowlands of Costa Rica and adjacent portions of Nicaragua.

The Panamanian endemics include *C. guntheriana, C. serpens,* and the dwarf, bifid-leaved *C. correae* and *C. verecunda.* The elusive *C. microphylla* is known only from the type and a few collections in western Panama. Not surprisingly, with their adjacent boundary and a contiguous mountain range, many species are common to Costa Rica and Panama including *C. brachyclada, C. dammeriana, C. deneversiana, C. matae, C. pittieri, C. pygmaea, C. scheryi, C. warscewiczii* and the bifid-leaved *C. amabilis, C. deckeriana, C. palmeriana, C. robertii,* and *C. sullivaniorum.*

Several species extend into South America including *C. allenii, C. amabilis, C. lucidifrons, C. pygmaea,* and *C. sullivaniorum* in Colombia and *C. deneversiana* in Ecuador.

Northern South America

There are surprisingly few chamaedoreas restricted to northern South America although several have been named from the area. Here are *C. latisecta, C. linearis,* and *C. smithii,* all of subgenus *Morenia.* The distinctive and commonly cultivated *C. fragrans* is found in the Huanuco Valley of Peru.

The Amazonian portions of Colombia, eastern Ecuador and Peru, northern Bolivia, and western Brazil have few chamaedoreas. Among them are *C. angustisecta,* noted for its attractively mottled leaf sheaths and petioles, and *C. pauciflora,* both of subgenus *Moreniopsis* and with powerfully fragrant staminate flowers. The widespread *C. pinnatifrons* is found from lowland to montane forests throughout the area.

Shared Species

With few exceptions, chamaedoreas native to the northern center of distribution in Mexico and Guatemala do not occur in the southern area of Costa Rica and Panama and vice versa. Relatively low Nicaragua and Honduras seem to have been an effective barrier isolating the two centers of distribution. These notable exceptions include *C. arenbergiana* from Mexico, Guatemala, Honduras, Costa Rica, and Colombia; *C. costaricana* from Panama to Honduras; *C. graminifolia* from Belize, Guatemala, and Costa Rica; *C. tenella* from Mexico and Costa Rica; *C. simplex* from Mexico, Guatemala, and Colombia; *C. stenocarpa* from Guatemala, Costa Rica, and Panama; *C. stricta* from Mexico, Guatemala, Costa Rica, and Panama; *C. woodsoniana* from the Panamanian/Colombian border to Mexico; and, of course, the widespread and ever-present *C. pinnatifrons* and *C. tepejilote.*

Palms of Middle Elevations

Not abundant in tropical lowland forest, *Chamaedorea* attains its greatest diversity in tropical montane forest and cloud forest from about 800–1,500 meters elevation. In fact in 19th- and early 20th-century German horticulture, chamaedoreas were called *bergpalme,* meaning literally mountain palm since they were commonly introduced to European gardens from tropical montane forests. At selected locations in such wet or moist forests it is not uncommon to find up to eight species of the genus growing in proximity. In fact, near El Valle, Cocle, in central Panama, ten species of *Chamaedorea* can be found although each is restricted to one of three distinct habitats. Eight, including *C. allenii, C. amabillis, C. lucidifrons, C. serpens, C. sullivaniorum, C. tepejilote, C. warscewiczii,* and *C. woodsoniana,* occur in wet forest on slopes and in ravines. *C. costaricana* is found in moist, seasonally dry forest while *C. correae* is restricted to windswept ridgetops in mossy, elfin cloud forest.

At times, *Chamaedorea* may be one of the most conspicuous elements of the forest understory. Before being wiped out by collectors, *C. sullivaniorum* occurred in virtual monocultures near El Valle, Panama. In areas of three to four meters square it covered the ground thickly, nearly to the exclusion of all other vegetation.

Although not as common or species-rich at low elevations, *Chamaedorea* may still be a conspicuous component of the forest at low altitudes. Heinzman and Reining (1988) reported densities of *C. elegans* and *C. oblongata* in undisturbed areas in Peten, Guatemala, an area depicted as lowland forest. *C. elegans* averaged 5,933 plants per hectare (2.5 acres) and *C. oblongata* 2,570 plants per hectare. Olmsted (1988) surveyed *C. seifrizii* in Quintana Roo in the Yucatan area of Mexico adjacent to Peten, Guatemala. Highest densities reported for this species were 195 plants per hectare. Kahn and Mejia (1991) reported that of 34 species of palms at a 0.5 hectare study site in Peru, *C. integrifolia* (= *C. pauciflora*), with 208 individuals, was the fifth most common overall and third most common understory species.

Chamaedorea has a remarkable altitudinal range as a genus. A few species occur at sea level while several species in Guatemala and Mexico including *C. nubium, C. parvisecta,* and others occur between 2,000 and 3,000 meters. The species from relatively high elevations in Mexico and Guatemala frequently occur in forests of *Quercus* (oak), *Liquidambar,* and *Alnus* (alder), attesting to their tolerance of cool temperatures. *C. costaricana* occurs at 2,300 meters in Panama and probably higher. In northern South America *C. pinnatifrons* approaches 2,700 meters elevation or, perhaps, even higher.

A few species of *Chamaedorea* have a wide altitudinal range. *C. tepejilote* occurs at sea level up to 1,600 meters in several locations and *C. pygmaea* is found as low as 200 meters and as high as 2,000 meters in eastern Panama. *C. linearis* in South America and *C. pochutlensis* from western Mexico have been reported from as low as 50 meters to over 2,000 meters.

The genus has a remarkable latitudinal range also, occurring from northeastern Mexico to northern Bolivia. Inland from the east coast of Mexico, *C. radicalis* occurs in mountain oak forest north of the Tropic of Cancer, as mentioned earlier, putting it outside the limit of the tropics. Not far to the south, *C. microspadix* is found approaching the tropical limit. Not surprisingly, both species are the most cold-hardy of the genus and easily tolerate several degrees below freezing. On the west coast of Mexico, *C. pochutlensis* approaches the tropical limit and is associated with pines and oaks. *C. pinnatifrons, C. tepejilote,* and *C. woodsoniana* have the widest range in the genus, occurring from Mexico to northern South America with the former found as far south as Bolivia.

Palms of Dense Primary Forests

With few exceptions, chamaedoreas are palms of the dense primary wet forest or cloud forest (see Plates 4 and 5). They will not survive and do not establish themselves in disturbed or secondary forest. They require shade, abundant moisture, and high humidity to become established and grow well. In fact, several of the species restricted to cloud forests where humidity is constantly high and temperatures moderate with little daily or seasonal fluctuation are the most difficult to cultivate. Those species characterized by their essentially stemless habit, thin and decurrent pinnae, and pinnate eophylls (Hodel and Uhl 1990b) are especially sensitive. These have such exacting growth requirements, difficult to reproduce artificially, that they are rare in cultivation and seldom do well.

Although the genus is most species-rich in wet premontane and montane forest with a developed canopy, there are several specialized and more open habitats that harbor their own distinct species of *Chamaedorea.* Among them are two inhabitants of dwarf habit with usually bifid leaves restricted to cloud forests in central Panama. *C. guntheriana* occurs in open, windswept cloud forest near the Continental Divide. There, the low, open, shrubby forest contains no tall trees and seems to be of an almost sclerophyllus nature. Many of the species occurring there have stiff, thick, leathery, grayish green leaves and a dwarf habit as does *C. guntheriana.* The area experiences a pronounced dry period from January through April and many individuals of *C. guntheriana* grow in nearly full sun.

Similarly, *C. correae* occurs in a mossy, elfin cloud forest on windswept ridges further to the west in central Panama. Although the forest there is wetter and trees are slightly more conspicuous than where *C. guntheriana* occurs, it is still fairly open and the more or less thick leaves and dwarf habit of *C. correae* reflect this.

Chamaedorea pittieri and *C. parvifolia,* species nearly identical in habit to each other but with vastly different flowers, occur sparingly in moist cloud forest with a well developed canopy at relatively high elevations in Panama and Costa Rica. Both occur in windy areas on the leeward side in the rain-shadow of the Continental Divide. Precipitation is less and the dry period more pronounced there than on the windward side but cloud cover is frequent. Their adaptation to the harsher environment is apparent in their short, nearly stemless habit and leaves with more or less thick, stiff, gray-green pinnae. Their short stems are often buried in the deep leaf litter and only with age elongate and become visible.

One species, *C. seifrizii,* is well adapted to more open forest or woodland and occurs in such habitat in the Yucatan Peninsula of southern Mexico and northern Guatemala and Belize. There it is often found growing in full sun on the fringes of cultivated areas in disturbed scrub or woodland. In fact, it is difficult to determine if these stands represent truly wild populations or remnants of groups cultivated by earlier peoples and long-since abandoned. The form with stiff, narrow leaflets seems to predominate, raising the possibility that, if cultivated, it was selected for its tolerance of full sun. In any case, it seems to be reproducing successfully in these exposed situations. Seed plantations of *C. seifrizii* in full sun in southern Florida and Hawaii demonstrate its tolerance or preference for high light. In fact, vigorous, robust-growing plants and maximum seed production are attained in high light. *C. seifrizii,* will

→

Plate 4. Chamaedorea in Habitat. A. *C. metallica* on limestone rock, *D. & R. Hodel 950,* Temazcal, Veracruz, Mexico; **B.** *C. pinnatifrons* in cloud forest, *D. & M. Hodel 694,* Monteverde, Costa Rica; **C.** Marianne Hodel stands with clustering form of *C. tepejilote* in lowland forest, *D. & M. Hodel 843,* Cayo, Belize; **D.** *C. tuerckheimii* in montane rain forest, *Hodel & Castillo 1010,* Alta Verapaz, Guatemala.

B
D

tolerate lower light and shade very well, especially the forms with broader pinnae. It is these broad-leaflet forms rather than the narrow-leaflet ones that are more common in denser forest in northern Guatemala.

Although not as tolerant of high light as *C. seifrizii,* there are other species that occur in open forest and tolerate similar situations. *C. graminifolia* and *C. adscendens* occur on limestone outcroppings in Belize and Guatemala. The vegetation on limestone is often more open than on surrounding, more fertile soils, resulting in a greater amount of light reaching the plant. *C. graminifolia* has leaves with very narrow, gray-green, drooping pinnae, features considered characteristic of some plants tolerant of high light. The leaves of *C. adscendens* are among the thickest and most leathery in the genus, reflecting its open, seasonally dry habitat.

Chamaedorea pochutlensis, with multiple stems and long-pinnate leaves from seasonally drier forests of western Mexico, tolerates nearly full sun in southern California, especially near the coast although it tends to yellow slightly in the hotter interior valleys of the greater Los Angeles basin. Similarly, *C. radicalis* and *C. microspadix* grow in full sun at the Huntington Gardens in San Marino, California. The former is especially tolerant of high light; this is not too surprising considering its native habitat of rather open, rocky, oak forest. The latter tends to yellow slightly but it comes from denser forest.

Chamaedorea costaricana with multiple stems and long-pinnate leaves can be a tall-growing palm, at least for *Chamaedorea,* and in its native habitat may occur in more open, disturbed forest. In these situations its stems easily emerge into higher light either above or amongst the rather open canopy. Not surprisingly, it tolerates a good deal of light along the immediate coast in southern California and would probably tolerate even higher light in areas with higher relative humidity such as Hawaii and Florida. In fact, we can generalize that plants will tolerate a higher level of light in areas with higher humidity. Chamaedoreas that may grow well in full sun in southern Florida or Hawaii where humidity is high will need some protection, even in coastal southern California, to prevent them from yellowing and detracting from their beauty. I have observed magnificent clumps of *C. tepejilote* in full sun at Tikal in northern Guatemala that showed only slight yellowing. However, in southern California this species yellows badly and even burns in full sun.

Although essentially palms of undisturbed, primary forests, some chamaedoreas are able to survive in secondary or disturbed forests. The widespread and often cultivated *C. tepejilote* is frequently found in such situations. Species with a cespitose habit (clustering stems), such as *C. costaricana,* seem to be more successful in surviving after the forest is cut or disturbed. Since species with a solitary stem have only one growing point, once this is cut or destroyed that individual is doomed. However, species with multiple stems will often resprout naturally from the base and produce new shoots and stems. Also, the multi-stemmed species with their stems of various heights produce their own microclimate. The larger stems with leaves provide some shade for the smaller, younger shoots and stems, enhancing the plant's ability to survive without the natural protective forest canopy. I have observed this phenomenon with *C. costaricana* in Panama and Costa Rica and *C. seifrizii* in Guatemala and Belize. In Guatemala we discovered several clumps of *C. brachypoda* in areas cleared of all vegetation except large *Orbignya* palms. The *Chamaedorea* had survived repeated cutting back to the ground and chewing by browsing cattle and had simply resprouted from their underground rhizomes and grew in the shade provided by the *Orbignya.*

Variety of Substrates

Chamaedoreas inhabit a wide array of substrates. Many are found on relatively rich soils, often of volcanic origin. These fertile soils may be of fairly heavy texture but have excellent structure, enhancing their aeration, porosity, and drainage. Soils of southern California, especially the coastal region, may be heavy also; unfortunately, these have poor structure and are thus poorly drained. Soils in the tropics are often covered by an organic layer of humus that is composed of leaf litter, twigs, small branches, logs, and mosses, lichens and other microflora in various states of decay. The leaf litter varies in depth from virtually no covering at all to 20 cm. In some instances, where it can collect and be held in place by the root buttresses of large canopy trees, the leaf litter may be 30 cm or more in depth and creates somewhat of a special habitat. Some chamaedoreas, notably *C. stenocarpa* and *C. stricta* from Guatemala, seem to be found most frequently in proximity to large trees whose buttresses trap and collect organic material. Both these species appear stemless but actually possess short, creeping stems that are submerged in the deep pockets of decaying leaf litter.

The leaf litter of the forest floor serves several functions. It helps to conserve moisture, supply a low but constant stream of nutrients from the decaying material, and provide an excellent medium for the germination of seeds and establishment of seedlings and young plants. High in organic matter, the leaf litter holds water and nutrients while at the same time it has good aeration, porosity, and drainage. In fact, most chamaedoreas have an extensive network of roots in the leaf litter although they also have deeper roots in the mineralized soil below. By covering the often steeply sloping ground and encouraging the development of an extensive root system, the leaf litter prevents soil erosion in areas subject to high rainfall.

A few chamaedoreas seem to be leaf-litter collectors. *C. rigida* from the Atlantic slope of Oaxaca, Mexico, has stiffly ascending, bifid leaf-blades that tend to collect falling leaves, twigs, and other organic material, funneling them into the crown of the plant [see Plate 5(C)]. *C. oreophila,* also from the Atlantic slope of Mexico, has pinnate leaves that perform much the same function.

Another substrate upon which chamaedoreas commonly occur is calcareous soils or limestone rock. Although they are found on other substrates, some species do exceptionally well often on pure limestone outcroppings with only a thin layer of leaf litter. *C. adscendens, C. ernesti-augustii, C. graminifolia, C. metallica, C. oblongata, C. neurochlamys, C. tepejilote, C. seifrizii,* and *C. geonomiformis* occur in such situations in southern Mexico, northern Guatemala, and Belize [see Plates 4(A) and 5(B)]. In Costa Rica *C. tepejilote* and *C. macrospadix* are found on limestone rocks on the Pacific slope and the same two species with *C. warscewiczii* occur in a similar situation on the Atlantic slope. Southern Florida has a substrate composed predominantly of calcareous rock and it is no surprise that these species perform exceptionally well there. Habitats on limestone rocks at low elevations can experience seasonally, very dry periods. *C. graminifolia* and *C. seifrizii,* are subjected to exceedingly dry conditions for several months each year without any apparent loss of vigor.

Although *C. tepejilote* occurs on many types of substrates, it will often exhibit distinct site preferences from one area to another. Chazdon (1985) pointed out that at La Selva in the Atlantic lowlands of Costa Rica, *C. exorrhiza* (= *C. tepejilote*) tends to be found only on the more nutrient-rich recent alluvial deposits of the Rio Puerto Viejo.

Of Epiphytes, Lithophytes, and Rheophytes

Although there are no true epiphytes in *Chamaedorea,* seeds of many species will often be deposited in crotches of tree trunks, on fallen logs, or other places above ground where leaf litter can collect and provide the proper environment for seeds to germinate and establish themselves. *C. pygmaea* has been reported at least once from Chiriqui, Panama, to occur epiphytically. I have observed numerous specimens of various *Chamaedorea* in Mexico, Belize, Guatemala, Costa Rica, and Panama that at least began their lives epiphytically although they eventually sent forth larger roots and became established in the ground.

In many situations chamaedoreas grow on solid rock, usually limestone, and thus could be considered partial lithophytes. Near Cerro Punta in Chiriqui, Panama, at over 2,000 meters elevation, we observed an enormous individual of *C. costaricana* with multiple stems to 15 meters long that was growing on a 10-meter round rock. Sometimes rock outcroppings are the only places where forest has not been destroyed and here the palms have survived, as is true of *C. quezalteca* near Coban in Guatemala. But true lithophytes they are not since they are not restricted to rocks.

Chamaedorea has at least one rheophyte, a plant restricted to growing along swiftly moving water courses where it is occasionally inundated. The appropriately named *C. cataractarum* was first collected in Oaxaca, Mexico, where it grows in cataracts in mountain streams [see Plate 5(A)]. Locally known as the river palm, it is confined to these water courses in Oaxaca and in the adjacent states of Chiapas and Tabasco. It is well adapted to such a habitat with its flexible petioles, leaves, and pinnae and low, creeping, branched stem securely anchored by a tenacious network of roots.

Pollination

Palms have often been considered to be largely wind-pollinated but recent studies indicate that they are probably insect-pollinated. While some species may be wind-pollinated, understory rain forest palms such as *Chamaedorea* are probably not. The lower reaches of the rain forest are structurally and floristically poorly adapted to wind pollination. Moore (1973c) suggested that "the abundance of pollen and multiplicity of stamens or excess of staminate flowers produced by many palms of the rain forest may well be an adaptation to insect depredation rather than to wind pollination."

Of understory palms in the American tropics, Schmid (1970) discounted wind as the pollinating mechanism for *Asterogyne martiana* and felt that flies transfer pollen, while Essig (1971) suggested that bees and beetles did the same for two species of *Bactris*. Essig (1973) implied that flies, bees, or beetles pollinated two palms of the rain forest understory in New Guinea, *Ptychosperma macarthurii* and *Hydriastele microspadix*. Croat (1978) noted that bees and weevils visited flowers of *C. wendlandiana* (= *C. tepejilote*) in Panama but apparently they did not collect pollen. Henderson (1986) observed bees, beetles, and flies collecting pollen at staminate flowers and observed that they were attracted to pistillate flowers of *C. costaricana* in Costa Rica.

That most species of *Chamaedorea* have flowers that are scented or aromatic seems to suggest that they are indeed insect-pollinated. Also, many have brightly colored flowers, usually yellow, orange, or red, and sticky pollen, again suggesting insect pollination (Fisher and Moore 1977). The flowers of *C. fragrans, C. pauciflora,* and *C. angustisecta* are noted for their sweet fragrance. Flowers of *C. deckeriana* have a distinct scent, somewhat resembling spicy anise. Others like *C. elatior, C. seifrizii, C. pochutlensis,* and *C. graminifolia* have floral odors that some consider to be offensive.

Flowers of most chamaedoreas seem to be at anthesis for several days, sometimes as long as ten (Henderson 1986). Fisher and Moore (1977) reported that some species of *Chamaedorea* have apparently lengthened their period of anthesis with multiple inflorescences that open in succession, one after the other. Multiple staminate inflorescences are a diagnostic feature of subgenus *Morenia* and also appear frequently in subgenus *Stephanostachys*. Some species are known for the short duration of anthesis of staminate flowers. *C. allenii* from Panama has an inflorescence with densely packed staminate flowers that are at anthesis for only a day or two at the most. They tend to disintegrate and rot away rapidly, one reason why these and staminate flowers of many *Chamaedorea* are infrequently found in the wild.

Some species, such as *C. oblongata, C. pochutlensis, C. seifrizii* and especially members of subgenus *Stephanostachys,* produce copious amounts of dry, light pollen that suggest they may be wind-pollinated. In fact, when staminate inflorescences of *C. tepejilote* are disturbed at anthesis they send forth great clouds of white pollen. Cook (1943a) and Barry (1967) noted these two types of *Chamaedorea* pollen, one sticky and rather heavy and the other dry and light. The implications of these two pollen types for growers of *Chamaedorea* who want to produce seeds on their plants is discussed under propagation in the chapter on culture.

Predators and Survival

Chamaedoreas may have several mechanisms that tend to enhance their survival. Uhl and Moore (1973) reported the occurrence of raphide crystals as common in the genus, especially in the various floral organs. The crystals are highly irritating to marauding predators, mainly insects, that may feed on flowers, again more evidence that insects frequent *Chamaedorea* flowers and may be involved in pollination.

Chamaedoreas in the wild often have leaves that exhibit chewing damage to some degree or another. I have observed this to be common and Ibarra-Manriquez (1988) reported that distinct patterns of leaf feeding characterize certain species of *Chamaedorea* in Veracruz, Mexico. Despite this, the apical bud is rarely attacked and destroyed. De Steven and Putz (1985) showed that of eight species of understory palms in Panama, only *C. wendlandiana* (= *C. tepejilote*) displayed no animal-caused damage. They speculated that the terminal shoot possesses some chemical repellents. Oyama (1991) and Oyama and Dirzo (1991) studied beetle-feeding on *C. tepejilote* (probably *C. alternans*) in Veracruz, Mexico. The beetle in their study exhibited species and gender preferences, desiring *C. tepejilote* over other plants including other species in the same genus. The beetle also preferred pistillate plants over staminate ones.

Flexible stems that do not break easily and have

→

Plate 5. Chamaedorea in Habitat. A. *C. cataractarum* (a rheophyte) along river, *D. & R. Hodel 923,* Chiapas, Mexico (note flood debris among petioles); **B.** *C. stolonifera* on pitted and fissured limestone rock, *D. & R. Hodel 933,* Chiapas, Mexico; **C.** *C. rigida,* a litter collector in cloud forest, *D. & R. Hodel 941,* Oaxaca, Mexico.

C

a tendency to put forth adventitious roots readily enhance the survival ability of chamaedoreas in the forest. Although falling litter may severely damaged leaves, stems are infrequently broken and, rather, falling objects will push them to the forest floor in a prostrate manner. Here they tend to root and the growing point will reestablish itself once again in an upward fashion. As mentioned earlier the species with multiple stems and thus multiple growing points have a further advantage in their ability to send up new shoots from the base even if the upper part is totally decimated.

The juice of fruits of some chamaedoreas is highly irritating, especially to areas of tender skin. Standley and Steyermark (1958) reported that mischievous boys in Guatemala would push through the crowded market places, spreading the juice on unsuspecting persons, much to the shoppers' discomfort. The irritating fruits, while not as strong as in other genera such as *Arenga, Gaussia, Caryota,* and *Ptychosperma,* probably provide a degree of protection from predators. At the same time, however, the fruit-rachilla color scheme of *Chamaedorea,* either black fruits on red or orange rachillae or red or orange fruits on orange, green or light colored rachillae, serves as an attractant to potential dispersal agents like birds or other small animals. Foster (1986), Gallina in Coates-Estrada and Estrada (1986), Snow and Snow (1978), Wheelwright et al. (1984), and Zona and Henderson (1989) reported that birds and/or mammals were probable dispersal agents of several species of *Chamaedorea.*

Researchers have recently investigated various ecological aspects of *C. alternans* (reported as *C. tepejilote*) in Veracruz, Mexico, including growth and reproduction (Oyama 1990, 1991, Oyama and Mendoza 1990), biomass allocation (Oyama and Dirzo 1988), and effects of defoliation on survival (Oyama and Mendoza 1990). Interested readers should consult these references for details.

CHAPTER 4

THE GENUS CHAMAEDOREA

Chamaedorea is named from the Greek *chamai,* meaning on the ground or low, and *dorea,* meaning gift, apparently in reference to the generally low habit of these plants and/or the showy fruits that are easily reached. Unfortunately, Willdenow did not specify the meaning when he named the genus. The proper taxonomic and nomenclatural citations for *Chamaedorea* are given below. Following these is a description of the genus.

Chamaedorea *Willd.,* Species Plantarum 4(2): 638 & 800. 1806 (conserved name). Type: *C. gracilis* Willd. (illegitimate name) **C. pinnatifrons** (Jacq.) Oerst. (*Borassus pinnatifrons*).

Morenia Ruiz & Pav., Florae Peruvianae et Chilensis. Prodromus. 150. 1794 (rejected name). Type: *M. fragrans* Ruiz & Pav. [= *C. linearis* (Ruiz & Pav.) Mart.].

Nunnezharia Ruiz & Pav., Florae Peruvianae et Chilensis. Prodromus. 147. 1794 (rejected name). ("*Nunnezharoa*"). Type: *N. fragrans* Ruiz & Pav. [= *C. fragrans* (Ruiz & Pav.) Mart.].

Nunnezia Willd., Species Plantarum 4: 890 & 1154. 1806. (= *Nunnezharia*).

Kunthia Humb. & Bonpl., Plantas Aequinoctiales 2: 127, t. 122. 1809. Type: *K. montana* Humb. & Bonpl. [= *C. linearis* (Ruiz & Pav.) Mart.].

Chamaedorea sect. *Chamaedorea* Mart., Historia Naturalis Palmarum 3: 307. 1849.

Chamaedorea sect. *Collinia* Liebm. ex Mart., Historia Naturalis Palmarum 3: 308. 1849.

Chamaedorea sect. *Stachyophorbe* Liebm. ex Mart., Historia Naturalis Palmarum 3: 309. 1849.

Stachyophorbe (Liebm. ex Mart.) Liebm. ex Klotzsch, Allg. Gartenzeitung 20: 363. 1852. Lectotype: *S. cataractarum* (Mart.) Liebm. ex Klotzsch (= *C. cataractarum* Liebm. ex Mart.).

Chamaedorea subgenus *Stephanostachys* Klotzsch, Allg. Gartenzeitung 20: 363. 1852.

Chamaedorea subgenus *Collinia* (Liebm. ex Mart.) H. A. Wendl., Index Palmarum 57. 1854.

Chamaedorea subgenus *Chamaedorea* (Mart.) H. A. Wendl., Index Palmarum 59. 1854.

Chamaedorea subgenus *Eleutheropetalum* H. A. Wendl., Index Palmarum 58. 1854.

Collinia (Liebm. ex Mart.) Liebm. ex Oerst., Vidensk. Meddel. Dansk. Naturhist. Foren. Kjoebenhavn 1858: 5. 1859. Not Rafinesque. Lectotype: *Collinia elegans* (Mart.) Liebm. ex Oerst. (= *C. elegans* Mart.).

Chamaedorea subgenus *Chamaedorella* Oerst., Vidensk. Meddel. Dansk. Naturhist. Foren. Kjoebenhavn 1858: 23. 1859.

Chamaedorea subgenus *Chamaedoropsis* Oerst., Vidensk. Meddel. Dansk. Naturhist. Foren. Kjoebenhavn 1858: 15. 1859.

Chamaedorea subgenus *Psilostachys* Oerst., Vidensk. Meddel. Dansk. Naturhist. Foren. Kjoebenhavn 1858: 23. 1859.

Dasystachys Oerst., Vidensk. Meddel. Dansk. Naturhist. Foren. Kjoebenhavn 1858: 25. 1859. Type: *D. deckeriana* (Klotzsch) Oerst. (*Stachyophorbe deckeriana* Klotzsch) [= *C. deckeriana* (Klotzsch) Hemsl.].

Eleutheropetalum (H. A. Wendl.) Oerst., Vidensk. Meddel. Dansk. Naturhist. Foren. Kjoebenhavn 1858: 6. 1859. Type: *E. ernesti-augustii* (H. A. Wendl.) Oerst. (= *C. ernesti-augustii* H. A. Wendl.).

Spathoscaphe Oerst., Vidensk. Meddel. Dansk. Naturhist. Foren. Kjoebenhavn 1858: 29. 1859. Type: *S. arenbergiana* (H. A. Wendl.) Oerst. (= *C. arenbergiana* H. A. Wendl.).

Stephanostachys (Klotzsch) Oerst., Vidensk. Meddel. Dansk. Naturhist. Foren. Kjoebenhavn 1858: 26. 1859. Type: *S. casperiana* (Klotzsch) Oerst. (= *C. casperiana* Klotzsch).

Kinetostigma Dammer, Notizbl. Konigl. Bot. Gart. Berlin 4: 171. 1905. Type *K. adscendens* Dammer [= *C. adscendens* (Dammer) Burret].

Chamaedorea sect. *Stephanostachys* Burret, Notizbl. Bot. Gart. Berlin-Dahlem 11: 160. 1933a.

Habit and Stems

Chamaedorea displays an amazing diversity of habits (see Plate 6). Plants are unarmed and may be very small, even dwarf, to moderate in size. Perhaps the smallest is *C. tuerckheimii* that has leaves less than 15 cm long and may flower and fruit when less than 20 cm tall. The largest are forms of *C. costaricana* that form clumps ten meters across and whose stems may attain 15 meters in length. *C. woodson-*

iana and forms of the widely distributed *C. tepejilote* have robust, solitary stems that may attain ten cm in diameter yet not the length of those of *C. costaricana.*

Even within a species, plants can vary in size considerably. Health and vigor of the plant, the amount of light, and natural variation all affect the size of individual plants. Large-growing, cespitose species, like *C. costaricana, C. quezalteca, C. pochutlensis,* and *C. graminifolia,* display a good deal of variation in stem diameter and leaf size. Newer stems toward the outside of the clump are usually more robust, vigorous, and have larger leaves than older ones toward the inside.

Plants are equally diverse in form. Most species are solitary but several are cespitose and have stems that multiply mostly by basal branching, as in *C. microspadix, C. costaricana, C. pochutlensis, C. fragrans,* and *C. hooperiana.* Some species, such as *C. tepejilote* and *C. elatior,* have solitary and cespitose forms. The basal shoots of species with cespitose habit may be erect immediately from the base of the plant or, as Tomlinson (1961) pointed out, may grow horizontally below or at ground level for some distance before turning upward.

Rarely, plants are rhizomatous, as in *C. brachypoda* and *C. rhizomatosa* or stoloniferous, as in *C. stolonifera.* A few species, such as *C. serpens* and some forms of *C. costaricana, C. elatior,* and *C. quezalteca,* have aerial branches. In at least one species, *C. cataractarum,* the stem branches in a dichotomous manner. Tomlinson and Moore (1966) and Fisher (1974) thoroughly investigated this unique habit.

Stems of most chamaedoreas are slender and range from one to three cm in diameter. However, the stem may be as small as six mm in diameter, as in *C. tenerrima,* or as large as ten cm in diameter, like those of some forms of *C. costaricana, C. tepejilote,* and *C. woodsoniana.* Stems are mostly erect and several meters tall with easily discernible internodes, as in *C. oblongata, C. glaucifolia, C. pochutlensis,* and *C. tepejilote.* There are no truly stemless chamaedoreas; however, some species, like *C. radicalis* and *C. brachyclada,* appear stemless although they actually possess a short, often curved stem with very congested nodes. Such stems are usually subterranean or at or near ground level and buried in the leaf litter, giving the plant a stemless appearance. Plants may appear stemless for their entire life or with age may form a short, visible stem, as in *C. pittieri, C. parvifolia, C. volcanensis,* and some forms of *C. radicalis.*

Several species, such as *C. palmeriana, C. chazdoniae, C. correae, C. pumila, C. castillo-montii,* and *C. sullivaniorum* have decumbent stems that creep along the ground in a prostrate manner before turning upward. These plants will often initially grow in an erect manner but soon fall over and become prostrate. Even species noted for their erect stature, like *C. oblongata,* sometimes have creeping stems in certain situations in the wild. Species with cespitose habit that form large clumps, such as *C. costaricana, C. quezalteca,* and *C. pochutlensis,* will frequently have their largest stems leaning outward, sometimes to the point of being decumbent or sprawling along the ground. The highly variable *C. elatior* from Mexico and Guatemala is the only truly scandent or climbing species in the genus. Its stems may attain 30 meters in length as they snake into overhead vegetation. Some of the large clustering species may have outward-leaning stems that are supported on the adjacent vegetation; however, this is passive rather than active climbing.

Stems of chamaedoreas are mostly clean, smooth, green, ringed at various distances, and usually variously and minutely white-spotted. The internodes of *C. oreophila* and *C. vulgata* are quite distinctly white-spotted. Many chamaedoreas have self-cleaning leaves that tend to fall on their own or can be easily removed after senescence. With others, old leaf bases are persistent and partially or wholly cover the stem until they weather away or the emerging inflorescences thrust them off. Most species with short stems, especially those appearing stemless, have very persistent leaf bases. Internodes may be elongated, to 40 cm or more in length, as in some of the larger species, such as *C. costaricana,* or very com-

→

Plate 6. Habit, Stems, and Leaves of Chamaedorea. A. Bifid leaves, *C. tenella,* cultivated, garden of B. Shannon, Hollywood, Florida; **B.** Stemless, pinnate-leaved *C. macrospadix* with inflorescences arising from leaf litter, *D. & M. Hodel 692A,* Monteverde, Costa Rica; **C.** Wide pinnae, *C. warscewiczii, Hodel & Grayum 702,* trail down to river, C.A.T.I.E., Turrialba, Costa Rica; **D.** Stems clustering by short creeping basal branches from which new shoots arise, *C. costaricana,* cultivated, Wahiawa Botanic Garden, Honolulu, Hawaii; **E.** Slender solitary upright stem and bifid leaves, *C. amabilis, Hodel 741 bis,* cultivated, Virginia Robinson Botanical Garden, Beverly Hills, California, *ex* El Valle, Panama; **F.** Solitary upright stem, numerous linear pinnae, *C. glaucifolia,* cultivated, Waimea Falls Park & Arboretum, Honolulu, Hawaii.

A
B
D
F

pacted and barely visible, to as little as two to three mm long in the dwarf, stemless species like *C. stenocarpa.* Leaf scars or nodes are pale or brown and may be shallow or prominent and swollen, and narrow or wide. When nodes are shallow and well spaced, the stem is greenish and smooth. When they are prominent, swollen, and congested, the stem is ridged, rough, and brownish.

In most species, the age and vigor of the plant and the amount of light greatly affect the length of the internode. Internodes tend to be longer on young, vigorous plants in relatively low light while they are often shorter on older, less vigorous plants and/or those in higher light. These conditions have much less effect on stems with naturally short or very short internodes.

Aerial prop roots supporting stems of chamaedoreas at the base are common. Some species, like *C. tepejilote* and *C. rojasiana,* are known for their production of aerial roots. Other species known for their decumbent habit, such as *C. correae, C. chazdoniae,* and *C. serpens,* will root along the stem, especially at the nodes. Most species not normally of a decumbent habit will still root along the stem if it has fallen or been knocked over and if humidity is high.

Leaves

Chamaedorea displays as much diversity in size and shape of leaves as it does in habit and stems (see Plate 6). Most species have regularly, less often variously, pinnate leaf blades. Some species, as in *C. ernesti-augustii, C. stolonifera,* and *C. fragrans,* have bifid leaves undivided except at the apex where they are bilobed or incised to various degrees. The leaf blades of *C. fragrans* are apically incised to well over one-half their length, those of *C. geonomiformis* to one-third to one-half, and those of *C. tenella* to less than one-third. *C. tuerckheimii* is the only species in the genus with a leaf that can be considered entire. Although the leaf margin is toothed, the blade is oval and undivided except infrequently when it is minutely notched at the apex.

There are several species, such as *C. adscendens, C. allenii,* and *C. volcanensis,* that may have pinnate or bifid leaves on one individual, a feature often related to age. Younger plants have bifid leaves, juvenile plants have variously pinnate leaves, and older plants have leaves completely and regularly pinnate. These cannot be considered strictly juvenile and adult forms, though, since these species frequently flower with bifid leaves. In other species, like *C. metallica, C. castillo-montii, C. dammeriana, C. pauciflora,* and *C. pygmaea,* the phenomenon appears to be less a function of age and more likely due to natural variation, with some of the adult population having a pinnate leaf or leaves in a crown of bifid ones. Still others, as in the highly variable and wide-ranging *C. pinnatifrons,* are predominantly pinnate-leaved although populations of bifid-leaved individuals exist.

Leaves vary greatly in size from species to species. As mentioned earlier, those of *C. tuerckheimii* are the smallest in the palm family. In contrast, leaves of *C. arenbergiana, C. tepejilote, C. woodsoniana,* and *C. carchensis* may be two to three meters in length. It should also be noted that the size of leaves within a species varies greatly and a number of factors affects it.

Although leaves emerge in an erect manner and often remain that way for some time, most usually become more or less horizontal eventually. A few species, like *C. brachyclada,* retain their erect, upswept leaves. Crowns may be very leafy and contain 15 or more leaves, as in *C. rigida* and *C. dammeriana.* In contrast, *C. brachyclada* usually does not have more than three good leaves. Within a species, the number of leaves in a mature plant is a good indication of its health and vigor. The more leafy the crown, the healthier the plant when compared with other individuals of the same species. Cultivated plants tend to have fewer leaves than their wild counterparts.

Leaf sheaths may be short or long and tubular, as in *C. oblongata,* or splitting opposite the petiole and tubular only around its base, as in *C. pumila.* Several species including *C. costaricana* and *C. quezalteca* have prominent ligules at the apex of the sheath on either side of the petiole. The ligules are diagnostic but often wither away and are most apparent on newly emerging leaves. Sheaths are green but become brown with age and persist on the stem and

→

Plate 7. Inflorescences of Chamaedorea. A–D Staminate; **A.** *C. geonomiformis, Hodel 792A,* cultivated, research collection, Los Angeles; **B.** *C. ernesti-augustii, Hodel 755A,* cultivated, research collection, Los Angeles; **C.** *C. glaucifolia, Hodel 807,* cultivated, garden of R. Palmer, Whittier, California; **D.** *C. oreophila, Hodel 765A,* cultivated, research collection, Los Angeles. **E** Pistillate; **E.** *C. alternans, D. & R. Hodel 918,* Catemaco, Veracruz, Mexico (note multiple arrangement at a node).

B
C
E

invariably are longitudinally striate-nerved. They may be thick, leathery, and durable, as in *C. hooperiana* and *C. radicalis,* or thin and papery as in *C. amabilis.* Several species have sheaths that are conspicuously costate due to a prominently raised median ridge extending from the petiole.

Petioles are variable in length, even among individuals of the same species. The amount of light greatly influences their length. *C. foveata* and *C. woodsoniana* have petioles that are densely pitted with irregular, conspicuous depressions giving living material a rough texture. Petioles are essentially three-sided, being flattened or grooved above and rounded below. Upper and lateral surfaces are green and often minutely white-spotted. Nearly all species have petioles with a yellow or pale green band below extending from the rachis onto the sheath. The band may be nearly indiscernible, as in *C. fragrans* and *C. brachyclada,* or prominent and distinct, as in *C. tepejilote.* Several species, such as *C. glaucifolia, C. pochutlensis,* and *C. costaricana,* have petioles, rachises, and sometimes sheaths variously covered with a more or less deciduous glaucous indument. Rachises are usually green, frequently minutely white-spotted, essentially two-sided and angled above, and rounded and yellow or pale green below.

Blades are various shades of green, the most typical being the rich greens of *C. tepejilote, C. elegans,* and *C. oblongata.* However, several species including *C. adscendens, C. metallica, C. oreophila,* and *C. radicalis* have bluish green blades. Others are gray-green, such as those of *C. glaucifolia* and *C. graminifolia.* Some species, like *C. sullivaniorum* and *C. pumila,* have iridescent green blades that seem to sparkle slightly when light is reflected at a critical angle. A few species, such as *C. sullivaniorum* and *C. queroana,* have slightly mottled blades. Surfaces of blades may be glossy, as in *C. oblongata,* or dull green like those of *C. microspadix.*

Although leaf blades are frequently the same shade of green on both sides, many are paler on the underside. Some species, such as *C. microspadix* and *C. nubium,* have various forms which may be distinctly grayish white on the underside. Blades also vary in their texture, some being thick and leathery, as in *C. adscendens, C. correae, C. guntheriana, C. pumila,* and *C. sullivaniorum.* However, others, such as *C. stenocarpa* and *C. amabilis,* are thin and papery.

Pinnae may be relatively few, less than ten on each side of the rachis, as in *C. oblongata* and *C. sartorii,* or numerous, to 25 or more on each side, as in *C. hooperiana, C. pochutlensis,* and *C. woodsoniana.* Pinnae may be short or long, narrow or wide, acuminate, and straight or sigmoid. Many species have pinnae less than 15 cm long while others, such as *C. woodsoniana, C. carchensis,* and forms of *C. tepejilote,* have pinnae that may exceed 60 cm in length. *C. arenbergiana, C. warscewiczii,* and *C. lucidifrons* have pinnae to 15 cm wide while those of *C. glaucifolia* may be less than one cm wide. Middle pinnae of the leaf blade are generally the longest; pinnae decrease slightly in length and sometimes width toward the base of the rachis. They decrease in length more noticeably toward the apex of the rachis with the apical pair of pinnae usually the shortest. However, the apical pair are often as wide or conspicuously wider than the largest middle pinnae or other pinnae combined, as in *C. rojasiana, C. verapazensis,* and some forms of *C. pinnatifrons.* A few species, such as *C. pygmaea,* have the apical pair of pinnae as long as the longest middle ones or, as in the striking *C. tenerrima,* they are conspicuously longer. The terminal pinnae of the climbing, vinelike *C. elatior* are usually strongly reflexed and hooklike, enabling the plant to climb into adjacent vegetation.

Pinnae are broadly lanceolate, nearly rhombic, as in some forms of *C. oblongata*; generally lanceolate, as in most species; or long-lanceolate or linear, like those of *C. glaucifolia* and *C. hooperiana.* Many species, such as *C. sartorii, C. oblongata,* and *C. warscewiczii,* have sigmoid pinnae while those of others may be straight or nearly so, as in *C. radicalis, C. pochutlensis,* and *C. glaucifolia.* Pinnae are usually regularly arranged on the rachis; however, those of *C. glaucifolia* and *C. klotzschiana* are aggregated in groups. Pinnae with their lower margins decurrent on the rachis distinguish several species (Hodel and Uhl 1990). Margins of pinnae of *C. brachyclada, C.*

→

Plate 8. Flowers of Chamaedorea. Cultivated, California. **A.** *C. elegans,* staminate, *Hodel 787,* research collection, Los Angeles; **B.** *C. metallica,* staminate, *Hodel 808,* research collection, Los Angeles; **C.** *C. pinnatifrons,* staminate, *Hodel 815,* garden of L. Rossten, Huntington Beach; **D.** *C. tenella, Hodel 714 bis,* research collection, Los Angeles, *ex* Osa Peninsula, Costa Rica; **E.** *C. tuerckheimii,* pistillate, *Hodel 750,* research collection, Los Angeles; **F.** *C. fragrans,* staminate, *Hodel 866,* research collection, Los Angeles; **G.** *C. microspadix, Hodel 759,* garden of R. Palmer, Whittier; **H.** *C. arenbergiana,* staminate, *Hodel 803,* garden of P. Sullivan, Ventura.

B
D
F
H

stenocarpa, C. scheryi, and *C. undulatifolia* are conspicuously decurrent along the rachis to the next leaflet while those of *C. cataractarum* and some forms of *C. tepejilote* are only briefly so.

Nervation of pinnae is quite variable between and within species. Some species, such as *C. oblongata* and *C. pochutlensis,* have only a prominent midrib with all the other nerves faint or inconspicuous. Others, like *C. sartorii, C. costaricana,* and *C. quezalteca,* have a prominent midrib, and one or more prominent primary nerves on either side of this, and numerous but inconspicuous secondaries. Still others including *C. allenii, C. linearis,* and *C. tepejilote* have several prominent primary nerves, conspicuous secondaries between each pair of primaries, and numerous and inconspicuous tertiaries. Primary nerves may be prominent, keeled, and drying a distinct yellow, as in *C. tepejilote, C. alternans,* and *C. volcanensis,* or low and drying pale, as in *C. oblongata* and *C. neurochlamys. C. woodsoniana* and *C. warscewiczii* have primary nerves that in the live state are conspicuous but not raised above although they are prominent and elevated below.

Even species with bifid blades exhibit a variety of nervation. *C. stolonifera* and *C. metallica* have several conspicuous primary nerves not noticeably raised. In contrast, *C. pumila, C. tuerckheimii,* and *C. sullivaniorum* have the primary nerves so prominent and elevated that the blade takes on a corrugated effect.

Inflorescences

Although not so readily apparent, *Chamaedorea* displays a large amount of floral variation that rivals that of stem and leaf. Inforescences vary greatly (see Plate 7), and may be interfoliar, as in *C. oreophila* and *C. geonomiformis,* or infrafoliar, as in *C. oblongata, C. brachypoda, C. fragrans, C. pochutlensis,* and *C. tepejilote.* In some, such as *C. glaucifolia, C. klotzschiana,* and *C. seifrizii,* they are infrafoliar but often emerge from among the old, persistent sheaths. Some species may have inflorescences that are inter- or infrafoliar or, as in *C. metallica, C. ernesti-augustii, C. neurochlamys, C. pinnatifrons,* and others, they may be interfoliar when they emerge but are infrafoliar in fruit.

Inflorescences are usually solitary at a node although subgenera *Morenia* (*C. linearis*) and *Moreniopsis* (*C. angustisecta, C. pauciflora*), and several species of subgenus *Stephanostachys,* including *C. alternans, C. deckeriana,* and *C. oreophila,* have several staminate inflorescences at a node. Fisher and Moore (1977) and Davis (1979) provided informative accounts of this phenomenon. Multiplicity of inflorescences effectively extends anthesis and increases the opportunities for pollination since individual inflorescences open consecutively, one after the other. In *C. alternans,* several pistillate inflorescences may occur at a node.

Inflorescences are usually erect to spreading, but frequently drooping when laden with fruits. They are usually branched to one (rarely two) order(s) or less often may be spicate or furcate, as in *C. allenii, C. deckeriana, C. tenella,* and *C. pauciflora.* Some species, such as *C. sullivaniorum* and *C. pygmaea,* have branched staminate but spicate pistillate inflorescences. The staminate inflorescence of each species generally has more rachillae than the pistillate one. In many species, such as *C. metallica, C. ernesti-augustii,* and *C. dammeriana,* inflorescences may be spicate, furcate, or have several branches. Although these cases reflect natural variation, in some species an inflorescence with more rachillae often appears to be a function of age. It is not uncommon, for example, to find the first inflorescences spicate in plants of *C. radicalis, C. volcanensis, C. cataractarum,* and *C. undulatifolia* among others. Then, at subsequent flowerings, furcate inflorescences appear and, later, inflorescences with numerous rachillae.

Peduncles are short, as in *C. fragrans,* or long, as in *C. macrospadix,* and usually slender but may be robust, as are those of *C. woodsoniana* and some forms of *C. tepejilote.* They normally emerge erect or spreading but the pistillate often become drooping or nodding when heavily laden with fruits. They are somewhat flattened basally and cylindric apically. Where exposed, they are greenish or less often pale or whitish at anthesis and red-orange, orange, or infrequently green in fruit.

Sheathing the peduncle are a relatively short, tubular prophyll with a tapering, bifid tip and one to several peduncular bracts that are essentially tubular, as in *C. oblongata,* or less frequently loosely sheathing or even slightly inflated, as in *C. robertii.*

→

Plate 9. Fruits of Chamaedorea. A. *C. nubium, Hodel & Castillo 899,* Alta Verapaz, Guatemala; **B.** *C. radicalis, Hodel 820,* cultivated, garden of R. Palmer, Whittier, California; **C.** *C. seifrizii,* cultivated, Miami, Florida; **D.** *C. microspadix,* cultivated, Lakewood, California.

B
D

Peduncular bracts are obliquely open and acuminate, acute-acuminate, or bifid apically. The apical portion of one overlaps the basal portion of the one just above it and they increase in length distally. The longest is usually the second-to-the-last or last visible one although the last visible bract may conceal a short rudimentary one. Bracts are membranous, as in *C. amabilis,* papery, as in *C. oblongata* and most others, or leathery, as in *C. robertii.* All are longitudinally striate nerved, a feature that is more evident when bracts are dry. At anthesis they are green, as in *C. pumila,* or brownish, as in *C. glaucifolia.* Rarely are they another color, such as the burgundy-red in *C. robertii.* Bracts are more or less persistent; however, in some species, such as *C. amabilis* and *C. deckeriana,* they will have become shredded or weathered away by the time fruits mature.

The rachises may be short, as in *C. stolonifera,* or long, as in *C. woodsoniana.* They are rounded or longitudinally angled and usually greenish at anthesis or less often whitish, as in subgenus *Morenia.* The rachis usually becomes red-orange or orange in mature fruit or rarely green, as in *C. radicalis.*

The rachillae or flower-bearing axes are usually greenish or pale and slender at anthesis and swollen or thickened and red-orange or orange in fruit. Rachillae of *C. radicalis* and subgenus *Morenia* are greenish in fruit. Rachillae may be erect, spreading, or drooping. Generally, the staminate rachillae of each species are more drooping and the pistillate stiffer or erect although the latter are often nodding when laden with fruits.

Flowers

Chamaedoreas have small or minute, usually no more than four to five mm in diameter, symmetrical, staminate or pistillate flowers densely or remotely arranged in spirals and sessile or partly sunken in elliptic or rounded pits in the axis (see Plate 8). Most species have flowers arranged in a solitary manner although subgenera *Moreniella* (*C. radicalis* and *C. microspadix*) and *Morenia* have staminate flowers paired or in short acervuli (lines). Less frequently, as in subgenus *Stephanostachys,* staminate flowers are contiguous in bud. Other species, such as *C. adscendens, C. robertii, C. volcanensis,* and *C. castillo-montii,* have staminate flowers that are densely arranged but not contiguous in bud. Several species of subgenus *Stephanostachys,* such as *C. allenii, C. oreophila,* and *C. deckeriana* as well as a few species in other subgenera, have pistillate flowers contiguous or nearly so. Generally, staminate flowers of a given species appear to be more densely arranged than their pistillate counterparts.

Flowers of some species, especially the staminate, are aromatic. Some find the odor of the staminate flowers of a few, such as *C. elatior, C. pochutlensis,* and *C. seifrizii,* overpowering or objectionable. Flowers of other species, such as *C. angustisecta, C. pauciflora,* and the aptly named *C. fragrans,* are delightfully sweet and fragrant. Those of *C. fragrans* are legendary, as Spruce (1869) noted: "My specimens, dried fourteen years ago, still give out their fine odour of mignonette with a dash of primrose when hot water is poured on them." The flowers of *C. deckeriana* have a mysterious, primeval scent, reminiscent of a dark, dense, wet forest.

The calyx is usually low and ringlike or cup-shaped; less frequently it is well developed and prominent in bud as in *C. crucensis, C. allenii,* and *C. volcanensis.* It is greenish or pale in color and three-lobed or -parted to various degrees, the three sepals being united basally or distinct and imbricate.

The staminate corolla may have valvate, free, and erect or apically spreading petals. Less frequently the petals are connate at their tips and there adnate to the pistillode and the corolla opens by lateral slits, as in subgenus *Chamaedorea* (*C. pinnatifrons*). Staminate flowers and the fruiting perianth of members of subgenus *Chamaedorea* are usually conspicuously ribbed when dry. Petals are usually greenish or yellow, or rarely orange, as in *C. fragrans,* or white, as in *C. microspadix, C. crucensis, C. volcanensis,* and subgenus *Morenia.* There are six stamens with short, broad or awl-shaped filaments and dorsifixed, included, or didymous or oblong anthers. Pistillodes vary and may be columnar, expanded basally, and sometimes expanded and/or trilobed apically.

The pistillate corolla usually has imbricate petals; less frequently they are valvate, as in subgenus *Eleutheropetalum* (*C. ernesti-augustii, C. metallica, C. stolonifera,* and *C. sartorii*), or rarely connate and opening through a terminal aperture, as in *C. elegans* (subgenus *Collinia*). Petals are usually yellow or greenish yellow, rarely orange, like those of subgenus *Eleutheropetalum* and *C. fragrans,* or whitish as, in subgenus *Morenia.* Staminodes may be present and toothlike or absent. Pistils are ovoid or globose, greenish, yellow, or pale in color with the styles lacking or rarely present and then short. Stigmas are sessile or nearly so, small, and recurved.

Fruits

Fruits are small and variable in shape, frequently among individuals of the same species, but are usually globose to oblong (see Plate 9). Rarely they are angled from mutual pressure, as in those species with densely packed or contiguous pistillate flowers, as is often the case with *C. arenbergiana, C. deckeriana, C. allenii, C. crucensis, C. zamorae,* and *C. oreophila.* The shape of the fruits of these species is dependent on the percent of fertilized flowers. Where few flowers are fertilized, fruits have sufficient space to develop and become more or less globose. However, where a high rate of fertilization has occurred, swelling fruits will be angled by mutual pressure of adjacent fruits as they develop.

Fruits are rarely curved, as in *C. angustisecta, C. falcifera, C. neurochlamys, C. brachypoda,* and some forms of *C. oblongata.* They are usually black, sometimes orange or red, or rarely yellow. *C. microspadix, C. radicalis,* subgenus *Morenia,* and some members of subgenus *Stephanostachys* (*C. deckeriana* and *C. oreophila*), have red or orange fruits. The *C. pinnatifrons* complex has fruits that mature yellow and then persist on the rachillae and age to dark brown or black, or progress through a yellow or orange stage to mature dark brown or black. Epicarps are generally smooth although those of *C. allenii* are rough; mesocarps are fleshy; and endocarps are thin. As in fruits, seeds are variable in shape, even among individuals of the same species, but are usually globose or ellipsoid. Seeds may be angled from mutual pressure as were their fruits in those species mentioned above with densely packed or contiguous pistillate flowers.

Eophylls are bifid although in a few species, such as *C. elegans, C. stenocarpa, C. undulatifolia, C. brachyclada,* and *C. scheryi,* they are pinnate (Hodel & Uhl 1990). Usually, seedling leaves will remain bifid for some time before becoming pinnate. However, in a few species, like *C. pachecoana, C. liebmannii, C. whitelockiana,* and *C. parvisecta,* seedling leaves will become pinnate at an early stage, perhaps as early as the third or fourth leaf. These species and those with pinnate eophylls often prove troublesome to identify in the field as they appear at a young stage to be a pinnate-leaved, dwarf species (with a reduced number of pinnae) distinct from its adult counterpart that has numerous pinnae. The problem is compounded when the plants tend to flower at a young stage and have much reduced inflorescences, thus appearing distinct from those of their nearby, older counterparts. Presenting a similar problem are species such as *C. dammeriana, C. volcanensis,* and *C. queroana* that tend to flower at a young stage and have reduced inflorescences while still retaining their bifid juvenile leaves. These are easily mistaken as distinct species from their older, pinnate-leaved counterparts with well developed inflorescences.

CHAPTER 5

THE SPECIES OF CHAMAEDOREA

Accounts of the individual species, including nomenclatural citations, descriptions, and discussions, are presented in the appropriate subgenus. A key to the subgenera of *Chamaedorea* is provided. The number in parentheses following the subgenus in the key is the page on which that subgenus and its species are treated.

Key to the Subgenera of *Chamaedorea*

Petals connate in flowers of both sexes. .... **I. Collinia** (p. 36)
Petals distinct in flowers of both sexes.
  Petals not thickened in fruit but distal portion deciduous and thin base only remaining within calyx in fruit.
    Petals of pistillate flowers bright orange, very briefly imbricate basally, cucullate and valvate apically; staminate inflorescences solitary at node; fruits black. .... **II. Eleutheropetalum** (p. 40)
    Petals of pistillate flowers white to yellow-green, imbricate basally, spreading apically; staminate inflorescences multiple at node; fruits red. .... **III. Morenia** (p. 50)
  Petals thickened and persistent in fruit.
    Staminate flowers with petals connate apically and there adnate to pistillode, corolla opening by lateral slits; perianth usually strongly nerved when dry. .... **IV. Chamaedorea** (p. 56)
    Staminate flowers with petals spreading or erect, not connate apically.
      Staminate flowers contiguous, densely crowded in 4–7 rows along axis or axes. .... **VIII. Stephanostachys** (p. 232)
      Staminate flowers separated, borne singly, or paired or in short lines or clusters.
        Staminate inflorescences 3–4 at nodes. .... **VII. Moreniopsis** (p. 228)
        Staminate inflorescences solitary at a node.
          Staminate flowers borne singly. .... **V. Chamaedoropsis** (p. 120)
          Staminate flowers paired or in short lines. .... **VI. Moreniella** (p. 222)

More work is needed on the subgenera of *Chamaedorea* so that they may be better circumscribed. In some instances the boundaries between the subgenera are somewhat blurred with certain species of one subgenus florally approaching species of another subgenus. For example, staminate flowers of *C. graminifolia, C. hooperiana, C. elatior,* and *C. glaucifolia,* are bright yellow, strongly and distinctly aromatic, only faintly nerved or nerveless when dry, and, at times, tend to spread slightly apically well after anthesis. Although I include them in subgenus *Chamaedorea* since at anthesis the staminate petals are connate apically and the corolla opens by lateral slits, they approach several cespitose species in subgenus *Chamaedoropsis* florally, namely *C. seifrizii* and *C. pochutlensis.* These latter species also have nerveless, bright yellow, aromatic, staminate flowers but the petals spread apically yet often only slightly so, the tips remaining incurved over the stamens and nearly adnate to the pistillode. Perhaps these are transitional species between the two subgenera and are representative of the ongoing evolutionary process.

I. Subgenus COLLINIA

Chamaedorea subgenus **Collinia** (Liebm. ex Mart.) H. A. Wendl., Index Palmarum 57. 1854.

Chamaedorea sect. *Collinia* Liebm. ex Mart., Historia Naturalis Palmarum 3: 308. 1849.
Collinia (Liebm. ex Mart.) Liebm. ex Oerst., Vidensk. Meddel. Dansk. Naturhist. Foren. Kjoebenhavn 1858: 5. 1859. Not Rafinesque.

Key Characters: petals connate in flowers of both sexes, corolla opening apically by a triangular aperture.

Distribution: one variable species, widely distributed on the Atlantic slope from eastern and southern Mexico to Guatemala and Belize; to about 1,400 m elevation; widely cultivated.

1. Chamaedorea elegans *Mart.,* Linnaea 5: 204. 1830. Type: Mexico, Veracruz, Barranca de Tioselo, *Schiede s.n.* (*1015*) [holotype M (photo!); isotypes BR (photo!), MO!]. Epithet: from the Latin *elegans* meaning elegant. **PLATES 8(A), 10, 125(D, E).**

Chamaedorea elegans var. *angustifolia* M. Martens & Galeotti, Bull. Acad. Roy. Sci. Bruxelles 10: 122. 1843. Type: Mexico, Veracruz, *Galeotti 4976* [holotype BR (photo!)].

Collinia elegans var. *angustifolia* (M. Martens & Galeotti) M. Martens & Galeotti, Bull. Acad. Roy. Sci. Bruxelles. 10: 122. 1843.

Chamaedorea humilis Liebm. ex Mart., Historia Naturalis Palmarum 3: 308. 1849. Type: Mexico, Oaxaca, *Liebmann s.n.* (later numbered *6586*) (holotype C!; isotype P!). Epithet: from the Latin *humilis* meaning low-growing.

Chamaedorea deppeana Klotzsch, Allg. Gartenzeitung 20: 362. 1852. Type: none designated.

Chamaedorea helleriana Klotzsch, Allg. Gartenzeitung 20: 362. 1852. Type: none designated.

Collinia deppeana (Klotzsch) Klotzsch, Allg. Gartenzeitung 20: 362. 1852.

Collinia elegans (Mart.) Oerst., Vidensk. Meddel. Dansk. Naturhist. Foren. Kjoebenhavn 1858: 6. 1859.

Collinia humilis (Liebm. ex Mart.) Oerst., Vidensk. Meddel. Dansk. Naturhist. Foren. Kjoebenhavn 1858: 6. 1859.

Nunnezharia elegans (Mart.) O. Kuntze, Revisio Generum Plantarum 2: 730. 1891.

Nunnezharia humilis (Liebm. ex Mart.) O. Kuntze, Revisio Generum Plantarum 2: 730. 1891.

Nunnezharia pulchella (Linden ex Hemsl.) O. Kuntze, Revisio Generum Plantarum 2: 731. 1891.

Chamaedorea pulchella Linden ex Hemsl., Bot. Mag. 130: t. 7959. 1904. Type: Cult., *Hemsley s.n.* (holotype K!). Epithet: from the Latin *pulchellus* meaning beautiful and little.

Kunthia deppii Hort. in Zucc., Allg. Gartenzeitung 2: 245. 1834.

Chamaedorea lindeniana Hort. in H. A. Wendl., Index Palmarum 12. 1854. Not H. A. Wendl.

Chamaedorea martiana Hort. in H. A. Wendl., Index Palmarum 12. 1854. Not H. A. Wendl.

Chamaedorea elegantissima Hort. in Kerch., Les Palmiers 239. 1878.

Neanthe bella O. F. Cook, Science 86: 120. 1937. Name of no botanical standing.

Neanthe elegans (Mart.) O. F. Cook, Science 86: 122. 1937. Name of no botanical standing.

Neanthe neesiana O. F. Cook, Science 86: 122. 1937. Name of no botanical standing.

Common Names: parlor palm, good-luck palm, neanthe bella, collinia.

Vernacular Names: pacaya, pacayito, xate, kiik–Guatemala; tepejilote, palma camedor, pesmilla, palmilla de hojas, angostas, palmita camedor, palma fina, tutchast—Mexico.

Habit: solitary, slender, erect or infrequently decumbent, to 2 m tall or more but flowering when very small and less than 30 cm tall, then appearing stemless.

Stem: 0.8–1.5 cm diam., green, densely ringed with prominent nodes, internodes 0.5–3 cm long.

Leaves: 5–8, spreading, pinnate; **sheath** 8–20 cm long, very obliquely open nearly to base and tubular only in lower ⅓, short ligule apically on either side of petiole, margins brownish and ragged, light green or whitish below margin, longitudinally green-striate-nerved; **petiole** 10–40 cm long, slender, grooved and green above, rounded and pale below; **rachis** 15–60 cm long, very slender, 4-sided, angled and green above, rounded below with narrow yellow band extending onto petiole; **pinnae** 11–21 on each side of rachis, 15–30 × 1–3 cm, linear to narrowly lanceolate, long-acuminate, contracted basally, thin, dark green, midrib prominent and pale, elevated or keeled above, 1–2 less prominent primary nerves on each side of midrib, secondaries numerous, faint.

Inflorescences: interfoliar, erect, shorter than or equal to or greatly exceeding leaves; **peduncles** 15–90 cm long, 5–9 mm wide at base, ± flattened, 4–6 mm wide at apex, rounded, green where exposed in flower, red-orange in fruit. **Staminate** with 4–7 **bracts**, uppermost exceeding peduncle, largest to 35 cm long, acuminate and bifid apically, fibrous or ± papery, longitudinally striate-nerved, brown in flower; **rachis** 1.5–20 cm long, longitudinally ridged or angled, green; **rachillae** 5–35, lower ones the longest, these to 15 cm long, becoming progressively shorter toward apex of rachis, 2 mm diam., spreading, simple or branched, sharply angled, green.

Pistillate similar to that of staminate but with 6–10 **bracts**; **rachis** slightly shorter, flexuous, orange in fruit; rachillae fewer in number and shorter than those of staminate, to 10 cm long, ± stiff, green in flower, red-orange in fruit.

Flowers: Staminate in remote to moderate spirals, 3 × 4 mm, depressed-globose, yellow, aromatic, nerved when dry, sessile or slightly sunken in elliptic depressions; **calyx** 0.75–1 × 2–2.5 mm, moderately lobed, green, **sepals** connate in basal ½, rounded apically; **petals** 2.5 × 2 mm, connate, corolla opening by a 3-angled pore apically, fleshy; **stamens** 1.5–2 mm long, **filaments** connate, whitish, **anthers** 0.75–1 mm long, entire, yellow; **pistillode** equal to or slightly exceeding corolla, 6-angled, flared slightly apically, pale yellow-green.

Pistillate in remote spirals, 3 × 2.75 mm, globose, yellow, nerved when dry, slightly sunken in elliptic depressions 1–1.5 mm long; **calyx** 1 × 2 mm, deeply lobed, green, **sepals** connate in basal ⅓, rounded to acute apically; **petals** 2–2.5 × 1.5–2 mm, connate, corolla opening by a 3-angled pore apically, thick, fleshy; **staminodes** lacking; **pistal** 2 × 1–1.5 mm, depressed-globose, green; **stigma** lobes separated, sessile, blunt, clear-colored.

Fruits: 4–7 mm diam., globose, black; seeds 3–6 mm diam., globose; eophyll pinnate.

Distribution: MEXICO. Chiapas. Hidalgo. Oaxaca. Puebla. San Luis Potosi. Tabasco. Veracruz. GUATEMALA. Huehuetenango. Alta Verapaz. Peten. BELIZE. Cayo. Moist or wet, dense forest on the Atlantic slope; to 1,400 m elevation; often on limestone.

Discussion: William Schiede discovered *C. elegans* in the Barranca de Tioselo, Veracruz, Mexico, and sent material to Martius (1830) who described and named it. Guillaumin (1923b) reported that Linden introduced it to European gardens in 1839, and it was cultivated at the Musee de Paris in France since then. *C. elegans* has been cultivated as early as 1900 in California (Reidel 1957).

Moore (1976) considered *Chamaedorea humilis* to be a synonym of *C. elegans*. Liebmann (1846) first mentioned *C. humilis* by name only when he listed *Collinia humilis* in the account of his Mexican collections. Martius (1849) actually described it a few years later as *Chamaedorea humilis* and placed it in section *Collinia*. Oersted (1859) stated that *C. humilis* was similar to *C. elegans* but differed in the longer, narrower leaflets and simply branched inflorescences. He established *Collinia* as a separate genus from *Chamaedorea*. Oersted also stated that, contrary to Liebmann's description of the species as stemless, plants grown in the botanical garden in Denmark did form a short trunk to 50 cm tall with internodes 2.5 cm long. Later, Burret (1933a) stated, citing Liebmann and Galeotti collections from Oaxaca, that sometimes the rachillae of *C. humilis* are branched, as in *C. elegans*, and are not simple as originally described. Burret maintained *Collinia* as a separate genus from *Chamaedorea* but considered the species *humilis* to be in the latter genus.

Chamaedorea elegans exhibits a wide degree of variability. Standley and Steyermark (1958) stated that it is difficult to find differences that are of specific significance when considering plants from the entire range. Cook (1937, 1943a) and Cook and Joyner (1938, 1939) considered the Guatemalan material distinct and referred to it under the invalid name *Neanthe bella*. In general, these appear to be more delicate, smaller, and of a more compact nature than plants from the northern part of the range in Mexico. Cook and Joyner (1939) referred to these latter plants from Mexico under the invalid name *Neanthe elegans*. Its more rapid rate of growth, generally larger size, and more open and rangy habit distinguish it. Contrary to this, commercial seedsmen use the name "neanthe bella" to refer to a form with broader leaflets from San Luis Potosi and Hidalgo near the northern end of the range in Mexico. They refer to Cook's *Neanthe elegans* and some similar forms of the Guatemalan material as "collinia."

The length of the peduncles and shape of the leaflets exemplify well the variability in *C. elegans*. Hemsley (1904) named and illustrated a form in *Curtis' Botanical Magazine* that Linden collected in Mexico and introduced to European gardens in 1885. It has drooping linear leaflets and erect spadices with long peduncles up to a meter in length that greatly exceed the leaves. Hertrich (1951, p. 23) and McCurrach (1960, p. 43) have a photograph of a plant of this form taken at the Huntington Botanical Gardens in San Marino, California, erroneously captioned as *C. graminifolia*. A form with broader leaflets and with inflorescences not exceeding the leaves occurs in Mexico and a form with narrow leaflets and inflorescences again not exceeding the leaves grows in Guatemala.

As one of the most popular and widely grown members of the genus, *Chamaedorea elegans* has been illustrated in the literature over the years, sometimes incorrectly. Bailey (1935, p. 732) and *Garden & Forest* (Anonymous, 1895) incorrectly illustrated and captioned *C. elegans* as *C. glaucifolia*. Nicholson (1900, p. 23, fig. 237) erroneously illustrated *C. elegans* as *C. graminifolia*. Krempin (1990, p. 93, 95) illustrated *C. elegans* but incorrectly captioned the photographs as *C. monostachys* and *C. radicalis*, respectively. Conversely, illustrations captioned as *C. elegans* (Hooker 1855c, Naudin 1858b) are actually *C. sartorii*. Hertrich (1951, p. 23) and Muirhead (1961, p. 49) have illustrations captioned as *Collinia elegans* at the Huntington Botanical Gardens in San Marino, California, but judging from the photographs and the accompanying descriptions in the text it is actually *C. pochutlensis*. In Everett (1981, p. 712) the names *C. elegans* and *C. erumpens* on the two bottom photos should be reversed.

Staminate and pistillate flowers of *C. elegans* are similar in size, shape, and color, and it is more difficult to distinguish the sexes than in other species of the genus. Both sexes have flowers with petals that are firmly united nearly to the top, resulting in a triangular opening at the apex of the corolla. In staminate flowers there is a large, greenish pistillode whose broadened tip obstructs and nearly closes the aperture. The sexes are easily confused since the pistillode is much larger and more prominent than the functional pistil of the pistillate flowers and a drop of nectar will often form at its tip. However,

Plate 10. Chamaedorea elegans. A. Cultivated, Hotel Posada Loma, Fortin de Las Flores, Veracruz, Mexico; **B.** *Hodel 849,* Tikal, Guatemala; **C.** *Hodel 767,* cultivated, garden of R. Palmer, Whittier, California; **D.** Pistillate plant, *D. & R. Hodel 946,* near Fortin de Las Flores, Veracruz, Mexico; **E, F** Cultivated, research collection, Los Angeles; **E.** Staminate inflorescence, *Hodel 787*; **F.** Fruits, *Hodel 767.*

→

B
D
F

staminate flowers are generally more densely arranged on the rachillae, slightly larger, and a deeper shade of yellow. Also, the apical opening in the corolla of the pistillate flowers is not obstructed since the pistil is of a much smaller size. Cook (1937, 1943a) and Cook and Joyner (1938, 1939) gave admirable and well illustrated yet somewhat rambling accounts of various aspects of *C. elegans* and its several forms. They included such topics as ecology, hand-pollination, floral and foliar variation, seed and fruit production, propagation, and adaptation for use in the interiors of homes and offices. Unfortunately, their nomenclature was outmoded and invalid.

Chamaedorea elegans is one of the most commonly cultivated palms in the world. Over 400 million seeds, nearly all from Mexico, are distributed annually throughout the world. It is a popular plant in the United States, Europe, Australia, and the Orient. It is commercially produced in vast numbers as a potted decorative item for indoor use in temperate regions or outdoors in subtropical areas. Slow growing and tolerant of relatively low light and dry air, it is well adapted for use in the interiors of homes and offices. While tough and durable, it maintains its slender, graceful, elegant habit for many years. In fact, Cook and Joyner (1939) and Cook (1943a) reported that specimens were kept in indoor situations for 40–50 years, flowering and fruiting regularly during that period and maintaining their vigor throughout. Since it is one of the few species of the genus with a pinnate eophyll, it is also extremely popular as a small seedling in dish gardens and terrariums.

In commercial production, *C. elegans* is normally potted with 10–15 or even more plants to a container to give a full, bushy, multi-stemmed look. Since the plants grow at various rates, several plants of differing heights compose the final product. *C. elegans* is very effective as a mass planting, as singles in a formal border, a foreground planting, or a tropical accent in a container or the ground. It is somewhat susceptible to mites and thrips and is considered a prime host of several injurious nematodes.

II. Subgenus ELEUTHEROPETALUM

Chamaedorea subgenus **Eleutheropetalum** H. A. Wendl., Index Palmarum 58. 1854.

Eleutheropetalum (H. A. Wendl.) Oerst., Vidensk. Meddel. Dansk. Naturhist. Foren. Kjoebenhavn 1858: 6. 1859.

Key Characters: petals distinct in flowers of both sexes, petals of pistillate flowers bright orange, thick, fleshy, very briefly imbricate basally, cuplike, valvate distally, not thickened in fruit but thin base only remaining within calyx; fruits black.

Distribution: four species on the Atlantic slope of eastern and southern Mexico, Guatemala, Belize, and Honduras; to about 1,300 m elevation; often on limestone; all four moderately common in cultivation.

Key to the Species of Subgenus *Eleutheropetalum*

Leaves pinnate.
  Leaves with a metallic blue sheen. .... **3. C. metallica**
  Leaves green, without metallic blue sheen. .... **4. C. sartorii**
Leaves bifid.
  Leaves with a metallic blue sheen. .... **3. C. metallica**
  Leaves green, without metallic blue sheen.
    Solitary habit. .... **2. C. ernesti-augustii**
    Stoloniferous habit, forming colonies. ... **5. C. stolonifera**

2. Chamaedorea ernesti-augustii *H. A. Wendl.,* Allg. Gartenzeitung 20: 73. 1852b. Type: Cult., *Wendland s.n.* [holotype GOET!; isotypes C!, G(photo!), L!, M(photo!)]. Epithet: honors Ernest August, ruler of Hannover from 1837–1851. **PLATES 2(B), 7(B), 11.**

Morenia ernesti-augustii (H. A. Wendl.) H. A. Wendl., Allg. Gartenzeitung 21: 3. 1853a.

Eleutheropetalum ernesti-augustii (H. A. Wendl.) Oerst., Vidensk. Meddel. Dansk. Naturhist. Foren. Kjoebenhavn 1858: 7. 1859.

Nunnezharia ernesti-augustii (H. A. Wendl.) O. Kuntze, Revisio Generum Plantarum 2: 730. 1891.

Chamaedorea glazioviana Drude ex Guillaum., J. Soc. Natl. Hort. France, Ser. 4, 24: 231. 1923b. Type: Cult., *Glaziou s.n.* (*9020*) (holotype P!;

→

Plate 11. Chamaedorea ernesti-augustii. A. Fruits, cultivated, garden of F. Ketchum, Huntington Beach, California; **B.** Juan Jose Castillo holds leaves of exceptional specimen, *Hodel & Castillo 879,* Izabal, Guatemala; **C, D** Cultivated, research collection, Los Angeles; **C.** Pistillate flowers, *Hodel 755B*; **D.** Staminate inflorescence, *Hodel 755A*; **E.** Fruiting plant on limestone rock, *D. & M. Hodel 879,* Toledo, Belize.

B
E

isotype C!). Epithet: honors Glaziou, collector of the type.

Chamaedorea simplicifrons Hort. in Heynh., Nomenclator Botanicus Hortensis 135. 1840.

Chamaedorea latifrons Hort. in H. A. Wendl., Index Palmarum 12. 1854.

Geonoma latifrons Hort. in H. A. Wendl., Index Palmarum 12. 1854.

Hyospathe elegans Hort. in H. A. Wendl., Index Palmarum 12. 1854. Not Mart.

Geonoma coralliflora Hort. in Chabaud, Les Palmiers de la Cote d'Azure 47. 1915.

Morenia corallifera Hort. in Ruffo, Le Palme de Villa Lucia 29. 1920.

Common Name: none.

Vernacular Names: guaya de abajo, guayita, cola de pescado, rabo de bobo, nesheshiptmil—Mexico.

Habit: solitary, slender, erect, to 2 m tall or more, sometimes flowering when nearly stemless.

Stem: 1–1.5 cm diam., green, smooth, slightly prominent nodes, internodes 0.5–3 cm long.

Leaves: 5–8, spreading, bifid; **sheath** 5–15 cm long, obliquely open above middle, cylindric basally, longitudinally striate-nerved; **petiole** 10–25 cm long, slightly channeled and green above, rounded and pale below; **rachis** 20–25 cm long, angled and green above, rounded below with a yellowish band extending onto sheath; **blade** 25–60 × 20–30 cm, broadly cuneate-obovate, deeply incised apically, shortly emarginate at tips, dull dark green above, dull green below, ± thick, margins toothed, 20–30 teeth per side, 12–18 primary nerves on each side of rachis, each terminating in a tooth, 2 secondaries between each pair of primaries, secondaries ending in teeth, primaries pale, dull, scarcely prominent below, prominently keeled above.

Inflorescences: interfoliar but sometimes infrafoliar in fruit, solitary. **Staminate** with **peduncle** to 30 cm long or more, erect, green where exposed in flower; **bracts** 4–7, lowest the smallest, uppermost slightly shorter than to exceeding peduncle, tubular, fibrous; **rachis** 15–20 cm long, straight, green; **rachillae** 15–25, lower ones the longest, these to 17 cm long, becoming progressively shorter toward apex of rachis, slender, diverging from rachis at a right angle, drooping, simple or rarely 1-branched, green in flower.

Pistillate erect, spicate, furcate or sometimes with 3–4 rachillae; **peduncle** to 70 cm long, green in flower, orange in fruit; **bracts** 5, similar to those of the staminate; if branched, **rachis** to 3 cm long, pale green in flower, red-orange in fruit; flower bearing portion or **rachillae** to 30 cm long, thick, erect, longitudinally angled and pale green in flower, rounded, thickened, red-orange or red in fruit.

Flowers: Staminate in moderate spirals, 2.5 × 3 mm, depressed-globose, yellow to orange, sunken in shallow depressions 1.5–2 mm long; **calyx** 1.25 × 2–2.5 mm, deeply lobed, clear-colored, **sepals** connate in basal ½, membranous, nerveless; **petals** 2 × 2–2.5 mm, connate briefly basally, free and valvate apically, thick, fleshy, cupped over stamens, nerveless; **stamens** 1.5–1.75 mm high, **filaments** shorter than anthers, 0.5–0.75 mm long, clear-colored, **anthers** 0.5–1 mm long, deeply bifid apically and basally, yellow; **pistillode** equalling petals, columnar, expanded apically into a truncate 6-angled cap, flared basally, clear-colored.

Pistillate in remote spirals, 2.5–3.5 × 2.5–3.5 mm, slightly depressed-globose, orange changing to brick-red, aromatic, sunken in prominent elliptic depressions; **calyx** 1.5–2 × 2.5–3 mm, deeply lobed, whitish, **sepals** imbricate in basal ⅓, acute apically, nerveless; **petals** 3.5 × 1.5–2 mm, very narrowly imbricate at base in bud though at anthesis appearing valvate throughout, thick, fleshy in upper portion, this usually deciduous as fruit develops, cup-shaped; **staminodes** prominent; **pistil** not exceeding petals, depressed-globose, of 3 nearly free carpels lightly connate centrally near base at anthesis, whitish, **stigma** lobes sessile, recurved, whitish.

Fruits: 15 × 8–10 mm, subglobose to ellipsoid, green to blue-green becoming black at maturity, remains of a single stigma at base, abortive carpels often adherent to perianth in fruit or pale depressions where abortive carpels may have torn away, mesocarp fleshy, thin, mucilaginous, aromatic; seeds 10 × 7 mm, ellipsoid.

Distribution: MEXICO. Chiapas. Oaxaca. Tabasco. Veracruz. GUATEMALA. Alta Verapaz. Huehuetenango. Izabal. Peten. BELIZE. Cayo. Stann Creek. Toledo. HONDURAS. Atlantida. Comayagua. Cortes. Yoro. Dense, wet forest on the Atlantic slope; mostly at low altitudes but to 1,000 m elevation; often on limestone.

Discussion: Linden discovered *C. ernesti-augustii* in Tabasco, Mexico, and introduced it to Europe in 1847, where Wendland (1852b) described and named it from cultivated plants. It was apparently widely cultivated in Europe in the latter half of the 19th century and first part of this century. Today, it is fairly common in gardens and collections in California, Florida, Hawaii, and Australia and, to some extent, in Europe. Seeds of this species, nearly all from Mexico, are handled commercially on a limited scale and distributed throughout the world.

Chamaedorea ernesti-augustii is very distinct and not likely to be confused with any other species of the genus with the exception, perhaps, of *C. geonomiformis.* In fact, McCurrach (1960, p. 43) and Krempin (1990, p. 92) illustrated and Standley and Record (1936) listed *C. ernesti-augustii* erroneously as *C. geonomiformis. C. geonomiformis* differs in its shorter and narrower, more shallowly incised leaves with rounded nerves; the few-branched staminate inflorescence; and the staminate flowers with the petals connate apically and the corolla opening by lateral slits. A species closely related to *C. ernesti-augustii, C. sartorii,* differs mainly in its leaves having several sigmoid pinnae on each side of the rachis and a pistillate inflorescence with more (4–7) rachillae and never spicate. Hybrids existing between *C. ernesti-augustii* and *C. sartorii* and *C. stolonifera* are discussed in the chapter on hybrids.

Chamaedorea ernesti-augustii is a handsome, striking, highly ornamental species that displays a significant amount of foliar variation. The form from Belize and northeastern Guatemala is noted for its exceptionally large leaves. An excellent choice as a tropical accent for containers for indoor use or for outdoors on a covered patio in subtropical and tropical areas, *C. ernesti-augustii* does equally well in the ground as a foreground subject or is very effectively used as a mass planting. Because it is such a handsome plant, though, it can easily stand alone and still be an effective landscape subject.

Unfortunately, *C. ernesti-augustii* and other species of subgenus *Eleutheropetalum* do not set fruits readily in cultivation without some human assistance. Hand-pollination is necessary since pollen tends to adhere to the anthers and natural insect pollinators are usually absent. Douglas (1988) gave an account of the hand-pollination techniques to ensure an adequate set of fruits on species of subgenus *Eleutheropetalum.* See the discussion on propagation in the chapter on culture for further information.

3. Chamaedorea metallica *O. F. Cook ex H. E. Moore,* Principes 10: 45, figs. 1–3. 1966. Type: Cult., *Moore 9373* (holotype BH!). Epithet: refers to the metallic blue-green sheen of the leaf. **PLATES 4(A), 8(B), 12, 13.**

Common Name: none.
Vernacular Name: metalica—Mexico.

Habit: solitary, slender, erect, 2–3 m tall.

Stem: 1.3–1.5 cm diam., green, minutely but distinctly white-spotted, nodes prominent, internodes 1.5–2.5 cm long, adventitious roots often appearing at base.

Leaves: 12–16, rather stiffly ascending or ascending-spreading with blades cupped upward toward apex, usually bifid but often variously pinnate, metallic-blue-green; **sheath** 7–8 cm long, tubular but split opposite petiole to within 1–2 cm of base; **petiole** 2.5–4 cm long, very slightly grooved and green above, rounded and pale below; **rachis** 19–32 cm long, angled and green above, rounded below with a yellowish band extending onto sheath; **blade** 20–30 × 15 cm, 10–12 cm long on upper margin, cuneate-obovate, incised apically to ⅓ its length, 8–11 primary nerves on each side of rachis, these prominent but not much elevated above, impressed below, 5–6 secondaries between each pair of primaries, minutely and irregularly toothed along margin at apex of each primary nerve especially toward base; or, sometimes blade pinnate, **pinnae** 3–8 on each side of rachis, 2.5–6.5 cm wide, sigmoid, 1–3-nerved, toothed along outer margin above middle.

Inflorescences: interfoliar but often infrafoliar in fruit, solitary, spreading-erect. **Staminate** with **peduncle** 10–25 cm long or more, pale or greenish in flower; **bracts** 3–4, lowermost 3 cm long, flattened, 2-edged, others 6–14 cm long, uppermost exceeding peduncle, all tubular, greenish, papery; **rachis** 4–6 cm long, pale or greenish in flower; **rachillae** 10–12, these 9–15 cm long, 1.5–2 mm wide at base, simple, ascending to erect, usually flexuous at apex, pale green in flower.

Pistillate spicate, sometimes furcate or with 3–4 rachillae; **peduncle** 12–26 cm long, green where exposed in flower, red-orange in fruit; **bracts** 6, similar to those of staminate, green in flower, brown in fruit; flower-bearing portion or **rachillae** 12–14 cm long, 4 mm diam., erect, pale green in flower becoming swollen to 7 mm diam. and red-orange in fruit.

Flowers: Staminate in moderate spirals, 3 × 4 mm, depressed-globose, dull purplish brown in bud, bright orange at anthesis, ± sunken in superficial elliptic depressions 1.5–2 mm long; **calyx** 1.25 × 2.25 mm, shallowly lobed, pale green to white, **sepals** connate in basal ¾, rounded to acute apically; **petals** 3 × 3 mm, connate in basal ½, valvate apically, bilobed, thick, fleshy, cupped over stamens; **stamens** nearly equalling petals, white, **filaments** 1.5–2 mm long, connate, adnate to pistillode, greenish, anthers essentially sessile on the filament tube, incurved, deeply bifid apically and basally, yellow; **pistillode** slightly exceeding petals, columnar with a broad flat 3-angled irregularly toothed cap apically, expanded basally, white.

Pistillate in remote spirals, 3.5 × 3 mm, globose, bright orange, sunken 1 mm in axis, aromatic; **calyx** 1–1.5 × 2.75 mm, lobed, pale green, **sepals** imbricate in basal ½, rounded apically; **petals** 2–3 × 1.5 mm, connate briefly basally, valvate apically, thick, fleshy, strongly cup-shaped; **staminodes** 6, nearly as long as pistil, white; **pistil** 1.5–1.75 × 1.5 mm, depressed-globose, of 3 carpels connate in lower half but readily separable, pale green, **stigma** lobes sessile, recurved, clear-colored.

Fruits: 12 × 9 mm, globose-ellipsoid when fresh, drying slightly smaller, dull black at maturity, epicarp smooth, mesocarp thin, fleshy, green with slender elongate or branched flat fibers appressed against thin endocarp; seeds 10 × 7 mm, brown, with 2 arcuately ascending raphe-branches.

Distribution: MEXICO. Veracruz. Oaxaca? Dense, wet forest on the Atlantic slope; 50–600 m elevation; usually on limestone.

Discussion: O. F. Cook collected *C. metallica* as long ago as 1905 at Cordoba, Veracruz, Mexico. Cook also collected it from cultivated plants in St. Louis, Missouri, in 1907 and at the Doheny Estate in Los Angeles, California, in 1912 and 1914. Apart from Cook's collections, no specimens existed in herbaria and it seemed to have escaped the attention of horticulturists and botanists until the 1960s. In 1960, the late Horace Anderson of Leucadia, California, sent seeds he had collected in Mexico to Nat DeLeon in Miami, Florida, who forwarded some to H. E. Moore, Jr. at Cornell. Plants resulting from these seeds matured in 1965 at Cornell, were hand-pollinated, and produced fruits in 1966. Moore (1966) named the species from these plants. Cook had recognized *C. metallica* as a distinct species and suggested the epithet which Moore adopted when he formally named the species.

In 1989, we found *C. metallica* growing on limestone rocks with *C. elatior, C. sartorii, C. oblongata,* and the cycad *Dioon spinulosum* near Temazcal,

Veracruz. Local residents reported it to be very common on an island in Lake Temazcal.

Despite being one of the most distinctive species of the genus due of its dark green, leathery leaves with a metallic sheen, *C. metallica* has been erroneously referred to *C. tenella* on several occasions in cultivation and the literature. In fact, Everett (1981, p. 73) illustrated and Lane (1988) listed *C. metallica* as *C. tenella. C. tenella,* close to and possibly not distinct from *C. geonomiformis,* is uncommon in cultivation. *C. tenella* differs from *C. metallica* in the green blades lacking the metallic sheen; the spicate staminate inflorescence; and the yellow staminate flowers with the petals connate apically and the corolla opening by lateral slits. Although normally with bifid leaves, it is not unusual for *C. metallica* to have leaves that divide with age into 3–8 pairs of pinnae.

Today, *C. metallica* is rather common and widespread in cultivation and appears in gardens and collections in Europe, Australia, the Orient, Hawaii, California, Florida, Central America, and elsewhere. Seeds are handled commercially from both wild and cultivated plants in Mexico and distributed worldwide.

Chamaedorea metallica is a handsome palm that, due to its dwarf habit and relatively slow rate of growth, makes an excellent subject for pots and tubs. It is a durable plant and relatively easy to grow. Extremely tolerant of low light, it performs exceedingly well indoors and is also very resistant to mites.

4. Chamaedorea sartorii *Liebm.* in Mart., Historia Naturalis Palmarum 3: 308. 1849. Type: Mexico, Veracruz, Barranca de San Francisco, *Liebmann s.n.* (some sheets later numbered *10801, 10802, 10803*) (holotype C!; isotypes GH!, MO!, P!, U!, US!). Epithet: honors Carl Sartorius, 19th-century German botanist. **PLATE 14.**

Chamaedorea aurantiaca Brongn., Rev. Hort. Ser. 3, 1: 86. 1847. Type: Mexico, *Ghiesbreght 32* (holotype P!; isotype F!). Epithet: from the Latin *aurantiacus,* meaning orange, in reference to the color of the flowers.

Morenia oblongata H. A. Wendl., Allg. Gartenzeitung 21: 3. 1853a.

Chamaedorea oblongata (H. A. Wendl.) H. A. Wendl., Index Palmarum 59. 1854. Not Mart.

Chamaedorea oblongata var. *conferta* H. A. Wendl., Index Palmarum 59. 1854. Not Mart.

Eleutheropetalum sartorii (Liebm.) Oerst., L'Amerique Centrale 1: 13, t. 3, figs. 1–6. 1863.

Nunnezharia aurantiaca (Brongn.) O. Kuntze, Revisio Generum Plantarum 2: 731. 1891.

Nunnezharia sartorii (Liebm.) O. Kuntze, Revisio Generum Plantarum 2: 731. 1891.

Eleutheropetalum sartorii var. *conferta* H. A. Wendl. ex Burret, Notizbl. Bot. Gart. Berlin-Dahlem 11: 729. 1933a.

Chamaedorea mexicana Hort. in Heynh., Nomenclator Botanicus Hortensis 135. 1840.

Chamaedorea hartwegii Hort. in Anonymous, Gard. Chron. 23 (New Series, pt. 2): 410. 1885.

Morenia sartorii Hort. in Ruffo, Le Palme di Villa Lucia 30. 1920.

Chamaedorea wobstiana Hort. in Linden, Catalogue general 117: 4. 1885.

Common Name: none.

Vernacular Names: tepejilote cimarron, tepejilote, tepejilotillo, tepejilote chiapaneco—Mexico.

Habit: solitary, slender, erect, to 3–4 m tall.

Stem: 0.8–1.6 cm diam., green, prominently ringed, internodes 3–30 cm long.

Leaves: 3–6, spreading, pinnate; **sheath** 11–23 cm long, tubular, obliquely open, longitudinally striate-nerved; **petiole** 17–45 cm long, slightly grooved and green above, rounded and pale below; **rachis** 30–50 cm long, ± prominently and sharply angled and green above, rounded and pale below; **pinnae** 5–10 on each side of rachis, 20–40 × 4–7 cm, lanceolate to long-lanceolate, long-acuminate, oblique basally, sigmoid, falcate, ± thick, deep lustrous green, a central midrib and 2 marginal primary nerves, these rounded, conspicuous, 3 secondaries on each side of midrib, tertiaries numerous, faint.

Inflorescences: interfoliar but often infrafoliar in fruit, solitary, erect, 30–70 cm long; **peduncles** 20–50 cm long, green in flower, orange in fruit; **bracts** 4–6, upper one the largest, to 25 cm long, slender, fibrous, longitudinally striate-nerved, brown in flower; **rachises** 2–15 cm long, green in flower, orange in fruit. **Staminate** with 30 **rachillae**, these 15–25 cm long, slender, pendulous, flexuous, green.

Pistillate with 4–8 **rachillae**, these 15–20 cm long, erect, becoming spreading in fruit, sharply angled and light green in flower, rounded, thicker, fleshy, and orange-red in fruit.

Flowers: Staminate in moderately dense spirals, 2.5 × 2.5–3 mm, depressed-globose, orange, slightly sunken in depressions 1.5–2 mm across; **calyx** 1.25 × 2 mm, deeply lobed, white, **sepals** connate in basal ½, rounded to acute apically, slightly membranous, nerveless; **petals** 2 × 2 mm, connate briefly basally, free and valvate apically, cupped over stamens, acute, thin, slightly

Plate 12. Chamaedorea metallica. Cultivated, California. **A, B, E, F** Research collection, Los Angeles; **C, D,** garden of R. Douglas, Walnut Creek. **A.** Spicate infructescence with mature fruits, *Hodel 817*; **B.** Pistillate plants (left, top), *Hodel 817,* and staminate plant (right), *Hodel 808*; **C, D.** Crown of leaves (note pinnate leaves in D); **E, F.** Staminate inflorescence and staminate flowers, *Hodel 808.*

B
D
F

fleshy, nerved on inside; **stamens** equalling pistillode, **filaments** white, 0.75–1 mm long, slightly connate basally, **anthers** 1–1.25 mm long, narrowly oblong, yellow; **pistillode** shorter than petals, columnar, flared abruptly apically, expanded basally, whitish.

Pistillate in remote spirals, 2.5 × 3 mm, globose, slightly sunken in rounded depressions 2–3 mm across, orange to red, slightly aromatic; **calyx** 1 × 2.5 mm, deeply lobed, pale green to white, **sepals** connate and/or imbricate briefly basally, rounded to acute apically, thin, nerveless; **petals** 2.5 × 1.5–2 mm, connate in basal ½, free and valvate apically, acute, cup-shaped, thick, fleshy, very slightly nerved on inside; **staminodes** 6, large, slightly membranous, free, inserted at base of pistil; **pistil** 1.5 × 1.25–1.5 mm, subglobose, pale green to white, **stigma** lobes separated, rounded, recurved, clear-colored.

Fruits: 9.5–12.5 × 7–8 mm, ellipsoid-ovoid, black, epicarp transparent, thin, mesocarp slightly fleshy, thin, green, mucilaginous, aromatic, endocarp fibrous-nerved; seeds 7–10 × 5–6 mm, ellipsoid.

Distribution: MEXICO. Veracruz. Puebla. Oaxaca. HONDURAS. Atlantida. Olancho. Wet or moist forest on the Atlantic slope; 100–1,300 m elevation; often on limestone.

Discussion: Liebmann discovered *C. sartorii* near Mirador, Veracruz, and described and named it in Martius (1849). It is close to *C. ernesti-augustii* but the latter species differs in the bifid leaves and spicate or few-branched pistillate inflorescence. Vegetatively, *C. sartorii* is similar to *C. oblongata* and *C. neurochlamys*, and sterile material is sometimes difficult to distinguish. Foliage of *C. oblongata* usually dries grayish, however, while that of *C. sartorii* and *C. neurochlamys* dries dark green, brown or black. *C. sartorii* and *C. oblongata* have pinnae with thicker texture than those of *C. neurochlamys*. When fertile, staminate flowers with petals free at the apex and orange, cup-shaped petals easily distinguish *C. sartorii* from *C. oblongata* and *C. neurochlamys*.

Grown in Europe since about 1850, *C. sartorii* is fairly widespread in cultivation today, occurring in gardens and collections in California, Florida, Hawaii, Australia, and elsewhere. In California, it has been cultivated since 1900 (Reidel, 1957). Hybrids between *C. sartorii* and *C. ernesti-augustii* and *C. stolonifera* are discussed in the chapter on hybrids.

Although some collectors malign *C. sartorii* as inferior in appearance to *C. oblongata*, the reputation may be unwarranted as *C. sartorii* is a handsome species with dark, lustrous green foliage. The leaves are darker green than those of *C. oblongata* and just as thick and leathery. *C. sartorii* can be used in the same situations in the landscape as *C. oblongata* and is equally attractive.

5. Chamaedorea stolonifera *H. A. Wendl. ex Hook. f.*, Bot. Mag. 118: t. 7265. 1892. Type: Cult., *Hooker* [holotype K (photo!)]. Epithet: from the Latin *stolo*, meaning a stolon, in reference to the stoloniferous habit. **PLATES 1(C), 5B, 15.**

Common Name: none.
Vernacular Name: none.

Habit: stoloniferous, 1–2 m tall, forming colonies of the same sex to 10 m across.

Stems: separated, new stems arising at a considerable distance from old ones, erect, 5–8 mm diam., green, prominently ringed, internodes 3–6 cm long.

Leaves: 5–8, spreading, bifid; **sheath** to 10 cm long, tubular, slightly oblique apically, longitudinally striate-nerved; **petiole** 5 cm long, flattened or faintly grooved and green above, rounded and pale below; **rachis** to 10 cm long, slightly angled and green above, rounded below with a yellowish band extending onto sheath; **blade** 30 × 25 cm, incised apically to ⅔ its length, lobes divergent, oblong-lanceolate, 25–30 × 9–10 cm, deep green, acute, outer margin toothed along upper half, teeth becoming more closely spaced apically, 10 prominent primary nerves on each side of rachis, these pale and keeled above, 2 secondaries between each pair of primaries, tertiaries numerous, inconspicuous, closely spaced.

Inflorescences: usually infrafoliar but sometimes interfoliar initially, solitary, erect-spreading, to 25–30 cm long; **peduncles** to 20 cm long, slender, erect, green in flower, orange in fruit; **bracts** 3–5, uppermost reaching or just extending onto rachis slightly, fibrous, lanceolate, acuminate, imbricate, green in flower, brown in fruit, longitudinally striate-nerved; **rachis** 1–2 cm long, green in flower, red-orange in fruit. **Staminate** with 2–7 **rachillae**, these 7–13 cm long, 1.5–2 mm diam., spreading and flexuous, more or less rounded but slightly longitudinally angled, light green, minutely white-spotted.

Pistillate with 2–4 **rachillae**, these to 10 cm long, slightly angled longitudinally and green in flower, becoming thickened, rounded, and orange in fruit.

Flowers: Staminate in ± dense spirals, 2.5–3 × 3.5–4.5 mm, strongly depressed-globose, flattened apically, brick-red to yellow-orange, slightly sunken; **calyx** 1 × 2.5–3 mm, obscurely lobed, pale green to white, membranous, sepals imbricate in basal ½–¾, rounded to acute apically; **petals** 2 × 3 mm, connate in basal ¾, free and valvate apically, strongly cup-shaped, slightly bilobed, acute, thick, ± fleshy; **stamens** shorter than pistillode, **filaments** connate in a fleshy broad 6-toothed ring, whitish tinged with orange near apex, **anthers** 1 mm long, on teeth of filament ring, divaricate, **pistillode** barely exserted beyond petals, columnar, truncate and abruptly flared apically, orange.

Pistillate in ± remote spirals, 2 × 3.5–4 mm, broadly obconical, red-orange, sunken in broadly elliptic pits; **calyx** 0.5 × 2.5–3 mm, deeply lobed, pale green, membranous, **sepals** imbricate

Plate 13. Chamaedorea metallica. Cultivated. **A.** Mass planting in garden of C. Graff, Miami, Florida; **B.** Different leaf forms, garden of R. Douglas, Walnut Creek, California.

in basal ¼–½, rounded to acute apically; **petals** 2.5 × 2.5, briefly imbricate basally, free and valvate apically, acute, cup-shaped, thick, fleshy; **pistil** 1–1.5 mm high, depressed-globose, whitish; **stigma** lobes sessile, recurved, clear-colored.

Fruits: 7–9 mm diam., globose, black.

Distribution: MEXICO. Chiapas. Wet forest on the Atlantic slope; 600–800 m elevation; usually on limestone.

Discussion: Wendland cultivated *C. stolonifera* at Herrenhausen at least as early as 1882 from material originating in Mexico. In 1882, Wendland sent plants to the Royal Botanic Gardens at Kew, England, where they flowered during the spring months. Joseph Hooker of Kew then named the species from these flowering plants. *C. stolonifera* seemed to have disappeared in the wild until Faustino Miranda rediscovered it in 1949 in Chiapas. The late Harold E. Moore, Jr. also collected it there in 1952. In 1989, we found several clumps of *C. stolonifera* growing on limestone rocks in a wet forested area of Chiapas. The clumps of one sex only were about 10 meters across and possessed stolons emerging from cracks and crevices in the limestone rocks. We returned with living staminate material and seeds to augment the few staminate plants recently produced in cultivation.

Chamaedorea stolonifera should be considered endangered. It is known from only a few collections and the remaining populations are unprotected and precariously near to cultivated areas. Also, commercial growers have recently expressed a desire to exploit the wild populations for propagative material, a situation that would certainly lead to the demise of the species in the wild.

Today, *C. stolonifera* is widely cultivated and appears in gardens and collections in California, Florida, Australia, and Europe. It is somewhat similar to *C. brachypoda* with which it is often confused in cultivation but the rhizomatous habit, thinner texture of the leaves, and the green flowers easily distinguish the latter species. Braun (1983, p. 63) illustrated *C. brachypoda* but labeled it erroneously as *C. stolonifera.*

Chamaedorea stolonifera is one of the most striking and distinctive members of the genus due to its exceedingly stoloniferous habit resulting in the formation of relatively dense clusters of stems through and out of which creep the spreading stolons. These stolons give rise to new plants often at a considerable distance from the original plant. Normally, stolons originate at basal nodes and move across the top of the soil but they have been observed originating at nodes as high up the main stem as 15 cm, arching downward and submerging below the soil, and then reemerging ½–1 meter distant. They have been known to cross under narrow sidewalks of concrete and brick and other such obstructions in this manner. The plant makes an attractive potted specimen and is especially striking when stolons have climbed over the lip of the pot and are holding plantlets at their ends suspended in midair, often with aerial roots. Sometimes stolons will even emerge through the drainage holes in the bottoms of pots. *C. stolonifera* is a neat and clean palm. It is self-cleaning, the senescing leaf, petiole, and sheath falling as a single unit, leaving no dry, brown leaf bases on the stem.

Chamaedorea stolonifera is readily propagated vegetatively by severing the stolons that have produced rooted plantlets at their ends and then excising and replanting or potting these plantlets. Perhaps this ease of vegetative propagation was one of the factors that contributed to the virtual disappearance of staminate plants from cultivation. Essentially, all plants in cultivation today are pistillate although it appears that several clones are present. These clones vary in the shape and size of the leaf, the serrations on the margin, and the angle at which the foliar lobes diverge. As more staminate plants become available, seed propagation will be possible.

By the early 1970s only three staminate plants were known to exist. Two of them, one in Delaware at Longwood Gardens and one in Florida, subsequently died. The third plant was in Richard Palmer's garden in Whittier near Los Angeles. Palmer obtained his plant in the 1960s from the late David Barry, Jr. of Los Angeles. It was very old and at the base bore visible evidence of numerous stolons having been removed at one time. It had ceased to send out new stolons and had become a single-stemmed, spindly plant, perhaps nearing the end of its natural life. In the early and middle 1980s, Palmer hand-pollinated his pistillate plants and gave pollen to Lou Hooper of La Habra, California. Palmer and Hooper were successful in setting fruits. In 1989,

Plate 14. Chamaedorea sartorii. A. Cultivated, garden of N. Bezona, Kona, Hawaii; **B, C.** Fruiting plant and infructescences, *D. & R. Hodel 947,* near Fortin de Las Flores, Veracruz, Mexico.

→

B

Hooper reported that at least three of his dozen or so seedlings grown from the seeds resulting from his pollination efforts produced staminate inflorescences. These and the recent rediscovery of *C. stolonifera* in the wild should serve to perpetuate the staminate of the species.

III. Subgenus MORENIA

Chamaedorea subgenus **Morenia**.

Morenia Ruiz & Pav., Florae Peruvianae et Chilensis. Prodromus 2: 150. 1794.
Kunthia Humb. & Bonpl., Plantas Aequinoctiales 2: 128, t. 122. 1809.

Key Characters: petals distinct in flowers of both sexes, spreading, white to yellow-green, imbricate basally, staminate flowers clustered in groups; staminate inflorescences multiple at a node; fruits red.

Distribution: three or perhaps more species confined to northern South America on the western Andean slope from northern Colombia to southern Ecuador and on the eastern Andean slope from southern Venezuela through Ecuador and Peru into Bolivia; 50–2,800 m elevation; only *C. linearis* appears to be cultivated.

Key to the Species of Subgenus *Morenia*

Pinnae broadly lanceolate, markedly expanded at middle, markedly narrowed apically and basally. .... **6. C. latisecta**
Pinnae linear to lanceolate, not markedly expanded at middle, not markedly narrowed apically and basally.
  Larger plants 2–10 m tall, 13 or more pinnae on each side of rachis. .... **7. C. linearis**
  Smaller plants less than 2 m tall, 8–9 pinnae on each side of rachis. .... **8. C. smithii**

6. **Chamaedorea latisecta** (*H. E. Moore*) *A. Gentry,* Ann. Missouri Bot. Gard. 73: 163. 1986. **Figure 1.**

Morenia latisecta H. E. Moore, Gentes Herb. 8: 203, fig. 87. 1949. Type: Colombia, Putamayo, *Schultes & Villareal 7676* (holotype BH!). Epithet: from the Latin *latisectus* meaning cut into broad sections in reference to the broad pinnae.
Common Name: none.
Vernacular Name: none.

Habit: solitary, erect, to 8 m tall.
Stem: green, ringed.
Leaves: pinnate, to 1.3 m long; **sheath** and **petiole** glabrous, striated; **rachis** at least 1 m long; **pinnae** 11 on each side of rachis, lower and median ones 40–60 × 9–12 cm, broadly lanceolate, contracted basally to 1.5 cm wide, falcate, tapered on lower margin to a slender acuminate tip, alternate to subopposite at intervals of mostly 6–8 cm along rachis, upper and terminal pinnae 25–40 × 3–5 cm, lanceolate, more closely spaced, ± evenly tapered to a shorter tip, all glabrous, prominently nerved, green above, green with yellowish nerves below.
Inflorescences: Staminate interfoliar, 6 per node, innermost pair developing first, outermost last; **peduncle** 25 cm long; **bracts** 3, lower one 12 cm long, upper one 20 cm long, exceeding peduncle, tubular with flaring apices; **rachis** 12 cm long; **rachillae** 50, these 3–7.5 cm long, simple, closely spaced.
Pistillate solitary, large; **peduncle** 40 cm long, erect; **bracts** 5, lowermost 3.5 cm long; **rachis** 27 cm long, pendulous; **rachillae** 30, these 12–20 cm long, simple, ascending with pendulous tips.
Flowers: Staminate borne spirally in pairs, depressed and angled in bud, yellow; **calyx** 0.6 mm high, lobed, lobes deltoid; **petals** 3 mm long, ovate, valvate, thick; **stamens** ½–⅔ as long as petals; **pistillode** conspicuous, 3-lobed.
Pistillate with **calyx** 3 mm high, deeply lobed, lobes entire or irregularly crenate, exceeding corolla remains.
Fruits: immature 12 × 10 mm, nearly globose, green.

Distribution: COLOMBIA. Putumayo. Dense, wet forest; 2,000–2,300 m elevation.

Discussion: The description is taken from Moore (1949) who described and named *C. latisecta* from material that R. E. Schultes collected in Colombia. The broadly lanceolate pinnae markedly narrowed at both ends distinguish it from all other species in subgenus *Morenia. C. latisecta* is known with certainty only from the type and does not appear to be cultivated. However, further study may show that it is but a variant of the *C. linearis* complex.

7. **Chamaedorea linearis** (*Ruiz & Pav.*) *Mart.,* Historia Naturalis Palmarum 2: 5. 1823. **PLATES 2(C), 3, 16.**

Morenia fragrans Ruiz & Pav., Florae Peruvianae et Chilensis. Prodromus 2: 150, pl. 32. 1794.

Plate 15. Chamaedorea stolonifera. A. Colony of one plant (same sex) on limestone rock, *D. & R. Hodel 933A,* Chiapas, Mexico; **B–E** Cultivated, research collection, Los Angeles; **B.** Potted specimen with stolons creeping out of container; **C.** Close-up of stolons (note adventitious roots); **D.** Pistillate flowers, *Hodel 757*; **E.** Staminate flowers, *Hodel 771* (in both D and E note thick, orange, cuplike petals and that petals are removed from center flowers).

→

C
E

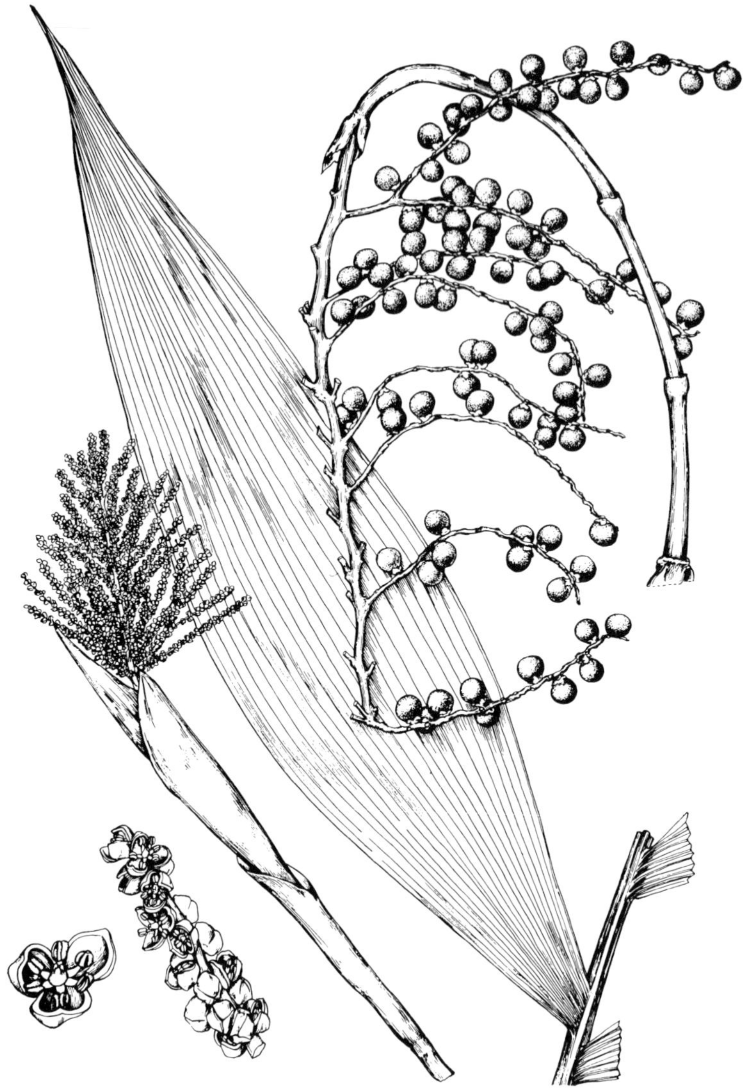

87. MORENIA LATISECTA. Pinna, staminate and pistillate spadices ×⅓; portion of staminate rachilla ×3; staminate flower ×4.

Figure 1. Chamaedorea latisecta. (Reprinted from *Gentes Herbarum* 3: 204. 1949. Courtesy of L. H. Bailey Hortorium.)

Not *C. fragrans* (Ruiz & Pav.) Mart. (based on *Nunnezharia fragrans* Ruiz & Pav.). Type: Peru, Huanuco, near Muna, *Ruiz & Pavon s.n.* [holotype MA (photo!)]. Epithet: from the Latin *fragrans* meaning scented pleasantly, in reference to the fragrant flowers.

Martinezia linearis Ruiz & Pav., Systema Vegetabilium Florae Peruvianae et Chiliensis 297. 1798. Type: Peru, Pozuzu, *Ruiz & Pavon s.n.* (*B15*) [holotype MA (photo!); isotype F!, G (photo!)]. Epithet: *linearis* is Latin for linear, in reference to the shape of the pinnae.

Kunthia montana Humb. & Bonpl., Plantas Aequinoctiales 2: 128, t. 122. 1809. Type: Colombia, San Pablo, *Humboldt & Bonpland s.n.* (holotype P). Epithet: from the Latin *montanus* meaning pertaining to the mountains, in reference to the habitat.

Morenia poeppigiana Mart., Historia Naturalis Palmarum 3(7): 161. 1837. Type: Peru, Huanuco, Huallaga, *Poeppig 1546* [holotype W (photo!)]. Epithet: honors Eduard Poeppig, collector of the type.

Morenia corallina Karst., Linnaea 28: 274. 1857. Type: Colombia, Santa Marta, *Karsten t. 171* (in Karsten, 1866!). Epithet: from the Latin *corallinus* meaning coral red, in reference to the color of the fruits. Not *Morenia corallina* Hort.

Morenia lindeniana H. A. Wendl., Bot. Zeitung Berlin 17: 17. 1859. Type: Cult., *Wendland s.n.* (holotype GOET?; isotype L!). Epithet: honors the horticulturist J. Linden of Belgium, discoverer of the species. Not *C. lindeniana* H. A. Wendl.

Nunnezharia corallina (Karst.) O. Kuntze, Revisio Generum Plantarum 2: 730. 1891.

Nunnezharia lindeniana (H. A. Wendl.) O. Kuntze, Revisio Generum Plantarum 2: 730. 1891.

Nunnezharia linearis (Ruiz & Pav.) O. Kuntze, Revisio Generum Plantarum 2: 730. 1891.

Nunnezharia montana (Humb. & Bonpl.) O. Kuntze, Revisio Generum Plantarum 2: 730. 1891. Not *C. montana* Liebm.

Nunnezharia morenia O. Kuntze, Revisio Generum Plantarum 2: 730. 1891.

Nunnezharia poeppigiana (Mart.) O. Kuntze, Revisio Generum Plantarum 2: 730. 1891.

Morenia linearis (Ruiz & Pav.) Burret, Notizbl. Bot. Gart. Berlin-Dahlem 11: 316. 1932.

Morenia montana (Humb. & Bonpl.) Burret, Notizbl. Bot. Gart. Berlin-Dahlem 11: 316. 1932.

Morenia caudata Burret, Notizbl. Bot. Gart. Berlin-Dahlem 13: 337. 1936a. Type: Ecuador, La Merced, *Schultze-Rhonhof 1862* (holotype B, destroyed). Epithet: from the Latin *caudatus* meaning excessively acuminate, in reference to the long-acuminate tips of the pinnae.

Morenia macrocarpa Burret, Notizbl. Bot. Gart. Berlin-Dahlem 13: 333. 1936a. Type: Peru, Huanuco, *Weberbauer 3600* (holotype B, destroyed). Epithet: from the Greek *macro* meaning large and *carpa* meaning fruit, in reference to the large fruits.

Morenia robusta Burret, Notizbl. Bot. Gart. Berlin-Dahlem 13: 335. 1936a. Type: Colombia, Caldas, *Dryander 20* (holotype B, destroyed). Epithet: from the Latin *robustus* meaning robust, in reference to the habit.

Chamaedorea polyclada Burret, Notizbl. Bot. Gart. Berlin-Dahlem 13: 341. 1936b. Type: Ecuador, Balao, *Preuss 1949* (holotype B, destroyed). Epithet: from the Greek *poly* meaning many and *clada* meaning branches, in reference to the numerous rachillae of the inflorescence.

Morenia microspadix Burret, Notizbl. Bot. Gart. Berlin-Dahlem 15: 33. 1940. Type: Ecuador, Rio Negro, *Schultze-Rhonof 2614* (holotype B, destroyed). Neotype (Bergmann 1989): Ecuador, Pastaza, *Bergmann & Pedersen 62566* (AAU). Epithet: from the Greek *micro* meaning small and the Latin *spadix,* in reference to the small spadix (inflorescence).

Chamaedorea poeppigiana (Mart.) A. Gentry, Ann. Missouri Bot. Gard. 73: 162. 1986.

Chamaedorea megaphylla A. Gentry, Ann. Missouri Bot. Gard. 73: 163. 1986. Type: Peru, Huanuco, La Divisiora, *Gentry et al. 29572* (holotype MO!). Epithet: from the Greek *mega* meaning large and *phyllus* meaning leaved, in reference to the large leaves.

Chamaedorea formosa Hort. in Bull, Catalogue 1876: 4. 1876.

Chamaedorea montana Hort. in Siebert & Voss, Vilmorin's Blumengartneri 1155. 1896. Not *C. montana* Liebm.

Common Name: none.

Vernacular Names: Cana de la vibora, Cana de San Pablo, corallina, turumuto—Colombia; ceja de la montana—Peru.

Habit: solitary, ± stout, occasionally slender, erect, infrequently decumbent, 2–10 m tall.

Stem: 2–8 cm diam., smooth, glabrous, glossy, light green, yellow, or mottled green, prominently ringed, internodes 2–20 cm long, base of stem with numerous aerial roots to 1.5 cm diam., these branched, orange-brown, prickly.

Leaves: 3–8, erect-spreading, pinnate, rarely bifid or irregularly pinnate, to 3.8 m long; **sheath** 40–125 cm long, tubular for 30–85 cm and forming a crownshaft, striated, green, brown-margined; **petiole** 7–23 cm long, to 3 cm diam., flattened or channeled and green above, rounded and pale below; **rachis** 90–230 cm long, flattened at base and prominently angled toward apex and green above, rounded below with a pale yellow band extending onto sheath; **blade** elliptic to oblong to ovate; **pinnae** 13–32 on each side of rachis, basal pinnae 12–60 × 1–7 cm, middle ones 16–85 × 2.5–12 cm, apical ones 20–50 × 3–10 cm, linear to lanceolate, slightly sigmoid to falcate, acuminate, slightly contracted basally and there calloused above at attachment to rachis, green above, paler below, dominant midrib and 3–4 prominent primary nerves on each side of this or 6–10 equally prominent primary nerves per pinna, nerves rounded and pale white to yellow below, angled above, 1–2 secondaries and tertiaries between each primary, these thin, filiform, glabrous below when young but with age having minute, deciduous, dark brownish, lepidote scales; or if blade bifid then to 1 m or more long, incised apically, lobes to 25 cm long, 50 or more prominent primary nerves on each side of rachis.

Inflorescences: Staminate infrafoliar, 3–18 per node, erect-spreading, glabrous, developing centrifugally, borne on a common hypopodium; **peduncle** 7–40 cm long, base semicrescent and flattened basally, 1-4 cm wide, oval apically, white or yellow in flower; **bracts** 3–10, prophyll bifid, 2–27 cm long, 2nd–8th bracts 3–38 cm, upper 2 bracts often rudimentary and concealed, lanceolate, tubular or loosely sheathing, thin-membranous or ± leathery, longitudinally striated to furrowed, splitting apically and becoming tattered and shredded, white shaded with green in flower, becoming dry, brownish; **rachis** 4–35 cm long, white or yellow in flower; **rachillae** 15–30, these 1–12 cm long, basal ones the longest, erect-spreading or slightly drooping, angled, straight or flexuous toward the apex, white in flower.

Pistillate infrafoliar, solitary, erect; **peduncle** 10–64 cm long, 1–2 cm diam., rounded, whitish where exposed in flower, drooping and green in fruit; **bracts** 2–7, similar to those of the staminate, drying brown by anthesis; **rachis** 2–50 cm long, straight, angled, whitish in flower, pale green in fruit; **rachillae** 5–53, these 5–22 cm long, all ± the same length or basal ones slightly shorter, whitish in flower, recurved, drooping and pale green in fruit.

Flowers: Staminate in groups of 2–3 or a few rarely solitary, 5 mm high, globose, whitish, sessile; **calyx** 1 mm high, lobed, membranous, **sepals** connate in basal ⅔, obtuse to acute apically; **petals** 2–3 × 2–2.5 cm, broadly-ovate to obovate, free to base, valvate, spreading apically, concave, acute, glossy white, fleshy-membranous, with 5 faint nerves; **stamens** 1.5–3 mm long, connate basally, white, **filaments** filiform, sigmoid, spreading, **anthers** 1.2 mm long, oblong, basifixed or dorsifixed, erect basally, divergent apically, 2-cleft; **pistillode** 2 × 0.5 mm, shorter than stamens, subglobose basally, cylindric-conical, split apically into 3 erect lobes, fleshy.

Pistillate in moderately dense or lax spirals, solitary, rotund-campanulate, short-cylindric, superficial; **calyx** 1 mm high, shallowly lobed, green, **sepals** connate in basal ¾–⁹⁄₁₀, erect, acute apically; **petals** 2.5–4 × 3 mm, ovate to oblong, membranous basally, acute, free, valvate, and reflexed apically, glossy white; **staminodes** 6 but sometimes absent, adnate to petals basally, appressed, filiform, barely 2 mm high, flattened apically and crested; **pistil** 2–3 × 2.5–3 mm, obovate-globose, slightly 6-lobed apically, pale green or whitish, **stigma** lobes 1 mm long, sessile, ± deltoid, erect to spreading, interior margin stigmatose-papillose, after swelling ± revolute.

Fruits: 1–2.5 cm diam., globose to elliptic-globose, orange to deep red, rugose, sometimes 3 developing per flower, stigmatic residue basal; seeds 7–10 × 8–14 mm, elliptic-globose, minutely but prominently papillose basally, endocarp brown with vestiges of vascular bundles slightly raised, endosperm hard, durable, solid, embryo small, conic, located on the dorsal side toward apex.

Distribution: VENEZUELA. COLOMBIA. ECUADOR. PERU. BOLIVIA. Dense, wet forest on the eastern Andean slope from Venezuela to Bolivia and western Andean slope from Colombia to southern Ecuador; 50–2,800 m elevation.

Discussion: The description is from Bergmann (1989) and supplemented from Ruiz and Pavon (1794, 1798), Humboldt and Bonpland (1809), Martius (1823, 1837), Wendland (1859), Burret (1936a, 1940) and Gentry (1986). *C. linearis* is a highly variable species in size, shape, and nervation of pinnae and size and shape of fruit. Various authors have proposed different segregates based on these variable characters. However, when considered over its entire range, the differences seem insignificant. The name of one of the segregates, *Morenia corallina* Karst., should not be confused with *Morenia corallina* Hort., a named applied to *C. ernesti-augustii* and *C. oblongata* in 19th-century European horticulture.

The floral description of *Kunthia montana* in Humboldt and Bonpland (1809) is somewhat troubling since they drew much attention to the hermaphroditic flowers, a situation unheard of in the genus. They were either working from a mixed collection or a misinterpretation of staminate flowers with prominent pistillodes as hermaphroditic.

Ramirez-Goyena (1911) reported *Kunthia montana* from Nicaragua although he is probably referring to *C. woodsoniana* or a robust form of *C. costaricana.*

Chamaedorea linearis is one of the largest-growing species in the genus. Only *C. tepejilote, C. woodsoniana,* and some forms of *C. costaricana* may have stems and leaves as large as those of *C. linearis. C. carchensis* has leaves that may be as large as those of *C. linearis* but its stems are smaller.

Although experiences cultivating members of subgenus *Morenia* are few, *C. linearis* appears to be a vigorously growing plant with stems attaining 2–3 meters in height in five years. In Hawaii, it is of relatively easy culture and makes an attractive specimen with its large, green leaves and ringed, lime-green nearly ivory trunk. It is especially attractive when staminate plants are flowering. The white, multiple inflorescences are showy and attractive as are the large, crimson fruits of pistillate plants. Recently, Lyon Arboretum in Honolulu, Hawaii, has distributed seeds and seedlings of *C. linearis* as *C. poeppigiana* under the number L-82.0023. Their plants were grown from seeds that Tim Plowman collected in Peru. Besides Hawaii, *C. linearis* is cultivated in California, Florida, and, perhaps, elsewhere. Although Linden introduced it to European gardens from Colombia in the 1850s, it has apparently disappeared from cultivation there.

8. Chamaedorea smithii *A. Gentry,* Ann. Missouri Bot. Gard. 73: 164. 1986. Type: Peru, Junin, *Foster & Smith 9171* (holotype MO!). Epithet: honors D. Smith, co-collector of the type.

Plate 16. Chamaedorea linearis. Cultivated. **A-D** *Hodel 831,* Lyon Arboretum (*82.023*), Honolulu, Hawaii, *ex* Peru; **A.** Flowering (staminate) plant; **B.** Stem with infructescences and red mature fruits; **C.** Staminate inflorescences (note multiple arrangement at a node); **D.** Immature fruits; **E.** Plant, Fairchild Tropical Garden, Miami, Florida. →

B
E

Common Name: none.
Vernacular Name: none.

Habit: solitary, slender, erect, to 2 m tall.

Stem: green, prominently ringed.

Leaves: 65–75 cm long, pinnate; **petiole** 9-15 cm long; **pinnae** 8–9 on each side of rachis, 15–29 cm long, lowermost less than 1 cm wide, some terminal and median ones to 5 cm wide, variable in shape, narrowly lanceolate-elliptic to linear, not sigmoid, 3–7 cm apart, long-acuminate, alternate or subopposite, drying grayish green with contrastingly tannish main veins below, ± minutely lepidote below, otherwise glabrous, 5–7 primary nerves, fewer in narrow basal pinnae, primaries integrating into and not very strongly differentiated from others.

Inflorescences: Staminate not seen. **Pistillate** with **peduncle** 30 cm long; **bracts** 4, lowermost attached near base, 8–9 cm long, second attached 4 cm above base, 17 cm long, third attached 13–14 cm above base, 31 cm long, uppermost 24–25 cm above base (7–9 cm below lowermost rachilla), 5–6 cm long, narrow, 1 mm wide, membranous; **rachis** 10–13 cm long, rather straight; **rachillae** 15–21, these to 3.5–5 cm long in flower, to 14 cm long in fruit, diverging at nearly right angle from rachis.

Flowers: Staminate not seen. **Pistillate** several mm apart, not at all sunken, scars actually slightly raised; **calyx** 1.5–2 mm high, lobed, **sepals** connate in basal ⅔; **petals** 1–1.5 mm long, broadly ovate, folded over bud?; **pistal** 1.5 mm high, thick-triangular ovoid, **stigma** lobes 1 mm long, sessile, recurved.

Fruits: 8–10 mm diam., globose, red, with persistent and conspicuous calyx cup.

Distribution: PERU. Junin. Dense, wet *Podocarpus* forest, 1,800 m elevation.

Discussion: The description is from Gentry (1986), who stated that *C. smithii* is endemic to the lower montane *Podocarpus* forests where it is locally very common near Rondayacu about 45 km south of San Ramon and north of Monobamba near the Tarma-Chanchamayo border. However, the forest remnants there are one of the last vestiges of this type of vegetation in Peru and, accordingly, the species must by regarded as highly threatened.

Gentry (1986) compared *C. smithii* to members of subgenus *Chamaedorea,* namely *C. boliviensis* and *C. lanceolata* (both = *C. pinnatifrons,* a highly variable species ranging from Mexico to Bolivia and Ecuador). Gentry also compared *C. smithii* to *C. pauciflora,* a lowland Amazonian species of subgenus *Moreniopsis.* Unfortunately, *C. smithii* is known only from the type, a fruiting pistillate collection, and staminate material is not available. However, the pyramidal pistillate inflorescence with its long, straight rachis and relatively short, curved fruiting rachillae, green fruits maturing dark red, and unribbed fruiting perianth suggest subgenus *Morenia.*

IV. Subgenus CHAMAEDOREA

Chamaedorea subgenus **Chamaedorea** (Mart.) H. A. Wendl., Index Palmarum 59. 1854. (in part, sect. b.)

Chamaedorea sect. *Chamaedorea* Mart., Historia Naturalis Palmarum 3: 307. 1849. (in part.)

Chamaedorea subgenus *Chamaedoropsis* Oerst., Vidensk. Meddel. Dansk. Naturhist. Foren. Kjoebenhavn 1858: 15. 1859. (in part.)

Chamaedorea subgenus *Chamaedorella* Oerst., Vidensk. Meddel. Dansk. Naturhist. Foren. Kjoebenhavn 1858: 23. 1859.

Chamaedorea subgenus *Psilostachys* Oerst., Vidensk. Meddel. Dansk. Naturhist. Foren. Kjoebenhavn. 1858: 24. 1859.

Key Characters: petals distinct in flowers of both sexes, thickened and persistent in fruit, staminate flowers with petals connate apically and basally and adnate apically to pistillode and corolla opening by lateral apertures; staminodes present or lacking; perianth usually strongly nerved when dry; fruits black, less often red-orange, yellowish, or dark brown.

Distribution: about 25 species widely or locally distributed on the Atlantic and/or Pacific slopes from Mexico to Ecuador, Bolivia, and Brazil; to nearly 3,000 m elevation; about ten species are cultivated.

Key to the Species of Subgenus *Chamaedorea*

Plants vinelike with apical pinnae usually reflexed, downward-pointing, hooklike. .... **10. C. elatior**
Plants not vinelike and apical pinnae not reflexed, downward-pointing, or hooklike.
  Leaves bifid.
    Cespitose habit. .... **30. C. serpens**
    Solitary habit.
      Leaf blades conspicuously toothed on margins, staminate rachillae erect-spreading.
        Leaf blades incised apically to ¼ their length. .... **9. C. amabilis**
        Leaf blades incised apically to ½ or more their length. .... **24. C. palmeriana**
      Leaf blades not conspicuously toothed, staminate rachillae pendulous.

Leaf blades obscurely nerved; staminate flowers not strongly nerved.
Leaf blades usually more than 15 cm long, incised apically to more than ⅓ their length, inflorescences usually branched. .... **13. C. geonomiformis**
Leaf blades usually less than 15 cm long, incised apically to less than ⅓ their length, inflorescences spicate. **32. C. tenella**
Leaf blades with ± prominent nerves; staminate flowers strongly nerved.
Leaf blades with 7 or fewer, smooth, low nerves above on each side of rachis, 3 secondaries conspicuous between each pair of primaries. .... **28. C. rojasiana**
Leaf blades with 8 or more, sharply keeled and elevated nerves above on each side of rachis, 2 secondaries between each pair of primaries. .... **31. C. simplex**
Leaves pinnate.
Cespitose habit.
Stems with rhizomes, forming loose colonies; pinnae drying plicate. .... **15. C. graminifolia**
Stems without rhizomes, forming ± tight clumps; pinnae not drying plicate.
Basal shoots emerging vertically from behind persistent, thick leaf sheaths; pinnae linear. .... **16. C. hooperiana**
Basal shoots not emerging as above; pinnae lanceolate, apical pair very broad.
Stems erect; apical pinnae 50 cm or more long. .... **22. C. murriensis**
Stems decumbent, rooting and sprouting along their length; apical pinnae to 30 cm long. .... **30. C. serpens**
Solitary habit.
Pinnae with slitlike windows at rachis, petiole densely ferrugineous-lepidote .... **12. C. ferruginea**
Pinnae lacking slitlike windows, petiole not densely ferrugineous-lepidote.
Pinnae aggregated in groups along rachis.
Pinnae linear. .... **14. C. glaucifolia**
Pinnae lanceolate. .... **17. C. klotzschiana**
Pinnae regularly spaced along rachis.
Pinnae straight or only slightly falcate.
Pinnae drying plicate; lower rachillae forked. .... **18. C. liebmannii**
Pinnae not drying plicate; lower rachillae simple.
Pinnae thick, with prominent midrib, other nerves ± obscure; inflorescences equalling or exceeding leaves. .... **20. C. macrospadix**
Pinnae thin, with several conspicuous nerves; inflorescences not exceeding leaves. .... **29. C. schiedeana**
Pinnae sigmoid.
Apical pinnae broad, usually as broad or broader than others combined.
Leaf sheaths long-open, splitting deeply opposite petiole. .... **33. C. verapazensis**
Leaf sheaths tubular.
Pistillate rachillae pendulous even in flower. .... **21. C. matae**
Pistillate rachillae erect or spreading.
Fruits red, yellow, or orange when soft ripe.
Fruits strongly sickle-shaped. .... **11. C. falcifera**
Fruits ± rounded or only slightly elongated. .... **27. C. pinnatifrons**
Fruits black when soft ripe.
Rachillae 10 or more. .... **22. C. murriensis**
Rachillae 3 or fewer. .... **28. C. rojasiana**
All pinnae ± similar in size, apical pair only slightly larger.
Fruits red, yellow, or orange when soft ripe.
Fruits ± globose or only slightly elongated. .... **27. C. pinnatifrons**
Fruits markedly elongated and reniform or sickle-shaped.
Fruits reniform or bilobed. .... **23. C. neurochlamys**
Fruits sickle-shaped. .... **11. C. falcifera**
Fruits black when soft ripe.
Pinnae thin with numerous, conspicuous nerves drying shiny yellow above and below. .... **19. C. lucidifrons**
Pinnae ± thick, not too conspicuously nerved above.
Pinnae with nerves prominently elevated and thickened below. .... **34. C. warscewiczii**
Pinnae with nerves not prominently elevated and thickened below.
Plants appearing stemless; pinnae 4–6 on each side of rachis, gray-green, less than 5 cm wide. **25. C. parvifolia**
Plants with visible, elongate stem; pinnae 7–10 on each side of rachis, green, more than 10 cm wide. .... **26. C. pedunculata**

9. Chamaedorea amabilis *H. A. Wendl. ex Dammer,* Gard. Chron., Ser. 3, 36: 245. 1904b. Type: Costa Rica, Alajuela, Rio Sarapiqui, *Wendland s.n.* (holotype K!; isotype L!). Epithet: from the Latin *amabilis* meaning lovely, attractive. **PLATES 6(E), 17.**

Nunnezharia amabilis (H. A. Wendl. ex Dammer) O. Kuntze, Revisio Generum Plantarum 2: 730. 1891.

Chamaedorea coclensis L. H. Bailey, Gentes Herb. 6: 236, fig. 123. 1943a. Type: Panama, Cocle, El Valle, *Allen 1815* (holotype MO!). Epithet:

named for Cocle Province in Panama where the type was collected.

Common Name: none.
Vernacular Name: none.

Habit: solitary, slender, erect but sometimes decumbent, 1–2 m tall.

Stem: 7–10 mm diam., conspicuously ringed, internodes 5 cm long, conspicuous adventitious prop roots at base to 10 cm high.

Leaves: 4–5, spreading, bifid, rich green; **sheath** 12 cm long, tubular, green, faintly longitudinally striate-nerved; **petiole** 3–12 cm long, flat and bright green above with a raised, light green, triangular knoblike appendage at base, rounded and with a ± indistinct pale band below; **rachis** 15–20 cm long, angled and green above, rounded and pale-banded below; **blade** 30–50 × 15–30 cm, broadly obovate, 15–25 cm wide at apex, incised apically to ⅕–¼ its length, thin-papery, glabrous, glossy below, cuneate at base, outer margins on upper half conspicuously serrate-dentate, 25 prominent nerves on each side of rachis.

Inflorescences: infrafoliar, ascending, slender; **peduncles** 10–15 cm long, 4 mm wide at base, 5 mm wide at apex, green in flower where exposed, orange in fruit; **bracts** 4, prophyll 1 cm long, 2nd bract 3 cm, 3rd 7 cm, distal one 10 cm and exceeding peduncle, papery-membranous, greenish but rapidly turning brownish or rotting to nearly transparent by anthesis, longitudinally striate-nerved, acuminate, bifid. **Staminate** with very short **rachis** to 1 cm long, green in flower; **rachillae** 3–7, 15–25 cm long, 3–4 mm diam., erect-spreading, green, minutely white-spotted.

Pistillate spicate or sometimes furcate; flower-bearing portion or **rachillae** 15–20 cm long, 5 mm diam., erect, green and minutely white-spotted in flower, swollen and reddish orange in fruit.

Flowers: Staminate in rather dense spirals, 2.5–3 × 2.5–3 mm, subglobose, slightly angled, obtuse or ± flattened, greenish at anthesis ageing to brown, aromatic, slightly sunken in rounded depressions 2 mm long; **calyx** 0.5–0.75 × 2–2.5 mm high, shallowly knotched, greenish ageing to brown, **sepals** connate basally, broadly rounded apically, nerveless; **petals** 2.5 × 3 mm, connate apically and there adnate to pistillode and corolla opening by very small, basal, lateral apertures, triangular, acute; **stamens** 1.5 mm high, connate basally, **anthers** 0.75 mm long; **pistillode** 2–2.5 mm high, obovoid, swollen apically, there 1 mm wide, green.

Pistillate in rather dense spirals, 2–5 mm distant, 2 × 3 mm, depressed-globose, green, swelling with age whether pollinated or not to 4 mm wide and becoming greenish yellow, slightly sunken in elliptic depressions; **calyx** 0.5 × 3–4 mm, shallowly lobed, green, **sepals** briefly connate basally, very broadly rounded apically, nerveless; **corolla** with a terminal triangular opening, 1.5 mm across, **petals** 2 × 3–3.5 mm, imbricate in basal ¾, very broadly truncate to acute or undulate apically, ± thickened, lime-green, white-margined, cup-shaped; **pistil** 1.5 × 2 mm, depressed-globose, dark green, **stigma** lobes sessile, short, black.

Fruits: 12 × 9 mm, globose to oblong, black, smooth, glossy.

Distribution: COSTA RICA. Alajuela. Cartago. Heredia. San Jose. PANAMA. Cocle. Colon. COLOMBIA. Choco. Dense, wet forest; 500–1,000 m elevation.

Discussion: Wendland discovered *C. amabilis* along the Rio Sarapiqui near the village of San Miguel in Costa Rica about the middle of the 19th century. He returned to Germany with living material and grew it in his greenhouses at Herrenhausen for many years. He worked up a description and name for it which were published a year after his death by his friend Udo Dammer (1904b). Dammer noted that none of Wendland's plants was alive by the early 1900s.

Chamaedorea amabilis is distributed in wet forests at middle elevations mainly on the Atlantic slope in Costa Rica and Panama. It also occurs sparingly in northwest Colombia. In Panama it occurs up to and just over the Continental Divide at appropriate elevations, such as at El Cope and El Valle.

One of the most striking members of the genus, *C. amabilis* is highly sought after by collectors who do not hesitate to strip great numbers of them from the forest which has led to their decline in the wild (Hodel 1988a). *C. amabilis* is cultivated in California, Hawaii, Costa Rica, and Australia, although it is not common in any of these places. Although Dammer (1904b) reported that all of Wendland's plants had died, the species may still survive in Europe.

Chamaedorea amabilis is surprisingly hardy considering its origin and has withstood 0°C (32°F) without sustaining damage in the Los Angeles area.

10. Chamaedorea elatior *Mart.,* Linnaea 5: 205. 1830. Not Mart., Historia Naturalis Palmarum 3: 157, t. 126, t. R, figs. 1 & 2s, t. Z2, fig. 12, t. Z4, figs. 6–10, t. Z6, figs. 4–33. 1837; not Hort. Type: Mexico, Veracruz, Jalapa, *Schiede s.n.* [holotype M!; isotypes BM (photo!), BR (photo!)]. Epithet: from the Latin *elatus* meaning tall, in reference to the long, tall stems. **PLATE 18.**

Chamaedorea scandens Liebm. in Mart., Historia Naturalis Palmarum 3: 308. 1849. Not Hort. Type: Mexico, Veracruz, El Mirador, *Liebmann s.n.* (some sheets later numbered *6580*) (holotype C!; isotypes F!, GH!, MO!, P!, U!, US!). Epithet: from the Latin *scandens* meaning climbing, in reference to the habit.

Chamaedorea affinis Liebm. in Mart., Historia Naturalis Palmarum 3: 308. 1849. Type: Mex-

Plate 17. Chamaedorea amabilis. A. Blades are nearly 40 × 30 cm, *D. & M. Hodel 741A,* El Valle, Panama; **B, C** Cultivated, research collection, Los Angeles, *ex* El Valle, Panama; **B.** Infructescence with mature fruits, *Hodel 741B bis*; **C.** Staminate inflorescences, *Hodel 741A bis*; **D, E.** Cultivated, Jardin Botanico Robert y Catherine Wilson, Costa Rica.

C
E

Fig. 45. *Chamaedorea falcifera.* a, leaf × ½; b, top of sheath × ½; c, staminate inflorescence × ½; d, section of staminate rachilla × 3; e, staminate calyx × 10; f, staminate flower × 10; g, staminate corolla × 10; h, staminate flower in vertical section × 10; i, stamens in side, back, and front views × 10; j, pistillate inflorescence × ½; k, fruit × 2; l, seed × 2; m, seed in vertical section × 2; n, base of fruit with abortive carpels × 8; o, abortive carpels in vertical section × 8; p, abortive carpels in cross section × 10; q, pistillate flower reconstructed × 5. Flowers drawn from reconstituted material with nerves less prominent than when dry. Illustrator: Marion Ruff Sheehan.

Figure 2. Chamaedorea falcifera. (Reprinted from *Principes* 2: 69. 1958.)

bal. Dense, wet lowland forest on the Atlantic slope; 40–300 m elevation; on limestone.

Discussion: The description is taken from Moore (1958), who described and named the species from material that Steyermark had collected in Guatemala. *C. falcifera* is noted for its curious, strongly sickle-shaped fruits; in fact, the shape of the fruits seems to be the only reliable character distinguishing it from *C. pinnatifrons.* Fruits are orange to red at maturity, similar to those of *C. pinnatifrons* and *C. neurochlamys.*

At least as typified by the type, *C. falcifera* is not in cultivation and material so referred is *C. neurochlamys. C. falcifera* is somewhat of an enigma

→

Plate 19. Chamaedorea geonomiformis. A. Staminate plant on limestone rock, *Hodel & Castillo 1023A,* Izabal, Guatemala; **B–D** Cultivated, Los Angeles; **B.** Fruits (immature), *Hodel 792B*; **C, D.** Staminate plant and staminate inflorescences, *Hodel 792A.*

C
E

ico, Oaxaca, *Liebmann s.n.* (some sheets later numbered *6576*) (holotype C!). Epithet: the Latin *affinis* means neighboring or allied to, in reference, apparently, to its similarity to *C. scandens.*

Chamaedorea montana Liebm. in Mart., Historia Naturalis Palmarum 3: 308. 1849. Type: Mexico, Oaxaca, *Liebmann s.n.* (some sheets later numbered *6574, 6575*) (holotype C!; isotypes P!, US!). Epithet: from the Latin *montanus* meaning growing in the mountains, in reference to the habitat.

Chamaedorea resinifera H. A. Wendl., Allg. Gartenzeitung 21: 179. 1853c. Type: Cult., *Wendland s.n.* [holotype GOET(photo!); isotypes C!, G(photo!), L!]. Epithet: from the Latin *resinifer* meaning resin-producing, probably in reference to the resinous smelling, highly aromatic staminate flowers.

Chamaedorea desmoncoides H. A. Wendl., Allg. Gartenzeitung 21: 177. 1853c. Type: Cult., *Wendland s.n.* [holotype GOET!; isotypes C!, G(photo!), L!]. Epithet: a Latin derivation meaning desmoncuslike, in reference to the climbing, vining stems similar to those of the genus *Desmoncus.*

Nunnezharia affinis (Liebm.) O. Kuntze, Revisio Generum Plantarum 2: 730. 1891.

Nunnezharia elatior (Mart.) O. Kuntze, Revisio Generum Plantarum 2: 730. 1891.

Nunnezharia desmoncoides (H. A. Wendl.) O. Kuntze, Revisio Generum Plantarum 2: 730. 1891.

Nunnezharia oaxacensis O. Kuntze, Revisio Generum Plantarum 2: 730. 1891.

Nunnezharia regia (Hort.) O. Kuntze, Revisio Generum Plantarum 2: 731. 1891.

Nunnezharia repens (Hort.) O. Kuntze, Revisio Generum Plantarum 2: 731. 1891.

Nunnezharia resinifera (H. A. Wendl.) O. Kuntze, Revisio Generum Plantarum 2: 731. 1891.

Chamaedorea elatior var. *bambusoides* H. A. Wendl. ex Dammer, Gard. Chron., Ser. 3, 38: 42. 1905c.

Chamaedorea elatior var. *desmoncoides* H. A. Wendl. ex Dammer, Gard. Chron., Ser. 3, 38: 42. 1905c.

Chamaedorea scandens var. *bambusoides* H. A. Wendl. ex Dammer, Gard. Chron., Ser. 3, 38: 42. 1905c.

Chamaedorea scandens var. *desmoncoides* H. A. Wendl. ex Dammer, Gard. Chron., Ser. 3, 38: 42. 1905c.

Chamaedorea bambusoides Gerome, Rev. Hort. 83: 571. 1911.

Chamaedorea bambusoides var. *graminifolia* Gerome, Rev. Hort. 83: 570. 1911.

Chamaedorea bambusoides var. *juncea* Gerome, Rev. Hort. 83: 571. 1911.

Chamaedorea regia Hort. in H. A. Wendl., Index Palmarum 11. 1854.

Chamaedorea repens Hort. in H. A. Wendl., Index Palmarum 11. 1854.

Chamaedorea resimpia Hort.?

Chamaedorea robusta Hort. in H. A. Wendl., Index Palmarum 11. 1854.

Anothea scandens O. F. Cook, Nat. Hort. Mag. 22: 135. 1943b. Name of no botanical standing.

Platythea graminea O. F. Cook, Nat. Hort. Mag. 26: 226. 1947b. Name of no botanical standing.

Common Name: none.

Vernacular Names: tepejilote, tepejilotillo, cola de gullo, junco de bejuco, junco—Mexico.

Habit: solitary or infrequently cespitose and then often branching up to a meter or more above ground, sprawling or subscandent, often arching, but nearly erect when young, to 20 m long or more, crawling up through taller vegetation or simply sprawling on ground.

Stem: 0.8–2 cm diam., green, smooth, nodes slightly prominent, internodes 10–30 cm long.

Leaves: 5–15, deeply bifid when young becoming progressively more pinnate up the stem until completely pinnate, spreading, arching; **sheath** to 60 cm long, elongate, tubular, persistent, green, minutely white-spotted, finely longitudinally striate-nerved; **petiole** 0–30 cm long, flat and green above, pale and rounded below with a slightly prominent light green ridge extending onto sheath; **rachis** 0.5–3 m long, sharply angled and green above, rounded and with a very faint green band below; **blade** to 3 m long in very robust individuals but normally 0.5–1.5 m long; **pinnae** 10–55 on each side of rachis, 20–45 × 1.5–5 cm, linear-lanceolate to linear, alternate basally, opposite apically, equally attenuate, strongly indurate-calloused at very narrowed base, deep green, progressively more reflexed and hooklike toward apex of blade and there downward-pointing, a prominent central midrib and numerous but inconspicuous nerves of lesser orders on each side of this.

Inflorescences: interfoliar, breaking through sheaths, erect; **peduncles** 10–20 cm long, flattened, 1.2 cm wide, 6 mm thick, erect, green where exposed; **bracts** 3–7, prophyll smallest, this to 10 cm long, others generally 10–15 cm long, often only upper one visible, others concealed by petiole and sheath, upper bract to 13 cm long, stout, fibrous, brown at anthesis, longitudinally striate-nerved, acute-acuminate, bifid, upper 2 exceeding peduncle; **rachises** 5–25 cm long, green, obtusely angled. **Staminate** with up to 35 **rachillae**, these to 25 cm long, rounded distally, 3 mm diam., lower ones flattened near base and there to 5 mm wide, simple, spreading, divaricate at right angles, obtuse, or ± reflexed.

Pistillate similar to staminate but larger, branched portion to 40 × 30 cm, **rachillae** few to many, to 30 cm long, simple, pale orange or salmon colored or brownish in mature fruit.

Flowers: Staminate in moderately dense spirals, 4 × 3.5–4 mm, globose but depressed apically in center, bright yellow, very aromatic, slightly sunken in depressions 1–1.5 mm diam.; **calyx** 1 × 2.5 mm, shallowly lobed, green, **sepals** connate nearly to top, broadly truncate apically, nerveless; **petals** 3.5–4 × 2.5–3.25 mm, obovate, yellow, connate basally and apically, adnate to pistillode apically and corolla opening by lateral slits, acute, slightly fleshy, nerveless; **stamens** 3 mm high, slightly exceeding pistillode, **filaments** 1.25–1.5 mm long, whitish, **anthers** 1.5–2 mm long, oblong, white; **pistillode** 2.5–3 mm high, columnar, upper half swollen, truncate apically, greenish.

Pistillate in remote spirals, 2.5–3 × 2.5–3 mm, globose, yellow, slightly sunken in superficial, elliptic depressions 1–2 mm long; **calyx** 1–1.25 × 2.3 mm, deeply lobed, green, **sepals** connate and or slightly imbricate basally, broadly rounded apically, lightly nerved on inside; **petals** 2.5 × 2.5 mm, imbricate in basal ¾, acute apically, yellow, fleshy, nerveless; **staminodes** absent; **pistil** 2.5 × 1.5–1.75 mm, globose, green, **styles** short or lacking, **stigma** lobes distinct, recurved.

Fruits: 8–11 mm diam., globose, black with a glaucous bloom, epicarp thin, slightly transparent, mesocarp fleshy, mucilaginous, greenish orange, endocarp membranous, slightly nerved; seeds 6–7 mm diam., globose, brownish.

Distribution: MEXICO. Chiapas. Oaxaca. Puebla. Veracruz. GUATEMALA. Alta Verapaz. Huehuetenango. Sacatepequez. San Marcos. Moist or wet, dense forest mostly on the Atlantic slope but occasionally on the Pacific slope in Guatemala; 100–1,500 m elevation.

Discussion: As the only truly vinelike and climbing member of the genus, *C. elatior* is very distinctive and not likely to be confused with any others. Its climbing, vinelike habit and reflexed, hooklike pinnae which enable it to cling to taller vegetation and grow to great lengths or heights clearly set it apart. One would think that with such a distinctive habit there would be no difficulty in the application of the correct name. However, this is not the case. Schiede discovered *C. elatior* near Jalapa, Mexico, and sent material to Martius (1830) who published its name and meager description without illustration. The original description is only a few lines in length and tells us next to nothing about the plant, not even whether it is climbing and vinelike or if its apical pinnae are reflexed and hooklike. Also, Schiede's type consists of an immature inflorescence and part of a leaf, either of which alone or together is hardly diagnostic.

A few years later, Martius (1837) gave another, and this time illustrated, account of *C. elatior* in one of the volumes of *Historia Naturalis Palmarum.* However, apparently the species he discussed and illustrated in his 1837 account was not *C. elatior* as he treated it in 1830 but rather another undescribed species that Karwinsky had recently introduced to European gardens from Mexico. Wendland (1853c) pointed out that Martius (1837) confused *C. elatior* with this recent introduction. Wendland recognized the error and named the recent introduction *C. karwinskyana* (= *C. pochutlensis*). However, the confusion did not end until as recently as the 1930s; many specimens of *C. karwinskyana* in collections in Europe were (and probably still are) labeled as *C. elatior* (Burret 1935).

I have examined Schiede's type of *C. elatior* from Munich and much to my consternation, it does not fit well with material bearing that name in herbaria and cultivation. Instead of the apical pinnae being strongly reflexed, those of the type are forward-pointing as in *C. karwinskyana.* If Schiede's type does, in fact, represent *C. karwinskyana* then this species (as would *C. pochutlensis*) would become a synonym of *C. elatior* and the climbing, vinelike plant that is so well known as *C. eliator* must be given a new name. The appropriate epithet would be *C. scandens* since it is the oldest synonym for the climbing, vining plant.

However, despite the differences in Schiede's type and the material commonly accepted as *C. elatior,* I am reluctant to propose nomenclatural changes at this time. It is not inconceivable that Schiede's type does indeed represent the climbing, vinelike plant. That plant's first pinnate leaves often do not display strongly reflexed, apical pinnae; in fact, all pinnae may be forward-pointing on young leaves. It could be that the leaf in Schiede's type is one of these young ones not displaying the reflexed apical pinnae. That Martius made no mention of the climbing, vining habit may simply be a reflection of the fact he did not see the plant in its whole and had only Schiede's words from which to derive an idea of its habit. Martius could easily have been confused since the original material he had to work from in 1830 was quite inadequate and this is reflected in his meager description. Upon seeing the more recent introductions in cultivation, Martius may have mistakenly took them to be *C. elatior* and used this cultivated material to expand descriptions of this species in *Historia Naturalis Palmarum.* Wendland (1853c) examined additional and, apparently, more typical material of Schiede's collection at Berlin (unfortunately, the Berlin material has since been destroyed) and came to the conclusion that Martius erred. Wendland felt that the true *C. elatior* was quite similar to *C. scandens.* He in turn named as *C. karwinskyana* the material that Martius (1837) described and illustrated as *C. elatior* in *Historia Naturalis Palmarum.*

Chamaedorea elatior is extremely variable throughout its range. The many names cited in synonymy were distinguished mainly on the number of pinnae or rachillae of the pistillate inflorescence and size of various parts. Standley and Steyermark (1958) stated that the variation is evident not only in the various stages through which the leaves go as they develop along the stem but in size of parts of plants from the same region. For example, it is possible to find elongate or very short petioles on the same plant. The number of pinnae and of rachillae of the pistillate inflorescence appears to vary sufficiently to make separation of the forms cited in synonymy not only difficult and unworkable but illogical.

That its distinctive habit has enhanced its notoriety and resulted in a fairly well documented history in cultivation is evidenced by the numerous synonyms and horticultural forms and frequent references to it in cultivation in Europe and in nursery and gardening catalogs. Linden sent material to Europe from Mexico in 1846 that Wendland (1853c) later named *C. desmoncoides*. The Southern California Acclimatization Association introduced it to southern California by the turn of the century as *C. bambusoides* (Guillaumin 1923b). Franceschi, who had a nursery near Santa Barbara, California, listed *C. desmoncoides* as early as 1900 in his catalog of plants (Guillaumin 1923b) and *C. bambusoides* was in the 1908 edition (Reidel 1957). It appeared in the Doheny collection in Los Angeles in the 1920s and 1930s probably from material that E. Howard, whom Doheny hired to procure plants for his estate, had collected in Mexico.

Widely cultivated, *C. elatior* appears in gardens and collections in Hawaii, California, Florida, Australia, Venezuela, and Europe. The form with a solitary, stout stem is the most commonly encountered while the branching, thin-stemmed form is rather uncommon. Some find the strongly smelling staminate flowers objectionable. Plants are difficult to place in the landscape. Without overhead support or taller vegetation such as a sturdy tree on which to climb, they will tend to sprawl along the ground in an unmanageable and potentially destructive manner. Since the stems can become quite heavy, this may damage other plants. It is a unique and interesting palm, though, and when well placed and displayed, such as on a tree or trellis against a wall or side of a building, it will provide an attractive accent.

11. Chamaedorea falcifera *H. E. Moore,* Principes 2: 68, fig. 45. 1958. Type: Guatemala, Izabal, *Steyermark 41640* (holotype F!). Epithet: from the Latin *falcifera* meaning bearing a scythe or sickle, in reference to the curious, sickle-shaped fruits. **FIGURE 2.**

Common Name: none.
Vernacular Name: none.

Habit: solitary, slender, 1–6 m tall.

Stem: 5–7 mm diam., internodes 1.5–3.2 cm long.

Leaves: 3–5, pinnate; **sheath** 6–9 cm long or more, tubular, briefly oblique apically and there striate-nerved; **petiole** 6–15 cm long or more; **rachis** 10–25 cm long, both petiole and rachis pale below; **pinnae** 3–4 on each side of rachis, end pair largest, these 14–21 cm long on upper margin and 5–10 cm along rachis with one side lower than other, lower pinnae progressively smaller, middle ones 11–20 × 2.5–5 cm, basal ones 7–15 × 1–3 cm, sigmoid, acute to acuminate, rich shining green above, paler dull green below, basal and middle ones with a prominent midrib and submarginal primary nerves, 1–3 secondaries on each side of midrib, at least midrib ± keeled, end pair with 4–7 primary nerves keeled above, these pale and shining below when dry, 2–4 secondaries between primaries, these inconspicuous.

Inflorescences: inter- or infrafoliar, arching, nodding, solitary; **peduncles** 15–21 cm long; **bracts** 3–4, these tubular, brown, lower of which is inserted 1 cm above flattened base, uppermost slightly exceeding peduncle in staminate plants; **rachises** 1–2 cm long. **Staminate** with 4–7 **rachillae**, these 6–15 cm long, pendulous, slender, green.

Pistillate with 4–8 **rachillae**, these 5–15 cm long, slender, ± stiff, orange in fruit.

Flowers: Staminate in moderately dense spirals, yellowish green, essentially superficial but leaving elliptic scars 2 mm long; **calyx** 0.75–2 mm, deeply lobed, strongly nerved when dry, **sepals** connate and/or imbricate nearly to apex, rounded or straight apically; **petals** valvate, connate apically and basally and adnate apically to pistillode and corolla opening by lateral slits, **stamens** subsessile, **anthers** entire apically, divided basally; **pistillode** equalling stamens, ± columnar, truncate apically.

Pistillate with **calyx** 0.7 mm high in fruit, shallowly lobed, strongly nerved; **petals** 2 mm high in fruit, imbricate, strongly nerved; **staminodes** absent; **pistil** not seen.

Fruits: 10–14 × 3–4 mm, sickle-shaped, acute apically, orange to red, essentially superficial on rachillae, mesocarp strongly fibrous, endocarp membranous; seeds 9 × 2 mm; perianth persistent at base of fruit, abortive carpels adherent to fruit at maturity.

Distribution: GUATEMALA. Alta Verapaz. Iza-

Plate 18. Chamaedorea elatior. A, B Veracruz, Mexico; **A.** Bifid juvenile leaf of plant on limestone rock, *D. & R. Hodel 949,* Temazcal; **B.** Plant climbing on tree branch, *D. & R. Hodel 921,* Catemaco; **C-F** Cultivated, research collection, Los Angeles; **C.** Cespitose form with basal and aerial branching, *Hodel 805*; **D-F** *Hodel 862*; **D.** Young plant with bifid and pinnate leaves; **E.** Terminal portion of adult leaf showing reflexed, downward-pointing pinnae; **F.** Staminate inflorescences. →

B
D
F

Fig. 45. *Chamaedorea falcifera.* a, leaf × ½; b, top of sheath × ½; c, staminate inflorescence × ½; d, section of staminate rachilla × 3; e, staminate calyx × 10; f, staminate flower × 10; g, staminate corolla × 10; h, staminate flower in vertical section × 10; i, stamens in side, back, and front views × 10; j, pistillate inflorescence × ½; k, fruit × 2; l, seed × 2; m, seed in vertical section × 2; n, base of fruit with abortive carpels × 8; o, abortive carpels in vertical section × 8; p, abortive carpels in cross section × 10; q, pistillate flower reconstructed × 5. Flowers drawn from reconstituted material with nerves less prominent than when dry. Illustrator: Marion Ruff Sheehan.

Figure 2. Chamaedorea falcifera. (Reprinted from *Principes* 2: 69. 1958.)

bal. Dense, wet lowland forest on the Atlantic slope; 40–300 m elevation; on limestone.

Discussion: The description is taken from Moore (1958), who described and named the species from material that Steyermark had collected in Guatemala. *C. falcifera* is noted for its curious, strongly sickle-shaped fruits; in fact, the shape of the fruits seems to be the only reliable character distinguishing it from *C. pinnatifrons.* Fruits are orange to red at maturity, similar to those of *C. pinnatifrons* and *C. neurochlamys.*

At least as typified by the type, *C. falcifera* is not in cultivation and material so referred is *C. neurochlamys. C. falcifera* is somewhat of an enigma

---

Plate 19. Chamaedorea geonomiformis. A. Staminate plant on limestone rock, *Hodel & Castillo 1023A,* Izabal, Guatemala; **B–D** Cultivated, Los Angeles; **B.** Fruits (immature), *Hodel 792B*; **C, D.** Staminate plant and staminate inflorescences, *Hodel 792A.*

D

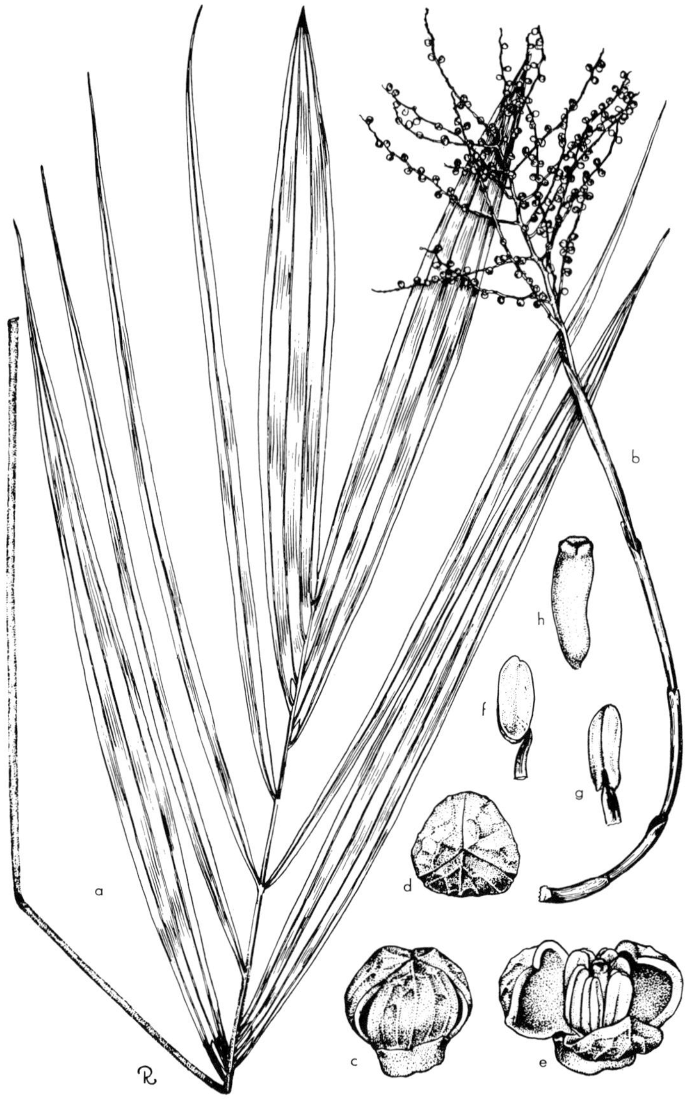

98. CHAMÆDOREA FERRUGINEA. a, leaf ×½; b, spadix ×½; c, d, e, staminate flower, side, top, expanded ×8; f, g, stamen ×10; h, pistillode ×12.

Figure 3. Chamaedorea ferruginea. (Reprinted from *Gentes Herbarum* 8: 232. 1951. Courtesy of L. H. Bailey Hortorium.)

and only the type and a few other collections represent it in herbaria. We have searched for it without success at the type locality. It is either very rare, extinct, or an atypical *C. neurochlamys* or *C. pinnatifrons.*

12. Chamaedorea ferruguinea *H. E. Moore,* Gentes Herb. 8: 236, t. 98. 1951. Type: Mexico, Oaxaca, *Schultes 905* (holotype BH!). Epithet: from the Latin *ferrugineus* meaning rusty or light brown with a mixture of red, in reference to the lepidote scales of this color densely covering the petiole. **Figure 3.**

Common Name: none.

Vernacular Names: ya-bo-ga-dau, ye-dza, be-yadau (fruit)—Mexico.

→

Plate 20. A–C, Chamaedorea geonomiformis. D, E, Chamaedorea glaucifolia. Cultivated. **A, B** Fairchild Tropical Garden, Miami, Florida; **A.** Staminate plant; **B.** Pistillate plant; **C.** Pistillate plant, *Hodel 792B,* research collection, Los Angeles. **D.** Virginia Robinson Botanical Garden, Beverly Hills, California; **E.** Glaucous petiole, research collection, Los Angeles.

B
C
E

Habit: solitary, to 4 m tall.

Stem: not seen.

Leaves: pinnate; **petiole** 40 cm long or more, 3 mm diam., slender, greenish and densely ferrugineous-lepidote; **rachis** 16–19 cm long, 3-sided, green, dark punctuate and sparsely lepidote; **pinnae** 3–6 on each side of rachis, basal pinnae to 30 × 1.5–2 cm with slitlike windows at rachis, middle ones to 31 × 5.5–7 cm, end pair to 20 cm long on the inner margin, 1.7–2.2 cm wide, 6.5–9 cm along rachis and with slitlike window there, all linear to linear-lanceolate, acuminate, dark green above, slightly paler below, basal pair 3–4-nerved, middle ones 1-nerved, end pair to 3–5-nerved, nerves pale but prominent.

Inflorescences: Staminate with **peduncle** 20 cm long; **bracts** 6, prophyll 2 cm long, fifth nearly equalling peduncle, exceeding incomplete sixth, tubular, longitudinally striated, bifid; **rachillae** 20, these 3–8 cm long, 0.5–1 mm thick, erect to spreading.

Pistillate not seen.

Flowers: Staminate 2.5 mm high, globose, sessile; **calyx** 0.5 mm × 2.5 mm, lobed; **petals** 2.5 mm long, valvate, connate apically and basally and adnate apically to pistillode and corolla opening by lateral slits, brown, prominently nerved and ± reticulate when dry; **stamens** shorter than petals, **filaments** very short; **pistillode** columnar, indistinctly 3-lobed with a prominent depression at apex.

Pistillate not seen.

Fruits: not seen.

Distribution: MEXICO. Oaxaca. Dense, wet forest; 950 m elevation.

Discussion: The description is from Moore (1951), who described and named *C. ferruginea* from only a staminate plant that R. E. Schultes had collected in 1939 in the district of Choapam in Oaxaca, Mexico. It is apparently known only from the type, whose fenestrate leaves with the slitlike windows at the rachis in the basal and apical pairs of pinnae are very distinctive. No other *Chamaedorea* has this character, a feature common with members of the genus *Reinhardtia.* In fact, vegetatively, especially in the fenestrate pinnae and densely ferrugineous-lepidote petiole, *C. ferruginea* is suspiciously similar to some *Reinhardtia* species known to occur in the area, raising the possibility of a mixed collection. The inflorescence of the type is similar to that of *Chamaedorea liebmannii,* a species also known to occur in the area.

According to Schultes, local inhabitants, who apparently eat the fruits, know this palm by the several Zapotec names listed earlier.

13. Chamaedorea geonomiformis *H. A. Wendl.,* Allg. Gartenzeitung 20: 1. 1852a ('*geonomaeformis*'). Type: Cult., *Wendland s.n.* [holotype GOET!; isotypes C!, L!, M!, W (photo!)]. Epithet: a Latin derivative meaning having a similar shape, form, or habit to palms in the genus *Geonoma.* **PLATES 1(B), 7(A), 19, 20(A-C).**

Nunnezharia geonomiformis (H. A. Wendl.) Hook. f., Bot. Mag. 100: t. 6088. 1874.

Chamaedorea fenestrata Hort. in H. A. Wendland, Index Palmarum 12. 1854.

Chamaedorea humilis Hort. in H. A. Wendl, Index Palmarum 12. 1854. Not (Liebm.) Mart.

Geonoma fenestrata Hort. in Dammer, Gard. Chron., Ser. 3, 36: 202. 1904b.

Geonoma humilis Hort. in Dammer, Gard. Chron., Ser. 3, 36: 202. 1904b.

Migandra O. F. Cook, Nat. Hort. Mag. 22: 142. 1943b. A name of no botanical standing, the species was the type for this genus but Cook did not actually make the transfer.

Common Name: necklace chamaedorea.

Vernacular Names: pacaya—Honduras; capucacapocha—Guatemala.

Habit: solitary, erect or occasionally decumbent, to 1.5 m tall, often flowering when stemless.

Stem: 0.5–1 cm diam., green, smooth, ringed, internodes 1–3 cm long, adventitious roots conspicuous at base.

Leaves: 5–10, erect-spreading, bifid, rich green above, green below; **sheath** 8–10 cm long, tubular, green, obliquely open above middle, longitudinally striate-nerved; **petiole** 2–15 cm long, flat and green above, rounded and pale below; **rachis** 11–22 cm long, angled and yellow-green above, rounded and yellowish below; **blade** 15–30 × 15 cm, incised apically ⅓–½ its length and there 10 cm wide, oblanceolate or oblong-elliptic, lobes acute, only shallowly and obscurely toothed toward apex, 9–12 nerves on each side of rachis, these obscure above, pale yellowish and prominent below, not keeled above, 2 secondaries between each pair of primaries, tertiaries numerous, faint.

Inflorescences: interfoliar but often infrafoliar in fruit, solitary, erect-spreading; **peduncles** 10–20 cm long, slender, 2–5 mm wide at base, 2–3 mm wide at apex, erect, green in flower where exposed, red-orange in fruit; **bracts** 5, tubular, closely sheathing, papery, drying brown by anthesis, longitudinally striate-nerved, uppermost shorter than to exceeding peduncle; **rachises** 2 cm long, green in flower, red-orange in fruit. **Staminate** with 3–6 **rachillae**, rarely spicate or with 2 rachillae and then usually only

→

Plate 21. Chamaedorea graminifolia Cayo, Belize. **A-D** *Hodel & Thomas 1132A* (staminate) *& 1132B* (pistillate); **A.** Plant in nearly full sun; **B.** Lou Thomas stands among colony in dense shade on steep limestone slope; **C.** Staminate inflorescence; **D.** Pistillate inflorescence; **E, F** *Hodel & Thomas 1128B*; **E.** Infructescence with immature fruits (note how inflorescence broke through persistent leaf sheaths); **F.** Stem with dense glaucous covering.

B
D
F

when flowering for one of first times, to 20 cm long, pendulous, slender, pale yellow.

Pistillate with 2–3 **rachillae,** rarely spicate and then usually only when flowering for one of first times, 5–15 cm long, erect-spreading, light or yellowish green in flower becoming thickened and orange-red in fruit.

Flowers: Staminate in rather dense spirals, 4 × 3.5–4 mm, globose-ovoid, yellowish, slightly sunken in elliptic depressions 3 mm long; **calyx** 0.5 × 3 mm, ringlike, shallowly lobed, not nerved when dry, yellow, membranous, **sepals** connate nearly to apex, broadly rounded apically; **petals** 3–4 × 4 mm, valvate, connate apically and basally and adnate apically to pistillode and corolla opening by lateral slits, membranous, nerveless or only faintly nerved; **stamens** 2–2.5 mm long, **filaments** equalling anthers, **anthers** 1.5 mm long, entire apically, briefly separated basally; **pistillode** 3 mm high, columnar, slightly narrowed apically.

Pistillate in moderate spirals, 1–2 × 2.5 mm, dome-shaped, yellow, slightly sunken in rounded depressions; **calyx** 0.75 × 2.5 mm, shallowly lobed, membranous, yellow, **sepals** imbricate and/or connate in basal ¾, broadly rounded apically, strongly nerved when dry; **petals** 2 × 3 mm, imbricate nearly to apex, acute, faintly nerved when fresh, strongly nerved when dry; **staminodes** not seen; **pistil** 2.5 × 2–2.5 mm, dome-shaped, yellow, **styles** short or lacking, **stigma** lobes slightly pointed, low, clear-colored.

Fruits: 8–12 mm diam., globose, blue-black, falling with perianth attached; seeds 6–9 mm diam.; abortive carpels adherent to fruit.

Distribution: MEXICO (without locality). GUATEMALA. Alta Verapaz. Izabal. Peten. BELIZE. Toledo. HONDURAS. Atlantida. COLOMBIA? Dense, wet forest on the Atlantic slope; 100–900 m elevation; usually on limestone.

Discussion: Warscewicz discovered *C. geonomiformis* in Guatemala and sent seeds of it to various gardens in Europe. By 1848, it was growing at Herrenhausen, in Van Houtte's garden at Hale and Allardt in Berlin, and at the Berlin Botanical Garden. Wendland (1852a) named it from his material cultivated at Herrenhausen. A popular plant in European stovehouses of the time, gardeners and collectors referred to it as the necklace palm because the long, pendulous, densely flowered rachillae bore a resemblance to a string of pearls.

Oersted (1859) felt that *C. geonomiformis* was sufficiently distinct and erected a subgenus, *Psilostachys,* for the species. In 1896 *C. geonomiformis* was reported at the Musee de Paris in France and Franceshi offered it as early as 1900 in California (Guillaumin 1923b).

Chamaedorea geonomiformis is very close to *C. tenella* and the latter species is perhaps not distinct. The more distinctly toothed margin of the leaves, smaller habit, spicate inflorescences, and more shallowly incised apex of the blade are used to distinguish *C. tenella.* There is a great deal of variation with the first two characters, although *C. tenella* generally seems to be a consistently smaller plant. However, as plants of *C. geonomiformis* become older and taller their leaves tend to become smaller. I have observed small-leaved forms of *C. geonomiformis* in cultivation that were referred to *C. tenella* yet had branched staminate inflorescences.

The inflorescences are also variable but those of *C. geonomiformis* are usually branched but sometimes spicate while those of *C. tenella* are always spicate. I have observed plants of *C. geonomiformis* with spicate inflorescences though these almost always occur when the plant is flowering for one of the first times. Later inflorescences are branched with the number of rachillae generally increasing with each subsequent inflorescence until about 4–6 branches have been attained.

The degree of the incision at the apex of the blade is perhaps the best character. Wendland stated that *C. geonomiformis* had blades incised ⅓–½ their length while those of *C. tenella* were more shallowly incised, only about ⅕–¼ their length. Moore (1966) suggested that *C. tenella* may simply be a high altitude form of *C. geonomiformis* yet we have collected material on the Osa Peninsula in Costa Rica at about 100 meters elevation that has smaller leaves and spicate inflorescences that matches up well with those described for *C. tenella.* Also, we collected material at nearly 1,000 meters elevation in Guatemala that corresponds to *C. geonomiformis.* In Mexico we collected material that matches up with *C. tenella* in Veracruz and Chiapas at about 800 meters elevation.

In summary, although the two are close, the larger habit, larger blades more deeply incised at the apex, and inflorescences usually branched distinguish *C. geonomiformis.*

Chamaedorea geonomiformis has been confused with *C. ernesti-augustii* in a few floristic accounts (Standley and Record 1936, perhaps Standley 1931). McCurrach (1960, p. 43) and Krempin (1990, p. 92) illustrated *C. ernesti-augustii* but erroneously captioned the photographs as *C. geonomiformis. C. geonomiformis* has been reported from Colombia only recently (Galeano and Bernal 1987) although

Plate 22. Chamaedorea hooperiana. Cultivated, California. **A.** Large clump, Ingwersen Nursery, Oceanside; **B.** Infructescence, *Hodel 690,* Huntington Botanical Garden (*43001*), San Marino.

→

I have not seen the specimens. One specimen at the Gray Herbarium documents *C. geonomiformis* from Mexico but without a specific locality.

Chamaedorea geonomiformis is rather common in gardens and collections in California and Florida. One of the most recent introductions came from material collected by the Huntington Botanical Gardens in Belize in 1977. Plants grown from these seeds have matured, set seeds, and produced second generation plants in California. *C. geonomiformis* is a choice and handsome plant for shady gardens in tropical and subtropical regions.

14. Chamaedorea glaucifolia *H. A. Wendl.*, Index Palmarum 64. 1854. Type: Cult., *Wendland s.n.* (holotype GOET!; isotype L!). Epithet: from the Latin *glaucus* meaning covered with a powdery gray material and *folius* meaning leaved, in reference to the glaucous leaf bases, petioles, and rachises. **PLATES 6(F), 7(C), 20(D, E).**

Nunnezharia glaucifolia (H. A. Wendl.) O. Kuntze, Revisio Generum Plantarum 2: 730. 1891.
Chamaedorea crucifolia Hort. in Crouch., Garden 1: 283. 1872.
Chamaedorea elegantissima Hort. in Kerch., Les Palmiers 239. 1878.
Chamaedorea glaucophylla Hort. in Siebert & Voss, Vilmorin's Blumengartnerei 1155. 1896.
Discoma glaucifolia O. F. Cook, Nat. Hort. Mag. 22: 137. 1943b. Name of no botanical standing.

Common Name: none.
Vernacular Name: none.

Habit: solitary, erect, to 5 m tall.

Stem: 2–3.5 cm diam., green, prominently ringed, internodes 25–35 cm long.

Leaves: 3–5, pinnate, erect, ascending, only later spreading; **sheath** 25–40 cm long, tubular, green, obliquely open apically and there ± pale but with brown ragged margin, slightly glaucous, leathery but drying grayish brown and ± woody, persistent, durable, longitudinally striate-nerved with a rounded ridge extending from petiole; **petiole** 15–40 cm long, slightly grooved and green above especially toward base, rounded and green and densely glaucous below; **rachis** 1–2 m long, angled and green above, rounded and green below and densely glaucous; glaucous covering on leaf sheath, petiole, and rachis is ± deciduous and tends to weather away with time; **pinnae** 50–70 on each side of rachis, 30–35 × 0.5–1.2 cm, narrowly linear to linear-lanceolate, clustered in groups of 2–4, borne in different planes and directions, long-acuminate, gray-green, prominent midrib and several much less prominent primary and secondary nerves on each side of this.

Inflorescences: infrafoliar, usually among old leaf sheaths, long-pedunculate, erect, solitary; **peduncles** to 50 cm long, ± stout, 7 mm diam. at apex, 1.2 cm diam. at base, erect or downward-arching when heavily laden with fruits, light green in flower, red-orange in fruit; **bracts** 6, prophyll 5 cm long, 2nd bract 7 cm, 3rd 11 cm, 4th 18 cm, 5th 20 cm, 6th 15–20 cm, uppermost exceeding peduncle, closely sheathing, acute, bifid, fibrous, longitudinally striate-nerved, light brown at anthesis. **Staminate** with **rachis** 10–20 cm long, obtusely angled, light green in flower; **rachillae** 15–30, these 25–30 cm long, 2 mm diam., obtusely and longitudinally angled, drooping, light green, minutely and faintly white-spotted.

Pistillate with **rachis** 15 cm long, longitudinally and obtusely angled, green in flower, red-orange in fruit; **rachillae** 12–30, lower ones the longest, these to 15 cm long, 2 mm diam., ± stiff, erect, spreading, obtusely and longitudinally angled, light green in flower, minutely and faintly white-spotted, red-orange in fruit.

Flowers: Staminate 2.5 × 3 mm, depressed-globose, yellow, slightly aromatic, slightly sunken in rounded depressions; **calyx** 1 × 1.75 mm, lobed, green, **sepals** imbricate in basal ½, triangular and acute apically; **petals** 2.5 × 2.5 mm, valvate, connate apically and basally and apically adnate to pistillode and corolla opening by lateral slits, acute, lightly nerved on inside; **stamens** 2 mm high, **filaments** 0.75–1 mm long, green, **anthers** 1.5 mm long, oblong, yellowish; **pistillode** 2.5 mm high, columnar, green.

Pistillate in rather lax spirals, 2 × 2 mm, globose, light yellow; **calyx** 1 × 2 mm, deeply lobed, nerveless, **sepals** imbricate in basal ⅓, rounded apically; **petals** 2 × 3 mm, imbricate nearly to apex, acute, nerveless; **pistil** 2 × 2 mm, globose, greenish, **styles** short or lacking, **stigma** lobes short, slightly recurved, barely exceeding petals.

Fruits: 7–10 mm diam., globose, black.

Distribution: MEXICO. Chiapas. Moist or wet forest on the Atlantic slope; 500–1,000 m elevation; often in exposed locations on limestone.

Discussion: Linden introduced *C. glaucifolia* to the Jardin Botanique de Bruxelles apparently from Chiapas, Mexico, about the middle of the 19th century. Wendland (1854) described and named the species, apparently from staminate material only, that was cultivated at the Brussels garden. Wendland stated that the species was collected in pine woods in New Granada near Chiapas. New Granada usually refers to present-day Colombia and Panama. However, Wendland and others also used it to refer to southern Mexico. In any case, collections document it at several localities in Chiapas.

Chamaedorea glaucifolia is a very distinctive species, not likely to be confused with any others and is easily recognized by its numerous linear pinnae

→

Plate 23. Chamaedorea hooperiana. Cultivated, California. **A.** Clustering habit with persistent woody leaf sheaths from behind which new shoots arise, research collection, Los Angeles; **B.** Pistillate flowers, *Hodel 804*, garden of P. Sullivan, Ventura; **C.** Staminate flowers, *Hodel 772* (holotype), garden of L. Hooper, La Habra; **D.** Leaf and pistillate inflorescence, *Hodel 804*, garden of P. Sullivan, Ventura.

B
C

borne in several planes and clustered in groups along the rachis. It has sometimes been erroneously referred to in cultivation as *C. graminifolia,* a cespitose species not in cultivation. *C. glaucifolia* is among the finest textured of all the species of the genus. The amount of glaucous indument on the rachis and petiole seems to vary considerably. Some individuals are thickly covered while others are only lightly covered or almost bare. The glaucous indument seems to be less than permanent, often shedding and weathering away with age, leaving only the newly emerged leaves with petioles and rachises covered in such a manner. There is also considerable variation in the size of pinnae and the diameter of stems.

Widely cultivated, *C. glaucifolia* appears in gardens and collections in Hawaii, California, Florida, Venezuela, and Australia. In California, it was apparently first offered in the trade in 1912 (Reidel 1957). It is relatively easy to grow and tolerates higher light than most members of the genus. Because of its unusual and fine texture, it is used very effectively as an accent in a container or planter. It also makes an interesting study in texture and is quite attractive as a mass planting through which other plants may be observed.

15. Chamaedorea graminifolia *H. A. Wendl.,* Index Palmarum 62. 1854. Type: Cult., *Wendland s.n.* [holotype GOET!; isotypes C!, L!, W(photo!)]. Epithet: from the Latin *gramen* meaning grass and *folius* meaning pertaining to the leaves, in reference to the narrow grasslike pinnae. **PLATE 21.**

Nunnezharia graminifolia (H. A. Wendl.) O. Kuntze, Revisio Generum Plantarum 2: 730. 1891.

Chamaedorea schippii Burret, Notizbl. Bot. Gart. Berlin-Dahlem 11: 1038. 1934. Type: Belize, Toledo, Jacinto Hills, *Schipp 1222, S-569* (holotype B, destroyed; isotypes BM!, F!, GH!, MICH!, MO!, S!). Epithet: honors William Schipp, collector of the type.

Common Name: none
Vernacular Name: chapai—Belize.

Habit: cespitose, erect to leaning, to 4 m tall, forming open clusters and colonies 1–5 m across by spreading rhizomes.

Stems: separated, 2–3 cm diam., green with dense glaucous covering, ringed, internodes to 30 cm long.

Leaves: 4–6, erect-spreading, pinnate, to 1.75 m long; **sheath** to 20–40 cm long, tubular for ¾ its length, obliquely open apically and there light green and longitudinally striate-nerved and slightly glaucous, ± thick and durable, green, brown-margined, drying brownish, persistent; **petiole** 15–25 cm long, slightly glaucous, ± rounded above and gray-green but slightly grooved near base, rounded and green below; **rachis** to 70–110 cm long, angled and green above, gray-green and rounded below with a faint yellow-green band extending onto sheath, slightly glaucous; **pinnae** 22–42 on each side of rachis, to 40 × 1.5–2.5 cm, narrowly linear to long lanceolate, straight, acute to acuminate, ± drooping apically, gray-green, regularly arranged except basal ones sometimes clustered irregularly, all except ± wider terminal pair with prominent midrib scarcely keeled above and 1 primary nerve on each side of this, pinnae drying slightly plicate.

Inflorescences: infrafoliar, emerging from persistent leaf bases, erect, solitary; **peduncles** 15–35 cm long, to 1 cm wide at base, 2–3 mm diam. at apex, greenish in flower, red-orange in fruit; **bracts** 5–7, prophyll to 2.5 cm and bifid, 2nd bract 6 cm, 3rd 11 cm, 4th 16 cm, 5th–7th to 20 cm, uppermost exceeding peduncle and often concealing a rudimentary bract, tubular, brownish in flower, acute to auminate, longitudinally striate-nerved, upper ones slightly inflated, apical one exceeding peduncle and extending onto rachis; **rachises** 5–10 cm long, greenish yellow in flower, red-orange in fruit. **Staminate** with 12–35 **rachillae**, these to 25 cm long, 1–2 mm diam., ± pendulous, green in flower.

Pistillate with 10–22 **rachillae**, these to 15 cm long, spreading to ascending or curved-ascending, ± stout and stiff, 3 mm diam., yellow-green and minutely white-spotted in flower, red-orange in fruit.

Flowers: Staminate in moderate spirals, 2 × 2–2.5 mm, depressed-globose, flattened apically, yellow, aromatic, superficial or only slightly sunken in elliptic depressions 1.5 mm long; **calyx** 0.5–0.75 × 1.5 mm, deeply lobed, green, **sepals** connate in basal ¼–½, broadly rounded apically, nerveless; **petals** 2.5–3 × 1.75 mm, valvate, connate apically and basally and adnate apically to pistillode, perhaps with age becoming free apically, broadly oblong, acute, ± fleshy, lightly nerved, more so when dry; **stamens** 1.5–1.75 mm high, **filaments** 1 mm long, **anthers** 0.75–1 mm long, dorsifixed, yellow; **pistillode** 2 mm high, columnar, broadened basally, 3-lobed, yellowish.

Pistillate in moderate spirals, 2 × 1.5–2 mm, ovoid to globose, yellowish, slightly sunken in elliptic depressions 1.75 mm long; **calyx** 0.5–0.75 × 1.75 mm, deeply lobed, greenish, **sepals** imbricate in basal ½, broadly rounded apically; **petals** 1.75 × 2.5 mm, imbricate nearly to apex, triangular, acute, ± erect apically, faintly nerved when dry; **staminodes** not seen; **pistil** 2 × 1.75 mm, ± globose or ovoid, green, **stigma** lobes very short, virtually sessile.

Fruits: 6–10 mm diam., globose, black; abortive carpels adherent to fruit.

Distribution: BELIZE. Cayo. Toledo. GUATEMALA. Alta Verapaz. MEXICO (Tehuantepec.) COSTA RICA? Moist or wet forest on the Atlantic slope; to 700 m elevation; on limestone.

Plate 24. Chamaedorea klotzschiana. Cultivated. **A.** Leaf with pinnae clustered in groups along rachis, *Hodel 802A,* research collection, Los Angeles; **B.** Plant, Fairchild Tropical Garden, Miami, Florida; **C.** Staminate inflorescence, *Hodel 802A,* research collection, Los Angeles; **D.** Infructescence with mature fruits, garden of P. Sullivan, Ventura, California.

CHAMAEDOREA
C
D

Discussion: Wendland described and named *C. graminifolia* from cultivated material that Linden supposedly introduced to European gardens from Guatemala. Strangely, the horticultural and botanical literature has given scant attention to it over the years, then often erroneously or only to note that it was a poorly known and understood species.

Staminate flowers of *C. graminifolia,* like those of *C. hooperiana, C. elatior,* and *C. glaucifolia,* are bright yellow, strongly aromatic, only faintly nerved or nerveless when dry, and, at times, tend to spread slightly apically well after anthesis. Although I include them in subgenus *Chamaedorea* since at anthesis the staminate petals are connate apically and the corolla opens by lateral slits, they approach several cespitose species in subgenus *Chamaedoropsis* florally, namely *C. seifrizii* and *C. pochutlensis.* These latter species also have nerveless, bright yellow, aromatic staminate flowers but the petals spread apically yet often only slightly so, the tips remaining incurved over the stamens and nearly adnate to the pistillode. Surprisingly, Wendland's type material of *C. graminifolia* at GOET has the staminate petals free apically yet when Wendland named and described the species, he did so stating that the petals were connate apically.

Although apparently never common in the wild in the past, *C. graminifolia* is now, due to habitat destruction, even more uncommon. Its existence in Costa Rica is somewhat tenuous; only three collections, none of which is recent, document it there. There is also a collection at the Bailey Hortorium from plants cultivated in the 1920s in California that purportedly originated from the Tehuantepec region of Mexico.

We found *C. graminifolia* at several locations in Belize, always at the summits of limestone hills where it forms open clumps of finely divided leaves to as much as five meters across and of the same sex. The forest in the area not on, but surrounding, the limestone hills has largely been destroyed, so it is not possible to determine if the species once inhabited lower, level areas. However, it appears that *C. graminifolia* is mainly restricted to the summits and adjacent upper slopes since we did not encounter it on the lower slopes even though there was often suitable habitat. The forest on limestone hills is quite open and the palms grow on solid limestone rock covered with a layer of leaf litter. During the extended dry season, the substrate can become exceedingly parched. In May, 1991, companion shrubs and trees were actually wilted due to the intense and prolonged dry spell yet *C. graminifolia* seemed to be thriving.

The numerous and narrow linear pinnae and stems forming rather open and wide-spreading clumps readily distinguish *C. graminifolia.* In many ways, it reminds one of a cespitose *C. glaucifolia.* It even has the characteristic glaucous covering on the leaf sheath, petiole, and rachis as does the latter species. Once they drop their persistent leaf sheaths, the stems of *C. graminifolia* are also covered with a waxy, glaucous material. The clump-forming habit of *C. graminifolia* is rather distinctive. Stems are well separated by rhizomes and often emerge from the ground a meter distant from the main stem.

Material grown as *C. graminifolia* is usually misidentified; the name is often applied to a form of *C. elegans* or to *C. seifrizii.* Hertrich (1951, p. 23) and McCurrach (1960, page 43) contain photographs of a form of *C. elegans* with long-pedunculate inflorescences incorrectly captioned as *C. graminifolia.* Muirhead (1961) listed *C. graminifolia* but is probably referring to *C. glaucifolia* or *C. seifrizii.* Krempin (1990, p. 92, 93) discussed and illustrated *C. graminifolia* but the description and photograph depict *C. seifrizii.*

Standley (1931) and Chittenden (1951) listed *C. graminifolia* for Costa Rica and, although it has been collected there a few times, their descriptions depict other species. Standley (1920) listed it from the Yucatan in Mexico although he expressed uncertainty about the application of the name and was probably referring to *C. seifrizii.* Standley and Calderon (1925) listed *C. graminifolia* for El Salvador where it is known as *cuiliote* or *cuiliote amargo* but the plant is probably *C. quezalteca.*

On the other hand, the name *C. schippii,* a synonym of *C. graminifolia,* is rather frequent in the literature and horticulture. Burret described and named it from material that William Schipp, who called it one of the most beautiful and elegant palms, had collected in the Jacinto Hills of southern Belize. However, this name is also usually misapplied in

→

Plate 25. Chamaedorea liebmannii. A. Mature plant, *D. & R. Hodel 926,* Chiapas, Mexico; **B.** Mature plant, *D. & R. Hodel 944,* Oaxaca, Mexico; **C, D** Mature plants, *Hodel & Castillo 1004A* (staminate) *& 1004B* (pistallate), Huehuetenango, Guatemala; **C.** Staminate; **D.** Pistillate.

B
D

horticulture, in this instance to hybrid plants in cultivation. Pure *C. schippii* (= *C. graminifolia*) is not cultivated, at least in southern California; plants so referred there are actually hybrids with *C. pochutlensis.* Krempin (1990, p. 92, 93) discussed and illustrated *C. schippii* but the description and photograph depict this hybrid.

Collectors have made this hybrid in California and perhaps elsewhere reportedly between *C. schippii* (= *C. graminifolia*) and *C. costaricana.* However, it is not *C. costaricana* but *C. pochutlensis* that was utilized as one of the parents with *C. schippii* (= *C. graminifolia*). Since *C. graminifolia* in cultivation is already hybridized and, therefore, not pure, the resulting offspring are in reality backcrosses with *C. pochutlensis.*

Local inhabitants in Belize refer to *C. graminifolia* as *chapai* and state that its leaves are boiled in water and the liquid used to treat high blood pressure and diabetes.

16. Chamaedorea hooperiana *Hodel,* Principes 35: 188. 1991. Type: Cult., *Hodel 772* (holotype BH!; isotypes CR!, HNT!, K!, MEXU!, MO!, NY!). Epithet: honors Louis Hooper of La Habra, California from whose garden the type originated. **PLATES 22, 23.**

Common Name: none.
Vernacular Name: none.

Habit: Cepitose, new lateral shoots emerging from tops of old dried persistent basal sheaths, forming fairly dense clumps to 3–4 m across, erect, leaning with age, to 4–5 m tall.

Stems: 2–2.5 cm diam., green, ringed, often covered with old leaf bases, internodes to 15 cm long.

Leaves: 5–7, erect-spreading, pinnate; **sheath** to 40–50 cm long, tightly clasping, obliquely open apically and there splitting deeply opposite petiole with age, rough-brown-margined, below this whitish and longitudinally striate-nerved, old sheaths persistent, drying brown, hard, durable, ± woody; **petiole** to 20–35 cm long, lime-green and grooved especially near base above, rounded and pale below; **rachis** to 0.8–1 m long, sharply angled and lime-green above, rounded below with a green or yellowish band extending onto sheath, attenuate apically; **pinnae** 20–26 on each side of rachis, lower and middle ones longest, these to 40 × 1.8 cm, becoming progressively shorter toward apex of rachis, regularly arranged, opposite to subopposite, flat off rachis, elongate-lanceolate, ± straight, only slightly falcate, long-acuminate, slightly contracted basally, ± thick, durable, a prominent pale midrib sharply angled above and 2 much less prominent primary nerves on each side of this, secondaries and tertiaries faint and inconspicuous, midrib prominent below, apical pair of pinnae slightly wider.

Inflorescences: infrafoliar, emerging from tops of dried persistent sheaths, erect. **Staminate** with **peduncle** to 30 cm long, 1.5 cm wide at base, erect-spreading, pale green where exposed; **bracts** 5–6, tubular, obliquely open apically and there bifid except largest which is long-acuminate and greatly exceeds peduncle and often conceals 1 or 2 smaller ones, longitudinally striate-nerved, brown in flower; **rachis** 20 cm long, light green; **rachillae** 40–45, these to 25 cm long, 1.75–2 mm diam., spreading-drooping, light green.

Pistillate with **peduncle** to 50 cm long, 1.5 cm wide at base, erect, pale or greenish in flower where exposed, reddish orange in fruit; **bracts** 6, similar to those of staminate inflorescence, brown and persistent in fruit; **rachis** to 22 cm long, green-yellow in flower, reddish orange in fruit; **rachillae** 40, lower ones longest, these to 22 cm long, apical ones to 10 cm long, spreading slightly, ± stiff, yellow-green in flower, reddish orange in fruit.

Flowers: Staminate in moderately dense spirals, 3–3.5 mm distant, 3–3.5 × 3–4 mm, subglobose, bright yellow, strongly aromatic, slightly sunken in elliptic depressions; **calyx** 1 × 2 mm, shallowly lobed, membranous, light green, **sepals** connate nearly to apex, broadly rounded apically; **petals** 3.5 × 3 mm, valvate, connate apically and basally and adnate apically to pistillode and corolla opening by lateral slits, perhaps petals later apically spreading slightly but remaining inwardly curved, broadly acute, margins ± thickened; **stamens** 2.5 mm tall, **filaments** 1–1.5 mm long, very pale green, nearly clear-colored, **anthers** 1 mm long, yellow changing to white; **pistillode** 2.75 mm tall, columnar, light green, darkened and narrowed apically.

Pistillate in rather remote spirals, 8 mm distant, 3 × 3 mm, globose, greenish yellow, slightly sunken in elliptic depressions; **calyx** 1–1.5 × 3 mm, prominently lobed, very light green or nearly yellow, **sepals** connate and or lightly imbricate briefly basally, broadly rounded apically; **petals** 2.5–2.75 × 3 mm, tightly imbricate nearly to apex, acute; **pistil** 3 × 3 mm, globose, light green, **stigma** lobes sessile, distinct but low, rounded.

Fruits: 7–8 mm diam., oblong-globose, black.

Distribution: MEXICO. Veracruz. Dense, wet forest; 1,000–1,500 m elevation.

Discussion: *C. hooperiana* presently exists in only a few collections in southern California. Pauleen Sullivan in Ventura, Lou Hooper in La Habra, Jack Ingwersen in Oceanside, and the Huntington Botanical Gardens in San Marino have mature plants in their gardens. For years these plants were unidentified and their origin uncertain. The common story was that the existing plants originated from one introduction from Central America. However, the Huntington's plant may have originated from a different introduction.

In December, 1989, during field work in Mexico, we found *C. hooperiana* in the wild in the Catemaco region of Veracruz, enabling us to provide a known

---

→

Plate 26. A, Chamaedorea liebmannii. B, Chamaedorea lucidifrons. C, Chamaedorea macrospadix. A. Pistillate inflorescence and infructescence with mature fruits, *D. & R. Hodel 944,* Oaxaca, Mexico (note twice-branched rachillae). **B.** Lou Hooper holds leaf and infructescence, *Hodel et al. 1110,* type locality, El Valle, Panama. **C.** On rocky slope, *D. & M. Hodel 692,* Monteverde, Costa Rica.

C

locality for it. Sullivan has plants of both sexes and produces seeds regularly that she has distributed to local palm collectors and hobbyists. I originally thought that this species was *C. karwinskyana* and plants of *C. hooperiana* may have been distributed as such. However, I have since examined the type of *C. karwinskyana* and realize that it can be included with *C. pochutlensis.*

Similar florally to *C. graminifolia* and vegatatively to *C. pochutlensis, C. hooperiana* can be distinguished in the manner in which it sends forth new shoots from the base of the plant. These emerge from the tops of the persistent, nearly woody, basal leaf sheaths. In addition, *C. hooperiana* has thicker, durable, nearly plasticlike leaves. Rhizomatous stems emerging some distance from the parent plant and soft, thin, narrowly linear pinnae distinguish *C. graminifolia.* Other distinguishing characters of *C. pochutlensis* include the staminate flowers with apically spreading petals and broader, softer pinnae.

A handsome species of easy culture, *C. hooperiana* is a vigorous and relatively fast grower. Its eventual size should be considered when placing it in the landscape. With age, it will form rather dense clumps several meters across. Stems toward the perimeter of the clump tend to lean outward gracefully, occupying even more space. It is more resistant to infestations of mites than *C. costaricana* and *C. pochutlensis* and is much superior as an indoor plant. In fact, it holds great promise for use in interior situations due to its tolerance of low light and low humidity and resistance to pests. Collectors in southern California have made hybrids between *C. hooperiana* and *C. pochutlensis.*

17. Chamaedorea klotzschiana *H. A. Wendl.,* Index Palmarum 63. 1854. Type: Cult., *Wendland s.n.* (holotype GOET!; isotype C!). Epithet: honors Wendland's friend and colleague, botanist J. Friedrich Klotzsch. **PLATE 24.**

Nunnezharia klotzschiana (H. A. Wendl.) O. Kuntze, Revisio Generum Plantarum 2: 730. 1891.

Common Name: none.
Vernacular Name: tepejilote—Mexico.

Habit: solitary, erect, 1–4 m tall.

Stem: 1.5–2.8 cm diam., green, smooth, prominently ringed, internodes 5–20 cm long.

Leaves: 4–6, pinnate, erect-spreading; **sheath** 25–35 cm long, tubular, obliquely open apically, fibrous, green, drying grayish brown and persisting on stem, durable, nearly woody, longitudinally striate-nerved with a rounded ridge extending from petiole; **petiole** 17–40 cm long, slightly channeled and green above, rounded and pale below; **rachis** 0.6–1 m long, slightly angled and green above, rounded below with a ± inconspicuous greenish stripe extending onto sheath; **pinnae** 12–20 on each side of rachis, 20–40 × 3.6–5 cm, narrowly lanceolate to lanceolate, ± sigmoid, clustered in groups of 2–4, groups 8–17 cm apart, long-acuminate, margins entire, lustrous green, midrib and 2 marginal primary nerves especially prominent below, 3 secondaries on each side of midrib, tertiaries numerous, faint.

Inflorescences: infrafoliar, sometimes emerging from top of or thrusting off persistent dried leaf sheaths, 40–70 cm long, long-pedunculate; **peduncles** 30–45 cm long, ± rounded but flattened basally, 1–2 cm wide at base, 0.7–1 cm wide at apex, erect and green in flower, downward-arching and red-orange in fruit; **bracts** 5–6, upper one largest, to 30 cm long and exceeding peduncle, fibrous, durable, acute-acuminate, ± tightly sheathing, longitudinally striate-nerved. **Staminate** with **rachis** 5–20 cm long, green in flower, downward-pointing; **rachillae** 12–20, these 20–30 cm long, 2 mm diam., drooping, green in flower.

Pistillate with **rachis** 5–20 cm long, erect, green in flower, red-orange in fruit; **rachillae** 12–20, these 15–25 cm long, 1.5–2 mm diam., ± stiff, spreading, green and erect in flower, red-orange and downward-pointing when heavily laden with fruits.

Flowers: Staminate in ± dense spirals, 3 × 3.5 mm, depressed-globose, flattened apically, greenish yellow, sunken in elliptic superficial depressions 1.5–2 mm long; **calyx** 1 × 2–2.5 mm, shallowly lobed, light green, nerveless, **sepals** connate in basal ¾, rounded apically; **petals** 3–3.5 × 3.5 mm, obovate, valvate, connate apically and basally and apically adnate to pistillode and corolla opening by lateral slits, thin, fleshy, acute, nerveless; **stamens** 3 mm high, **filaments** 0.75–1.25 mm long, greenish yellow, **anthers** 1.75–2 mm long, oblong, yellow; **pistillode** 2.5–3 mm high, columnar, truncate apically, greenish yellow.

Pistillate in lax spirals, 3 × 2–2.5 mm, ovoid-globose, yellow, in rounded superficial depressions 1.5–2 mm wide; **calyx** 1 × 2.5 mm, lobed, green, **sepals** connate in basal ¼, triangular and rounded apically; **petals** 2.5 × 2.5–3 mm, obovate to scarcely circular, imbricate nearly to apex, and there straight and only slightly pointed, nerveless; **staminodes** lacking or 6, small, separated, inserted at base of pistil, slightly triangular; **pistil** 3 × 2–2.5 mm, ovoid-globose, yellow-green, **stigma** lobes sessile, separated, angled, slightly elevated and recurved, just exceeding petals.

Fruits: 9–11.5 × 7–9 mm, globose-ovoid, black, occasionally 2–3 in one perianth, epicarp thin, slightly transparent, mesocarp slightly fleshy, green, mucilaginous, endocarp membranous, nerved; seeds 6–9 × 5.5–7 mm, globose, brownish.

Distribution: MEXICO. Veracruz. Dense, wet forest; 500–1,250 m elevation.

Discussion: Wendland (1854) described and named *C. klotzschiana* from material growing at Herrenhausen. It is apparently restricted to the state

Plate 27. Chamaedorea macrospadix. Costa Rica. **A.** Staminate plant, *Hodel & Grayum 701A,* along trail down to river, C.A.T.I.E., Turrialba (note open leaf sheaths); **B.** Fruiting plant on limestone rocks, *D. & M. Hodel 712B,* above Neily, Puntarenas.

of Veracruz in Mexico. *C. klotzschiana* is one of the most distinct members of the genus and is easily recognizable by its lanceolate pinnae clustered in groups of 2–4 along the rachis. *C. glaucifolia* also has its leaflets clustered in groups along the rachis but the more numerous (40–80 on each side), narrowly linear pinnae in several planes distinguish it from *C. klotzschiana*.

Because of its attractive foliage, *C. klotzschiana* is widely cultivated and much sought after by collectors. It is found in gardens and collections in California, Florida, Hawaii, and Australia.

18. Chamaedorea liebmannii *Mart.*, Historia Naturalis Palmarum 3: 308. 1849. Type: Mexico, Oaxaca, *Liebmann s.n.* (some sheets later numbered *6582*) (holotype C!; isotypes P!, US!). Epithet: honors Frederick Liebmann, collector of the type. **PLATES 25, 26(A).**

Collinia elatior Liebm., Overs. Kongel. Danske Vidensk. Selsk. Forh. Medlemmers Arbeider 1846: 7. 1846. Name only.

Chamaedorea lepidota H. A. Wendl., Allg. Gartenzeitung 21: 138. 1853b. Type (syntypes): Mexico, Oaxaca, *Galeotti 4975* (F!, GOET!) and *4989E* [BH!, BR!, GOET!, P!, W(photo!)] and Cult., *Wendland s.n.* [C!, G(photo!), GOET!]. Epithet: from the Greek *lepidotus* meaning covered with small scales, in reference to the type of indument often found on the leaves.

Collinia elatior Liebm. ex Oerst., Vidensk. Meddel. Dansk. Naturhist. Foren. Kjoebenhavn 1858: 6. 1859. New name only based on *Chamaedorea liebmannii* Mart.

Nunnezharia lepidota (H. A Wendl.) O. Kuntze, Revisio Generum Plantarum 2: 730. 1891.

Nunnezharia liebmannii (Mart.) O. Kuntze, Revisio Generum Plantarum 2: 730. 1891.

Chamaedorea liebmannii var. *lepidota* H. A. Wendl. ex Dammer, Gard. Chron. Ser. 3, 36: 246. 1904b.

Chamaedorea aequalis Standl. & Steyerm., Publ. Field Mus. Nat. Hist., Bot. Ser. 23: 196. 1947. Type: Guatemala, Huehuetenango, *Steyermark 51666* (holotype F!). Epithet: is a Latin word meaning equal, perhaps in reference to the equally tapering, long-acuminate pinnae although the authors did not specify the application of the name.

Chamaedorea velutina Hort. in H. A. Wendl., Index Palmarum 13. 1854.

Lophothele ramea O. F. Cook, Nat. Hort. Mag. 22: 142, fig. 14. 1943b. Name of no botanical standing.

Common Name: none.
Vernacular Name: none.

Habit: solitary, slender, erect, infrequently decumbent, to 4 m tall, sometimes flowering when stemless.

Stem: 2 cm diam., nodes only slightly thickened, internodes to 10 cm long but usually shorter.

Leaves: 4–9, pinnate, erect-spreading, sometimes covered with minute whitish lepidia; **sheath** 10–13 cm long, tubular, striate-nerved, subauriculate apically; **petiole** 12–27 cm long, ± triangular, slightly channeled and green above, rounded and pale below; **rachis** 30–45 cm long, sharply angled and green above, rounded and green below with a yellow band extending onto sheath; **pinnae** 13–18 on each side of rachis, 16–30 × 1.5–3 cm, linear to narrowly lanceolate, long-acuminate, constricted and plicate basally, alternate or slightly subalternate, gleaming-velutinous, lower somewhat reflexed and upper often confluent, a prominent midrib and 2–4 primary nerves on each side of this, secondaries numerous, slightly visible.

Inflorescences: inter- or infrafoliar, spreading; **peduncles** 20–45 cm long; **bracts** 5–7, upper one to 25 cm long, fibrous, thin; **rachises** 2–9 cm long. **Staminate** with up to 26 **rachillae**, these to 25 cm long, ± pendulous toward apex, lower ones forked or branched, green in flower.

Pistillate with 18–24 **rachillae**, these to 20 cm long, angled or ribbed, erect-spreading, lower ones sometimes branched, green in flower, red-orange in fruit.

Flowers: Staminate in ± dense spirals, 3 × 2.5 mm, ± globose, yellow, in superficial elliptic depressions 1.5 mm long; **calyx** 1 × 1.5 mm, shallowly lobed, pale green, membranous, nerveless, **sepals** connate in basal ¾, rounded or straight apically; **petals** 2.5–3.5 × 1.5–2.5 mm, valvate, connate apically and basally and adnate apically to pistillode and corolla opening by lateral slits, obovate, thin, slightly fleshy, angled apically, nerved on inside; **stamens** 2.5 mm high, **filaments** 1.5–2 mm long, orange, slightly connate basally, **anthers** 1.5 mm long, narrowly oblong, dorsifixed, yellow; **pistillode** 2.5 mm high, columnar, attenuate, rounded apically.

Pistillate in spirals, 3 × 1.5 mm, ovoid, yellow, in superficial elliptic depressions 1.5 mm long; **calyx** 1.25 × 1.5 mm, deeply lobed, green, slightly membranous, strongly nerved on inside, **sepals** connate in basal ⅓, acute and brown-margined apically; **petals** 3 × 1.25 mm, connate briefly basally, imbricate apically nearly to top, long-ovate, erect and ± recurved, margins angled, thin, strongly nerved; **staminodes** lacking or rarely 3 or 6; **pistil** 1.75 × 1 mm, subglobose, green, attenuate slightly apically, **stigma** lobes sessile, greenish, slightly separated, erect, rounded.

→

Plate 28. A, B, Chamaedorea matae. C, Chamaedorea neurochlamys. A, B. Leaf, infructescence (immature fruits) and plant, *D. & R. Hodel 715,* Osa Peninsula, Costa Rica. **C.** Fruiting plant, cultivated, Wahiawa Botanic Garden (*77.493*), Honolulu, Hawaii, *ex* Toledo, Belize.

C

Fruits: to 10 mm diam., globose, black, epicarp slightly transparent, mesocarp thin, green, mucilaginous, endocarp slightly membranous, nerved; seeds 6.5 mm diam. globose; fruiting perianth nerved.

Distribution: MEXICO. Chiapas. Oaxaca. Puebla. Veracruz. GUATEMALA. Huehuetenango. Dense, wet forest on the Atlantic slope; 1,200–1,800 m elevation.

Discussion: Martius named and described *C. liebmannii* from material that Liebmann had collected near Teotalcingo and Petlapa in Oaxaca, Mexico. Liebmann (1846) suggested a name for this species (*elatior*) but did not publish it formally. When Martius (1849) formally named the species, the prior use of the epithet *elatior* in *Chamaedorea* precluded its use, and Martius gave it a new name in honor of Liebmann.

Although not presently cultivated, *C. liebmannii* was grown in Europe by about 1850 from material that Galeotti had introduced from Oaxaca. Wendland (1853b) named and described *C. lepidota* from Galeotti's material. Guillaumin (1923b) reported that *C. lepidota* was cultivated at the Musee de Paris in France since about 1850 and Belgian horticulturists knew it as *C. velutina.*

Chamaedorea liebmannii occurs at relatively high elevations, usually in oak and/or pine forest. In Mexico, it grows with *C. woodsoniana* near the summits of Volcan San Martin and Volcan Santa Martha in Veracruz. In Oaxaca, we found it with *C. woodsoniana, C. queroana,* and *C. rigida,* while in Chiapas, we observed it with *C. nubium* and *C. rojasiana.* In Huehuetenango in adjacent Guatemala, we found it with *C. carchensis, C. quezalteca, C. simplex,* and *C. verapazensis.*

Dried herbarium material consisting of only leaves and fruits is similar to and may be difficult to distinguish from forms of *C. elegans* with narrow pinnae. This similarity to *C. elegans* prompted Liebmann and Oersted to place *C. liebmannii* in *Collinia* with *C. elegans.* However, the smaller habit, connate pistillate petals, and recurved stigma lobes distinguish *C. elegans.* Of course, if staminate flowers are available, the two species are relatively easy to distinguish. Staminate petals of *C. elegans* are connate nearly to the apex and the corolla opens by a terminal pore. Also, the ranges of the two species do not greatly overlap. *C. elegans* is most common below 1,200 m elevation while *C. liebmannii* occurs above 1,200 m. Plants of both species have pinnate leaves at a very early age. *C. elegans* has pinnate eophylls while the leaves of *C. liebmannii* become pinnate as early as the third seedling leaf.

In their descriptions of *C. aequalis,* Standley and Steyermark (1947, 1958) included material from Solola and Quetzaltenango on the Pacific slope of Guatemala with substantially larger dimensions than the type from the Atlantic slope of Huehuetenango. The extraneous material from the Pacific slope has a nerveless fruiting perianth and is best referred to *C. whitelockiana* or is an unnamed species. Although *C. aequalis* is included as a synonym of *C. liebmannii,* only the range of dimensions of the type is included in the description here.

19. Chamaedorea lucidifrons *L. H. Bailey,* Gentes Herb. 6: 244, fig. 127. 1943a. Type: Panama, Cocle, *Allen 1814* (holotype MO!; isotype BH!). Epithet: from the Latin *lucidus* meaning shining or clear and *frons* meaning leaf, apparently in reference to the shining lower surface of the pinnae. **PLATE 26(B).**

Common Name: none.
Vernacular Name: none.

Habit: solitary, slender, erect, to 3 m tall.

Stem: 2–3 cm diam., green, ringed, internodes 5–10 cm long.

Leaves: 3–5, erect-spreading, pinnate; **petiole** to 45 cm long, ± flattened and green above and slightly grooved near base, green and rounded below; **rachis** 65 cm long, angled and green above, rounded and green below with a pale indistinct yellow band extending to sheath; **pinnae** 7 on each side of rachis, 30–35 × 8–12 cm, oblong-lanceolate, alternate, strongly sigmoid, long-acuminate, contracted basally, glossy, lucid particularly on lower surface, strongly nerved with 7–10 conspicuous and prominent primary nerves, these drying keeled and pale above, keeled and shining yellow below, secondaries nearly as conspicuous as primaries, tertiaries faint.

Inflorescences: Staminate not seen. **Pistillate** emerging from bare stem well below the leaves; **peduncle** to 60 cm long, slender, 1–1.5 cm wide at base, erect; **bracts** 6–8, tightly sheathing, brownish and ± splitting in fruit; **rachis** 2–4 cm long; branched portion 22 cm long and nearly as broad; **rachillae** 7–11, these 10–25 cm long, slender, 1 mm diam., simple, glabrous, upwardly s-curved, flexuous, orange in fruit.

Flowers: not seen.

Fruits: 5–6 × 4–5 mm, ellipsoid-globose, black, irregularly and shallowly wrinkled when dry; perianth nerved, small, lobes spreading.

→

Plate 29. Chamaedorea neurochlamys. A, B Cayo, Belize; **A.** Mature plant, *D. & M. Hodel 842*; **B.** Marianne Hodel and young plant, Rio Frio Cave; **C.** Infructescence with mature fruits, cultivated, Jardin Botanico Robert y Catherine Wilson, Costa Rica; **D.** White apex of leaf sheath with green nerves, cultivated, research collection, Los Angeles.

B
D

Distribution: PANAMA. Cocle. COLOMBIA? Dense, wet forest near the Continental Divide; 600 m elevation.

Discussion: The description is from Bailey (1943a, 1943b) who named and described *C. lucidifrons* without having seen good flowers of either sex, and is supplemented from our collection at the type locality. *C. lucidifrons* is very distinctive vegetatively because of its broadly sigmoid pinnae with conspicuous and prominent nerves, the latter which resemble, in many respects, those of *C. tepejilote.* The pinnae are among the broadest of any *Chamaedorea*; only those of some forms of *C. arenbergiana* and *C. warscewiczii* approach the pinnae of *C. lucidifrons* in width.

Staminate flowers are not known for *C. lucidifrons* so its placement in subgenus *Chamaedorea* is tentative. However, the nerved fruiting perianth seems to suggest that it is appropriately placed here. Probably closest to the highly variable *C. pinnatifrons,* the broadly sigmoid, thin pinnae markedly contracted basally and with numerous nerves visible above and below, the inflorescences emerging well below the leaves, and the slender, s-curved, upward-pointing fruiting rachillae distinguish *C. lucidifrons.* Apparently not cultivated, *C. lucidifrons* is not widely distributed, occurring only sparingly at the type locality near El Valle. A Kalbreyer collection from Colombia is tentatively referred to *C. lucidifrons.*

20. Chamaedorea macrospadix *Oerst.,* Vidensk. Meddel. Dansk. Naturhist. Foren. Kjoebenhavn 1858: 20. 1859. Type: Costa Rica, *Oersted s.n.* (later numbered *6568*) (holotype C!). Epithet: from the Greek *macro* meaning long or large and the Latin *spadix* meaning the flower stalk (inflorescence). **PLATES 6(B), 26(C), 27.**

Nunnezharia macrospadix (Oerst.) O. Kuntze, Revisio Generum Plantarum 2: 730. 1891.

Common Name: none.
Vernacular Name: none.

Habit: solitary, erect or sometimes decumbent, to 4 m tall but flowering when stemless.

Stem: 1.5–2 cm diam., green, prominently ringed, internodes 6 cm long.

Leaves: 3–7, erect-ascending, pinnate; **sheath** 15–20 cm long, splitting deeply opposite petiole and very open, clasping completely only toward base, durable, auriculate, brown-margined apically and below this whitish, longitudinally striate-nerved; **petiole** 30–65 cm long, slightly grooved and green above, rounded and pale below; **rachis** to 60 cm long, angled and green above, rounded below with a pale yellow band extending onto sheath; **pinnae** 8–13 on each side of rachis, to 30 × 5 cm, lanceolate, acuminate, regularly spaced, ± straight, only slightly sigmoid or falcate, contracted and plicate basally, ± thick, a very prominent midrib with two much-less-prominent primary nerves and three secondaries on each side of this, end pinnae wider.

Inflorescences: interfoliar, erect, sometimes spreading when heavily laden with fruits; **peduncles** equalling or exceeding leaves, to 1 m long or more, 1.5 cm wide at base, 5 mm wide at apex, erect to spreading, light green or yellowish in flower, light orange in fruit; **bracts** 6–9, 25–30 cm long, tubular, rather tightly clasping, obliquely open apically, acuminate, bifid, brown in flower, longitudinally striate-nerved; **rachises** 7–10 cm long, obtusely angled longitudinally, green in flower, red-orange in fruit. **Staminate** with 10–25 **rachillae**, these 20–35 cm long, pendulous, mostly simple although lower ones may be forked, green.

Pistillate with 8–15 **rachillae**, these to 20–25 cm long, spreading and ± stiff but drooping when laden with fruits, flexuous, green in flower, red-orange in fruit.

Flowers: Staminate in rather dense spirals, 2 × 1.5 mm, tapered apically and basally, ovoid, greenish; **calyx** low, lobed, **sepals** connate basally, rounded apically; **petals** valvate, connate apically and basally and adnate apically to pistillode and corolla opening by lateral slits, conspicuously nerved when dry.

Pistillate globose and greenish in young bud; **calyx** lobed, **sepals** imbricate basally; **petals** imbricate nearly to apex; **pistil** ± globose, greenish-yellow, **stigma** lobes sessile, separated.

Fruits: to 10 × 6 mm, oblong and slightly curved to obovoid or subglobose, black.

Distribution: COSTA RICA. Alajuela. Cartago. Limon. Puntarenas. San Jose. Dense, moist or wet forest on the Atlantic and Pacific slopes; 100–1,300 m elevation; often on limestone.

Discussion: Oersted (1859) described and named *C. macrospadix* from material he collected in the vicinity of Turrialba in Costa Rica. One of the most wide-ranging and variable chamaedoreas in Costa Rica, *C. macrospadix* occurs on the Atlantic and Pacific slopes from the Cordillera de Tilaran nearly to the Panamanian border. In fact, it would be surprising if *C. macrospadix* does not appear in collections from Panama in the future. It still occurs in the Turrialba area and can be seen along the trail down to the river at C.A.T.I.E. where it grows with *C. warscewiczii* and *C. tepejilote.* Also, we found it near San Miguel along the Rio Sarapiqui occurring with the same two species and in the Talamanca region near Panama. On the Pacific slope of Costa Rica we observed it in western Puntarenas near Monteverde and in southeastern Puntarenas in the Golfo Dulce region where it commonly occurs on

---

Plate 30. Chamaedorea palmeriana. Pistillate plant, *D. & M. Hodel 726B,* Chiriqui, Panama.

limestone outcroppings. In these situations it seems to remain virtually stemless for some time before, if ever, forming a stem and also has fewer pinnae and rachillae. Allen (1956) referred to it as *Chamaedorea* sp. *Allen 6742.*

Chamaedorea macrospadix is easily recognizable by its very long-pedunculate inflorescences which often exceed the leaves and which in young or stemless plants appear to arise at or near ground level. Also, as Allen (1956) noted, the pinnae have a satiny-glossy sheen on their undersides, similar to that found on the pinnae of *Neonicholsonia watsoniana.* A similar species with long-pedunculate inflorescences, *C. pedunculata,* differs in the broadly rhombic pinnae and staminate flowers rhombic- or diamond-shaped.

Chamaedorea macrospadix may appear in collections in the near future from seeds the International Palm Society Seed Bank distributed in 1987 and 1988 from our collections at C.A.T.I.E. (*Hodel 635, 701*) and the Golfo Dulce (*Hodel 629, 631, 712*).

21. Chamaedorea matae *Hodel,* Principes 35: 75. 1991a. Type: Costa Rica, Rincon de Osa, *Mata 497* (holotype CR!). Epithet: honors G. Mata, collector of the type. **PLATE 28(A, B).**

Common Name: none.
Vernacular Name: none.

Habit: solitary, slender, erect, to 3 m tall.

Stem: 1.5–2 cm diam., green, smooth, ringed, internodes 5–12 cm long.

Leaves: 4–7, spreading, pinnate; **sheath** to 20 cm long, tubular, obliquely open apically, longitudinally striated; **petiole** to 15–25 cm long, grooved and green above, rounded and pale below; **rachis** 50–70 cm long, angled and green above, rounded with a yellowish band below extending onto sheath; **pinnae** 4–5 on each side of rachis, basal ones to 30 × 9 cm, broadly lanceolate, strongly sigmoid, acuminate, thick, leathery, 5–6 prominent primary nerves, elevated above, drying yellow and prominent below, end pair conspicuously wider, to 15 cm wide, 10–12-nerved.

Inflorescences: interfoliar, erect to nodding or arching, solitary; **peduncles** to 50 cm long, 7–10 mm diam., greenish in flower and red-orange in fruit where exposed; **bracts** 5, prophyll to 6 cm long, 2nd and 3rd bracts to 15 cm, 4th to 23 cm, 5th to 26 cm, ± loosely sheathing, brownish in flower and fruit, bifid, acuminate, fibrous, becoming tattered; **rachises** 3–8 cm long, green in flower, red-orange in fruit. **Staminate** with up to 15 **rachillae**, to 25 cm long, pendulous, green in flower.

Pistillate with 7–10 **rachillae**, to 15–20 cm long, green and pendulous in flower, red-orange and pendulous in fruit.

Flowers: Staminate in moderately dense spirals 1–2 mm apart, 3 × 2.5 mm, ± ovoid to barrel-shaped, leaving superficial elliptic scars 2.5–3 mm long; **calyx** 2.5 × 0.75 mm, scarcely lobed, strongly nerved when dry, **sepals** connate nearly to apex, rounded or straight apically; **petals** 3.5 × 2.5–3 mm, valvate, connate apically and basally and apically adnate to pistillode and corolla opening by lateral slits, strongly nerved when dry; **stamens** shorter than pistillode, **anthers** 1.5 mm long; **pistillode** 2.5–2.75 mm high, columnar.

Pistillate leaving ± rounded superficial scars 2 mm long; **calyx** 0.75 mm high in fruit, scarcely lobed, strongly nerved; **petals** 2 × 3 mm in fruit, imbricate basally, broadly rounded apically, dark-centered, brown-margined, strongly nerved; **pistil** not seen.

Fruits: 7–8 × 5–6 mm, ± ovoid, black.

Distribution: COSTA RICA. Puntarenas. PANAMA. Herrera. Dense, moist or wet forest on the Pacific slope; to 400 m elevation.

Discussion: *C. matae* is close to *C. warscewiczii* but the latter species has double the number of pinnae with unelevated, smooth nerves above and although downward-pointing, curved rather than pendulous pistillate rachillae. *C. matae* is not cultivated except for the few plants in the research collection in Los Angeles.

22. Chamaedorea murriensis *Galeano,* Principes 31: 143. 1987. Type: Colombia, Antioquia, Cieneguetas, *Bernal et al. 705* (holotype COL). Epithet: derived from the location where the species was discovered, the *corregimiento de Murri* in Colombia. **FIGURE 4.**

Chamaedorea macroloba Galeano, Brittonia 38: 60, fig. 1. 1986. Not Burret.

Common Name: nonc.
Vernacular Name: none.

Habit: solitary or cespitose, erect, 2–3.5 m tall.

Stem: 2–3 cm diam., green.

Leaves: 7–10, pinnate; **sheath** 16 cm long, tubular, green, striated; **petiole** 5–6 cm long; **rachis** 42–50 cm long; **blade** obovate; **pinnae** 2–4 on each side of rachis, apical ones very large, 50 × 18–20 cm, occupying a little more than ½ rachis, elliptic, slightly toothed toward apex, 10–11-nerved, other pinnae 25–32 × 5–7 cm, lanceolate, sigmoid, acuminate, 4–7 cm apart, 1 prominent primary nerve per pinna, this yellowish, several secondaries and tertiaries, primaries total 13 on each side of rachis.

Inflorescenses: simply branched. **Staminate** with **peduncle** 72 cm long, 3 mm wide; **bracts** 6, upper one 20 × 1 cm, tubular, membranous to papery, slightly striated; **rachis** 7–8 cm long; **rachillae** 14, these to 30 cm long, 1 mm wide, pendulous, rigid.

Pistillate green in flower, red with mature fruits; **peduncle** 60–

→

Plate 31. Chamaedorea palmeriana. A. Fruiting plant, *Hodel et al. 972,* Braulio Carrillo National Park, Costa Rica; **B.** Staminate inflorescence with erect rachillae, *Hodel 726A bis,* cultivated, Los Angeles, *ex* Chiriqui, Panama; **C, D** Chiriqui, Panama; **C.** Pistillate plant, *D. & M. Hodel 726B*; **D.** Group of small plants, some just beginning to fruit (note machete, upper left).

C
D

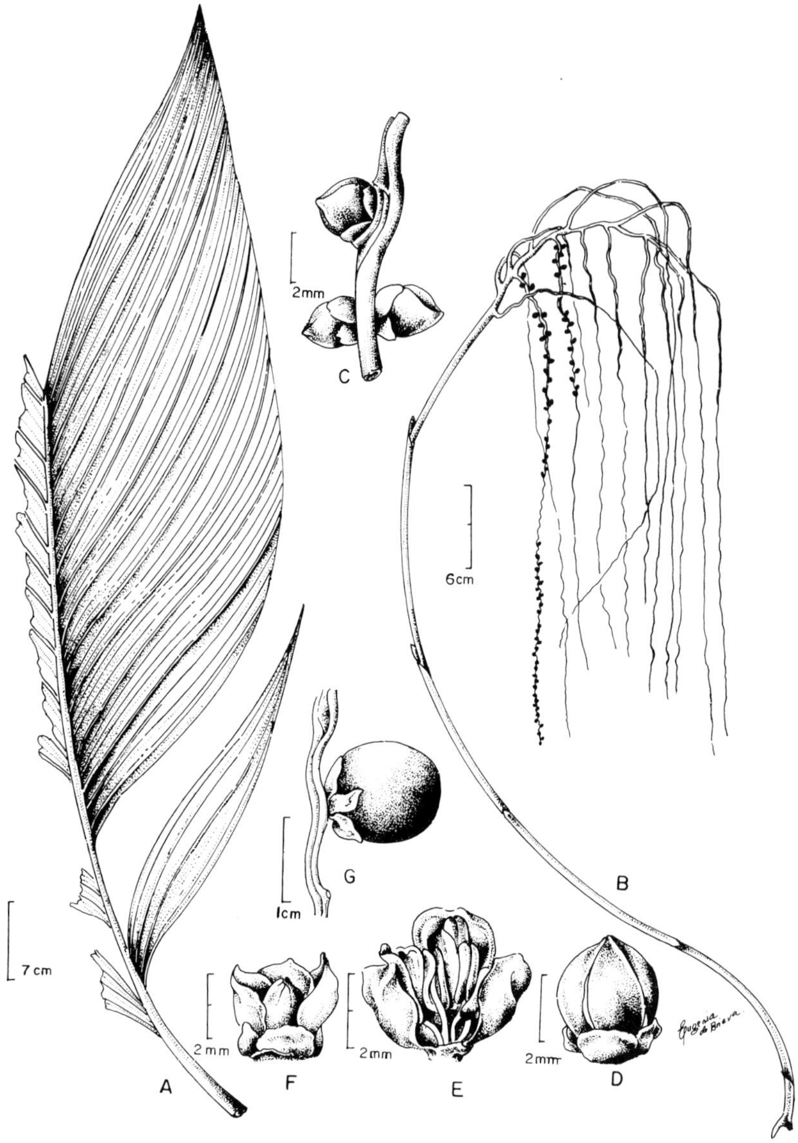

Figure 4. **Chamaedorea murriensis.** (Illustration by Eugenia de Brieva; courtesy of Gloria Galeano. This illustration originally appeared in *Brittonia* 38: 61. 1986.)

70 cm long, 4 mm wide; **bracts** 6, similar in shape and size to those of staminate; **rachis** 6–13 cm long; **rachillae** 11–14, these 16–20 cm long, rigid, 1.5 mm diam.

Flowers: Staminate in rather lax spirals, 3–3.5 × 2.5–3 mm, ± globose, dark when dried; **calyx** 1–1.5 mm × 2.5–3 mm, deeply lobed, **sepals** connate in basal ¼, broadly rounded apically; **petals** 3 × 1.5–2 mm, valvate, connate apically and basally and adnate apically to pistillode and corolla opening by lateral slits, lanceolate, thick; **stamens** 2.5 mm high, **filaments** 1.75–2 mm long, **anthers** 2 mm long, dorsifixed, straight, oblong, white when fresh; **pistillode** 2.5–3 mm high, columnar, truncate.

Pistillate in lax spirals, 2.5 × 3–3.5 mm, depressed-globose, dark when dried, superficial; **calyx** 1.5 × 2.5 mm, lobed, **sepals** imbricate in basal ½, broadly rounded apically; **petals** 2.5 × 2–2.5 mm, broadly ovate, imbricate, thick, acute and erect apically; **pistil** 2.2 mm high, ellipsoid, briefly 3-lobed apically.

Fruits: 12 × 9–10 mm, subglobose, black, stigmatic residue basal; seeds 10 × 8 mm, subglobose, brown.

Plate 32. A, B, Chamaedorea palmeriana. C, Chamaedorea parvifolia. A. *Hodel et al. 1119,* Chiriqui, Panama (pinnate leaf near hand is of *C. scheryi*); **B.** *Hodel et al. 972,* Braulio Carrillo National Park, Costa Rica. **C.** Virtually stemless pistillate plant with leaves, inflorescences arising from leaf litter, *Hodel & Grayum 965B,* Copey, San Jose, Costa Rica.

C

Distribution: COLOMBIA. Antioquia. PANAMA. Darien. Dense, wet forest; 700-1,500 m elevation.

Discussion: The description is from Galeano (1986), who stated that *C. murriensis* is related to *C. warscewiczii* but differs in the leaves obovate in outline instead of ovate and the very large apical pinnae occupying more than half the rachis. Not cultivated, *C. murriensis* is apparently rare in the wild, being known only from the type locality and a few collections in the Darien region of adjacent Panama.

23. Chamaedorea neurochlamys *Burret,* Notizbl. Bot. Gart. Berlin-Dahlem 11: 744. 1933a. Type: Guatemala, Alta Verapaz, *Tuerckheim 4045* [holotype B, destroyed (photo!); isotype US!]. Epithet: from the Greek *chlamys* meaning covering and *neuro* meaning nerved; however, Burret did not specify its application. **PLATES 28(C), 29.**

Docanthe alba O. F. Cook, Nat. Hort. Mag. 22: 95, fig. 9, 1943a. Name of no botanical standing.

Common Name: none.

Vernacular Names: pacaya, pacayo, chilac—Guatemala.

Habit: solitary, slender, erect, 1.5–4.5 m tall.

Stem: 1.5–2.5 cm diam., smooth, green, ringed, internodes 5–15 cm long.

Leaves: 3–5, pinnate, dull dark green above, glossy green below; **sheaths** 18 cm long, tubular, obliquely open apically and there whitish when fresh and green longitudinally striate-nerved; **petiole** 15–29 cm long, lightly grooved and green above, rounded and pale yellow below; **rachis** 45–65 cm long, angled and green above, rounded below with a pale yellow or whitish band extending onto sheath; **pinnae** 6–8 on each side of rachis, lower central ones longest, these 23–33 × 5–6.5 cm, uppermost pinnae slightly broader, 14–16 cm long on upper margin, 5-nerved, lower pinnae smaller, narrowly rhombic-sigmoid, regularly spaced and remote except lower 2–3 ± closer, acuminate, narrowed basally, prominent midrib and submarginal primary nerves shining and slightly keeled toward base above and pale and shining below, 4–5 slightly less prominent secondaries on each side of midrib, tertiaries fine and numerous.

Inflorescences: inter- or infrafoliar, solitary; **peduncles** 40–60 cm long, 1.5 cm wide at base, 6–9 mm wide at apex, erect and greenish yellow or whitish in flower, arching and downward-curving slightly and red-orange in fruit; **bracts** 5–6, tubular, appressed, fibrous, brownish in flower, often fallen away in fruit, longitudinally striate-nerved, acuminate, slender, uppermost exceeding peduncle. **Staminate** with **rachis** 4–7 cm long, green; **rachillae** 15–20 or more, these 15–20 cm long, slender, pendulous, light green.

Pistillate erect in flower, nodding in fruit; **rachis** 6–10 cm long, green in flower, red-orange in fruit; **rachillae** 10–25, these 9–14 cm long, slender and ascending but drooping when heavily laden with fruits, yellow-green in flower, red-orange in fruit.

Flowers: Staminate in rather dense spirals, 2 mm high, yellowish, scarcely sunken; **calyx** 0.5 mm high, shallowly lobed; **petals** valvate, connate apically and basally and apically adnate to pistillode and corolla opening by lateral slits, prominently nerved when dry; **stamens** with short **filaments**, **anthers** not bifid apically; **pistillode** longitudinally 3-angled, truncate apically.

Pistillate in lax spirals, 2–3 × 2.5 mm, globose, greenish yellow, scarcely sunken; **calyx** 0.75–1 × 2–2.5 mm, lobed, yellowish, **sepals** connate in basal ½–⅔, broadly rounded apically; **petals** 2.5 × 2.5 mm, imbricate nearly to apex, acute, strongly nerved when dry; **staminodes** not seen; **pistil** 2–3 × 2 mm, subglobose, yellow, **styles** short or lacking, **stigma** lobes short, recurved, clear-colored.

Fruits: 10 × 5 mm, subreniform or sometimes nearly bilobed or sickle-shaped, yellow to bright orange-red when soft ripe but ageing to dark brown when fully ripe; abortive carpels adhering to fruit at maturity.

Distribution: MEXICO. Campeche. Chiapas. Quintana Roo. GUATEMALA. Alta Verapaz. Huehuetenango. Izabal. Peten. BELIZE. Cayo. Stann Creek. Toledo. HONDURAS. Atlantida. Comayagua. Cortes. La Paz. Ocotepeque. Dense, wet, lowland forest on the Atlantic slope; 0–400 m elevation; occasionally on limestone.

Discussion: Burret (1933a) described and named *C. neurochlamys* from material that Tuerckheim collected near Cubilguitz, Alta Verapaz, Guatemala in 1915. It is fairly common in Belize and known there as "monkey-tail." Because of its red, subreniform fruit it has been confused in cultivation with *C. concolor* (= *C. pinnatifrons*) and the elusive *C. falcifera.* The Mexican forms of *C. pinnatifrons* (referred to as *C. concolor*) have only recently been introduced to cultivation although material of *C. pinnatifrons* from Venezuela has been in gardens and collections since the middle 1970s. As mentioned earlier, *C. falcifera* is known only from the type and a few collections and is not, at least in its typical form, in cultivation. Future work may, in fact, show it to be nothing more than a variant of *C. neurochlamys.* Differences among these three species are summarized below.

C. falcifera	*C. pinnatifrons*	*C. neurochlamys*
3–4 pinnae/side; fruit falciform; leaf sheath pale or only briefly white at apex	1–9 pinnae/side; fruit globose; leaf sheath pale or only briefly white at apex	7–8 pinnae/side; fruit subreniform; leaf sheath distinctly white at apex

Plate 33. Chamaedorea pedunculata. A, B. Staminate plant, inflorescence, *D. & M. Hodel 708A* (holotype), near Jardin Botanico Robert y Catherine Wilson, Costa Rica; **C.** Cultivated, Ho'omuluhia Botanic Garden, Honolulu, Hawaii.

C

Stevenson (1974) discussed *C. concolor* in south Florida but based on his description of the plants as having 5–7 pinnae on each side of the rachis, it seems probable that the species in question is *C. neurochlamys.* Vegetatively, *C. neurochlamys* is somewhat similar to the variable and widely cultivated *C. oblongata.* The green apex of the leaf sheath, free and spreading staminate petals, and black fruits distinguish *C. oblongata.*

Krempin (1990, p. 94) discussed and illustrated *C. neurochlamys* but the description and photograph depict *C. tepejilote.*

Chamaedorea neurochlamys is a handsome and attractive palm but is especially striking when bearing infructescences heavily laden with bright orange or red fruits. It is cultivated in Hawaii, California, Florida, and, perhaps, elsewhere. In 1977, the Seed Bank of the International Palm Society and the Huntington Botanical Gardens in San Marino, California distributed seeds of *C. neurochlamys* erroneously as *C. falcifera* from collections made by the latter institution in Belize.

24. Chamaedorea palmeriana *Hodel & Uhl,* Principes 34: 122. 1990c. Type: Panama, Chiriqui, *D. & M. Hodel 726A* (holotype BH!; isotype PMA!). Epithet: honors Richard W. Palmer who supported and encouraged my work in palms, especially that of *Chamaedorea.* **PLATES 30, 31, 32(A, B).**

Common Name: none.
Vernacular Name: none.

Habit: solitary, slender, erect but often creeping or decumbent, to 1 m high.

Stem: 7–10 mm diam., smooth, green, ringed, internodes 5–10 cm long, often with aerial roots at base and rooting along its length.

Leaves: 7 or sometimes more, spreading, bifid; **sheath** 5–7 cm long, tubular, green, minutely white-spotted, obliquely open apically, faintly longitudinally striate-nerved; **petiole** to 15 cm long, flattened and green above, rounded and green below; **rachis** 8–12 cm long, angled and green above, rounded below with a faint pale yellow or light green band extending onto sheath; **blade** to 25 × 25 cm, deeply bifid apically to more than ½ its length, ± thin, ± papery, bright emerald green, lobes to 15 × 9 cm, corrugated with 12–14 prominent nerves at top of each ridge above, exterior margins toothed.

Inflorescences: interfoliar, erect; **peduncles** to 20–30 cm long, slender, 5–7 mm wide at base and there ± flattened, 2–3 mm diam. at apex and there rounded, greenish in flower, orange in fruit; **bracts** 4–5, prophyll 5 cm long, 2nd–4th bracts 15 cm long, uppermost not exceeding peduncle, tightly sheathing, longitudinally striate-nerved, acuminate, bifid, green in flower. **Staminate** with rachis 1–5 cm long, flexuous, greenish in flower; **rachillae** 2–7, these to 15 cm long, 1–1.5 mm diam., erect-spreading, longitudinally ridged, green.

Pistillate spicate or rarely with 2–3 rachillae; **rachis** or flower-bearing portion 15–20 cm long, erect, 1.5–2 mm diam. and greenish in flower, thickened to 2.5 mm diam. and orange in fruit.

Flowers: Staminate in moderately dense spirals, 2.5 × 2.5 mm, angular-globose, greenish apically, yellowish basally, slightly sunken; **calyx** 0.5 × 1.75 mm, deeply lobed, brownish, **sepals** connate only briefly basally, broadly rounded apically; **petals** 2.5 × 2–2.5 mm, valvate, connate apically and basally and adnate apically to pistillode and corolla opening by small basal apertures, these elliptic, 0.5 mm long, orange-brown margin around apertures, rounded-triangular, acute; **stamens** included; **pistillode** 2.5 mm high, columnar.

Pistillate in rather dense spirals, 3–4 mm apart, 2 × 2 mm, depressed-globose, pale green, slightly sunken; **calyx** to 0.5 × 2 mm, shallowly lobed, **sepals** very briefly imbricate or connate basally, broadly rounded apically; **petals** 2–2.5 × 3 mm, imbricate basally, spreading apically, broadly rounded, acute, greenish or yellowish, 3 mm long in fruit, petals and sepals very lightly nerved on inside; **pistil** globose-subglobose, green, rounded, acute.

Fruits: when immature 12 × 6 mm, ellipsoid, green, closely spaced on rachillae, 12 × 8–10 mm when mature, ovoid to globose, black.

Distribution: PANAMA. Chiriqui. Veraguas. COSTA RICA. Alajuela. Cartago. Heredia. Limon. San Jose. Dense, wet forest on the Atlantic slope often up to and just over the Continental Divide; 450–1,800 m elevation.

Discussion: *C. palmeriana* is relatively widespread, occurring in very wet forest from Veraguas Province in Panama westward to at least the upper reaches of the Rio Sarapiqui in Costa Rica. It occurs mainly on the Atlantic slope in Costa Rica and Panama. However, in Panama where the Continental Divide is at a lower elevation, *C. palmeriana* is sometimes found just over it on the Pacific slope.

Chamaedorea palmeriana is closest to *C. amabilis* but the blades bifid apically to no more than ¼ their length, double the number of nerves on each side of the rachis (20–25 rather than 12–14), and the globose staminate flowers distinguish the latter species.

An attractive species often flowering when no more than 30 cm tall, *C. palmeriana* is noted for its deeply bifid, corrugated, emerald green leaves. In the wild or when well grown, it is a very leafy plant, often

---

Plate 34. Chamaedorea pinnatifrons. A. Cultivated, Fairchild Tropical Garden, Miami, Florida; **B.** Staminate plant, cultivated, Waimea Falls Park & Arboretum (*75s2237*), Honolulu, Hawaii, *ex* Venezuela; **C.** Staminate plant, *D. & M. Hodel 694A,* Monteverde, Costa Rica; **D.** Staminate inflorescence, *Hodel 815,* cultivated, garden of L. Rossten, Huntington Beach, California.

B
D

bearing 7–10 leaves in a handsome and compact crown. Unfortunately, like other species of the genus from wet, cool, relatively high areas, *C. palmeriana* is somewhat difficult to cultivate and has only recently been introduced to a few gardens in southern California.

25. Chamaedorea parvifolia *Burret,* Notizbl. Bot. Gart. Berlin-Dahlem 11: 746. 1933a. Type: Costa Rica, San Jose, Copey, *Tonduz 11794* (holotype B, destroyed); Neotype (here designated): Costa Rica, San Jose, *Hodel & Grayum 965A* (BH!; isoneotype CR!). Epithet: from the Latin *parvus* meaning little or small and *folium* meaning leaf, in reference to the small leaves. **PLATE 32(C).**

Common Name: none.
Vernacular Name: none.

Habit: solitary, slender, erect, initially stemless but eventually to 2 m tall.

Stem: 1.5 cm diam., wholly or partially submerged in leaf litter, prominently ringed, internodes short, to 3 cm long, often with persistent leaf bases.

Leaves: 5–8, erect-spreading, pinnate; **sheath** to 15 cm long, obliquely open apically, densely and conspicuously longitudinally striated; **petiole** to 15 cm long or more; **rachis** 23 cm long, green above, white-banded below; **blade** 35 cm long, elliptic; **pinnae** 4–5 on each side of rachis, middle and upper ones largest, these to 16 × 3–4 cm, lower ones shorter, 7.5 × 2.5–2.8 cm, lowest even smaller, lanceolate, slightly sigmoid, long-acuminate, gray-green with a bluish cast above, paler below, ± thick, rather scattered along rachis although basal ones are closer, 2–3 prominent nerves per pinna, secondaries and tertiaries not too conspicuous, apical pair wider, to 8 cm wide, 3-nerved.

Inflorescences: Staminate erect, solitary; **peduncle** to at least 35 cm long, moderately slender, 3 mm wide at base and flattened; **bracts** 4, 3–4 mm diam. at base, briefly contracted apically, closely sheathing, uppermost briefly exceeding or equalling peduncle; **rachillae** up to 5, these to 10 cm long, spreading.

Pistillate similar to staminate but with 3 **rachillae**, these 7–8 cm long, moderately slender, nearly upright, longitudinally angled, orange-red in fruit.

Flowers: Staminate in rather dense spirals, 3.5 × 2.5 mm, ovoid, tapered apically and basally, drying black, superficial or only slightly sunken; **calyx** 0.75–1 × 2–2.5 mm, lobed, **sepals** connate in basal ½, broadly rounded apically; **petals** 4 × 2.5–3 mm, valvate, connate apically and basally and apically adnate to pistillode and corolla opening by lateral and elliptic apertures, acute, lightly but conspicuously nerved; **stamens** 2.5 mm high, **filaments** short, **anthers** 2 × 0.75 mm; **pistillode** 3–3.5 mm high, columnar, slender.

Pistillate with **calyx** very lightly nerved in fruit, suborbicular, nearly outspread; **corolla** not much larger than calyx, **petals** imbricate, free, suborbicular, broadly rotund.

Fruits: in loose spirals, superficial on the rachillae, small, nearly globose, longer than broad, black, immature fruit ellipsoid.

Distribution: COSTA RICA. Alajuela. Cartago. Puntarenas. San Jose. Moist forest, usually on the Pacific slope near the Continental Divide; 1,500–2,700 m elevation.

Discussion: The description is from Burret (1933a), who described and named *C. parvifolia* from a pistillate collection only, and is supplemented from our collections near the type locality. Tonduz's type was destroyed and duplicates have not been found. Since the species is distinctive in habit and habitat and many collections of it appear in several herbaria, it seems appropriate to retypify this taxon here.

Standley (1937) reported *C. parvifolia* to be a common palm of the oak (*Quercus*) forests of Canton de Dota in Costa Rica. He also reported it from the region of San Ramon.

Strangely, *C. parvifolia* seems very close in habit to *C. pittieri.* Both are attractive plants and have short, prominently ringed stems with very congested nodes; thick, durable, heavily striated sheaths; small leaves with more or less thick, grayish green pinnae with a bluish cast; and long-pedunculate, few-branched, stiffly erect inflorescences. In fact, when I first saw *C. parvifolia* near Copey, Canton de Dota, I assumed it was *C. pittieri.* Fortunately, I was able to collect good staminate material, and the apically connate petals with the corolla opening by lateral slits confirm its placement in subgenus *Chamaedorea* and set it well apart from *C. pittieri.* Still, the apparent similarities of the two in all aspects except staminate flowers are remarkable. *C. parvifolia* does not appear to be cultivated.

26. Chamaedorea pedunculata *Hodel & Uhl,* Principes 34: 131. 1990c. Type: Costa Rica, Puntarenas, San Vito de Coto Brus, *D. & M. Hodel 708A* (holotype BH!; isotype CR!). Epithet: from the Latin *pedunculatus* meaning provided with a peduncle, in reference to the long, conspicuous peduncle. **PLATE 33.**

Common Name: none.
Vernacular Name: none.

Plate 35. Chamaedorea pinnatifrons. A, B, D-F Costa Rica; **C** Panama. **A.** Fruiting plant, *Hodel et al. 967,* Braulio Carrillo National Park; **B.** *Hodel & Grayum 966,* Copey, San Jose; **C.** *D. & M. Hodel 728,* Chiriqui; **D.** Infructescence with mature fruits, *D. & M. Hodel 694B,* Monteverde; **E.** Infructescence with nearly mature fruits, *Hodel et al. 967,* Braulio Carrillo National Park; **F.** *D. & R. Hodel 723,* near La Selva.

→

B
D
F

Habit: solitary, erect or decumbent, to 2–3 m tall.

Stem: 2.5 cm diam., green, ringed, internodes 7.5 cm long.

Leaves: 4–6, erect-spreading, pinnate; **sheath** 25 cm long, splitting opposite petiole and obliquely long-open, clasping tightly only in basal ½; **petiole** 30–35 cm long, slightly grooved and green above, rounded and pale-banded below; **rachis** 1 m long, slender and attenuate apically, sharply angled and green above, rounded and green below; **pinnae** 7–8 on each side of rachis, lower middle the largest, these to 35–37 × 12–13 cm, becoming progressively smaller toward apex of rachis, broadly rhombic-lanceolate, regularly arranged, opposite or subopposite, slightly sigmoid, ± downward-cupped, long-acuminate with drooping tips, narrowly contracted basally with 8–10 nerves, these not very prominent above, more prominent with an equal number of secondary nerves below.

Inflorescences: interfoliar, erect, becoming spreading in flower or fruit, long-pedunculate; **peduncle** to 1 m long or slightly more, 2–2.5 cm wide at base, 1 cm wide at apex, green or pale in flower, orange in fruit; **bracts** 7–8, tightly sheathing, obliquely open apically, drying brown by anthesis; **rachises** 6–10 cm long, green in flower, red-orange in fruit. **Staminate** with 20 **rachillae**, these to 30 cm long, slender, pendulous, green.

Pistillate with 12–15 **rachillae**, these to 20–25 cm long, spreading, greenish yellow in flower, red-orange in fruit.

Flowers: Staminate in fairly dense spirals but not contiguous in bud, 3 × 2.5 mm, depressed-diamond-shaped, greenish yellow; **calyx** low, green; **petals** valvate, connate apically and basally and apically adnate to pistillode and corolla opening by lateral slits, longitudinally striate-nerved.

Pistillate not seen.

Fruits: 7–8 mm diam., obovoid-globose, black.

Distribution: COSTA RICA. Puntarenas. Dense, wet forest on the Pacific slope; 1,000 m elevation.

Discussion. *C. pedunculata* is similar to and perhaps not distinct from *C. macrospadix* but the long-lanceolate pinnae and ovoid staminate flowers distinguish the latter species. Not cultivated and collected only at the type locality, *C. pedunculata* is uncommon in forest remnants near Jardin Botanico Robert y Catherine Wilson where it grows with *C. pinnatifrons, C. crucensis,* and *C. brachyclada.*

27. Chamaedorea pinnatifrons (*Jacq.*) *Oerst.,* Vidensk. Meddel. Dansk. Naturhist. Foren. Kjoebenhavn 1858: 14. 1859. Epithet: from the Latin *pinnatus* meaning bearing pinnae or leaflets and *frons* meaning leaf. **PLATES 4(B), 8(C), 34-36.**

Borassus pinnatifrons Jacq., Plantarum Rariorum Horti Caesarei Schoenbrunnensis 2: 65, t. 247–248. 1797. Type: Venezuela, Caracas, *Bredemeyer s.n.* (*297*) [holotype W, destroyed? (photo!); isotypes BR (photo!), F!].

Chamaedorea gracilis Willd., Mem. Acad. Roy. Sci. Hist. (Bruxelles) 1804: 41. 1807. Type: Venezuela, Caracas, *Bredemeyer s.n.* [holotype W, destroyed? (photo!)]. Epithet: is a Latin word meaning thin or slender, in reference to the habit.

Martinezia lanceolata Ruiz & Pav., Systema Vegetabilium Florae Peruvianae et Chilensis 1: 297. 1798. Type: Peru, Chinchao, *Ruiz & Pavon s.n.* (*B20*) [holotype MA (photo!); isotype F!, FI (photo!), G (photo!), M (photo!)]. Epithet: from the Latin *lanceolatus,* apparently in reference to the lanceolate shape of the pinnae.

Chamaedorea concolor Mart., Historia Naturalis Palmarum 3: 160. 1837. Type: Mexico, Veracruz, Misantla, *Schiede 1018* [holotype M!; isotypes BH (fragments!), BR (photo!), GOET!]. Epithet: the Latin *concolor* means uniform in color, perhaps in reference to both surfaces of the leaf.

Chamaedorea lanceolata (Ruiz & Pav.) Kunth, Enumeratio Plantarum 3: 172. 1841.

Chamaedorea conocarpa Mart., Voyage dans l'Amerique Meridionale 6, t. 6, fig. 1, t. 16B. 1844. Type: Bolivia, Cumbrecilla, *d'Orbigny 42* (holotype P). Epithet: from the Latin *conus* meaning cone and *carpus* meaning fruit.

Hyospathe montana Mart., Voyage dans l'Amerique Meridionale 6, t. 6, fig. 1. 1844. Type: Bolivia, Cochabamba, *d'Orbigny 40* (holotype P). Epithet: from the Latin *montanus* meaning growing in the mountains.

Chamaedorea lindeniana H. A. Wendl., Allg. Gartenzeitung 21: 139. 1853b. Type: Cult., *Wendland s.n.* [holotype GOET!; isotypes C!, G (photo!), M (photo!)]. Epithet: honors 19th-century Belgian plant collector, J. Linden.

Chamaedorea bartlingiana H. A. Wendl., Index Palmarum 60. 1854. Type: Cult., *Wendland s.n.* [holotype GOET!; isotypes C!, G (photo!), L!]. Epithet: honors F. Bartling, botanist at Goettingen University.

Chamaedorea brevifrons H. A. Wendl., Index Palmarum 61. 1854. Type: Cult., *Wendland s.n.* [holotype GOET!; isotypes C!, L!, W (photo!)]. Epithet: from the Latin *brevis* meaning short and *frons* meaning leaf.

Plate 36. Chamaedorea pinnatifrons. A, B. Fruiting plant and infructescence, *D. & R. Hodel 917,* Catemaco, Veracruz, Mexico; **C.** Fruiting plant, *Hodel & Castillo 990,* Quetzaltenango, Guatemala; **D, E.** Infructescence and fruiting plant, *Hodel & Castillo 889,* Alta Verapaz, Guatemala.

→

B
E

Chamaedorea flavovirens H. A. Wendl., Index Palmarum 60. 1854. Type: Cult., *Wendland s.n.* [holotype GOET!; isotypes C!, G (photo!)]. Epithet: from the Latin *flavus* meaning yellow and *virens* meaning green, apparently in reference to the yellow-green leaves.

Chamaedorea bracteata H. A. Wendl., Bot. Zeitung Berlin 17: 29. 1859. Type: Cult., *Wendland s.n.* (holotype GOET!; isotype L!). Epithet: from the Latin *bracteatus* meaning having bracts, probably in reference to the large, loose, conspicuous peduncular bracts.

Chamaedorea bifurcata Oerst., Vidensk. Meddel. Dansk. Naturhist. Foren. Kjoebenhavn 1858: 13. 1859. Type: Costa Rica, Aguacate, *Oersted s.n.* (later numbered *6547*) (holotype C!). Epithet: from the Latin *bi* meaning two and *furcatus* meaning forked, in reference to the inflorescence with two rachillae.

Chamaedorea membranacea Oerst., Vidensk. Meddel. Dansk. Naturhist. Foren. Kjoebenhavn 1858: 22. 1859. Type: Nicaragua, Segovia, *Oersted s.n.* (holotype C!). Epithet: from the Latin *membraneceus* meaning membranous, apparently in reference to the leaves.

Chamaedorea pacaya Oerst., Vidensk. Meddel. Dansk. Naturhist. Foren. Kjoebenhavn 1858: 12. 1859. Type: Costa Rica, Jaris, *Oersted s.n.* (later numbered *6555*) (holotype, C!). Epithet: the vernacular name for several species of *Chamaedorea.*

Nunnezharia geonomoides Spruce, J. Linn. Soc. Bot. 11: 122. 1869. Type: Peru, Campana, *Spruce 67* (holotype K!). Epithet: meaning bearing a resemblance to palms of the genus *Geonoma.*

Chamaedorea geonomoides (Spruce) Drude in Mart., Flora Brasiliensis 3: 531. 1882a.

Chamaedorea lanceolata var. *littoralis* Drude in Mart., Flora Brasiliensis 3: 530. 1882a. Type: Brazil, near Bahia, *Blanchet 2465* [holotype?; isotype BR (photo!)]. Epithet: is a Latin word meaning pertaining to the sea-shore, probably in reference to it growing along inland waterways (rivers).

Nunnezharia bartlingiana (H. A. Wendl.) O. Kuntze, Revisio Generum Plantarum 2: 730. 1891.

Nunnezharia bifurcata (Oerst.) O. Kuntze, Revisio Generum Plantarum 2: 731. 1891.

Nunnezharia bracteata (H. A. Wendl.) O. Kuntze, Revisio Generum Plantarum 2: 730. 1891.

Nunnezharia brevifrons (H. A. Wendl.) O. Kuntze, Revisio Generum Plantarum 2: 730. 1891.

Nunnezharia concolor (Mart.) O. Kuntze, Revisio Generum Plantarum 2: 730. 1891.

Nunnezharia conocarpa (Mart.) O. Kuntze, Revisio Generum Plantarum 2: 730. 1891.

Nunnezharia flavovirens (H. A. Wendl.) O. Kuntze, Revisio Generum Plantarum 2: 730. 1891.

Nunnezharia lanceolata (Ruiz & Pav.) O. Kuntze, Revisio Generum Plantarum 2: 730. 1891.

Nunnezharia membranacea (Oerst.) O. Kuntze, Revisio Generum Plantarum 2: 730. 1891.

Nunnezharia pacaya (Oerst.) O. Kuntze, Revisio Generum Plantarum 2: 730. 1891.

Nunnezharia pinnatifrons (Jacq.) O. Kuntze, Revisio Generum Plantarum 2: 730. 1891.

Chamaedorea boliviensis Dammer, Notizbl. Konigl. Bot. Gart. Berlin 6: 262. 1915. Type: Bolivia, Pando, Alto Rio Acre, *Ule 115* (holotype B, destroyed). Epithet: derived from the name of the country where the type was collected.

Chamaedorea depauperata Dammer, Notizbl. Konigl. Bot. Gart. Berlin 6: 263. 1915. Type: Peru or Brazil, Alto Rio Acre, Sao Francisco, *Ule 9155b* [holotype B, destroyed; isotype F (fragments!), MG]. Epithet: from the Latin *depauperatus* meaning undeveloped or reduced, apparently referring to the habit.

Chamaedorea minor Burret, Notizbl. Bot. Gart. Berlin-Dahlem 11: 2. 1930. Type: Venezuela, Lara, *Saer 427* (holotype B, destroyed). Epithet: the Latin *minor* means small, apparently in reference to the habit.

Chamaedorea concinna Burret, Notizbl. Bot. Gart. Berlin-Dahlem 11: 747. 1933a. Type: Colombia, Santander, *Kalbreyer 418* (holotype B, destroyed); Neotype: Colombia, Santander, Toledo, *Bernal & Galeano 1354* (COL; isoneotypes HUA, NY). Epithet: from the Latin *concinnus* meaning neat, pretty, or elegant.

Chamaedorea dryanderae Burret, Notizbl. Bot. Gart. Berlin-Dahlem 11: 754. 1933a. Type: Colombia, Valle, *Dryander s.n.* (holotype B, de-

Plate 37. Chamaedorea rojasiana. Quetzaltenango, Guatemala. **A.** Bifid leaves, *Hodel & Castillo 991*; **B.** Pinnate leaves, staminate plant, *Hodel & Castillo 984A*.

→

stroyed). Epithet: honors E. Dryander, collector of the type.

Chamaedorea heilbornii Burret, Notizbl. Bot. Gart. Berlin-Dahlem 11: 750. 1933a. Type: Ecuador, Tandapi, *Heilborn 75A* (holotype B, destroyed; isotype S!). Epithet: honors O. Heilborn, collector of the type.

Chamaedorea herrerae Burret, Notizbl. Bot. Gart. Berlin-Dahlem 11: 748. 1933a. Type: Peru, Valle de Santa Ana, *Herrera 3641* (holotype B, destroyed). Epithet: honors F. Herrera, collector of the type.

Chamaedorea holmgrenii Burret, Notizbl. Bot. Gart. Berlin-Dahlem 11: 749. 1933a. Type: Ecuador, Rio Tenguel, *Holmgren 75* (holotype B, destroyed; isotype S!). Epithet: honors J. Holmgren, collector of the type.

Chamaedorea hoppii Burret, Notizbl. Bot. Gart. Berlin-Dahlem 11: 755. 1933a. Type: Ecuador, Archidona, *Hopp 1042* (holotype B, destroyed). Epithet: honors W. Hopp, collector of the type.

Chamaedorea kalbreyeriana H. A. Wendl. ex Burret, Notizbl. Bot. Gart. Berlin-Dahlem 11: 753. 1933a. Type: Colombia, Santander, *Kalbreyer 1969* (holotype B, destroyed). Epithet: honors W. Kalbreyer, collector of the type.

Chamaedorea macroloba Burret, Notizbl. Bot. Gart. Berlin-Dahlem 11: 757. 1933a. Type: Bolivia, Cochabamba, *Steinbach s.n.* (holotype B, destroyed).

Chamaedorea micrantha Burret, Notizbl. Bot. Gart. Berlin-Dahlem 11: 749. 1933a. Type: Guatemala, Alta Verapaz, *Tuerckheim A8* (holotype B, destroyed). Epithet: from the Greek *micranthus* meaning small-flowered.

Chamaedorea rhombea Burret, Notizbl. Bot. Gart. Berlin-Dahlem 11: 753. 1933a. Type: Costa Rica, Rancho de Anchiote, *Pittier 374* (holotype B, destroyed; isotype BR!). Epithet: from the Latin *rhombeus* meaning rhombic, in reference to the shape of the pinnae.

Chamaedorea oerstedii O. F. Cook & C. B. Doyle, Nat. Hort. Mag. 18: 168. 1939. ('*oerstedi*'). Name only, based on *C. pacaya* Oerst. Epithet: honors Danish botanist A. Oersted.

Chamaedorea aguilariana Standl. & Steyerm., Publ. Field Mus. Nat. Hist., Bot. 23: 196. 1947. Type: Guatemala, Quetzaltenango, *Standley 86890* (holotype F!). Epithet: honors D. J. Aguilar who assisted the authors.

Chamaedorea concolor Hort. in H. A. Wendl., Index Palmarum 11. 1854. Not Mart.

Chamaedorea lindeniana Hort. in H. A. Wendl., Index Palmarum 13. 1854. Not H. A. Wendl.

Chamaedorea scandens Hort. in H. A. Wendl., Index Palmarum 15. 1854. Not Liebm.

Chamaedorea weberbaueri Dammer ex Burret, Notizbl. Bot. Gart. Berlin-Dahlem 11: 757. 1933a. Name only and of no botanical standing.

Common Name: none.

Vernacular Names: molinillo—Venezuela; nucua-ene, cana brava, chontilla, chontilla blanca—Ecuador; pacaya, chilak, molinillo—Guatemala; palma rabo de bobo, cola de pescado, tepejilotillo—Mexico.

Habit: solitary, erect or occasionally decumbent, 0.5–4 m tall, sometimes flowering when stemless.

Stem: 0.5–3 cm diam., green, smoothed, prominently ringed, internodes 1–18 cm long, often with orange-brown aerial roots at base.

Leaves: 4–10, erect-spreading, pinnate, rarely bifid, 0.8–1.5 m long; **sheath** 6–40 cm long, tubular, obliquely open apically and there brown-margined and briefly auriculate, whitish below this, longitudinally striate-nerved; **petiole** to 5–55 cm long, grooved and green above, rounded and pale below; **rachis** to 70 cm long, angled and green above, rounded below with a pale band extending onto sheath; **pinnae** (2–)4–8 on each side of rachis, middle pinnae largest, these 15–40 × 2.5–12 cm, basal ones 8–37 × 1–7 cm, apical ones 15–35 × 4–15 cm, elliptic to lanceolate or rhombic-lanceolate, sigmoid, falcate, acuminate, a dominant midrib and 2–3 primary nerves on each side of this or 3–7 prominent primary nerves above, angled or keeled above, light green to yellow below, numerous secondaries and tertiaries, end pinnae wider with up to 8 primary nerves, or blade with pinnae mostly confluent with a very dominant end pair and only 1–2 small, basal pinnae; if **blade** bifid then 25–75 × 15–30 cm, incised apically for ⅓–⅔ its length, lobes 5–15 cm wide, pinnately ribbed with 8–14 primary nerves on each side of rachis, acute to acuminate, cuneate basally.

Inflorescences: inter- or infrafoliar, solitary, erect but nodding when heavily laden with fruits; **peduncles** 10–70 cm long, 1–1.5 cm wide at base and there flattened, 0.5–1 cm wide at apex and there rounded, erect but horizontal or downward-curving when laden with fruits, green or pale in flower where exposed, red-orange in fruit; **bracts** 3–7, prophyll 2–4 cm long, longest peduncular bract 15–50 cm long, tubular, tightly appressed or ± loose, long-lanceolate, acuminate, fibrous, green or brownish in flower, brownish in fruit, obliquely open apically, uppermost exceeding peduncle and extending onto rachis and/or often concealing a rudimentary one; **rachises** to 15 cm long, straight or flexuous, greenish yellow in flower, red-orange in fruit. **Staminate** with

→

Plate 38. Chamaedorea rojasiana. A. Staminate plant, *Hodel & Castillo 906A,* Quetzaltenango, Guatemala; **B.** Fruiting plant (immature fruits) in fog and mist, *D. & R. Hodel 925,* Chiapas, Mexico.

(3–)5–21 **rachillae**, these 5–28 cm long, 2 mm diam., pendulous, simple or rarely lower ones forked, green in bud, green to yellow-orange in flower.

Pistillate with (2–)5–15 **rachillae**, these 4–24 cm long, erect-spreading, ± stiff, simple or rarely lower ones forked, green to yellow in flower, becoming somewhat pendulous and red-orange in mature fruit.

Flowers: Staminate in moderately dense spirals, depressed-subglobose to ovoid, 3.5 × 3.5 mm, apically flattened, greenish to yellow, aromatic, superficial or slightly sunken in elliptic to rounded depressions to 1.5 mm long; **calyx** 0.5–1 × 2.5–3.5 mm, shallowly lobed, green to yellow-orange, **sepals** imbricate and/or connate in basal ⅔, rounded apically, nerved; **petals** 2–3 × 2–3 mm, valvate, petals connate apically and basally and apically adnate to pistillode and corolla opening by lateral and basal apertures, densely and strongly nerved; **stamens** to 2 mm high, **filaments** to 1.5 mm long, **anthers** to 1.25 mm long, oblong, dorsifixed; **staminodes** 3, 6 or lacking, small, triangular; **pistillode** to 2.5 mm high, columnar, longitudinally angled.

Pistillate in moderate to lax spirals, to 2 × 3.5–4 mm, ovoid to subconical or even shieldlike, flattened apically, greenish to yellowish, sunken in rounded or elliptic depressions to 3 mm long; **calyx** 1 × 4 mm, lobed, greenish or yellow, **sepals** connate and/or imbricate in basal ½, broadly obtuse apically, nerved when dry; **petals** 1–1.75 × 1.5–2.5 mm, connate basally, imbricate nearly to apex and there angled or rounded, nerved when dry; **pistil** 1.5–2 × 1.25–1.75 mm, ± globose to ovoid, **stigma** lobes sessile, ± reflexed.

Fruits: to 20 mm long and/or 4 mm wide or diam., ellipsoid to globose, green when immature, maturing and soft-ripening to yellow, orange, or red then ageing to black; fruiting perianth nerved.

Distribution: BOLIVIA. ECUADOR. BRAZIL. PERU. COLOMBIA. VENEZUELA. PANAMA. COSTA RICA. NICARAGUA. HONDURAS. EL SALVADOR. GUATEMALA. MEXICO. Dense, moist or wet forest on both the Atlantic and Pacific slopes; to 2,700 m elevation or perhaps higher; occasionally on limestone.

Discussion: As interpreted here, *C. pinnatifrons* is the most widespread and variable species in the genus and is most appropriately referred to as a species complex. The variability manifests itself in size and shape and number of parts of the various organs, including stems, leaves, pinnae, inflorescences, flowers, and fruits. While one can pick out apparently distinctive segregates from different regions, especially at altitudinal or latitudinal extremes, the variation seems insignificant (or perhaps overwhelming!) when considered over its entire range. The complex is fraught with countless intermediate forms that defy classification into known taxa. Failure to recognize the wide distribution and variability has resulted in the innumerable synonyms listed here.

While several workers have alluded to the possibility that many, if not all, of the taxa in subgenus *Chamaedorea* from South America could be included in *C. pinnatifrons,* no one has proposed the inclusion of such taxa as *C. pacaya* from Costa Rica and Panama and *C. concolor* from Mexico. Although admittedly a somewhat novel approach, I can see no good argument not to include them in the *C. pinnatifrons* complex, and I feel that, considering the tremendous variability, inclusion is their most appropriate disposition.

Perhaps leaves and inflorescences are the most apparent examples of variation in *C. pinnatifrons.* A collection near La Selva, Costa Rica at about 50 m elevation had leaves with only 2–3 pinnae on each side of the rachis with the end pair extremely broad, over twice as wide as the others combined. As one moves higher in elevation above La Selva, plants will have 5–7 pinnae on each side of the rachis with the end pair still larger but not as dramatically larger as the collections with less numerous pinnae. Chazdon (1987) made note of this gradation of leaf forms in *C. pinnatifrons* along a transect from La Selva upward into the Zona Protectora and Braulio Carrillo National Park although she referred to the species in question as *C. warscewiczii.*

In South America, bifid-leaved forms of *C. pinnatifrons* infrequently occur. In fact, a separate species, *C. geonomoides,* was even named from one of these bifid-leaved forms. However, the bifid leaf does not seem to be a good character in the *C. pinnatifrons* complex. Bifid-leaved forms occur near or with forms having nearly the entire blade bifid except for one or two small basal pinnae. Still other forms exist where the apical half of the blade is bifid but several or more pinnae compose the basal half. Although the completely pinnate leaf is most common, every conceivable combination, from bifid to eight or nine pinnae on each side of the rachis and combinations thereof, can be found in *C. pinnatifrons.*

Similarly, the number of rachillae varies greatly and does not appear to be a reliable character. Oersted named *C. bifurcata* from Copey in Costa Rica,

---

Plate 39. Chamaedorea schiedeana. A. Flowering plants, cultivated, Huntington Botanical Gardens, San Marino, California; **B, C** Near Huatusco, Veracruz, Mexico; **B.** Staminate plant, *D. & R. Hodel 952A*; **C.** Infructescences, *D. & R. Hodel 952B*; **D.** Leaf, staminate inflorescence (scale = 15 cm), *Hodel 758,* cultivated, garden of R. Palmer, Whittier, California.

→

C
D

basing it on the forked inflorescence. However, additional collections from the area show inflorescences with three, four, five, or even more rachillae, sometimes even on the same plant. Collections from South America exhibit a similar phenomenon. It seems that larger, robust, more vigorous forms have more and larger rachillae than smaller ones.

Further study in the *C. pinnatifrons* complex, especially in the area of correlating size and number of parts with environmental parameters, like altitude, precipitation, temperature, and soil and forest type, is certainly needed and may prove to be most rewarding. Perhaps it will cast some light on the taxonomic boundaries within the *C. pinnatifrons* complex.

Bredemeyer discovered *C. pinnatifrons* nearly 200 years ago in mountain forest near Caracas, Venezuela where it is still fairly common today. It was cultivated as early as 1789 as *C. gracilis* at the Imperial Gardens of Schoenbrunn at Vienna from seeds that Bredemeyer had sent from Venezuela. Wendland grew it at Herrenhausen by 1850 also as *C. gracilis* (Guillaumin 1923b).

Various collectors have introduced *C. pinnatifrons* to cultivation over the years but usually under names that are placed in synonymy here. Wendland described and named several species from cultivated material that others, especially Linden, had introduced to European gardens. Among them are *C. bartlingiana, C. bracteata, C. brevifrons, C. flavovirens,* and *C. lindeniana.*

The names of two other synonyms of *C. pinnatifrons* also figure prominently in horticulture. One, *C. concolor,* originally named from Mexico, is of common but erroneous usage in California and Florida palm gardens and collections, and probably elsewhere. Material so cultivated is *C. oblongata* or less frequently *C. neurochlamys.* Krempin (1990, p. 90) illustrated *C. oblongata* but erroneously captioned the photograph as *C. concolor. C. oblongata* has black fruits, leaf sheaths green throughout, and, being in subgenus *Chamaedoropsis,* has free and spreading staminate petals and nerveless fruiting perianth. Although in the same subgenus as *C. concolor* (= *C. pinnatifrons*), *C. neurochlamys* differs in the markedly white apex of the leaf sheath with green, longitudinal nerves and the subreniform, bilobed, or even sickled-shaped fruits. Stevenson (1974) in Florida and Jones (1984) in Australia list *C. concolor* as being cultivated in their areas yet they are probably actually referring to *C. neurochlamys.*

The other name, *C. pacaya,* is often but erroneously applied to *C. tepejilote. Pacaya* is a vernacular term for several species of *Chamaedorea* and also refers to the unopened staminate inflorescences of *C. tepejilote* that are used as food, especially in Mexico, Guatemala, Honduras, and El Salvador.

Still the name of another synonym, *C. bartlingiana,* is occasionally misapplied to material of a multi-stemmed species with long-pinnate leaves in gardens and collections in southern California. However, the misnamed plant has staminate flowers with apically free petals and is best referred to *C. pochutlensis.*

Chamaedorea pinnatifrons is presently cultivated in Hawaii, California, Florida, Australia, and perhaps elsewhere. Most of these cultivated plants originated through the efforts of August Braun in Caracas, Venezuela. In the middle 1970s, Braun distributed seeds that he had collected in the mountains behind Caracas to various collectors, botanic gardens, and the Seed Bank of the International Palm Society. In California and Hawaii, plants grown from seeds that Braun sent in 1975 have matured and produce fruits on a regular basis.

More recently, the Seed Bank distributed seeds of *C. pinnatifrons* in early 1988 of our collection from near Monteverde, Costa Rica (*Hodel 694* and *641* as *C. pacaya*) and in early 1990 from Veracruz, Mexico (*Hodel 917* as *C. concolor*).

Fruits of the *C. pinnatifrons* complex exhibit an interesting and characteristic color scheme during the maturation process. Fruits are green when immature, yellow or orange or sometimes even red when mature and soft ripe, and then often age to brown or black. They may be mature and ripe at least as early as the yellow stage, though, since fruits collected in that condition from near San Vito and La Selva in Costa Rica germinated in the greenhouse in Los Angeles. Fruits that were black from near Monteverde and San Vito in Costa Rica and Cerro Horqueta and Boquete in Panama also germinated. In fact, one of the characters that sets *C. pinnatifrons* off from other similar taxa in subgenus *Chamaedorea* with pinnate leaves, like *C. macrospadix, C.*

→

Plate 40. Chamaedorea serpens. A. Large clump on steep side of ravine, *D. & M. Hodel 745* (holotype), El Valle, Panama; **B.** Young plant, *Hodel et al. 1111,* El Valle, Panama; **C.** *Hodel 745 bis,* cultivated, research collection, Los Angeles, *ex* El Valle, Panama.

C

verapazensis, and *C. warscewiczii,* is the yellow-orange stage of the fruit-maturation process. Fruits of the other species mature from green to black without passing through a yellow-orange stage. Other distinguishing characters of the *C. pinnatifrons* complex include the inflorescences not exceeding the leaves, the sigmoid pinnae with elevated or keeled nerves above, and the globose to ovoid fruits.

Chamaedorea pinnatifrons grows in primary or secondary forest and is often quite locally abundant. Balslev and Barfod (1986) reported that it is one of the most common understory palms of the eastern lowlands of Ecuador. I have observed this ubiquitous species throughout Central America and Mexico and have found that with a few exceptions, if the forest supports any chamaedoreas at all, one will always seemingly be *C. pinnatifrons.* There are several local names for it. Standley and Steyermark (1958) listed *chilak, pacaya,* and *molinillo,* the latter name relating to the fact that the base of the stem with the cluster of stiff roots was used to stir coffee and other hot liquids.

28. Chamaedorea rojasiana *Standl. & Steyerm.,* Publ. Field Mus. Nat. Hist., Bot. Ser. 23: 205. 1947. Type: Guatemala, Quetzaltenango, *Steyermark 33479* (holotype F!). Epithet: honors Ulises Rojas, formerly director of the Jardin Botanico of Guatemala, on and near whose *finca* many of the known specimens were collected. **PLATES 37, 38.**

Common Name: none.

Vernacular Names: pacaya, molinillo—Guatemala.

Habit: solitary, slender, erect, 1–1.5 m tall though sometimes flowering when stemless.

Stem: 6–8 mm diam., green, smooth, ringed, internodes to 5.5–7 cm long, slender prop roots basally.

Leaves: 3–6, spreading, pinnate or less often bifid; **sheath** 7–12 cm long, 8–14 mm thick, oblique apically, prominently striate-nerved; **petiole** to 2.5–5 cm long or sometimes longer but shorter than sheath; **rachis** 12–18 cm long when pinnate, 7.5–8.5 cm long when bifid; **pinnae** 2–4 on each side of rachis, terminal pair largest, this to 13–17 cm long on upper margin, 4–6 cm along rachis, with 4–6 primary nerves on each side of rachis and 1–3 secondaries between them, basal ones smallest, these to 13 × 2.5 cm, each with a central midrib and 2 pairs of slightly less conspicuous primary nerves on each side, falcate-acuminate, sigmoid, thin, concolorous or slightly paler below, if bifid then blade to 20 × 20 cm, apically incised to more than ½ its length, lobes broadly diverging, acuminate, 7-nerved.

Inflorescences: interfoliar; **peduncles** 11–20 cm long, very slender, scarcely 2 mm thick near apex, erect or spreading; **bracts** 3–4, tubular, appressed, scarcely more than 3 mm broad, striate-nerved. **Staminate** spicate or with 2–3 **rachillae**; **rachis** or flower bearing portion 7–20 cm long, 4 mm thick, densely flowered to base, nodding or pendulous, greenish yellow in flower.

Pistillate spicate or furcate; **rachis** or **rachillae** 7–9 cm long, erect and yellow in flower, spreading and orange in fruit.

Flowers: Staminate in dense spirals, 3–3.5 × 1.75–2 mm, ovoid, yellowish, strongly nerved when dry, deeply sunken; **calyx** 0.5–1 × 2 mm, lobed, connate and/or imbricate in basal ½, **sepals** rounded apically; **petals** 3.5–4 × 1.5–2 mm, valvate, petals connate apically and basally and adnate apically to pistillode and corolla opening by lateral slits, acute; **stamens** shorter than pistillode, **filaments** short, **anthers** entire apically; **pistillode** 3.5 mm high, columnar, truncate.

Pistillate in moderately dense spirals, 1 × 2–3 mm, shieldlike, yellow, deeply sunken in oblong or oval depressions 2.5 mm long; **calyx** 0.25 × 2.5 mm, obscurely lobed, ringlike, membranous, ± transparent, **sepals** imbricate and/or connate nearly to top, straight apically; **petals** imbricate nearly to apex, irregularly rounded, prominently striate-nerved; **pistil** 1 × 2 mm, depressed-globose, yellowish, **styles** short or lacking, **stigma** lobes thick, blunt, separated, erect.

Fruits: drying 7–9 mm long, subglobose or ovoid-ellipsoid, orange at first then maturing purple-black.

Distribution: GUATEMALA. Quetzaltenango. San Marcos. Solola. Suchitepequez. MEXICO. Chiapas. Dense, moist or wet forest on the Pacific slope of Guatemala and Atlantic slope of Mexico; 1,200–2,600 m elevation.

Discussion: The description is from Standley and Steyermark (1947, 1958) and supplemented from our own collections. *C. rojasiana* appears related to *C. geonomiformis* and *C. simplex,* the latter with which bifid-leaved forms have been confused. *C. geonomiformis* differs in the nerveless or only obscurely nerved perianth of both sexes of flowers when dry, the branched inflorescences, and the thicker blade apically incised no more than ½ its length. *C. simplex* differs in the lack of conspicuous secondary nerves between the primaries, the primary nerves keeled above, and the superficial pistillate flowers. *C. rojasiana* is a handsome species that unfortunately is not in cultivation except for a few plants in the research collection in Los Angeles.

Chamaedorea rojasiana has an unusual distribution; it occurs on the Pacific slope of Guatemala and the Atlantic slope of Chiapas, Mexico. It is to be expected on the Pacific slope of Mexico near Guatemala.

29. Chamaedorea schiedeana *Mart.,* Linnaea 5: 204. 1830. Type: Mexico, Veracruz, Xalapa, *Schiede s.n.* [holotype M (lost?); isotypes BM (photo!), BR

Plate 41. Chamaedorea simplex. Staminate plant, *Hodel & Castillo 887,* Alta Verapaz, Guatemala.

(photo!), GOET!]. Epithet: honors German doctor Wilhelm Schiede, collector of the type. **PLATE 39.**

Kunthia xalapensis Otto & A. Dietr. Allg. Gartenzeitung 1: 1. 1833. Type: none designated.
Nunnezharia schiedeana (Mart.) O. Kuntze, Revisio Generum Plantarum 2: 731. 1891.
Chamaedorea gracilis Hort. in H. A. Wendl., Index Palmarum 15. 1854. Not Willd.
Chamaedorea speciosa Hort. in H. A. Wendl., Index Palmarum 15. 1854.

Common Name: none.

Vernacular Names: cuiliote varon, pacaya—Mexico.

Habit: solitary, slender, erect, 3–4 m tall.

Stem: 2.5–3 cm diam., smooth, green, ringed, internodes 5–10 cm long.

Leaves: 3–5, erect-spreading, pinnate; **sheath** to 20 cm long, tubular, obliquely open apically and there with a short ligule on either side, green but often persistent and then becoming brown, longitudinally striate-nerved; **petiole** 10–35 cm long, slightly grooved and green above, rounded and light green below; **rachis** to 60–70 cm long, angled and green above, rounded below with a light green band extending onto sheath; **pinnae** 10–14 on each side of rachis, lower-middle ones largest, these to 30 × 3–4 cm, narrowed basally, basal ones shorter and ± reflexed, upper ones shorter and apical pair sometimes confluent and wider, lanceolate to broadly or oblong-lanceolate, alternate or subopposite, thin-papery, sigmoid, drooping and acuminate apically, a prominent midrib and 2–3 conspicuous primary nerves on each side of this, secondaries inconspicuous.

Inflorescences: infrafoliar, erect or spreading, solitary; **peduncles** 20–60 cm long, slender, 0.5–1 cm wide at base and ± flattened, 5 mm diam. at apex and rounded, greenish in flower, red-orange in fruit; **bracts** 6–7, tubular, acuminate, bifid, brown in flower, fibrous, longitudinally striate-nerved, uppermost largest, this to 23 cm long and exceeding peduncle and extending onto rachis; **rachises** 5–25 cm long, light green in flower, red-orange in fruit. **Staminate** with 20–45 **rachillae**, these 15–20 cm long, 2 mm diam., ± drooping, green in flower.

Pistillate with 15–20 **rachillae**, these 15–25 cm long, 2.5 mm diam., ± stiff, ± parallel, greenish yellow or yellowish in flower, red-orange in fruit.

Flowers: Staminate in remote or ± dense spirals, 3 × 2.5–3 mm, ± globose, ± flattened apically, yellow, aromatic, slightly sunken in superficial and elliptic depressions to 1.5 mm long; **calyx** 1 × 2 mm, shallowly lobed, light green ageing to yellowish, brown-margined, **sepals** connate in basal ⅔, rounded apically, slightly membranous, lightly nerved on inside; **petals** 2–3 × 1.5–2.5 mm, valvate, connate apically and basally and apically adnate to pistillode and corolla opening by lateral slits, obovate, slightly thickened and fleshy, acute, nerved on inside; **stamens** 2–2.5 mm high, **filaments** 1.5–2 mm long, pale or greenish, ± thickened at base, **anthers** 1.5 mm long, linear-oblong, erect, light yellow; **pistillode** 2.5 mm high, columnar, truncate apically, light green.

Pistillate in remote spirals, 3 × 3 mm, globose, yellow, sunken in superficial and elliptic depressions 1.5–2 mm long; **calyx** 1 × 2.5 mm, deeply lobed, greenish yellow ageing to yellow, **sepals** connate in basal ⅓, rounded to acute apically, thin, nerved on inside; **petals** 2–3 × 2.5–3.5 mm, imbricate nearly to apex, widely oblong or orbicular, concave, slightly thickened and fleshy, erect and acute apically; **staminodes** 3–6, small, separated, slightly triangular, white; **pistil** 2–2.5 × 2 mm, depressed-globose, green, **stigma** lobes sessile, small, angled, slightly elevated, sometimes recurved, not exceeding petals.

Fruits: 6–9 mm diam., globose, black, epicarp thin, slightly transparent, mesocarp slightly fleshy, mucilaginous, green, endocarp membranous, nerved; seeds 4.5–6.5 mm diam., globose.

Distribution: MEXICO. Oaxaca. Puebla. Veracruz. Dense, wet forest on the Atlantic slope; 900–1,600 m elevation; often on limestone.

Discussion: Schiede discovered *C. schiedeana* in Xalapa, Veracruz, Mexico in the late 1820s and sent material to Martius (1830) who described and named it. Guillaumin (1923b) reported that it was cultivated in Europe as early as 1833 and was at the Musee de Paris, France in 1837. Today, it is found in collections and gardens in California where it is often referred to erroneously as *C. oblongata* or *C. concolor* (= *C. pinnatifrons*). Staminate flowers with apically free and spreading petals distinguish *C. oblongata* while yellow-orange or red fruits distinguish *pinnatifrons*. Also, *C. oblongata* and *C. pinnatifrons* have fewer pinnae, no more than eight on each side of the rachis, while *C. schiedeana* has 10–13 pinnae per side.

Burret (1933a) expressed doubt about some of the cited accounts and illustrations in the literature that have been referred to *C. schiedeana* including Otto and Dietrich (1833), Link (1834), Zuccarini (1834), and Martius (1839, 1845, 1849).

Chamaedorea schiedeana is similar in habit to and can be utilized in the same manner as *C. oblongata*. It is slightly more rangy with somewhat longer leaves than this latter species, though, and so some discretion is necessary when placing it in the garden or landscape. It is susceptible to infestations of mites but frequent spraying of the undersides of the leaves with a strong stream of water should control these pests.

30. Chamaedorea serpens *Hodel,* Principes 35: 77. 1991a. Type: Panama, Cocle, *D. & M. Hodel 745* (holotype BH!; isotype PMA!). Epithet: is a Latin word meaning creeping and rooting, in reference to the stems. **PLATE 40.**

Plate 42. Chamaedorea simplex. Staminate plant, *Hodel & Castillo 1003A,* Huehuetenango, Guatemala.

→

Common Name: none.
Vernacular Name: none.

Habit: cespitose, often branching up to 1 m up stem, decumbent, to 1.5 m tall, forming clumps 1–2 m across.

Stem: 5–10 mm diam., rooting and sprouting at nodes, sprawling and twisting along ground and through adjacent vegetation, smooth, green, prominently ringed, internodes 5–10 cm long.

Leaves: 4–5, erect-spreading, pinnate or rarely bifid; **sheath** to 20 cm long, tubular, obliquely open apically, green, longitudinally striated; **petiole** to 20 cm long, flat and green above, rounded and pale below; **rachis** to 25 cm long, angled and green above, rounded and with a pale yellow or light green band below extending onto sheath; **pinnae** 2–5 on each side of rachis, all but apical pair 12–18 × 2.5–4 cm, lanceolate, sigmoid, acuminate, narrowed basally, 4–5 prominent primary nerves above, 1 secondary between each pair of primaries, tertiaries numerous, faint, apical pair of pinnae (or if bifid) 27–30 × 8–12 cm, slightly sigmoid, acuminate, 9–12 prominent nerves above, these 1–2 cm apart, all pinnae ± thick, glossy green.

Inflorescences: inter- or infrafoliar, often emerging from persistent sheaths, erect; **peduncles** 15–25 cm long, 5–10 mm wide at base and there flattened, 1.8–3 mm diam. at apex and there ± rounded, pale to green in flower; **bracts** 4–5, prophyll 2 cm long, 2nd bract 8 cm, 3rd 12 cm, 4th 15 cm and exceeding peduncle, 5th 4 cm and sometimes concealed by the larger 4th, long-lanceolate, obliquely open apically, finely longitudinally striate-nerved, papery, greenish to brown in flower, lower ones acute, upper ones acute-acuminate; **rachises** to 3 cm long, greenish in flower. **Staminate** with 6–10 **rachillae**, these to 10 cm long, slightly drooping.

Pistillate with 10 **rachillae**, these to 15 cm long, erect-spreading, greenish in flower.

Flowers: Staminate in ± dense spirals, 1–2 mm apart, 2–2.5 × 1.8–2.5 mm, ± globose, greenish yellow in flower, drying brownish, only slightly sunken in superficial elliptic depressions 3 × 1.5 mm; **calyx** 0.8–1 × 1.8–2 mm, deeply lobed, only lightly nerved, **sepals** connate only briefly at base, rounded to acute apically; **petals** 2.5 × 2 mm, valvate, connate apically and basally and adnate apically to pistillode and corolla opening by basal and lateral apertures, acute, ± thin, lightly longitudinally striate-nerved; **stamens** 1.25–2 mm high, **anthers** 2 mm long, sessile or nearly so, longitudinally bilobed; **pistillode** 2–2.5 mm high, columnar, very slender, slightly flared apically.

Pistillate in dense spirals, 1–1.25 × 2 mm, subconic, nearly shieldlike, yellowish, sunken in elliptic to rounded depressions 2 mm long; **calyx** 0.75–1 × 2 mm, lobed, nerved when dry, **sepals** imbricate in basal ½, truncate to broadly rounded apically; **petals** 1.5 × 1.5 mm, imbricate nearly to apex, straight or acute; **pistil** 1 × 1 mm, depressed-ovoid to globose, **styles** short or lacking, **stigma** lobes recurved, short, not exceeding petals.

Fruits: not known.

Distribution: PANAMA. Cocle. Panama. Dense, wet forest and cloud forest at or near the Continental Divide; 800–1,000 m elevation.

Discussion: *C. serpens* is one of the most unusual members of the genus with its sprawling, decumbent, slender stems rooting and sprouting at the nodes along their length. The stems appear to grow upright until about a meter tall at which point they tend to fall over. In this manner, they form a rather loose colony of tangled stems several meters across and a meter high. The only other member of the genus approaching it in the branching habit of the stems is a form of *C. elatior* from Veracruz and Oaxaca, Mexico, that is easily distinguished by its long, vining, climbing stems and leaves with 10 or more often reflexed pinnae on each side of the rachis.

Florally, *C. serpens* is close to *C. pinnatifrons* and related species, like *C. warscewiczii* and *C. murriensis.* However, *C. serpens* is amply distinct in its creeping stems rooting and sprouting at the nodes and generally much smaller overall habit.

Apparently not cultivated except for a few plants in the research collection in Los Angeles, *C. serpens* occurs in wet forest and cloud forest at about 1,000 meters elevation in western central Panama. It is not a common plant; we found it on the sides of steep ravines near El Valle where it occurs in dense forest often shrouded in clouds.

31. Chamaedorea simplex *Burret,* Notizbl. Bot. Gart. Berlin-Dahlem 11: 758. 1933a. Type: Guatemala, Alta Verapaz, *Tuerckheim 2487* (holotype B, destroyed; isotypes L!, M!, US!). Epithet: the Latin *simplex* means simple, in reference to the bifid leaves. **PLATES 41-43.**

Common Name: none.
Vernacular Name: none.

Habit: solitary, slender, erect to decumbent, to 2.5 m tall.

Stem: 5–12 mm diam., green, shining, ringed, internodes 2.5–7 cm long, aerial roots basally.

Leaves: 6, spreading, bifid; **sheath** 7–9 cm long, tubular, obliquely and briefly open apically for 1.5–2 cm, glabrous, green, conspicuously and densely longitudinally striated; **petiole** 2.5–14 cm long, usually shorter than sheath, slender, green; **rachis** 2–5 cm long; **blade** bifid, incised apically for ¾ its length or more, as much as 20 cm long on upper margin, 5 cm wide, lobes broadly divergent, sigmoid, short-acuminate, curved along outer margin, ± thick, dark green above, paler below, 6–7 primary nerves on each side of rachis, these prominently keeled above, 2 thinner much less conspicuous secondaries between each pair of primaries, these visible above and below.

Inflorescences: interfoliar, solitary. **Staminate** spicate, to 50–60 cm long; **peduncle** to 15–25 cm long, less than 1 mm wide

→

Plate 43. Chamaedorea simplex. Fruiting plants, Guatemala. **A.** *Hodel & Castillo 1003B,* Huehuetenango; **B.** With immature fruits, Huehuetenango; **C.** *Hodel & Castillo 1008,* Alta Verapaz.

B

just below flowering portion, erect; **bracts** at least 2–3 and probably several more, closely sheathing, 2 mm diam., glabrous, longitudinally striated, obliquely open apically, lanceolate, acute, upper bract exceeding peduncle and extending onto flower-bearing portion; **rachis** or flower-bearing portion to 25 cm long, pendulous.

Pistillate spicate, 30–50 cm long, erect-spreading in flower, nodding in fruit; **peduncle** 20–35 cm long, erect in flower, spreading in fruit; **sheaths** similar to those of staminate but upper one apparently not exceeding peduncle; **rachis** or flower-bearing portion 8–9 cm long, erect-spreading in flower, becoming nodding and orange in fruit.

Flowers: Staminate very densely arranged but not contiguous in bud, ± sunken in elliptic depressions 3 mm long; **calyx** 1.5 mm high, shallowly lobed, densely nerved when dry, **sepals** connate in basal ⅔, broadly rotund apically; **petals** 2–3 mm long, valvate, connate apically and basally and adnate apically to pistillode and corolla opening by lateral slits, densely nerved when dry; **stamens** with **filaments** 0.75 mm long, **anthers** 1 mm long or a little more, 0.5 mm wide, oblong, dorsifixed; **pistillode** 1.5 mm high, columnar.

Pistillate in lax spirals, 2.5–3 × 3–3.5 mm, ± globose, sunken in elliptic depressions 2.5–3 mm long; **calyx** 1–1.5 × 3 mm, shallowly lobed, densely nerved when dry, **sepals** connate in basal ¾–⅚, broadly rounded apically; **petals** imbricate nearly to apex, free apically, suborbicular; **staminodes** lacking; **pistil** ovoid, **stigma** lobes pointed, slightly curved.

Fruits: 9–12 × 7.5–9 mm, rounded apically, ellipsoid or oblong, black, shining, seeds 5 mm diam., subglobose; abortive carpels adherent to fruit; perianth prominently nerved when dry.

Distribution: GUATEMALA. Alta Verapaz. Baja Verapaz. Huehuetenango. MEXICO. Chiapas. COLOMBIA? Dense, wet forest on the Atlantic slope; 1,300–1,500 m elevation; often on rocky slopes.

Discussion: The description is from Burret (1933a), Standley and Steyermark (1958), and our own collections. Burret stated that *C. simplex* is one of the most distinctive and elegant members of the genus. The leafy crown of thick, bifid, deeply incised and wide-spreading blades and spicate inflorescences in both sexes characterize this attractive plant. According to Standley and Steyermark (1958), *C. simplex* is one of the neatest and handsomest of all the Guatemalan *Chamaedorea,* of very distinctive appearance and easily recognizable even in the sterile state. After seeing it in the wild in Alta Verapaz and Huehuetenango in Guatemala, I agree fully with the above assessments of its handsomeness. Apparently, the only cultivated plants are in the research collection in Los Angeles. A Kalbreyer collection from Antioquia, Colombia is tentatively referred to *C. simplex.*

32. Chamaedorea tenella *H. A. Wendl.,* Gartenflora 29: 102. 1880. Type: Cult., *Wendland s.n.* (holotype GOET?, lost or destroyed); Neotype: Provenance unknown, *Kreuzpointner 1188* (M!). Epithet: from the Latin *tenellus* meaning delicate or slender, in reference to the habit. **PLATES 2(D), 6(A), 44, 45.**

Nunnezharia tenella (H. A. Wendl.) Hook. f., Bot. Mag. 107: t. 6584. 1881.

Common Name: none.
Vernacular Name: none.

Habit: solitary, slender, erect or decumbent, to 2 m tall.

Stem: 4–5 mm diam., green, ringed, internodes 2 cm long.

Leaves: 5–7, spreading, bifid; **sheath** 4 cm long, tubular, obliquely open apically, green, longitudinally striate-nerved; **petiole** 2–8 cm long, green and flattened or faintly grooved above, rounded and pale below; **rachis** 13–15 cm long, slightly angled and pale green above, rounded and yellow below; **blade** to 20 × 10 cm, oblong-oval or elliptic, 6 cm wide at apex and there incised apically ⅕–¼ its length, brilliant green, lustrous, acuminate, nearly entire or toothed along lateral margin especially toward apex, 10–12 obscure nerves above on each side of rachis, these pale yellow and more prominent below, an inconspicuous secondary between each of two primaries, tertiaries numerous but extremely faint.

Inflorescences: interfoliar but often infrafoliar in fruit, spreading but becoming downward-pointing when laden with fruits, spicate; **peduncles** 15–20 cm long, very slender, 1 mm diam., erect, slightly flattened basally, rounded apically, green to yellowish in flower, red-orange in fruit; **bracts** 4, prophyll 1.5 cm long, 2nd bract 2.5–4 cm, 3rd 5–6 cm, 4th 7–8 cm, sometimes a rudimentary 5th one to 1 cm long concealed by larger 4th one, membranous to thin-papery, slightly fibrous, finely longitudinally striate-nerved, acuminate, bifid, uppermost exceeding peduncle but apex soon rotting off so to appear less than peduncle, brownish in flower and becoming shredded or fallen away in fruit. **Staminate** with **rachis** or flower-bearing portion to 15 cm long, 1 mm diam., pendulous, green becoming yellowish at anthesis.

Pistillate with **rachis** or flower-bearing portion to 8 cm long, 2.5 mm diam., erect or spreading and green in flower, red-orange, swollen, fleshy, and downward-pointing in fruit.

Flowers: Staminate in very dense spirals but not contiguous, 4 × 3.5–4 mm, obovoid-globose, flattened apically, bright yellow to golden yellow, slightly sunken in circular depressions 2.5 mm diam.; **calyx** 0.5–0.75 × 2 mm, shallowly lobed, greenish yellow or yellow with brown margins, **sepals** connate in basal ¾, rounded apically; **petals** to 4.5 × 4–4.5 mm, valvate, connate apically and basally and apically adnate to pistillode and corolla opening by basal and lateral slits, obovate, acute, membranous, lightly nerved, more so when dry; **stamens** 1.5–2 mm high, **filaments** 1.5–1.75 mm long, yellowish, **anthers** 1.25–1.5 mm long, oblong, yellow; **pistillode** columnar, 3.5–4 × 0.75 mm, truncate apically, yellow.

Pistillate in ± remote to moderate spirals, 1.5 × 2–2.5 mm,

Plate 44. Chamaedorea tenella. Fruiting plant, 40 cm tall, *D. & R. Hodel 919,* Veracruz, Mexico.

subglobose or dome-shaped, yellow, sunken in elliptic depressions 2–2.5 mm diam.; **calyx** 0.75 × 2–2.5 mm, lobed, faintly nerved, greenish yellow or yellow, **sepals** connate briefly but mostly imbricate in basal ½, broadly rounded apically, slightly fleshy; **petals** 1.5–2 × 4–4.5 mm, imbricate nearly to apex and there truncate, fleshy, nerved; **staminodes** 6, small, separated; **pistil** 1.5 mm high, subglobose, greenish yellow, **styles** very short or lacking, **stigma** lobes separated, recurved, pale.

Fruits: 6–12 mm diam., ± globose, black.

Distribution: MEXICO. Chiapas. Veracruz. COSTA RICA. Puntarenas. Dense, wet forest; 100–1,000 m elevation; sometimes on limestone.

Discussion: Wendland (1880) described and named *C. tenella* from material cultivated at the Jardin Botanique at Zurich, Switzerland that Ortgies had introduced from Mexico in 1877. It was rare in cultivation by 1900 (Dammer 1904b) and had since largely disappeared from horticulture and botany until recently, when collections from Mexico and Costa Rica reestablished its existence and extended its range considerably. It is still very rare in cultivation today and most material so named is *C. metallica* or *C. geonomiformis.*

Chamaedorea tenella has an unusual bicentric distribution that is not without parallel in other groups (Grayum and de Nevers 1988). It occurs on the Atlantic slope of Veracruz and Chiapas in southern Mexico and then again in southeastern Costa Rica in the Golfo Dulce region on the Pacific slope. Although it is apparently generally distributed in both areas, it is certainly not common in either. Unfortunately, because of habitat destruction in Mexico and Costa Rica and the fact that collectors eagerly seek *C. tenella* due to its highly ornamental qualities, its continued existence in the wild is somewhat tenuous. Grayum and de Nevers (1988) consider it to be a rare and endangered species throughout its entire range.

In Costa Rica, we found it to be rather rare and to occur sparingly in dense lowland rainforest on the Osa Peninsula in the Golfo Dulce. Plants from this area have flowered in the greenhouse here in Los Angeles and we were fortunate to have good material of both staminate and pistillate plants. We also found *C. tenella* in wet forest at about 800 m elevation in Chiapas where it was growing on limestone and in the Catemaco area of Veracruz where it was on volcanic soils at a similar elevation. Plants from Chiapas tend to have stiffer, narrower blades more shallowly incised at the apex than their counterparts from Veracruz.

Chamaedorea tenella appears close to *C. geonomiformis* and is perhaps not distinct. Generally, the larger habit and leaves, the untoothed blade incised apically to more than ⅓ its length, and the usually branched inflorescences distinguish *C. geonomiformis.* Standley (1937) listed *C. geonomiformis* from Costa Rica although the description of the plant with spicate inflorescences depicts *C. tenella* or *C. deckeriana.*

Recent collections from Veracruz, Mexico have resulted in *C. tenella* becoming established in a few gardens and collections in California and Florida.

33. Chamaedorea verapazensis *Hodel & Castillo,* Phytologia 68: 390. 1990. Type: Guatemala, Baja Verapaz, *Hodel & Castillo 898A* (holotype BH!; isotype AGUAT!). Epithet: derived from the range of this species, originally found only in the Departments of Alta Verapaz and Baja Verapaz, Guatemala. **PLATE 46.**

Common Name: none.
Vernacular Name: none.

Habit: solitary, slender, decumbent then briefly erect, 1–2 m tall.

Stem: 8–10 mm diam., green, ringed, internodes 2–5 cm long, rooting along its length.

Leaves: 5–6, erect-spreading, pinnate, bright green; **sheath** to 15 × 1.3–1.5 cm, tubular, obliquely long-open apically, green, longitudinally striate-nerved; **petiole** 5–18 cm long, flattened and green above, rounded and green below; **rachis** 12–22 cm long, angled and green above, rounded below and with a pale green or yellowish band extending onto sheath; **blade** to 40 × 28 cm; **pinnae** 2–5 on each side of rachis, basal ones to 17 × 4.5 cm, lanceolate, strongly sigmoid, acuminate, 4–6 prominent primary nerves above, central one slightly more prominent, 1 secondary between each pair of primaries, tertiaries faint, inconspicuous, apical pinnae largest, to 15–22 × 5–9 cm, sigmoid, acuminate, exterior margin toothed toward apex, 7–10 prominent primary nerves above, 2 secondaries between each pair of primaries.

Inflorescences: interfoliar; **peduncles** to 55 cm long, 3–4 mm wide at base and flattened, 1–3 mm diam. at apex and rounded, straight, erect-spreading, pale or light greenish in flower, orange in fruit where exposed; **bracts** 4–5, prophyll to 6 cm long, 2nd bract to 18 cm, 3rd to 35 cm, 4th to 30 cm, and extending onto rachis and concealing rudimentary 5th bract, closely sheathing, acuminate and obliquely open apically, greenish in flower, greenish or brown in fruit, longitudinally striate-nerved. **Staminate** exceeding leaves; **rachis** to 7 cm long, greenish in flower; **rachillae**

Plate 45. Chamaedorea tenella. A. On limestone rock, *D. & R. Hodel 930,* Chiapas, Mexico; **B.** Fruiting plant on red clay, *D. & M. Hodel 714,* Osa Peninsula, Coata Rica; **C.** Staminate inflorescence, *Hodel 714A bis,* cultivated, Los Angeles, California, *ex* Osa Peninsula, Costa Rica.

→

C

7–10, these to 17 cm long, slender, 1 mm diam., pendulous, greenish yellow in flower.

Pistillate about equalling leaves; **rachis** to 3 cm long, pale green or greenish yellow in flower, orange in fruit; **rachillae** 2–5, these to 10 cm long, rigid, pale green or greenish yellow in flower, orange in fruit.

Flowers: Staminate in 4 moderately dense rows, 2.5–3 mm diam., globose, only slightly sunken in superficial elliptic depressions 2–2.5 mm long; **calyx** 1.5–2 × 2.5 mm, shallowly lobed, distinctly nerved when dry, **sepals** connate in basal ¾, straight or broadly rounded apically; **petals** 2.5–3 × 2.5 mm, valvate, connate apically and basally and adnate apically to pistillode and corolla opening by lateral apertures, broadly ovate, acute, thick, fleshy, distinctly nerved when dry; **stamens** 1.5–2 mm high, **filaments** 1.0 mm long, adnate basally to petals, **anthers** 0.75–1 × 0.5 mm, elongate; **pistillode** 2.5 × 1 mm, broadly columnar, tip flat.

Pistillate: in 4 ± lax and irregular rows, 2.5 mm diam., ± globose, sunken in elliptic depressions 2–2.5 mm long; **calyx** 1.5 × 2.5 mm, deeply lobed, strongly nerved when dry, **sepals** connate in basal ½, rounded apically; **petals** 2–2.5 × 2.5–3 mm, briefly connate basally, imbricate nearly to apex, rounded apically, fleshy, strongly nerved when dry; **staminodes** lacking; **pistil** 1.5–2 mm diam., globose, 3-angled, **styles** short or lacking, **stigma** lobes recurved, pointed, brownish.

Fruits: to 11 × 9 mm, ± globose, green maturing black.

Distribution: GUATEMALA. Alta Verapaz. Baja Verapaz. El Progreso. Huehuetenango. Zacapa. Dense, wet forest on the Atlantic slope; 1,700–2,100 m elevation.

Discussion: *C. verapazensis* is close to the highly variable *C. pinnatifrons,* from which it may not be distinct. However, the fruits maturing from green through a yellow or orange stage and then becoming black, the more tubular leaf sheaths, and the curved peduncles shorter than the leaves distinguish *C. pinnatifrons. C. verapazensis* may be confused with *C. rojasiana.* The smaller habit, more tubular leaf sheaths, and fewer rachillae distinguish the latter species. Other than a few plants in the research collection in Los Angeles, *C. verapazensis* is apparently not cultivated.

34. Chamaedorea warscewiczii *H. A. Wendl.,* Bonplandia 10: 37. 1862. Type: Cult., *Wendland s.n.* (holotype GOET!; isotype K!). Epithet: honors Josef Warscewicz, who introduced the species to cultivation. **PLATES 6(C), 47.**

Nunnezharia warscewiczii (H. A. Wendl.) O. Kuntze, Revisio Generum Plantarum 2: 731. 1891.

Chamaedorea latipinna L. H. Bailey, Gentes Herb. 6: 244, fig. 128. 1943a. Type: Panama, Bocas del Toro, Chiriqui Lagoon, *von Wedel 2275* (holotype MO!; isotype BH!). Epithet: from the Latin *lati* meaning broad and *pinna* meaning leaflet.

Common Name: none.
Vernacular Name: none.

Habit: solitary, erect, infrequently decumbent, to 5 m tall.

Stem: 1.5–3.5 cm diam., smooth, green, ringed, internodes 5–15 cm long.

Leaves: 4–8, pinnate, erect-spreading; **sheath** 15–25 cm long, tubular, obliquely open apically, green with a whitish margin briefly apically, a yellow ridge extending from petiole, drying brownish and persistent, longitudinally striate-nerved; **petiole** 15–35 cm long, slightly grooved and green above, rounded and pale below; **rachis** 35–50 cm long, 2–3 mm diam. apically and strongly attenuate, angled and green above, round and with a yellow band below extending onto sheath; **blade** oval or elliptic; **pinnae** 5–7 on each side of rachis, 25–37 × 2–13 cm, broadly sigmoid, rhombic, alternate, short-acuminate, ± leathery, contracted basally and there conspicuous because of 5–9 curved primary nerves, these smooth above and not much elevated but sharply keeled and elevated below, less distinct secondaries between each pair of primaries, end pair of pinnae confluent and ± broader basally.

Inflorescences: interfoliar but often infrafoliar in fruit, erect but becoming nodding or drooping when heavily laden with fruits; **peduncles** 30–60 cm long, 1–1.5 cm wide at base and ± flattened, 3–5 mm wide at apex, erect-arching, greenish white or yellowish in flower, red-orange in fruit; **bracts** 5–6, tubular, fibrous, brownish in flower and in fruit and often falling away, tightly sheathing, acuminate, bifid, longitudinally striate-nerved, uppermost longest and exceeding peduncle; **rachises** 5–15 cm long, light green in flower, red-orange in fruit. **Staminate** with 15–25 **rachillae**, these 20–30 cm long, pendulous, glabrous, light green in flower.

Pistillate with **rachillae** 10–15, these to 25 cm long, slender, ± spreading and light green in flower, downward-curving and orange in fruit.

Flowers: Staminate in rather lax spirals, 2–3 × 1.5–1.7 mm, ovate-globose, flattened apically, greenish yellow, only slightly sunken in shallow rounded depressions; **calyx** 0.5–1.5 mm, shallowly lobed, nerved, **sepals** imbricate basally, rounded apically; **petals** valvate, connate apically and basally and apically adnate to pistillode and corolla opening by lateral slits, acute-acuminate, nerved.

Pistillate in rather lax spirals, 2 × 1.5–1.7 mm, ovoid-globose in bud; **calyx** 0.5–0.75 × 1.5–1.75 mm across, ringlike, shallowly lobed, nerved, **sepals** connate nearly to apex, straight apically; **petals** imbricate nearly to apex, acute-acuminate, nerved; **pistil**

Plate 46. Chamaedorea verapazensis. Wet cloud forest, Guatemala. **A.** Staminate plant, *Hodel & Castillo 898A* (holotype), Baja Verapaz (note open leaf sheaths and long-pedunculate, straight inflorescences); **B.** Fruiting plant, *Hodel & Castillo 999,* Huehuetenango; **C.** Young plant, *Hodel & Castillo 893,* Alta Verapaz; **D.** Pistillate plant, *Hodel & Castillo 898B,* Baja Verapaz. (Note few-branched rachillae of pistillate plants.)

C
D

2 × 1.7 mm, ovoid-globose, **styles** short or lacking, **stigma** lobes short, recurved.

Fruits: 13 × 7 mm, ovoid to globose, tapered basally; seeds sharply pointed at site of attachment.

Distribution: PANAMA. Cocle. Colon. Darien. Panama. Veraguas. COSTA RICA. Alajuela. Cartago. Guanacaste. Heredia. Limon. San Jose. Dense, wet forest on the Atlantic slope up to and over the Continental Divide; 0–750 m elevation; sometimes on limestone.

Discussion: The description is from Wendland (1862) and supplemented from Bailey (1943a) and our own collections in Costa Rica and Panama. Warscewicz collected *C. warscewiczii* in Costa Rica and introduced it to European gardens by the middle of the 19th century. Wendland (1862) described and named it from cultivated plants at Herrenhausen.

Chamaedorea warscewiczii is similar to *C. matae* and some forms of the highly variable *C. pinnatifrons.* The very broad apical pinnae and pendulous (even in flower) pistillate rachillae distinguish *C. matae.* The erect (in flower) pistillate rachillae, thicker and rounded fruiting rachillae, and fruits that mature from a green through a yellow, orange, or red stage before ageing to brown or black distinguish *C. pinnatifrons.* Also, the pinnae of *C. warscewiczii* have conspicuous and smooth but not keeled nerves above and sharply keeled nerves below, features lacking in *C. matae* and *C. pinnatifrons. C. warscewiczii* is quite distinctive in fruit; the fruiting rachillae are slender and all uniformly curved in the same direction.

I have observed *C. warscewiczii* in Costa Rica along the trail down to the river at C.A.T.I.E. near Turrialba and below San Miguel along the Rio Sarapiqui. In both places it grows with *C. tepejilote* and *C. macrospadix.* In Braulio Carrillo National Park we observed it growing with *C. pumila.* In Panama, we found it on Cerro Campana with *C. tepejilote.* Bailey (1943a) referred to the populations on Cerro Campana as *C. flavovirens.*

Chamaedorea warscewiczii is a handsome species due to its leafy crown and broadly rhombic and sigmoid pinnae. It was reintroduced to cultivation in the late 1970s and fruiting plants occur at Lyon Arboretum in Hawaii (*L-78.796*). The International Palm Society Seed Bank distributed seeds in 1987 and 1988 from collections in Costa Rica (*Hodel 636, 702*) as *C. flavovirens.*

V. Subgenus CHAMAEDOROPSIS

Chamaedorea subgenus *Chamaedoropsis* Oerst., Vidensk. Meddel. Dansk. Naturhist. Foren. Kjoebenhavn 1858: 15. 1859. (in part.)

Key Characters: petals distinct in flowers of both sexes and thickened and persistent in fruit, staminate flowers borne singly with petals spreading or erect and not connate apically.

Distribution: widely or locally distributed on the Atlantic and Pacific slopes from Mexico to northern South America; to nearly 3,000 m elevation. Of the 40 or so taxa of subgenus *Chamaedoropsis* that have been validly published, about 20 are in cultivation.

Key to the Species of Subgenus *Chamaedoropsis*

Leaves bifid or entire but not pinnate.
  Cespitose habit.
    Stems with rhizomes. .... **37. C. brachypoda**
    Stems without rhizomes but spreading and multiplying by lateral, basal shoots.
      Leaf blade lobes narrowly divergent, margin toothed; staminate flowers orange. .... **47. C. fragrans**
      Leaf blade lobes broadly divergent, margin ± entire; staminate flowers yellow. .... **52. C. nubium**
  Solitary habit.
    Leaf blades entire. .... **74. C. tuerckheimii**
    Leaf blades bifid.
      Leaf sheaths tubular; stem elongate, either erect or decumbent.
        Leaf blades with 20 or more nerves on each side of rachis. .... **44. C. deneversiana**

Plate 47. Chamaedorea warscewiczii. A. Fruiting plant, *D. & M. Hodel 725,* Cerro Campana, Panama; **B.** Fruiting plant, *Hodel & Grayum 702,* along trail to river, C.A.T.I.E., Turrialba, Costa Rica; **C.** Staminate plant, *Hodel 832,* cultivated, Lyon Arboretum (*78.796*), Honolulu, Hawaii, *ex* Costa Rica; **D.** Frank Hausman holds plant cultivated in his garden, Homestead, Florida, *ex* Costa Rica.

B
D

Leaf blades with 10 or fewer nerves on each side of rachis.
Petioles rugose. .... **73. C. tenerrima**
Petioles not rugose.
Petioles gray, densely white-spotted.
Leaf blades incised apically ¾ their length. .... **41. C. correae**
Leaf blades incised apically ½ their length.
Leaf blades leathery, pliable, velvety-textured; inflorescences equalling or exceeding leaves, staminate rachillae pendulous. .... **35. C. adscendens**
Leaf blades stiff, plasticlike, gray-green; inflorescences shorter than leaves. .... **48. C. guntheriana**
Petioles green or only scarcely white-spotted.
Nerves prominently elevated above, leaf blades appearing corrugated. .... **76. C. verecunda**
Nerves conspicuous but not prominently elevated above, leaf blades flat.
Stem erect; crown of 7–15 leaves; staminate rachillae pendulous; staminate flowers with prominent calyx; fruits oval, narrowed at both ends. .... **43. C. dammeriana**
Stem decumbent; crown of 4–5 leaves; staminate rachillae spreading; staminate flowers with low calyx; fruits globose. .... **40. C. chazdoniae**
Leaf sheaths open; plants appearing stemless or stem very short.
Nerves prominently elevated above, leaf blades appearing corrugated.
Leaf blades leathery, pliable, iridescent and/or mottled blue-gray-green.
Leaf blades with 7 or fewer primary nerves on each side of rachis. .... **51. C. minima**
Leaf blades with 9 or more primary nerves on each side of rachis.
Leaf blades with 9–13 primary nerves on each side of rachis. .... **59. C. pumila**
Leaf blades with 15 or more primary nerves on each side of rachis. .... **72. C. sullivaniorum**
Leaf blades not leathery or pliable, not iridescent, only forest green.
Calyxes of both sexes well developed and prominent; staminate flowers cream-colored. .... **39. C. castillo-montii**
Calyxes of both sexes low, ringlike; staminate flowers whitish tinged with green. .... **65. C. robertii**
Nerves perhaps conspicuous but not prominently elevated above, leaf blades ± flat, not corrugated.
Leaf blades with 12 or more primary nerves on each side of rachis.
Leaf blades thin-papery; 10 or more staminate rachillae. .... **60. C. pygmaea**
Leaf blades thick, durable; less than 10 staminate rachillae.
Petioles short or lacking; leaf blades stiff. .... **64. C. rigida**
Petioles elongate; leaf blades pliable. .... **71. C. stricta**
Leaf blades with 10 or fewer primary nerves on each side of rachis.
Leaf blades ± mottled, nerves not drying shining yellow; fruiting perianth orange, showy; staminate rachillae erect to spreading. .... **61. C. queroana**
Leaf blades not mottled, nerves drying shining yellow; fruiting perianth brown, not showy; staminate rachillae drooping. .... **77. C. volcanensis**
Leaves pinnate.
Cespitose habit.
Stems with rhizomes, forming well separated, loose colonies. .... **63. C. rhizomatosa**
Stems without rhizomes, multiplying by lateral, basal shoots, forming tight clusters.
Leaf sheaths with prominent ligules at apex.
Ligules long-lanceolate, 5 cm long or more, brown, fibrous. .... **62. C. quezalteca**
Ligules triangular, short, less than 3 cm long, green with brown margin. .... **42. C. costaricana**
Leaf sheaths lacking ligules at apex.
Pinnae 18 or fewer on each side of rachis.
Pinnae straight, only a midrib prominent; peduncles less than 9 cm long; rachillae short, erect, stiff in fruit, not undulate. .... **67. C. seifrizii**
Pinnae sigmoid, a prominent midrib and conspicuous primary nerves; peduncles greater than 10 cm long; fruiting rachillae downward-pointing, often markedly undulate. .... **52. C. nubium**
Pinnae 20 or more on each side of rachis. .... **58. C. pochutlensis**
Solitary habit.
Leaf sheaths tubular; stem generally elongate, either erect or decumbent, if stem short or lacking then inflorescences equalling or usually exceeding leaves.
Stem short or lacking; inflorescences equalling or usually exceeding leaves. .... **35. C. adscendens**
Stem elongate; inflorescences usually shorter than leaves.
Pinnae 20 or more on each side of rachis. .... **80. C. woodsoniana**

→

Plate 48. Chamaedorea adscendens. On limestone rock, *Hodel & Castillo 1101A* (staminate) *& 1101B* (pistillate), Peten, Guatemala. **A.** Staminate plant with long-pedunculate inflorescence, pendulous rachillae, yellow flowers; **B.** Pistillate plant; **C.** Spicate infructescence with mature fruits; **D.** Densely white-spotted petiole.

C
D

Pinnae 13 or fewer on each side of rachis.
Petioles rugose; pinnae distinctly auriculate basally, terminal pair of pinnae broadly flared at right angles to rachis. . . . . **73. C. tenerrima**
Petioles not rugose but green or gray-green; pinnae not auriculate basally, terminal pair of pinnae not broadly flared at right angles to rachis.
Petioles gray-green and/or densely white-spotted.
Plants large; stem 2–3 cm diam. . . . . **78. C. vulgata**
Plants small, stem 5–8 mm diam.
Pinnae thin-papery, soft, dark green, terminal pair not as large as other combined. . . . . **50. C. microphylla**
Pinnae thick, ± stiff, ± gray-green, terminal pair much larger than others combined.
Pinnae with prominently elevated, pale or white nerves; staminate flowers maturing simultaneously throughout axis. . . . . **41. C. correae**
Pinnae obscurely nerved; staminate flowers maturing markedly and progressively from base to apex of axis. . . . . **48. C. guntheriana**
Petioles green, not densely white-spotted.
Leaf rachis strongly attenuate and filiform apically; staminate rachis strongly flexuous; fruiting peduncle and rachillae filiform, pendulous. . . . . **46. C. fractiflexa**
Leaf rachis not attenuate and filiform apically; staminate rachis straight; pistillate or fruiting rachillae not filiform but erect or spreading.
Pinnae thin-papery.
Pinnae 6 cm or more wide, drying with several prominent yellow nerves; inflorescences borne well below the leaves. . . . . **68. C. selvae**
Pinnae less than 5 cm wide, drying without prominent yellow nerves; inflorescences interfoliar or just below leaves.
Staminate rachillae pendulous; pistillate rachillae thick, blunt-tipped. . . . . **56. C. parvisecta**
Staminate rachillae erect; pistillate rachillae slender, spinose-tipped. . . . . **79. C. whitelockiana**
Pinnae thick- or medium-textured.
Stem decumbent, creeping. . . . . **40. C. chazdoniae**
Stem erect.
Pinnae 11 or more on each side of rachis; rachillae usually markedly undulate. . . . . **69. C. skutchii**
Pinnae 9 or fewer on each side of rachis; rachillae not markedly undulate.
Pinnae thick, ± leathery; 3–5 leaves in crown; inflorescences infrafoliar.
Pinnae 5–9 on each side of rachis, generally more than 5 cm wide. . . . . **53. C. oblongata**
Pinnae 3–5 on each side of rachis, generally less than 5 cm wide. . . . . **55. C. paradoxa**
Pinnae medium-textured; 7 or more leaves in crown; inflorescences interfoliar. . . . . **43. C. dammeriana**
Leaf sheaths long-open; stem short or lacking.
Pinnae decurrent on rachis; eophyll pinnate.
Pinnae not decreasing markedly in length toward apex of rachis. . . . . **60. C. pygmaea**
Pinnae decreasing markedly in length toward apex of rachis.
Margins of pinnae undulate; fruiting rachillae strongly curved, hooklike. . . . . **75. C. undulatifolia**
Margins of pinnae not undulate; fruiting rachillae ± straight or only slightly curved.
Pistillate inflorescences spicate or forked; generally small plants less than 70 cm tall. . . . . **70. C. stenocarpa**
Pistillate inflorescences with 3 or more rachillae; generally large plants 70 cm tall or more.
Leaves erect-ascending; 20 or more pinnae on each side of rachis; pistillate inflorescences with 60–100 short, stiff, filiform rachillae. . . . . **36. C. brachyclada**
Leaves spreading; fewer than 20 pinnae on each side of rachis; pistillate inflorescences with 3–15 thick rachillae. . . . . **66. C. scheryi**
Pinnae not decurrent on rachis; eophyll bifid.
Petioles rough-textured due to being densely pitted with minute, irregular depressions. . . . . **45. C. foveata**
Petioles ± smooth.
Pinnae 15 or more on each side of rachis; newly emerged petioles densely covered with black tomentum. . . . . **38. C. carchensis**
Pinnae 12 or fewer on each side of rachis; petioles lacking black tomentum.
Pinnae thick, ± stiff, blue- and/or gray-green. . . . . **57. C. pittieri**
Pinnae thin or medium-textured.
Leaf sheaths thick, durable, triangular in cross-section, arranged in a triangular manner. . . . . **49. C. lehmannii**
Leaf sheats ± thin or medium in thickness, rounded, not arranged in a triangular manner.

→

Plate 49. Chamaedorea adscendens. On limestone rock, Peten, Guatemala. **A.** Staminate inflorescence, *Hodel & Castillo 1101A*; **B.** Bifid-leaved plant; **C.** Young plant.

B

Pinnae 2–4 on each side of rachis; nerves prominently elevated and dark green; inflorescences of both sexes spicate. .... **39. C. castillo-montii**
Pinnae 7 or more on each side of rachis; nerves not prominently elevated; inflorescences of both sexes not spicate (or only one sex spicate).
Pinnae to 2 cm or less in width, ± obscurely nerved, sometimes minutely auriculate basally, end pair attached at right angles to rachis. .... **54. C. pachecoana**
Pinnae 3 cm or more in width, conspicuously nerved, not auriculate basally, end pair forward-pointing.
Pinnae ± mottled green, nerves drying dark; staminate rachillae erect or spreading; pistillate inflorescences spicate or forked; fruiting perianth bright orange, showy. .... **61. C. queroana**
Pinnae green and not mottled, nerves drying pale or yellowish; staminate rachillae pendulous; pistillate inflorescences with 5 or more rachillae; fruiting perianth brownish, not showy. ... **77. C. volcanensis**

35. Chamaedorea adscendens (*Dammer*) *Burret,* Notizbl. Bot. Gart. Berlin-Dahlem 11: 737. 1933a. Epithet: is a Latin word meaning ascending, apparently in reference to the inflorescences although the author did not specify the application of the name. **PLATES 48, 49.**

Kinetostigma adscendens Dammer, Notizbl. Konigl. Bot. Gart. Berlin 4: 172. 1905a. Type: Guatemala, Alta Verapaz, *Tuerckheim 8770* (holotype B, destroyed; isotype US!).

Tuerckheimia adscendens Dammer in J. D. Smith, Enumeratio Plantarum Guatemalensium 7: 53. 1905. Name only.

Common Name: none.
Vernacular Name: none.

Habit: solitary, erect, infrequently decumbent, to 2.5 m tall, flowering when appearing stemless.

Stem: 1 cm diam., often creeping and buried in the leaf litter, covered with persistent sheaths, green where exposed, ringed, internodes 2–2.5 cm long.

Leaves: 6, spreading, pinnate or infrequently bifid; **sheath** to 6 cm long, tubular, becoming obliquely long-open apically, thick, a rounded costa extending from petiole, gray-green; **petiole** 7 cm long or more, channeled and grayish above, rounded and grayish below; **rachis** to 16 cm long, angled and grayish above, rounded and grayish below; apex of sheath, petiole and rachis densely white-spotted; **pinnae** 2–6 on each side of rachis, middle ones largest, these to 16 × 2.5–3.5 cm, lower ones to 8 × 1.2 cm, obovate-lanceolate, long-acuminate, thick, leathery, velvet-textured, gray- and/or metallic-blue-green, contracted basally, regularly or unevenly spaced, ± obscurely nerved, a prominent but scarcely elevated midrib pale below and numerous inconspicuous nerves of lesser orders, terminal pair often much broader and 2–9-nerved, or sometimes one side undivided; or blade bifid, to 25–30 × 15–20 cm, incised apically to ½ its length, 6–9-nerved.

Inflorescences: infrafoliar, penetrating sheaths at nodes, ascending to erect in flower, ascending to spreading in fruit, sometimes appearing to arise from leaf litter, exceeding leaves; **peduncles** 40–70 cm long, slender, 1.5 cm wide at base, green where exposed in flower, red-orange in fruit; **bracts** 8–9, tubular, bifid or acuminate, brownish in flower, longitudinally striated, uppermost exceeding peduncle. **Staminate** with **rachis** to 10 cm long, green in flower; **rachillae** 2–10, these to 20 cm long, slender, pendulous, green in flower, lower ones sometimes forked.

Pistillate spicate; **rachis** or flower-bearing portion to 15 cm long, green in flower, red in fruit.

Flowers: Staminate in dense spirals but not contiguous in bud, 2–2.5 × 2–2.5 mm, ± globose, bright yellow, ageing to cream-colored, ± superficial on axis; **calyx** 0.5–1 × 2–2.5 mm, shallowly lobed, pale, membranous, **sepals** connate nearly to apex and there ± truncate; **petals** 2–3 × 1.5–2.5 mm, ovate, valvate, free nearly to base, spreading apically, acute; **stamens** 1–1.5 mm tall, **filiments** short or nearly lacking, **anthers** 1–1.25 mm long, entire apically; **pistillode** just exceeding stamens but shorter than petals, columnar.

Pistillate in moderate spirals, 2–2.5 × 1.5–1.75 mm, ovoid-globose, only slightly sunken; **calyx** 0.75–1 × 1.75–2.25 mm, ± deeply lobed, **sepals** connate and/or imbricate in basal ¼–½, broadly rounded apically, obtuse; **petals** 2–2.5 × 2 mm, briefly connate basally, imbricate nearly to apex, broadly rounded, acute-obtuse, minutely black-spotted apically, very faintly nerved on outside when dry; **staminodes** lacking or present; **pistil** 1.5–2 × 1.5–1.75 mm, conic, **stigma** lobes sessile, recurved.

Fruits: 8 mm diam. or long, globose or ovoid, black; seeds 6 mm long; abortive carpels adherent to corolla in fruit.

Distribution: GUATEMALA. Alta Verapaz. Peten. BELIZE. Toledo. Moist or wet forest on the Atlantic slope; to 700 m elevation; on limestone.

Discussion: Dammer (1905a) described and named *C. adscendens* from material that Tuerckheim had collected near Cubilguitz, Alta Verapaz, Guatemala. Dammer erected a new genus, *Kinetostigma,* for the species and later Burret (1933a) formally transferred it to *Chamaedorea. C. adscendens* is a remarkable palm in habit and habitat. Its durable, leathery leaves are perhaps the thickest in the genus, and their velvety texture and blue-gray-green color are attractive features. Although leaves are usually pinnate, bifid-leaved individuals are not uncommon. Young plants with bifid leaves are similar

Plate 50. Chamaedorea brachyclada. A. Marianne Hodel holds upright leaves, *D. & M. Hodel 705A,* San Vito de Coto Brus, Puntarenas, Costa Rica; **B, C.** Cultivated, Lyon Arboretum, Honolulu, Hawaii (note long-pedunculate inflorescences curving upward from base; **D, E** Inflorescences, cultivated, Los Angeles; **D.** Staminate inflorescence; **E.** Bottlebrushlike flowering portion of pistillate inflorescence.

B
METRIC
INCHES
E

to *C. metallica* except the blades are more deeply incised apically.

The habitat of *C. adscendens* is unusual also. Like *C. graminifolia,* the species appears restricted to the summits of steep limestone hills. In fact, the two species occasionally occur together. Only a thin layer of leaf litter covers the limestone rock and the area can experience a rather lengthy dry period. In 1991 during the height of the dry season, we found *C. adscendens* in southern Peten on solid limestone rock. While other dicotyledonous shrubs and trees were actually wilting from the lack of water, *C. adscendens* seemed to be unaffected. *C. adscendens* is rare in the wild; only a few collections in several herbaria document it, and we searched for it in the wild on four occasions before finally finding it in Guatemala. *C. adscendens* does not occur at the summit of each hill in a given area; it is often quite localized on a particular hill. Although we searched a half-dozen hills within a small area, we found *C. adscendens* only on one of these and then it was confined to only a small portion of the summit. It is difficult to explain this unusual distribution.

Staminate inflorescences at anthesis are densely flowered, resembling such species as *C. arenbergiana* and *C. tepejilote* of subgenus *Stephanostachys*; however, flowers are clearly distinct in bud. Although Moore (1963, 1976) indicated that *C. adscendens* was cultivated, it is doubtful if this is still true. Plants cultivated at Fairchild Tropical Garden under this name and originally collected in the Jacinto Hills in Belize died several years ago. Apparently the only plants in cultivation now are the few small ones in the research collection in Los Angeles.

36. Chamaedorea brachyclada *H. A. Wendl.,* Gartenflora 29: 101. 1880. Type: Cult., *Wendland s.n.* (holotype GOET!). Epithet: from the Greek *brachy* meaning short and *cladus* meaning shoot, in reference to the very short stem or nearly stemless habit. **PLATE 50.**

Nunnezharia brachyclada (H. A. Wendl.) O. Kuntze, Revisio Generum Plantarum 2: 730. 1891.

Common Name: none.
Vernacular Name: none.

Habit: solitary, erect, appearing stemless, to 2.5 m tall.

Stem: 2–3 cm diam., buried and creeping in leaf litter, closely and prominently ringed, internodes to 5–10 mm long.

Leaves: 3, erect, pinnate, 1–2 m long; **sheath** to 35 cm long, obliquely open and splitting deeply opposite petiole, clasping in a tubular manner only briefly near base, dark green; **petiole** 40–80 cm long, erect, slightly flattened and green above or grooved only near base, rounded and dark green below; **rachis** sharply angled and green above, rounded and green below, rachis and petiole with a covering of minute, rough, white, glistening spots especially at rachis-petiole junction; **pinnae** 20–30 on each side of rachis, largest to 30 × 2–3 cm, linear-lanceolate, ± straight but conspicuously falcate, acuminate, lower margin decurrent on rachis, dark green, 5-nerved with underside fairly rough.

Inflorescences: infrafoliar, long-pedunculate, arcuate, often appearing to rise from litter of forest floor; **peduncles** to 60 cm long, 1 cm wide at base, 5–7 mm diam. at apex, green and ± rounded in flower, reddish orange in fruit; **bracts** 5–9, prophyll to 2 cm long, 2nd bract 2 cm, 3rd 3 cm, 4th 6 cm, 5th 10 cm, 16th 12–15 cm, 7th, 8th, and 9th 15–20 cm, tubular, brown in flower, ± papery, longitudinally striate-nerved, acute-acuminate, bifid; **rachises** 10–12 cm long, green in flower, reddish orange in fruit. **Staminate** with 40–50 **rachillae**, lowermost longest, these to 10 cm long, uppermost shortest, to 5 cm long, 0.5 mm diam., perpendicular to rachis, spreading, slender, filiform, green.

Pistillate bottle-brushlike, **rachillae** 60–100, these 3–5 cm long, filiform, 0.5 mm diam., flexuous but ± stiff, simple or sometimes forked, 4–7-flowered, green in flower, reddish orange in fruit.

Flowers: Staminate in loose spirals, 3–3.5 × 2 mm, ovoid, green, slightly sunken; **calyx** 1 × 1.5 mm, shallowly lobed, **sepals** connate in basal ⅔, broadly rounded and brown-margined apically; **petals** 3 × 1.5–2 mm, ovate, valvate, connate in basal ½, free, acute and slightly recurved apically, thickened and lighter green medially on inside; **stamens** 2 mm high, **filaments** yellowish, **anthers** yellow, 1.5 mm long; **pistillode** 2 mm high, columnar, yellowish.

Pistillate in remote spirals, 3 × 2 mm, ovoid-globose, yellow-green, ± superficial; **calyx** 1 × 1.5–2 mm, deeply lobed, **sepals** sharply rounded and dark-margined apically; **petals** 3–3.5 × 2 mm, oval, imbricate nearly to apex and there free, erect; **pistil** 2 × 2 mm, globose, slightly 3-angled, dark green, **styles** short or lacking, **stigma** lobes slender, recurved, pointed, whitish, well below petal tips.

Fruits: 3–5 mm diam., globose, black; seeds 2–4 mm diam., globose; eophyll pinnate.

Distribution: PANAMA. Chiriqui. COSTA RICA. Puntarenas. San Jose. Moist or wet forest on the Pacific slope; 1,000–1,400 m elevation.

Discussion: Zahn collected seeds of this palm in Chiriqui in Panama and sent them to the famous plantsman Veitch in Europe prior to 1880. Wendland (1880) then described and named *C. brachyclada* from a cultivated pistillate plant he had obtained from Veitch. It was apparently lost to

→

Plate 51. Chamaedorea brachypoda. Cultivated. **A.** Large clump, Foster Botanic Garden (*65.1347*), Honolulu, Hawaii; **B.** Clump in garden of L. Hooper, La Habra, California; **C.** Carol Graff holds stem of large-leaved form (she collected it in western Honduras) in her garden, Miami, Florida.

C

cultivation until the 1960s when Robert G. Wilson established his garden at San Vito in Costa Rica near the Panamanian border. Wilson's garden included some native forest with populations of *C. brachyclada.* Moore (1971) noted it there as *Chamaedorea* sp. # 2. Wilson established and cultivated *C. brachyclada* in his garden and from these plants and wild plants in the adjacent forest he distributed seeds to fellow palm enthusiasts and botanical gardens. Through Wilson's efforts, collectors have grown *C. brachyclada* in California, Florida, and Hawaii since the late 1960s and early 1970s.

Although originally collected in Chiriqui, Panama, *C. brachyclada* is very rare there; in fact, I have seen only one collection of it from this area. Much of Chiriqui has been extensively deforested as have adjacent areas of Costa Rica. Grayum and de Nevers (1988) noted that this species is clearly threatened and endangered since suitable habitat is extremely scarce. As recently as 1987, though, a healthy, reproducing population of this palm was existing in forest remnants next to Wilson's garden.

A very distinctive species, the stemless habit, few (3–4), erect, long-pinnate leaves, numerous and essentially straight pinnae with lower margins decurrent, long petioles, and basal, long-pedunculate, arcuate inflorescences with numerous, short rachillae easily distinguish *C. brachyclada.* The pistillate inflorescence somewhat resembles a bottle brush with its slender, sturdy peduncle terminating in numerous, short, densely placed, stiff rachillae. Individuals will often flower and fruit when less than a meter tall. In these cases, the dimensions of the plant parts are much reduced.

Chamaedorea brachyclada forms a more or less natural group with *C. scheryi, C. pygmaea, C. stenocarpa,* and *C. undulatifolia* (Hodel and Uhl 1990b). They all share the virtually stemless habit, pinnae with lower margins decurrent, long-pedunculate inflorescences arising from the base, and pinnate eophylls. In addition, they usually occur at middle elevations in wet or cloud forest situations.

Chamaedorea brachyclada is difficult to cultivate, even in tropical locales. It does best in a moderate or slightly cool tropical climate. *C. brachyclada* appears to be particularly sensitive to low humidity. In California, leaves tend to yellow- and brown-tip due to the low humidity and high mineral content of the water. In these situations, the crown is often much reduced, being composed of only 2–3 leaves, and the plants take on a rather poor appearance. In addition, it is very susceptible to infestations of mites and thrips in cultivation, especially in areas with low relative humidity. Because of its pinnate eophyll, it is attractive even as a 1–2 leaf seedling.

37. Chamaedorea brachypoda *Standl. & Steyerm.,* Publ. Field Mus. Nat. Hist., Bot. Ser. 23: 198. 1947. Type: Guatemala, Izabal, *Steyermark 39185* (holotype F!). Epithet: from the Greek *brachy* meaning short and *podo* meaning foot, perhaps in reference to its rhizomatously branching stems or short petioles or peduncles. **PLATES 51, 52(A, B).**

Common Name: none.
Vernacular Name: pacaya—Guatemala.

Habit: cespitose and colonial with long, very slender, underground stems (rhizomes) forming dense clumps 1–2 × 3–4 m.

Stems: 5–7 mm thick, pale green, smooth, ringed, internodes 5–8 cm long.

Leaves: 5–8, spreading, bifid; **sheath** 15 cm long, tubular, 8 mm broad, obliquely open apically, drying brown, persistent, longitudinally striate-nerved; **petiole** 5–10 cm long, slightly grooved and green above, rounded and pale below; **rachis** 10–12 cm long, angled or ridged above, rounded and pale below; **blade** to 30 × 22 cm, narrowing to 20 cm wide at apex, incised apically to ½ its length, cuneate-acute basally, lobes 15–30 × 11 cm, divergent at angle of 90 degrees or more, short-acuminate, thin, pale below, margin toothed, 11–13 primary nerves on each side of rachis, prominent and keeled above, slightly falcate, 1 secondary nerve between each pair of primaries, tertiaries numerous and faint.

Inflorescences: infrafoliar, breaking through persistent leaf sheaths at nodes just below living leaves; **peduncles** 3–10 cm long, 3–4 mm wide at base and apex, ± stiff and ascending, green in flower, orange in fruit; **bracts** 6, prophyll and 2nd bract 0.5–0.8 cm long, 3rd 1.5 cm, 4th 2.5 cm, 5th 5 cm, 6th 4 cm and exceeding peduncle, all 4–5 mm broad, tubular, papery, acuminate, bifid, brownish in flower, longitudinally striate-nerved; **rachises** 1–2 cm long, green in flower, orange in fruit. **Staminate** with 4–8 **rachillae,** these 8–15 cm long, 1.5–2 mm diam., green. **Pistillate** with 4–8 **rachillae,** these 6 cm long, ± stiff, erect, parallel or slightly spreading, green in flower, orange in fruit.

Flowers: Staminate in fairly dense spirals, 3.5 × 3 mm, ob-

→

Plate 52. A, B, Chamaedorea brachypoda. C–E, Chamaedorea carchensis. A, B Cultivated, California; **A.** Pistillate inflorescence, *Hodel 827,* research collection, Los Angeles; **B.** Infructescence (immature fruits), garden of L. Hooper, La Habra. **C–E** In montane rain forest, *Hodel & Castillo 888A* (staminate) *& 888B* (pistillate), Alta Verapaz, Guatemala; **C.** Infructescence with maturing fruits; **D.** Older petioles covered with black indument, a diagnostic feature of *C. carchensis*; **E.** Young staminate inflorescence arising from base of plant.

B
C
E

ovoid-subglobose, greenish tinged with yellow, slightly sunken; **calyx** 1.5 × 2 mm, ± shallowly lobed, green, **sepals** connate in basal ⅔, broadly acute apically; **petals** 3 × 3 mm, ovate, valvate, connate in basal ⅔, adnate to stamen filaments and pistillode in a short stipe, free but incurved apically at anthesis and later opening more, acute, fleshy; **stamens** 1–1.5 mm long, appressed around pistillode, **filaments** very short, **anthers** 2-cleft apically; **pistillode** 3 × 2 mm, broadly columnar or ± swollen at middle, yellowish, terminating in a short nipplelike, 3-angled, greenish yellow apex.

Pistillate in ± loose spirals, 3 × 2–2.5 mm, ovoid or depressed-globose, greenish yellow, slightly sunken; **calyx** 1–1.5 × 2.5 mm, deeply lobed, green, brown-margined, **sepals** connate in basal ⅓–½, broadly rounded apically; **petals** 3 × 2–2.5 mm, ovate, imbricate to ½–⅔ to apex, free, erect or slightly recurved, acute apically, slightly thickened and ± fleshy; **staminodes** toothlike; **pistil** 2 × 2–2.5 mm, depressed-globose, green, **styles** short or lacking, **stigma** lobes short, recurved, thick, pale.

Fruits: 5–10 mm long, ellipsoid, black.

Distribution: GUATEMALA. Izabal. HONDURAS. Wet lowland forest on the Atlantic slope; to 150 m elevation.

Discussion: Standley and Steyermark described and named *C. brachypoda* from material the latter collected in Izabal. Much of the forest in Izabal is gone now and *C. brachypoda* is rare there. In 1989 we found a small population surviving in the shade of an *Orbignya* palm in a pasture. Apparently, the plant did not flower here as cattle periodically ate the stems down to the ground. Although the plants would resprout from rhizomes, cattle were not absent from the area long enough for the plants to grow up again and flower. Fortunately, Carol Graff (pers. comm.) of Miami, Florida recently discovered *C. brachypoda* in adjacent, less disturbed areas of Honduras.

With bifid leaves and densely rhizomatous stems, *C. brachypoda* is an attractive and popular ornamental. It is found in collections of many palm enthusiasts and botanical gardens in Hawaii, California, Florida, Costa Rica, Australia, and perhaps elsewhere. Despite its popularity, it is often confused with *C. stolonifera* in cultivation and is sometimes grown under that name. However, the stoloniferous habit, thicker stems, thicker textured leaves, and orange, hooded petals easily distinguish *C. stolonifera.*

Chamaedorea brachypoda is handsome as a potted specimen and can be maintained this way for many years. It produces a tremendous number of rhizomes and some will often emerge from drainage holes in the bottom of the pot. *C. brachypoda* is easy to propagate by divisions of the stems and/or rhizomes or by seeds. However, since most individuals in cultivation are pistillate, it is unusual to find seeds and plants are normally propagated vegetatively.

Chamaedorea brachypoda looks its best in a tropical climate. It is a superb performer in Hawaii, plants there forming dense, bushlike clusters several meters across with rich green leaves devoid of any yellowing or marginal necrosis or burning. On the contrary, plants in California tend to have somewhat yellowish leaves often with marginal necrosis. *C. brachypoda* is somewhat susceptible to infestations of mites.

38. Chamaedorea carchensis *Standl. & Steyerm.*, Publ. Field Mus. Nat. Hist., Bot. Ser. 23: 199. 1947. Type: Guatemala, Alta Verapaz, *Standley 90160* (holotype F!). Epithet: derived from the area where the type was collected, along the Rio Carcha near San Pedro Carcha, Guatemala. **PLATES 52(C-E), 53.**

Common Name: none.
Vernacular Name: none.

Habit: solitary, erect or briefly decumbent, flowering when appearing stemless, to 3 m tall.

Stem: 3–5 cm diam., green, ringed with swollen and brown leaf scars, internodes 1–5 cm long, covered with persistent leaf sheaths, basal portion of stem more densely ringed, curved, at or below leaf litter or ground.

Leaves: 3–4, ascending, pinnate, to 3 m long; **sheath** 10–30 cm long, 2–5 cm diam., obliquely long-open, tubular only near base, indurate, narrowed to petiole, persistent; **petiole** 1 m long or slightly more, nearly 1 cm diam., flat or grooved above, rounded below, densely covered with ± deciduous black tomentum; **rachis** 1–1.7 m long, angled and green above, rounded and green below, covered near the base with deciduous black tomentum; **pinnae** 15–20 on each side of rachis, middle ones 50 × 3.5–4 cm, linear-lanceolate, straight, not sigmoid, equally attenuate-acuminate, drooping, regularly placed, alternate, remote, gradually narrowed basally to 10–15 mm wide, slightly indurate, conspicuously 1-nerved and with 2 more slender submarginal nerves, secondaries very slender and inconspicuous, the terminal ones scarcely wider or in young leaves twice as wide as lower pinnae, both to 27 cm long.

Inflorescences: infrafoliar, erect, breaking through persistent sheaths; **peduncles** 35–100 cm long, stout, to 4 cm wide at base and there flattened, to 1.5 cm diam. at apex, green in flower, red-orange in fruit where exposed; **bracts** 6–8, prophyll to 8 cm long, 2nd bract to 12 cm, 3rd 15 cm, 4th 20 cm, 5th and 6th 25 cm, 7th 30 cm, 8th 10 cm and concealed by 7th, tubular, oblique apically, bifid, acute to obtuse, green in bud, brownish in fruit,

Plate 53. Chamaedorea carchensis. Juan Jose Castillo holds stem and infructescence, *Hodel & Castillo 888B,* montane rain forest, Alta Verapaz, Guatemala.

PROPERT
BOSTO
RED

densely striate-nerved; **rachises** to 25 cm long, stout, green in flower, red-orange in fruit. **Staminate** with 20–40 **rachillae**, these to 15 cm long, ± drooping, green in bud.

Pistillate with **rachillae** 18–40, these 6–25 cm long, stout, arcuate-erect to straight and parallel, stiff, green in flower, red-orange in fruit.

Flowers: Staminate only in immature bud, ± globose, green.

Pistillate 2 × 2 mm, globose, scarcely sunken in orbicular or broadly oval depressions 1–1.2 mm long; **calyx** 2 mm across, thickened in age, deeply lobed, **sepals** imbricate and/or connate in basal ¼–½, round apically, obtuse; **petals** 3 mm long, round-ovate, imbricate, obtuse, thickened in age.

Fruits: 10 × 9 mm, ± globose, black.

Distribution: GUATEMALA. Alta Verapaz. Huehuetenango. MEXICO. Chiapas. Wet, often rocky forest on the Atlantic slope; 900–1,350 m elevation.

Discussion: The description is from Standley and Steyermark (1947, 1958) who described and named *C. carchensis* from a pistillate collection only, and is supplemented from our own collections near the type locality and in Huehuetenango. Its initially stemless habit, indurate leaf sheaths, and straight pinnae remind one of *C. radicalis,* although *C. carchensis* grows much larger and is vastly different in flowers and fruits. Young plants have deeply bifid leaves not too dissimilar from juvenile leaves of *C. elatior. C. carchensis* is probably closest to *C. woodsoniana* but the more robust stem, more prominently nerved pinnae, massive staminate inflorescence with up to 100 rachillae, and larger, more elongated fruits distinguish the latter species. Also, *C. woodsoniana* develops a visible, elongated, aboveground stem much more rapidly than *C. carchensis.* The petioles of *C. carchensis* are densely covered with black tomentum that tends to weather away but remnants of which are usually noticeable on dried herbarium specimens and is diagnostic for the species. The leaves of *C. carchensis* are among the longest in the genus. The species is apparently not cultivated.

39. Chamaedorea castillo-montii *Hodel,* Phytologia 68: 397. 1990b. Type: Guatemala, *Hodel & Castillo 868A* (holotype BH!; isotype AGUAT!). Epithet: honors Juan Jose Castillo Mont, co-collector of the type and Curator of the Herbarium, Falcultad de Agronomia, Universidad de San Carlos, Guatemala. **PLATE 54.**

Common Name: none.
Vernacular Name: none.

Habit: solitary, briefly decumbent then erect to 1 m tall, often flowering when appearing stemless.

Stem: 1.5–2.5 cm diam., creeping, with adventitious roots along its length, green, conspicuously ringed, basal portion often covered with persistent leaf sheaths, internodes 1.5–2 cm long.

Leaves: 12–15, erect-spreading, bifid, less often variously pinnate; **sheath** 10–15 cm long, obliquely long-open apically, tubular only in basal ¼, green, minutely white-spotted, longitudinally striate-nerved, margins becoming rough, brown; **petiole** to 30 cm long, channeled or flattened and green above, rounded and green below, minutely white-spotted; **rachis** to 20 cm long, angled and green above, pale green below; **blade** to 38 × 27 cm, incised apically to more than ½ its length, interior margins to 22 cm long, entire, exterior margins to 38 cm long, toothed, lobes acute-acuminate, 30 × 10 cm, diverging at 65 degrees, rich green above, paler below, 11–13 prominent primary nerves on each side of rachis, these conspicuously elevated, 2 less prominent secondaries between each pair of primaries, tertiaries numerous; or blade variously pinnate with 2–5 pinnae on each side of rachis, all except apical pair 20 × 1.5 cm, linear-lanceolate, falcate, acuminate, a prominent midrib flanked by less prominent primary nerves, apical pair of pinnae very broad, more than twice as broad as others combined, 9–10 nerved.

Inflorescences: interfoliar, erect-spreading, spicate; **peduncles** erect, to 25 cm long, 5 mm wide at base and ± flattened, 2 mm wide at apex and rounded, very pale green where exposed or whitish where concealed in flower, orange where exposed or pale green where concealed in fruit; **bracts** 5, prophyll 2–2.5 cm long, 2nd bract 7–10 cm, 3rd 10–12 cm, 4th 14–18 cm long and extending onto rachis and concealing rudimentary 5th bract, tubular, acute-acuminate, bifid, green and minutely white-spotted in flower, brownish in fruit, longitudinally striate-nerved. **Staminate** with **rachis** to 30 cm long, pendulous, pale greenish yellow in flower.

Pistillate with **rachis** to 10 cm long, rigid, erect to spreading, pale greenish yellow in flower, orange in fruit.

Flowers: Staminate borne in six alternating rows and very closely appressed but not contiguous, 0.5 mm apart, attaining anthesis first at apex of rachis then progressing toward base, 1.5–1.75 × 2 mm, subglobose, whitish, superficial, leaving elliptic scars 2.5 mm long; **calyx** well developed and prominent in bud, 1.25–1.5 × 2–2.5 mm, shallowly lobed, membranous, **sepals** connate in basal ¾, broadly rounded to straight apically, tips inflexed forming a flat rim; **petals** 2.5–3 × 2.5 mm, ovate, valvate, free nearly to base and there briefly connate, rounded apically, fleshy; **stamens** equalling or slightly exceeding petals at anthesis, 1.75–2 mm high, **filaments** large, 1.5–2 × 0.5 mm, broadly columnar, ± terete, flared basally, fleshy, **anthers** short, 0.5–0.75 mm long, thin, exserted and crowded in a cluster above open petal tips, dorsifixed near middle, elliptic, versatile, sagittate basally; **pistillode** 1.5–1.75 × 0.75 mm, broadly columnar, flared basally, only slightly so apically, tip flat.

Pistillate borne in four loose rows, moderately dense, 0.5–2.0 →

Plate 54. Chamaedorea castillo-montii. In wet lowland forest on limestone rocks, *Hodel & Castillo 868A* (staminate) (holotype) *& 868B* (pistillate), Izabal, Guatemala. **A.** Staminate plant with spicate inflorescence; **B.** Staminate inflorescence with pendulous rachis; **C.** Plant with pinnate leaves; **D.** Spicate infructescence with just maturing fruits.

C
D

mm apart, 1 × 1.75 mm, depressed-globose, whitish, sunken for ½ their height in prominent elliptic depressions 2.5 mm long; **calyx** 0.5 × 2 mm, thin, membranous, shallowly lobed, **sepals** connate in basal ¾, broadly rounded apically; **petals** 1 × 1.75 mm, deltoid, imbricate in basal ½, valvate apically, rounded or slightly acute, thin, membranous; **staminodes** lacking; **pistil** 0.75–1 × 0.75–1.5 mm, ovoid-pyramidal, **stigma** lobes sessile, large, fleshy, ± triangular each with a distal longitudinal groove.

Fruits: 13 × 7 mm, oblong, narrowed at both ends, black.

Distribution: GUATEMALA. Izabal. Alta Verapaz. Wet forest on the Atlantic slope; 600–1,000 m elevation; usually on limestone.

Discussion: *C. castillo-montii* is highly localized and known only from a few collections at the type locality in Izabal and one other collection from Alta Verapaz. It is not uncommon in the forest at the type locality on Cerro San Gil overlooking the Atlantic Ocean and occurs there from about 600 m elevation up to the summit at 900 m. Most individuals have bifid leaves; about ten percent of the population, however, have a mixture of bifid and variously pinnate leaves. On Cerro San Gil, *C. castillo-montii* often occurs on limestone rocks and grows with *C. nationsiana, C. geonomiformis,* and *C. stenocarpa.* In Alta Verapaz, it grew with *C. tenerrima* and *C. tuerckheimii.*

Chamaedorea castillo-montii is unusual among Guatemalan members of the genus in its bifid leaves with prominently raised nerves, spicate inflorescences, and densely crowded but not contiguous staminate flowers with spreading petals. It appears close to *C. robertii* from Panama and Costa Rica with which it shares these features. However, the narrower and longer blade incised apically to no more than half its length, the more narrowly divergent lobes, green-tinged staminate flowers, and the low, ringlike calyxes distinguish *C. robertii.* Among Guatemalan members of subgenus *Chamaedoropsis, C. castillo-montii* is closest to *C. volcanensis* from western Guatemala. However, the branched inflorescences and generally pinnate leaves (if bifid then with much less prominent primary nerves not greatly elevated above) distinguish *C. volcanensis.* Other than a few plants in the research collection in Los Angeles, *C. castillo-montii* is not cultivated.

40. Chamaedorea chazdoniae *Hodel,* Principes 35: 73. 1991a ('*chazdonii*'). Type: Costa Rica, Heredia, Braulio Carrillo National Park, *Chazdon 205* (holotype CR!). Epithet: honors collector of the type, Robin Chazdon.

Common Name: none.
Vernacular Name: none.

Habit: solitary, decumbent then briefly erect to 1 m tall.

Stem: 7 mm diam., smooth, green, ringed, internodes to 4 cm long, rooting along its length.

Leaves: 4–5, spreading, pinnate or less often bifid; **sheath** to 7 cm long, tubular, briefly obliquely open apically, green, finely longitudinally striate-nerved; **petiole** to 7 cm long, 2–2.5 mm diam., longitudinally striated, flattened and green above, rounded and pale below; **rachis** to 7 cm long, angled and green above, rounded below with a pale band extending onto sheath; **blade** if pinnate to 30 × 20 cm with 2 **pinnae** on each side of rachis, apical pair largest, these to 18 × 5 cm, broadly lanceolate, contracted basally, acuminate, slightly sigmoid, exterior margin remotely and shallowly toothed, 5–6 prominent primary nerves above, 1 secondary between each pair of primaries, tertiaries numerous, faint, primaries and secondaries drying yellow especially below, basal pair of pinnae to 13 × 2.5 cm with 1–2 prominent primary nerves and 1–2 secondaries on each side of primaries; if blade bifid then to 20 × 16 cm, incised apically to ¾ its length, lobes broadly diverging, 18 × 5 cm, 5–7-nerved.

Inflorescences: interfoliar, perhaps infrafoliar in fruit; **peduncles** erect, to 15 cm long, 2–3 mm wide at base, 1–1.5 mm diam. at apex, greenish in flower, orange in fruit; **bracts** 4–5, prophyll 5 mm long, 2nd bract 1.5 cm, 3rd 4 cm, 4th 7 cm, tubular, greenish in bud, membranous, becoming brown and tattered at anthesis, finely longitudinally striated. **Staminate** with 5–10 **rachillae,** these to 11 cm long, slender, spreading, greenish in flower.

Pistillate: forked or with 3 **rachillae** to 8 cm long, erect, greenish in flower, orange in fruit.

Flowers: Staminate in loose spirals, 1–3 mm apart in bud, 2 × 2 mm, ± ovoid, 1 mm wide at apex, cream-colored, superficial, leaving elliptic scars 1–1.5 mm long; **calyx** 0.5 × 2 mm, lobed, membranous, **sepals** connate in basal ½–¾, broadly rounded apically; **petals** 2 × 2–2.5 mm, deltoid, valvate, free to base, erect, acute or slightly rounded; **stamens** short, **filaments** short or lacking, **anthers** bilobed, lobes elliptic; **pistillode** 1.25–1.75 mm, columnar, reddish and slightly lobed apically.

Pistillate in loose spirals, 2–3 mm apart, 2 × 1.5 mm, ovoid, superficial, leaving rounded scars 1.25 mm long; **calyx** 0.5 × 1.5 mm, scarcely lobed, membranous, **sepals** connate nearly to apex, straight apically; **petals** 1.75 × 2 mm, imbricate basally, acute or rounded apically; **staminodes** present; **pistil** 1.75–2 × 1.5 mm, ovoid, greenish, **styles** short or lacking, **stigma** lobes flat, recurved, darkened.

Fruits: 8–10 mm, globose, black.

Distribution: COSTA RICA. Heredia. Limon. Wet forest on the Atlantic slope; 700–1,100 m elevation.

Discussion: *C. chazdoniae* is known from only a

Plate 55. Chamaedorea correae. A–F In moist to wet cloud forest, Panama; **A.** *Hodel et al. 1113,* El Valle (note two small basal pinnae); **B, C** *Hodel et al. 1118,* Veraguas; **B.** Crown of leaves; **C.** Spicate pistillate inflorescences; **D, E** *Hodel et al. 1113,* El Valle; **D.** Decumbent habit with inflorescence emerging from stem well below crown of leaves; **E.** Leaves; **F.** Plant of decumbent habit and inflorescences, *Hodel et al. 1118,* Veraguas; **G.** Young plant, cultivated, Lyon Arboretum, Honolulu, Hawaii.

B
C
F
G

few collections from the Atlantic slope of Costa Rica in Braulio Carrillo National Park on Volcan Barba and outside the park on Volcan Irazu. It is most closely related to *C. correae* from Panama but the gray-green, thick leaves with more nerves on each side of the rachis, gray petioles, one to three ascending and densely flowered staminate rachillae, and green to yellow staminate flowers distinguish *C. correae*. *C. chazdoniae* is similar to *C. dammeriana* but the larger, erect habit, more numerous leaves with more nerves, pendulous staminate rachillae, densely placed and globose staminate flowers with a prominent calyx and stamens equalling the pistillode, and ovoid fruits distinctly narrowed at both ends distinguish the latter species.

Chazdon (1987) referred to *C. chazdoniae* as *Chamaedorea* sp. "chiquita." The epithet honors Chazdon who not only collected the type but has made significant contributions to our knowledge of the palm flora of Braulio Carrillo National Park. *C. chazdoniae* is apparently not cultivated.

41. Chamaedorea correae *Hodel & Uhl,* Principes 34: 125. 1990c. Type: Panama, Cocle, *Knapp & Dressler 3801* (holotype MO!). Epithet: honors Mireya Correa, botanist and professor at the University of Panama. **PLATE 55.**

Common Name: none.
Vernacular Name: none.

Habit: solitary, decumbent with prostrate portion longer than erect portion, to 2–3 m long, briefly erect to 1 m tall.

Stem: 5–10 mm diam., creeping, rooting at nodes where touching ground, green, prominently ringed, internodes 5–10 cm long.

Leaves: 4–5, erect-spreading, bifid or sometimes pinnate; **sheath** to 15 cm long, tubular, obliquely open apically, light green, longitudinally striate-nerved; **petiole** to 10 cm long, flat and gray-green above, gray-green and rounded below; **rachis** 5–15 cm long, angled and gray-green above, rounded below and with a pale yellow or light green band extending onto sheath; rachis, petiole, and upper part of sheath densely but minutely white-spotted; **blade** 15–25 × 20–30 cm, incised apically to ¾ its length, lobes broadly divergent, 15–25 × 4–12 cm, lanceolate, slightly sigmoid, dull green or gray-green, ± thick, acuminate, 8–10 primary nerves above, exterior margin toothed toward apex, or infrequently blade pinnate with a pair of small basal pinnae, these 8–12 × 1.5–3 cm, lanceolate, sigmoid, acuminate, narrowed basally, 2–3 prominent nerves above.

Inflorescences: infrafoliar, emerging well below or behind leaves, erect-ascending; **peduncles** 10–15 cm long, erect and green to greenish yellow in flower, erect or spreading and red-orange in fruit; **bracts** 5–6, prophyll 5 mm long, 2nd bract 1 cm, 3rd 2–3 cm, 4th 4 cm, 5th 6–8 cm, 6th 8–10 cm, tubular, abruptly flared apically, longitudinally striate-nerved, acute-acuminate, bifid, uppermost not exceeding peduncle. **Staminate** with 2–3 **rachillae** or rarely spicate, to 15–20 cm long, 1.5 mm diam., erect, finely longitudinally striated.

Pistillate spicate or less often forked; **bracts** similar to those of staminate; **rachis** or flower-bearing portion to 15–20 cm long, 2 mm diam., finely longitudinally striated, erect and greenish yellow in flower, erect or spreading and red-orange in fruit.

Flowers: Staminate rather densely arranged, 2 × 2.5 mm in bud, subglobose, greenish yellow, just prior to anthesis 2.5 × 2.5 mm, yellowish, ± superficial; **calyx** 2.5–3 mm across, membranous, lobed, **sepals** connate in basal ½–¾; **petals** 2–2.5 × 2.5 mm, rounded-triangular, valvate, only briefly connate basally, spreading apically, acute, obscurely nerved; **stamens** with **filaments** very short, **anthers** 0.75–1.25 mm long, flush against base of pistillode; **pistillode** 1.5–2 mm high, columnar, broadly lobed apically, flared basally and there adnate to filaments, green or yellowish.

Pistillate rather densely arranged, 2 × 2 mm, ovoid-globose, greenish yellow, ± superficial; **calyx** 1–1.25 × 2.5 mm, deeply lobed, green, fleshy, **sepals** connate in basal ¼; **petals** 2–2.5 × 2 mm, long-triangular, imbricate basally, spreading apically, acute; **pistil** 2–2.5 × 2 mm, globose, pale or greenish, **styles** short, **stigma** lobes flattened, recurved, pointed.

Fruits: 5–8 mm long, ellipsoid-globose, black.

Distribution: PANAMA. Cocle. Colon. Panama. Veraguas. Wet forest mainly on the Atlantic slope at or near the Continental Divide; 800–1,000 m elevation.

Discussion: With creeping stems that root at the nodes and thick, somewhat leathery, gray-green leaves, *C. correae* is one of the most distinctive members of the genus. It inhabits wind-swept, elfin cloud forest near or at the Continental Divide where its long stems creep through the moss-laden litter of the forest floor or are supported on low shrubs. The inflorescences appear on the stem well below or in back of the leaves, often as much as a meter or so. The bright red pistillate inflorescences are quite showy and persist long after the fruits have fallen.

Leaves of *C. correae* from near El Valle, Cocle usually have a broad terminal pair of pinnae with a few, small basal pinnae while those from Veraguas are bifid. At El Cope between El Valle and Veraguas, leaves are intermediate and may be either bifid or pinnate with a large terminal pair and small basal pinnae. *C. correae* is close to *C. guntheriana* but the smaller leaves with more narrowly divergent lobes, shorter peduncles, and flowers attaining anthesis in a markedly progressive manner up the axis distin-

---

Plate 56. Chamaedorea costaricana. A. Large clump in ravine, *Hodel & Grayum 698,* montane rain forest, Tapanti, Costa Rica; **B.** Along stream, *D. & M. Hodel 693A,* cloud forest, Monteverde, Costa Rica.

guish *C. guntheriana*. *C. correae* may also be confused with *C. chazdoniae* but the thinner, green blades with fewer nerves, green petioles, and spreading staminate rachillae with cream-colored flowers distinguish the latter species. *C. correae* has only recently been introduced to a few collections in California, Florida, Hawaii, and Australia.

42. Chamaedorea costaricana *Oerst.*, Vidensk. Dansk. Meddel. Naturhist. Foren. Kjoebenhavn 1858: 19. 1859 ("*costaricanas*"). Type: Costa Rica, Cartago, Turrialba, *Oersted s.n.* (later numbered *6543*) (holotype C!). Epithet: a derivative of Costa Rica, where the species was discovered. **PLATES 6(D), 56, 57, 58(A, B).**

Nunnezharia costaricana (Oerst.) O. Kuntze, Revisio Generum Plantarum 2: 730. 1891.

Chamaedorea biolleyi Guillaum., Bull. Mus. Hist. Nat. (Paris) 28: 543. 1922. Type: Cult., *Biolley s.n.* (*17376*) [holotype P!; isotypes GH!, US (photo!)]. Epithet: honors the collector of the type, P. Biolley.

Chamaedorea linearia L. H. Bailey, Gentes Herb. 6: 249, fig. 131. 1943a. Type: Panama, Chiriqui, El Boquete, *Pittier 2922* (holotype US!). Epithet: from the Latin *linearis* meaning linear, apparently in reference to the linear-lanceolate pinnae.

Chamaedorea seibertii L. H. Bailey, Gentes Herb. 6: 238. 1943a. Type: Panama, Chiriqui, *Seibert 218* (holotype MO!; isotype BH!). Epithet: honors the collector of the type, R. J. Seibert.

Omanthe costaricana O. F. Cook, Science 90: 298–299. 1939. Name of no botanical standing.

Common Name: none.
Vernacular Name: pacaya—Costa Rica.

Habit: cespitose by means of short horizontal stems at or near ground level, mostly erect but at times eventually leaning and becoming decumbent, forming rather dense or open clumps 3–8 × 3–6 m.

Stems: 2–6 cm diam., smooth, green, prominently ringed, internodes 5–30 cm long.

Leaves: 4–6, erect-spreading, pinnate, 1–2 m long; **sheath** 20–60 cm long or more, tubular, green, longitudinally striate-nerved, apex with a ligule on each side of petiole, these initially lanceolate and membranous and later deciduous or rotting away to leave a triangular base 1–2 cm long; **petiole** to 35 cm long, grooved near base and green above, rounded and pale below; **rachis** to 125 cm long, angled and green above, rounded below and with a pale band extending onto sheath; **pinnae** 20–26 on each side of rachis, 25–40 × 2.5–5 cm, linear-lanceolate to lanceolate, slightly sigmoid or falcate or nearly straight, alternate or subopposite, long-acuminate, contracted basally and there with a ± raised light green bump above, a whitish midrib prominent above and below with 2 prominent primary nerves on each side of this, 2 secondaries between each of these.

Inflorescences: infrafoliar, erect-spreading; **peduncles** 20–45 cm long, 1–6 cm wide at base, 0.5–1 cm wide at apex, green and erect in flower, reddish orange and nodding in fruit; **bracts** 5–8, tubular, longitudinally striate-nerved, papery, brownish in flower, becoming shredded with age and ± deciduous, uppermost exceeding peduncle; **rachises** 10–20 cm long, green in flower, reddish orange in fruit. **Staminate** with 15–30 **rachillae**, these 20–30 cm long, 2 mm diam., flexuous and ± pendulous or drooping, green to yellow in flower.

Pistillate with 10–20 **rachillae**, these 20–35 cm long, 3 mm diam., very slightly longitudinally ridged, green and spreading in flower, reddish orange and drooping when heavily laden with fruits.

Flowers: Staminate in moderate or ± dense spirals, 2.5–3.5 × 2.5–3 mm, oblong to depressed-globose, greenish yellow at anthesis ageing to brownish orange, slightly sunken; **calyx** 1 × 2–2.5 mm, deeply lobed, yellowish ageing to brownish orange, **sepals** free nearly to base and only briefly connate there, broadly rounded apically; **petals** 3 × 2–3 mm, ovate to deltoid, valvate, free nearly to base, spreading apically, ± fleshy, acute and slightly recurved apically; **stamens** 1–1.5 mm high, **filaments** short, **anthers** 1 mm long, brownish; **pistillode** to 2 mm high, columnar, slightly tapered apically, pale yellow.

Pistillate in ± lax spirals, 3–3.5 × 2–2.5 mm, ovoid-globose, yellow, slightly sunken; **calyx** 1 × 2 mm, deeply lobed, greenish yellow, brown-margined, **sepals** connate only briefly basally, broadly rounded apically; **petals** 3 × 2.5–3 mm wide, triangular, imbricate nearly to apex, free, erect, acute apically; **pistil** to 3 × 2 mm, depressed-obovoid, green apically, **styles** short or lacking, **stigma** lobes short, recurved, thickened, dark.

Fruits: 7–10 mm diam., globose to subglobose, green maturing black or purplish black; seeds 5–8 mm diam., globose, brownish, endocarp with distinctive s-shaped grooves.

Distribution: COSTA RICA. Alajuela. Cartago. Guanacaste. Heredia. Puntarenas. San Jose. PANAMA. Chiriqui. Cocle. Herrera. Los Santos. Veraguas. HONDURAS. Santa Rosa. NICARAGUA. Chontales. Granada. Jinotega. Madruz. Managua. Matagalpa. COLOMBIA? Moist or wet forest, mainly on the Pacific slope up to or near the Continental Divide but occasionally on the Atlantic slope; 600–2,300 m elevation.

Discussion: Oersted (1859) described and named

→

Plate 57. Chamaedorea costaricana. A. Infructescence (immature fruits), cultivated, Ho'omuluhia Botanic Garden (*78.256*), Honolulu, Hawaii; **B.** Staminate plant, *D. & M. Hodel 693A*, cloud forest, Monteverde, Costa Rica; **C.** Ligule at apex of leaf sheath, a diagnostic feature of *C. costaricana* (compare with that of *C. quezalteca*), *Hodel 762,* cultivated, research collection, Los Angeles; **D.** Handsome plant cultivated in garden of J. Wright, San Diego, California.

B
D

C. costaricana from material he collected in mountain forests about 1,000–1,500 m elevation near Turrialba, Costa Rica during his travels in Central America in the middle of the 19th century. Today, the native forest in this area is virtually gone, having been destroyed long ago for agricultural purposes. A few stands of *C. costaricana* still exist, though, not far from Turrialba in forest remnants on the slope of Volcan Barva; we collected it in 1990 near Turrialba on the road to Moravia de Chirripo.

Chamaedorea costaricana is a most variable species throughout its rather wide range. Different species have been proposed based principally on size and number of parts and prominence of nerves of the pinnae. The description attempts to account for the extremes in size of stems, leaves, pinnae, and inflorescences. Guillaumin (1922) described and named *C. biolleyi* from material cultivated at the Musee de Paris since 1900 from seeds that Biolley had sent from Costa Rica. Later, Bailey (1943a) described and named several species of *Chamaedorea* from Panama that were collected by various members of the Missouri Botanical Garden in support of the *Flora of Panama.* Two of these, *C. linearia* and *C. seibertii,* were of interest since their descriptions depicted plants close to *C. costaricana* although the names have not been applied to material in cultivation.

In 1987 and 1990, I had the opportunity to visit various localities in Costa Rica and Panama and to observe *C. costaricana* in the wild and in cultivation. In Panama I visited the type localities of *C. linearia* and *C. seibertii.* In each location I made collections that corresponded well with Bailey's descriptions of these species. Comparison of these with Bailey's types, Oersted's type of *C. costaricana,* Guillaumin's type of *C. biolleyi,* and with various other collections from Panama to Honduras show that, essentially, the differences are of size only.

Chamaedorea costaricana is widespread and well represented by collections throughout its range. Galeano and Bernal (1987) reported *C. linearia* from Colombia although they stated it is single-stemmed and did not mention the ligules at the apex of the leaf sheath. Moore (1975) reported it from Monteverde in northwestern Costa Rica as *Chamaedorea* sp. # 2. I have also collected *C. costaricana* in forest remnants near there and at several other localities in Costa Rica and Panama.

Generally, there seems to be a relationship between altitude and rainfall and size of plants. It seems that smaller forms occur in lower altitudes with lower and perhaps more seasonal rainfall while larger forms occur at higher altitudes where rainfall is greater and less seasonal and evaporation less. Above Cerro Punta in western Panama we observed several individuals of a very large form of *C. costaricana.* Its stems were quite stout, six cm in diameter, but were so long, over 15 m, that they could not support themselves and were snaking up into the branches of taller trees. Bailey's *C. linearia* typifies the large end of the spectrum with plants being larger in every part. At the small end is Bailey's *C. seibertii.* The typical *C. costaricana* that Oersted described represents the middle of the spectrum. In his key to *Chamaedorea,* Bailey (1943b) distinguished between *C. seibertii* and *C. linearia* by the former having straight pinnae that were not sigmoid or falcate. The straightness or curvature of pinnae appears to be a variable character. The presence of the two ligules at the apex of the leaf sheath is a diagnostic character unifying all these forms.

The ligules are lanceolate and membranous and more or less deciduous, at least in their upper portion. The lower or basal portion is not as membranous and is often persistent as a triangular tooth or auricle one to two cm long after the upper portion has fallen or rotted away. Since the upper portion is membranous and rots away rather rapidly, the overall visibility of this appendage is much reduced with time. Therefore, it is best viewed on leaf sheaths on the apical portions of stems near new, emerging leaves.

The ligules or the triangular bases that are a remnant of them are visible on Bailey's type specimens of *C. seibertii* and *C. linearia.* The presence of this appendage is unsurpassed as a diagnostic character, as it is apparent even on young seedlings. Cook (1939, 1943a) gave admirable accounts of *C. costaricana* that he referred to as *Omanthe costaricana.* The latter name has no botanical standing since Cook did not provide a Latin description. In the same accounts, Cook mentioned the ligules at the apex of the leaf sheath but he referred to them as antiligules.

→

Plate 58. A, B, Chamaedorea costaricana. C, D Chamaedorea dammeriana. A. Large clump fills ravine, cultivated, San Diego Zoo, California; **B.** Large stems of plant on humus-covered rock, *D. & M. Hodel 736A,* Cerro Punta, Chiriqui, Panama. **C, D** In lowland rain forest, *Hodel 954B,* Talamanca, Limon, Costa Rica; **C.** Infructescences with mature fruits; **D.** Fruiting plant (note very leafy crown).

D

Hertrich (1951) referred to *C. costaricana* under Cook's invalid name.

Chamaedorea costaricana has been cultivated for ornament in San Jose, Costa Rica at least since the late 1800s and probably earlier, the forested slopes of Volcan Barva and the Meseta Central serving as a ready and accessible source of plants and seeds. Today, handsome clumps are found throughout the city. It is also commonly found in towns and villages to the north of San Jose along National Route 9 that winds past Volcan Barva to the Atlantic lowlands. Here, such as at the towns of Heredia and Barva, it is seen as a large hedge, screen, or as single specimens in many of the yards and residences. From Costa Rica, it has found its way throughout the horticultural world and is one of the more widespread members of the genus in cultivation, being found in gardens and collections in California, Florida, Hawaii, Australia, and elsewhere.

Surprisingly, much of the material in California cultivated under the name *C. costaricana* is actually *C. pochutlensis.* A hybrid exists in California and perhaps elsewhere between *C. graminifolia* and what was reported as *C. costaricana.* However, it is not *C. costaricana* but *C. pochutlensis* that was utilized as one of the parents with *C. graminifolia. C. pochutlensis* is very similar to *C. costaricana* in habit but the absence of the two ligules at the apex of the leaf sheath distinguishes the former. Also, material in California and Hawaii grown as *C. woodsoniana* is actually *C. costaricana. C. woodsoniana* differs in its stout, solitary habit with robust stems 5–10 cm in diameter, lack of persistent ligules at the apex of the sheath, heavily nerved, straight pinnae, and larger, ellipsoid fruits. Krempin (1990, p. 97) illustrated *C. costaricana* but erroneously captioned the photograph as *C. woodsoniana.* Krempin (p. 90) also discussed and illustrated *C. costaricana* but the description and photograph seem to depict *C. microspadix.*

Chamaedorea costaricana is closest to *C. quezalteca* and the latter species is perhaps not distinct. Rare in cultivation outside its native range, the fibrous, persistent ligules, five to ten cm long, distinguish *C. quezalteca.*

Seeds of *C. costaricana* are handled on a commercial basis and distributed worldwide to nurseries, greenhouses, and other growers but to a much lesser extent than those of *C. elegans* or even *C. seifrizii.* The majority of seeds originates from cultivated plants in Costa Rica. *C. costaricana* is used mainly as an outdoor landscape subject in subtropical regions like Florida and southern California. It forms, with age, large handsome clumps that are effective backgrouind plants or screens for shady or partially sunny locations. Also, well grown individuals make attractive specimens for pots or tubs where their rich green, smooth, ringed, bamboolike stems and long-pinnate leaves can be appreciated fully.

Although it is grown to some extent for indoor landscape use, *C. costaricana* is susceptible to infestations of mites under these conditions. Better choices for indoors would be *C. pochutlensis* or *C. seifrizii,* both more resistant to mites.

43. Chamaedorea dammeriana *Burret,* Notizbl. Bot. Gart. Berlin-Dahlem 11: 737. 1933a. Type: Costa Rica, San Carlos, *Koschny s.n.* [holotype B, destroyed except for a tray of fruits (photo!); isotype GOET!). Epithet: honors German botanist Udo Dammer. **PLATE 58(C, D).**

Chamaedorea variabilis H. A. Wendl. ex Burret, Notizbl. Bot. Gart. Berlin-Dahlem 11: 726. 1933a. Name only.

Chamaedorea wedeliana L. H. Bailey, Gentes Herb. 6: 247, fig. 131. 1943a. Type: Panama, Bocas del Toro, *von Wedel 719* (holotype MO!; isotype BH!). Epithet: honors H. von Wedel, collector of the type.

Common Name: none.

Vernacular Names: sirik, palma bonita—Costa Rica.

Habit: solitary, erect, slender, 1–2 m tall.

Stem: 7–8 mm diam., green, prominently ringed, internodes to 10 cm long.

Leaves: to 15, spreading, not close but well spaced, pinnate, variously pinnate, or rarely bifid; **sheath** to 15 cm long, tubular, densely longitudinally striate-nerved, obliquely truncated apically; **petiole** 10–20 cm long, flat and/or grooved and green above, rounded and pale below; **rachis** to 30 cm long, angled and green above, rounded below with a white band extending onto petiole; **blade** 30 cm long or more, obovate; **pinnae** to 7 on each side of rachis, to 20 × 4.5 cm, oblong-lanceolate, slightly sigmoid, alternate, contracted basally, a midrib and 2–3 prominent primary nerves on each side of this, or variously pinnate with smaller basal pinnae and a broad terminal pair to 10–15 cm wide; or rarely bifid and then incised apically ½–⅔ its length, lobes oblong, upper margin obscurely toothed, 7–nerved.

→

Plate 59. Chamaedorea deneversiana. Jim Benzie holds stem of large specimen, *Hodel et al. 1114,* montane rain forest, Veraguas, Panama.

Inflorescences: interfoliar, to 45 cm long; **peduncles** to 30 cm long; **bracts** 5, prophyll to 2 cm long, others 5–10 cm long, tubular. **Staminate** with **rachis** to 4 cm long, green; **rachillae** 2–6, to 25 cm long, slender, pendulous, green in bud, slightly angled longitudinally.

Pistillate forked or with 3 rachillae, infrequently spicate; **rachillae** or flower-bearing portion to 20 cm long, straight, orange in fruit.

Flowers: Staminate in dense spirals, 0.5 mm apart, 2 × 2.5 mm, depressed-globose, leaving elliptic scars 2.5–3 mm long; **calyx** 0.5 × 2.25 mm, prominent, deeply lobed, membranous, **sepals** connate in basal ¼, rounded apically; **petals** 1.75 × 1.75 mm, ovate, valvate, free nearly to base, sharply acute, nerveless; **stamens** 1 mm high, equalling pistillode, **filaments** short, flared basally, **anthers** oval; **pistillode** 1–1.25 mm high, broadly columnar.

Pistillate in loose spirals; **calyx** lobed; **petals** imbricate, rotund.

Fruits: to 1.5 cm long, ± ovoid, narrowed at both ends, black.

Distribution: COSTA RICA. Alajuela. Guanacaste. Heredia. Limon. Puntarenas. PANAMA. Bocas del Toro. Colon. Panama. Wet or moist forest on the Atlantic and Pacific slopes; 200–700 m elevation.

Discussion: The description is from Burret (1933a) who described and named *C. dammeriana* from pistillate material that Koschny collected near San Carlos in Alajuela and staminate material that Wendland collected near the Rio Sarapiqui in Heredia. Bailey (1943a) and my own collections in Limon near the Panamanian border supplement the description. *C. dammeriana* is one of the most wide ranging and variable species in Costa Rica, occurring on both the Atlantic and Pacific slopes from one end of the country to the other. Strangely, it has been collected only infrequently in Panama and then only in the Atlantic lowlands although it occurs in relative abundance in nearby, adjacent regions of Costa Rica.

Chamaedorea dammeriana may be confused with *C. chazdoniae* but the much smaller, decumbent habit, much less numerous leaves with fewer nerves, spreading staminate rachillae with ovoid flowers lacking a prominent calyx and the stamens shorter than the pistillode, and globose fruits distinguish the latter species.

Chamaedorea dammeriana is a very handsome plant with a leafy crown holding 12–15 or more leaves. Attractive, bright orange, few-branched or spicate pistillate infructescences with shiny, black fruits emerge from behind leaf sheaths. According to Burret (1933a), Koschny's label stated that the species was known as *palma bonita*; however, this may merely be a description of the plant and not a common name. Indigenous peoples in the Talamanca region of Limon Province in Costa Rica refer to *C. dammeriana* as *sirik* and maintain that it has magical powers to heal various ailments. Medicine men place a leaf of the species near to and pointed at the ill person. A transfer of "healing power" occurs when the medicine man blows onto the leaf and toward the afflicted individual.

Pistillate inflorescences are somewhat variable; although usually forked or with three rachillae, they may be spicate when a plant first flowers. I observed a population in the Talamanca region of Limon that had inflorescences that were spicate, forked, or had three rachillae all on the same plant. Staminate specimens from Panama may have as many as six rachillae while those from Costa Rica usually have two to three rachillae. Immature fruits are narrowly elongate but become more rounded as they mature.

Leaves are similarly variable; usually blades have a very broad pair of terminal pinnae and a few smaller basal ones. At other times the blade may have up to seven pinnae, all about the same size, on each side of the rachis. Still less frequently, blades may be bifid although this usually occurs with young plants flowering for the first times; later, the blades will usually divide. Michael Grayum made a series of collections from the same population in Guanacaste that encompassed leaf blades ranging from bifid to completely pinnate with seven pinnae on each side of the rachis. Pistillate inflorescences of this population also varied from spicate to having three rachillae.

44. Chamaedorea deneversiana *Grayum & Hodel,* Principes 35: 133. 1991. Type: Panama, San Blas, *de Nevers et al. 5553A* (holotype MO!; isotype CAS!). Epithet: honors Gregory C. de Nevers, collector of the type and most of the known specimens of this species. **PLATES 59, 60.**

Common Name: none.
Vernacular Name: none.

Habit: solitary, decumbent to erect, 1.5–3 m tall.

Stem: 1.3–1.6 cm diam., smooth, green, ringed, internodes 5–12 cm long, rooting along its length.

→

Plate 60. Chamaedorea deneversiana. In montane rain forest, *Hodel et al. 1114,* Veraguas, Panama. **A, B.** Young plants; **C.** Leaf; **D.** Infructescence (immature fruits) emerges from well below the leaves (note decumbent habit).

C
D

Leaves: 4, spreading, bifid, rarely pinnate; **sheath** 19–26 cm long, obliquely open apically, green; **petiole** 15–45 cm long, ± flat and green above, rounded, rough, and gray-green below; **rachis** 35–60 cm long, angled and gray-green above, rounded, rough, and gray-green below, yellowish below when dry; **blade** 50–80 × 12–25 cm, oblong, incised apically ⅙–¼ its length, bright glossy green, margins coarsely toothed toward apex, rarely with few small basal pinnae, 20–24 primary nerves on each side of rachis, these sharply angled above, slender, sigmoid, prominent below, ± erose-margined.

Inflorescences: infrafoliar, borne well below the leaves. **Staminate** with **peduncle** 40 cm long, erect; **bracts** 6; **rachis** 5 cm long, slightly flexuous; **rachillae** 10, these 25–33 cm long, drooping.

Pistillate with **peduncle** 18–60 cm long, erect; **bracts** 5–6; **rachis** 2–3 cm long, slightly flexuous, orange in fruit; **rachillae** 3–6, these 10–30 cm long, drooping and yellow to orange in fruit.

Flowers: Staminate 1.1–1.3 × 1.8–1.9 mm; **calyx** 0.3–0.4 mm high, lobed, **sepals** connate in basal ½; **petals** valvate in bud, apparently free at anthesis, nerveless or obscurely nerved; **stamens** with **anthers** 0.6 mm long, ± reniform; **pistillode** 0.4 mm high, stout, anvil-shaped.

Pistillate 2.2–2.5 mm long, suburceolate; **calyx** 0.5–0.6 × 2.5–2.8 mm, lobed, nerved, **sepals** connate and/or imbricate basally; **petals** obscurely nerved externally, prominently 15-nerved within, constricted toward base where connate in a tube 0.2–0.3 mm high; **staminodes** lacking; **pistil** 1.5–1.9 mm long, **styles** short, **stigma** lobes separated.

Fruits: 6.8–7.5 × 5.6–5.8 mm, broadly ellipsoid, green to yellow (presumably immature).

Distribution: PANAMA. San Blas. Vergaguas. ECUADOR. Carchi. Cotopaxi. Pichincha. COSTA RICA? Wet forest on the Atlantic slope; to 850 m elevation.

Discussion: *C. deneversiana* is known from only a few collections from wet forests in Panama and Ecuador. It bears a resemblance to *C. amabilis,* a species occurring in similar habitats from Costa Rica to the Choco region of Colombia. In fact, the two species are very similar vegetatively, the former appearing as a giant-leaved form of the latter. However, in addition to being a smaller palm with smaller leaves and in a different subgenus, *C. amabilis* has spicate or infrequently furcate pistillate inflorescences and erect staminate rachillae. *C. deneversiana* may also appear similar to bifid-leaved forms of *C. murriensis* from Colombia. Leaf blades of the two are about the same size but the latter has a greater number of rachillae (staminate: 14 vs. 10; pistillate 11–14 vs. 4–6) and is in a different subgenus.

The limited distribution of *C. deneversiana* may be more apparent than real. A collection from western Costa Rica, *Grayum et al. 5000,* matches up very well with *C. deneversiana* and is tentatively referred here, differing only slightly in the number of rachillae and by its pinnate leaves. However, the leaves are nearly bifid, having large and dominant terminal lobes and only a pair of small, narrow basal pinnae.

45. Chamaedorea foveata *Hodel,* Phytologia 68: 403. 1990c. Type: Mexico, Oaxaca, *D. & R. Hodel 939A* (holotype BH!; isotype MEXU!). Epithet: from the Latin *fovea* meaning pit, in reference to the petioles being densely covered with minute pits. PLATE 61.

Common Name: none.
Vernacular Name: none.

Habit: solitary, erect, to 2.5 m tall, sometimes flowering when appearing stemless.

Stem: 2.5–3 cm diam., green, ringed, internodes 2–4 cm long, basal portion subterranean, curving.

Leaves: 3–5, spreading, pinnate, to 1.25 m long; **sheath** to 30 cm long, 2.5–4 cm diam., splitting deeply opposite petiole to ½ its length, tubular only in lower ½, dark green, minutely but conspicuously white-spotted, thick, durable, longitudinally striate-nerved when dry; **petiole** 20–40 cm long, robust, 1 cm diam., densely covered with minute irregular pits giving living material a rough texture and drying to narrow nearly elliptic contiguous fissures, very slightly flattened and only faintly grooved near base and green above, rounded and paler below; **rachis** to 60 cm long, angled and green above, rounded below with a yellow-green band extending onto sheath; **pinnae** 12–16 on each side of rachis, middle ones longest, these to 35 × 4–5 cm, terminal pair of pinnae 20 × 3.5–4.5 cm, basal pair 27 × 3.5–4.5 cm, lanceolate, nearly straight, only slightly sigmoid, falcate, acuminate, 5 prominent primary nerves above, 1 secondary between each pair of primaries in living material, secondaries inconspicuous above when dry, tertiaries numerous, all nerves drying distinctly pale yellow below.

Inflorescences: interfoliar but occasionally infrafoliar in fruit, erect, long-pedunculate; **peduncles** to 85 cm long, 1.5 cm wide at base and there flattened, 5–7 mm wide at apex, light green and remotely white-spotted where exposed in flower, orange in fruit; **bracts** 6, prophyll 8 cm long, 2nd bract 20 cm, 3rd 30 cm, 4th 38 cm, 5th 35 cm, 6th very short and rudimentary or to 15 cm long, equalled or exceeded by 5th one, tubular, acuminate, obliquely long-open, green to brown in flower, minutely white-spotted, longitudinally striate-nerved. **Staminate** with **rachis** 3 cm long, greenish yellow in flower; **rachillae** 8, to 25 cm long, spreading, greenish yellow in flower.

Pistillate with **rachis** 4–6 cm long, green or yellow-green in flower, orange in fruit; **rachillae** 3–7 or perhaps more, to 25 cm

Plate 61. Chamaedorea foveata. A. Pistillate plant, *D. & R. Hodel 939B,* pine-oak cloud forest, Oaxaca, Mexico; **B, C.** Pistillate inflorescence and plant, *Hodel 939B bis,* cultivated, research collection, Los Angeles.

C

long, 5 mm diam. at base, 1–1.5 mm diam. at apex, stiffly erect, greenish in flower, orange in fruit, slightly undulate when dry.

Flowers: Staminate in four moderate spirals, 1–3.5 mm apart, in bud 2 × 2.5 mm, dome-shaped, at anthesis 2.5 mm diam., globose, slightly sunken in elliptic depressions 3 mm long; **calyx** 1 × 3 mm, not adnate to sides of pit, lobed, thin, membranous, brown-margined, **sepals** connate in basal ½, rounded apically; **petals** 2 × 2 mm, valvate, ovate, free to base, spreading slightly apically, acute, fleshy but thin and membranous when dry, not nerved or only obscurely so, margins thickened; **stamens** 1–1.25 mm high, **filaments** 0.5–1 mm long, wider basally, slender, terete, awl-shaped apically, pale, **anthers** 1.5–2 × 0.5 mm, dorsifixed near middle; **pistillode** 1.5–2 × 0.5 mm, shorter than petals, columnar, enlarged in middle, definitely barrel-shaped, pale, tip 3-lobed, flat.

Pistillate in four loose spirals, 2–4 mm apart, 3.5 × 3 mm diam., globose-ovoid, yellow, superficial or only slightly sunken in elliptic depressions 2.5–3 mm long; **calyx** 0.5 × 2.5–3 mm, lobed, pale, membranous, **sepals** connate in basal ½, rounded apically; **petals** 2 × 2.5 mm, broadly ovate, briefly imbricate basally, broadly acute apically, fleshy, thicker distally, faintly nerved when dry; 1–3 **staminodes** present; **pistil** 2.5 × 2.5 mm, globose-ovoid, whitish, exceeding the petals, **styles** short or lacking, **stigma** lobes erect or slightly recurved, blunt, fleshy, whitish.

Fruits: ± globose, black.

Distribution: MEXICO. Oaxaca. Chiapas. Moist pine-oak forest on the Pacific and Atlantic slopes; 1,400–1,900 m elevation.

Discussion: *C. foveata* is unusual in its petioles that are densely covered with minute but conspicuous, irregularly shaped depressions giving living material a rough texture. It shares this feature with *C. woodsoniana* from the Atlantic slope. The larger habit, leaves, and inflorescences, much more numerous pinnae with prominently elevated nerves below, and much more numerous rachillae (the staminate with up to 100) distinguish *C. woodsoniana*. *C. foveata* appears closest to *C. vulgata* from the Pacific slope of Guatemala. The densely white-spotted but smooth petioles, larger but fewer pinnae with conspicuous secondary nerves, and the more or less superficial staminate flowers with the more prominent calyx distinguish *C. vulgata*. Other than a few plants in the research collection in Los Angeles, *C. foveata* is not cultivated.

46. Chamaedorea fractiflexa *Hodel & Castillo,* Principes 35: 6. 1991. Type: Guatemala, Quetzaltenango, *Hodel & Castillo 912A* (holotype BH!; isotype AGUAT!). Epithet: from the Latin *fractiflexus* meaning zigzag, in reference to the staminate rachis. **PLATE 62.**

Common Name: none.
Vernacular Name: none.

Habit: solitary, erect to decumbent, to 2 m tall.

Stem: 5–8 mm diam., smooth, green, internodes 3–7 cm long, often covered with persistent leaf sheaths.

Leaves: 3–8, spreading, pinnate; **sheath** to 12 cm long, tubular, obliquely open apically, green, longitudinally striate-nerved; **petiole** to 5 cm long, flat and green above, rounded and pale below; **rachis** to 20 cm long, angled and green above, rounded below with a pale band extending onto sheath; **pinnae** 5–8 on each side of rachis, lanceolate to broadly lanceolate, to 13 × 3.5 cm, regularly placed, opposite, sigmoid, long-acuminate apically, contracted basally, a prominent midrib and 2 primary nerves on each side of this, a secondary between each pair of primaries, tertiaries numerous, faint, end pinnae sometimes broader, to 6 cm wide, 4–5-nerved.

Inflorescences: interfoliar, slender, spreading in flower, pendulous in fruit; **peduncles** to 30 cm long, very slender, 2 mm wide at base and ± flattened, 0.5–1.5 mm diam. at apex and there ± filiform, green in flower, pendulous and orange apically in fruit; **bracts** 5–6, prophyll 2 cm long, 2nd bract 4.5 cm, 3rd and 4th 9 cm, 5th 7 cm, often a rudimentary 6th concealed by 5th, ± loosely sheathing, brown in flower and fruit, acute to acuminate, thin-papery, finely longitudinally striate-nerved. **Staminate** with **rachis** to 3 cm long, strongly flexuous, green; **rachillae** 5, these to 7 cm long, each attached at an "elbow" of rachis, green and slightly drooping in flower.

Pistillate with **rachis** to 1 cm long, green in flower, orange in fruit; **rachillae** 3, these to 7 cm long, green in flower, downward-pointing and orange in fruit.

Flowers: Staminate ± densely arranged, 1–2 mm apart, in bud 2–2.5 × 1–1.5 mm, bullet-shaped, at anthesis 2.5 × 2.5 mm, obovate; **calyx** to 0.75 × 1.5–2 mm, shallowly lobed, **sepals** connate and/or slightly imbricate nearly to apex, broadly rounded to straight apically, 0.5 mm wide; **petals** 3 × 1 mm, long-ovate, valvate nearly to base, erect apically, acute, lightly nerved on inside; **stamens** 1.5–1.75 mm high, **filaments** 0.5–1 mm long, darkened, **anthers** 1 mm long, oblong, bilobed, spreading apically and basally, dorsifixed, yellowish; **pistillode** 1.5–2.5 mm high, columnar, slightly lobed and widened apically, finely longitudinally striated.

Pistillate: loosely arranged, 3–4 mm apart, 1 × 1.5–1.75 mm, ± globose; **calyx** 0.75 × 2 mm, deeply lobed, obscurely nerved, **sepals** connate in basal ¼, acute apically; **petals** 1.5 × 1.5–2 mm, lightly imbricate basally, erect apically, acute, obscurely nerved; **pistil** 1.5 × 1 mm, ovoid.

Fruits: 8 × 7 mm, ± globose, black.

Distribution: GUATEMALA. Quetzaltenango. San Marcos. MEXICO. Chiapas. Wet forest on the Pacific slope; 2,000–2,900 m elevation.

Discussion: *C. fractiflexa* is rare, known only from a few collections in Guatemala and one in Mexico. It is closest to *C. parvisecta* but the more or less straight staminate rachis, thicker, ascending pistil-

Plate 62. Chamaedorea fractiflexa. In cloud forest, Quetzaltenango, Guatemala. **A.** Fruiting plant of decumbent habit, *Hodel & Castillo 905*; **B.** Infructescences (immature fruits, note filamentlike peduncles), *Hodel & Castillo 912B*; **C.** Staminate plant, *Hodel & Castillo 912A* (holotype).

→

C

late peduncle, densely flowered rachillae, and ascending fruit-bearing rachillae distinguish the latter species. *C. fractiflexa* is not cultivated except for a few plants in the research collection in Los Angeles.

47. Chamaedorea fragrans (*Ruiz & Pav.*) *Mart.*, Historia Naturalis Palmarum 2: 4, t. 3, figs. 1–2. 1823. Epithet: is the Latin word for fragrant, in reference to the sweet-smelling staminate flowers. **PLATES 8(F), 63.**

Nunnezharia fragrans Ruiz & Pav., Systema Vegetabilium Florae Peruvianae et Chilensis 294. 1798. Type: Peru, *Ruiz & Pavon s.n.* (*B18*) [holotype MA (photo!); isotypes F!, FI (photo!), G (photo!), M!].

Chamaedorea pavoniana H. A. Wendl. ex Dammer, Gard. Chron., Ser. 3, 36: 246. 1904b. Type: Cult., *Wendland s.n.* (holotype K?). Epithet: honors Spanish botanist of the late 1700s and early 1800s, Jose Antonio Pavon.

Chamaedorea ruizii H. A. Wendl. ex Dammer, Gard. Chron., Ser. 3, 36: 246. 1904b. Type: Cult., *Wendland s.n.* [holotype K?; isotype P (photo!)]. Epithet: honors Spanish botanist of the late 1700s and early 1800s, Hipolito Ruiz Lopez.

Chamaedorea cataractarum Hort. in Guillaum., J. Soc. Natl. Hort. France, Ser. 4, 24: 235. 1923b. Not Mart.

Chamaedorea verschaffeltii Hort. in Kerch., Les Palmiers 240. 1878 ("*verschaffelti*").

Chamaedorea gratissima Hort. in Linden, Catalogue 13. 1896.

Common Name: none.

Vernacular Names: sangapilla, siasia, chutasllium—Peru.

Habit: cespitose, erect, sometimes decumbent, forming clumps 2 × 3–6 m.

Stems: 0.5–1.5 cm diam., green, ringed, internodes 5–15 cm long.

Leaves: 4–7, spreading, bifid; **sheath** 10–15 cm long, tubular, mottled green, drying brown, persistent, longitudinally striate-nerved; **petiole** 2–5 cm long, slightly furrowed and green above, rounded and light green below; **rachis** to 20 cm long, angled and green above, rounded and with a faint light green band below extending onto petiole; **blade** 40–50 × 15–20 cm, narrowed to 6–10 cm wide at apex, incised apically to ¾ its length, cleft narrowly v-shaped, lobes 35–45 × 6–8 cm, linear-rhombic, ± straight, only slightly falcate, acuminate, green, glabrous, ± stiff, papery, rough or raspy texture, rich dark green, outer margin lightly toothed apically, 12 primary nerves on each side of rachis, one secondary between each pair of primaries, tertiaries numerous, very faint.

Inflorescences: infrafoliar, often bursting through old, persistent leaf sheaths, erect in flower, nodding in fruit; **peduncles** 5–10 cm long, 3–4 mm wide at base, 2–3 mm wide at apex, greenish in flower, orange in fruit; **bracts** 3–5, prophyll 8 mm long, 2nd bract 1 cm, 3rd 1.5 cm, 4th 2.5–3 cm, 5th 4 cm and less than to exceeding peduncle, tubular, papery-membranous, compressed, short-acuminate, bifid, brown in flower, longitudinally striate-nerved; if branched, **rachises** 0.5–3 cm long, greenish in flower, orange in fruit. **Staminate** with 2–6 **rachillae**, or much less frequently spicate, these 10–15 cm long, slender, 2–2.5 mm diam., spreading, flexuous, glabrous, ± rounded but sometimes slightly furrowed, green.

Pistillate with **rachillae** similar to those of staminate only orange and nodding in fruit.

Flowers: Staminate in dense or ± remote spirals, 4 × 3–3.5 mm, ovoid, bright orange with fragrance of a delicate herbal soap, slightly sunken; **calyx** 0.75–1 × 2.5–3 mm, lobed, greenish orange, **sepals** connate basally, rounded and dark-margined apically; **petals** 4 × 2–2.5 mm, ovate, valvate, free and spreading apically, acute, thick, fleshy; **stamens** ½ as high as petals, **filaments** united briefly basally, whitish, membranous, fleshy, **anthers** inwardly curved and touching pistillode; **pistillode** 3.5–4 mm high, columnar, deeply 3-parted, whitish tinged with orange.

Pistillate in ± remote spirals, 4 × 2.5–3 mm, bullet-shaped in bud, oblong-obovoid at anthesis, orange, slightly sunken; **calyx** 1 × 3 mm, deeply lobed, ± thick, green, **sepals** connate only briefly basally, broadly rounded to acute and dark-margined apically; **petals** 4 × 2.5 mm, ovate, valvate, free nearly to base, spreading and erect or slightly recurved apically, thick, fleshy, acute; **staminodes** present or lacking, if present then conspicuous, nearly equalling pistil, ovate, orange; **pistil** 2 × 2–2.5 mm, depressed-globose, whitish, 3-parted, **stigma** lobes sessile, very short, brownish, acute-rounded.

Fruits: 12 mm long, globose-ellipsoid, black, lustrous; seeds black, raphe laxly reticulate.

Distribution: PERU. Cuzco. Huanuco. Junin. San Martin. Moist or wet forest on the Atlantic slope; 400–1,000 m elevation.

Discussion: Ruiz and Pavon (1798) described and named *C. fragrans,* one of the earliest names in the genus, initially as a *Nunnezharia.* Later, Martius (1823) made the transfer to *Chamaedorea.* Spruce (1871) gave a fine account of this species as he found it during his travels in South America. It was apparently widely distributed in Amazonian Peru along the eastern slopes of the Andes especially between the Huallaga and Ucayali Rivers. It formed rather large colonies under the canopy of tall trees. The orange-colored, extremely fragrant flowers were said

---

→

Plate 63. Chamaedorea fragrans. Cultivated, California. **A.** Clump, *Hodel 818,* garden of L. Hooper, La Habra; **B.** Clump in Virginia Robinson Botanical Garden, Beverly Hills; **C, D** In garden of L. Hooper, La Habra; **C.** Immature fruits; **D.** Staminate inflorescence, *Hodel 818*; **E.** Pistillate flowers, *Hodel 822,* garden of R. Palmer, Whittier.

B
D
E

to perfume the forest for quite a distance. Peruvian girls, who called it *sangapilla,* would place the flowers in their hair, put them under their pillows, or use them to decorate small crosses that marked paths in the forest. Although much of the forest in this area has been destroyed, colonies of *C. fragrans* have been observed in their native habitat as recently as the late 1980s.

Chamaedorea fragrans appeared in cultivation in Europe on several occasions in the 18th and early 19th centuries. Linden introduced it there from Peru erroneously under the name *C. cataractarum* and Verschaffelt had it in his nursery catalog as *C. verschaffelti.* This latter name was applied to material cultivated at the Berlin Botanical Garden about 1888. Dammer (1904b), based on Wendland's notes, treated some of these introductions as separate species and described and named *C. pavoniana* and *C. ruizii.* Guillaumin (1923b) reported material cultivated under this latter name at the Musee de Paris, France in 1903. Burret (1933a) seemed to concur that these latter two species were somewhat distinct although very close to *C. fragrans.* The main differences seemed to be in the size and nervation of leaves, characters that appear to be variable. Unfortunately, I have not been able to locate the types of *C. pavoniana* and *C. ruizii.*

The staminate flowers of *C. fragrans* must rank among the most attractive in the genus. The snow-white stamens incurved to touch the pistillode are a perfect complement to the intense orange petals. Added to this is the sweet, heady fragrance, conjuring up images of a delicate, herbal soap.

Chamaedorea fragrans is fairly widespread in cultivation, appearing in gardens and collections in Hawaii, California, Florida, and Australia. Although both staminate and pistillate plants appear to be in cultivation, fruit is rarely, if at all, produced. Although scarce, seeds germinate readily; however, it is possible to propagate *C. fragrans* vegetatively by dividing the clustered stems with care. When well grown, this species is very attractive and sure to draw admiring attention with its neat, clustered stems and stiff dark green, v-shaped leaves. It is very striking as a potted specimen and can be maintained in such a manner for many years. *C. fragrans* is susceptible to mites and tends to brown-tip rather badly when subjected to low humidity and/or prolonged cool spells.

48. Chamaedorea guntheriana *Hodel & Uhl,* Principes 34: 126. 1990c. Type: Panama, Panama, *D. & M. Hodel 746* (holotype BH!; isotype PMA!). Epithet: honors Bill Gunther of Del Mar, California who has encouraged and supported my work on *Chamaedorea.* **PLATES 64, 65, 66(A-C).**

Common Name: none.
Vernacular Name: none.

Habit: solitary, slender, decumbent to 3 m long then briefly erect to 1 m tall.

Stem: 5–7 mm diam., smooth, green, ringed, conspicuously and minutely white-spotted, internodes to 2–4 cm long.

Leaves: 4–5, spreading, pinnate or less often bifid; **sheath** to 9 cm long, tubular, obliquely open apically, longitudinally striate-nerved; **petiole** to 9 cm long, ± rounded or slightly flattened and gray-green above, rounded and gray-green below; **rachis** to 12 cm long, angled and greenish above, rounded and greenish below with a faint yellow band extending onto sheath; petiole above and below and rachis above gray-green and densely white-spotted; **pinnae** 2–4 on each side of rachis, apical pair largest (or if bifid), to 19 × 3.5 cm with 5 conspicuous primary nerves, basal ones to 12 × 2.5 cm, lanceolate, falcate, acuminate, narrowed basally, stiff, ± thick and leathery, 2–3 prominent nerves, all pinnae with numerous secondaries, these ± faint, pinnae drying heavily striated.

Inflorescences: infrafoliar, stiff and ascending, to 15–20 cm long. **Staminate** forked or with 3 rachillae, or infrequently spicate; **peduncle** to 5–6 cm long, 2.5–4 mm wide at base and there ± flattened, 1.5–3.5 mm diam. at apex and there rounded, ascending, greenish in flower; **bracts** 4, prophyll 5 mm long, 2nd bract 1 cm, 3rd 2–3 cm, 4th 4 cm, tubular basally, ± inflated apically, brownish at anthesis, leathery, acute-acuminate, bifid; **rachis** or flower-bearing portion if spicate or **rachillae** to 10–12 cm long, to 2 mm diam., green, ascending.

Pistillate spicate; **peduncle** similar to that of staminate but brownish or dull orange in fruit; **bracts** similar to those of staminate but greenish at anthesis; **rachis** or flower-bearing portion to 6 cm long, 2.5 cm diam., erect, greenish in flower, becoming thickened and dull orange in fruit.

Flowers: Staminate in moderately dense spirals, maturing basally first and then attaining anthesis progressively toward the apex of axis, 2.5–3 × 2–2.5 mm, oblong to bullet-shaped, bright yellow, slightly sunken; **calyx** 1 × 2.5 mm, shallowly lobed, ring-like, pale green or yellowish, **sepals** connate and/or imbricate nearly to apex, broadly rounded apically; **petals** 2–2.5 × 2 mm, valvate, spreading, free nearly to base, acute, ± thick; **stamens** 1–1.5 mm high, **filaments** short, clear-colored, **anthers** 0.5 mm long, bilobed, brownish; **pistillode** 2–2.5 mm high, columnar, expanded basally and there 0.8 mm diam., whitish, apically 0.5 mm diam. and there yellow ageing to red.

Plate 64. Chamaedorea guntheriana. Cultivated, California. **A, B, D** *Hodel 856,* Research collection, Los Angeles, *ex* Panama; **C** Virginia Robinson Botanical Garden, Beverly Hills; **A.** Staminate inflorescences (note flowers maturing progressively from base to tip of rachillae); **B.** Staminate inflorescences; **C.** Young plant; **D.** Crown of leaves.

→

C
D

Pistillate in moderately dense spirals, 2.5–3 × 1.5–2 mm, bullet-shaped, yellow, slightly sunken; **calyx** 0.75 × 2.5–3 mm, shallowly lobed, ringlike, thickened, pale green, **sepals** connate in basal ⅔, rounded apically; **petals** 3 × 2.5–3 mm, imbricate nearly to apex, acute to mucronate apically; **pistil** 3 × 2 mm, ovoid, pale yellow-green, exceeding petals, **stigma** lobes sessile, short, recurved, clear-colored.

Fruits: 6 mm diam., globose, black.

Distribution: PANAMA. Panama. Windswept, moist, relatively open, dwarf cloud forest at or near the Continental Divide; 900–1,000 m elevation.

Discussion: *C. guntheriana* is a most unusual palm in habitat and habit. It occurs sparingly in dwarf, windswept, relatively open cloud forest at the Continental Divide in central Panama. This is not the typical, dripping-wet cloud forest festooned with epiphytes that one encounters in other parts of Panama and Costa Rica. Rather, certain elements of it seem to be of an almost sclerophyllous nature. The low, open, shrubby forest contains no tall trees. Many of the species occurring there have thick, grayish green leaves as does *C. guntheriana.*

Another aspect in which *C. guntheriana* is unusual is the manner in which the staminate flowers attain anthesis; this occurs in a very pronounced, progressive manner successively from the basal end of the rachis or rachillae to the distal or apical end.

With grayish green, thick, stiff pinnae, *C. guntheriana* is similar to *C. correae* but the slightly larger leaves and the staminate flowers attaining anthesis more or less simultaneously distinguish the latter species. Other than a few plants in the research collection in Los Angeles, *C. guntheriana* is not cultivated.

49. Chamaedorea lehmannii *Burret,* Notizbl. Bot. Gart. Berlin-Dahlem 11: 857. 1933a. Type: Guatemala, Alta Verapaz, *Lehmann 1374* (holotype B, destroyed); Neotype (here designated): Guatemala, Baja Verapaz, *Hodel & Castillo 897* (BH!; isoneotype AGUAT!). Epithet: honors F. C. Lehmann, collector of the type. **PLATE 66(D-F).**

Common Name: none.

Vernacular Names: chiquilote, guite—Guatemala.

Habit: solitary, erect, to 2 m tall, often flowering when stemless or nearly so.

Stem: 1–2 cm diam., green, densely ringed, internodes to 5 cm long, often covered with persistent leaf sheaths.

Leaves: 5–7, erect-spreading, pinnate; **sheath** 12 cm long, obliquely open for ½ its length, attenuate apically, thick, strongly striate-nerved, closely imbricate and arranged in 3 planes in a triangular manner; **petiole** 15–20 cm long, moderately stout, flat or grooved near base and green above, rounded and pale below; **rachis** to 45 cm long, angled and green above, rounded and pale below; **pinnae** 8 on each side of rachis, narrowly lanceolate, 20 × 2 cm or more, apical pair wider, to 3.5 cm wide, basal ones shorter and not as narrow, regularly disposed, straight, inconspicuously if at all sigmoid, equally attenuate apically and basally, a prominently keeled midrib and 1–2 prominent slender primary nerves on each side of this, secondaries inconspicuous.

Inflorescences: interfoliar, ascending, exceeding leaves, long-pedunculate; **peduncles** 30–40 cm long or more, straight, green in flower where exposed, orange in fruit where exposed; **bracts** 5–7, prophyll 2 cm long, others from 4–17 cm, tubular, slender, 5 mm diam.; **rachises** to 3 cm long, green in flower, orange in fruit. **Staminate** with 6 **rachillae** to 18 cm long, erect, green in flower.

Pistillate with 2–5 **rachillae** or rarely spicate, these 8–11 cm long, green in flower, rather stout, orange-red in fruit.

Flowers: Staminate in well separated loose spirals, 3.5 mm high, green, superficial; **calyx** short, ringlike, lobed, **sepals** connate in basal ¾, rounded apically; **petals** valvate, free apically, oblong or obovate-oblong, moderately thick; **stamens** with **filaments** ± robust, **anthers** short-oblong, not divided apically and as long as filaments; **pistillode** columnar.

Pistillate in remote spirals, subglobose, dull green, superficial; **calyx** short, deeply lobed, **sepals** connate and/or imbricate in basal ¼–⅓, broadly rounded apically; **petals** elliptic, imbricate, obtuse; **staminodes** present; **pistil** ± ovoid, **styles** short or lacking, **stigma** lobes separated, recurved.

Fruits: 7 mm diam. when dry, globose, black; seeds 6–7 mm diam., globose; abortive carpel adherent to corolla in fruit.

Distribution: GUATEMALA. Alta Verapaz. Baja Verapaz. El Progreso. Zacapa. Moist or wet forest and cloud forest on the Atlantic slope; 1,400–2,600 m elevation.

Discussion: The description is from Burret (1933a) and supplemented from Standley and Steyermark (1958). According to the latter authors, *C. lehmannii* bears a superficial resemblance to pinnate-leaved forms of *C. adscendens.* However, *C. lehmannii* appears closest to *C. pittieri* and occurs in a similar habitat as the latter species. *C. pittieri* differs in the strongly basally contracted, thicker, stiffer, leathery, velvety green pinnae with a bluish or grayish cast, leaf sheaths not arranged in a triangular manner, drooping staminate rachillae, more or less densely placed staminate flowers, and large, pistillate petals.

Lehmann's holotype was apparently destroyed and no isotypes have been found. Since collections of this species are in several herbaria, it seems appropriate to retypify this taxon here.

→

Plate 65. Chamaedorea guntheriana. A-C. In windswept, dwarf cloud forest on Continental Divide, *Hodel et al. 1127,* Panama (note bifid and pinnate leaves).

C

Apparently not cultivated, *C. lehmannii* has its closely imbricate leaf sheaths arranged in a triangular manner, much like those of the famous triangle palm *Neodypsis decaryi.*

50. Chamaedorea microphylla *H. A. Wendl.,* Bot. Zeitung Berlin 17: 102. 1859. Type: Cult., *Wendland s.n.* (holotype GOET!; isotype K!). Epithet: from the Greek *micro* meaning small and *phyllus* meaning leaf, in reference to the small leaves. **PLATE 67.**

Nunnezharia microphylla (H. A. Wendl.) O. Kuntze, Revisio Generum Plantarum 2: 730. 1891.

Common Name: none.
Vernacular Name: none.

Habit: solitary, slender, erect, to 1 m tall.

Stem: 1.5 cm diam., smooth, densely white-spotted and ± glaucous, internodes 4–7 cm long.

Leaves: 3–5, erect-spreading, pinnate; **sheath** 7–9 cm long, tubular, briefly open apically, moderately thick, longitudinally striate-nerved; **petiole** 10 cm long, flattened and grooved toward base and green above, rounded and green below; **rachis** to 25 cm long, angled and green above, rounded below with a narrow pale band extending onto sheath; rachis, petiole, and upper part of sheath minutely white-spotted; **pinnae** 7–9 on each side of rachis, middle ones 10–17 × 3–4 cm, narrowly lanceolate to oblong-lanceolate or long-trapezoid, lowest pinnae lanceolate and long-acuminate, to 9 × 1 cm, ± reflexed, sigmoid, falcate, abruptly short-acuminate, strongly contracted basally, dull dark green, not decurrent, alternate or subopposite, downward-cupped so drying wrinkled, thin, nearly transparent when dry, middle and upper ones regularly spaced, lower ones closer together, prominent midrib as well as 2 primary nerves on each side of this and several secondaries, upper pair confluent with 3 prominent nerves, interior margin 7–9 cm long.

Inflorescences: infrafoliar, erect-spreading; **peduncles** 15–40 cm long, 4–5 mm wide at base, 1–2 mm diam. at apex, green in flower where exposed; **bracts** 5–7, prophyll to 1.5 cm long, 2nd bract 2 cm, 3rd 6 cm, 4th–6th 9–12 cm, 7th 9 cm long and equalling or exceeding peduncle, tubular, thin-papery, brownish at anthesis, shredded or fallen away in fruit, bifid, acute-acuminate, longitudinally-striated. **Staminate** with **rachis** 5–10 cm long, flexuous, longitudinally angled, green, nodding; **rachillae** 12–15, 10–18 cm long, narrowly divergent from rachis, spreading basally, drooping apically, longitudinally angled, simple or lower rachillae forked, green.

Pistillate with **rachis** 3–7 cm long, flexuous, longitudinally angled, green in flower, probably orange or red in fruit; **rachillae** 7–12, these 4–8 cm long, slender, ± straight but flexuous, strongly undulate when dry, narrowly divergent from rachis, spreading, simple or some of lower ones forked, green in flower, probably orange or red in fruit.

Flowers: Staminate in loose spirals, 2.5–3 × 2–2.5 mm, ovoid, yellow, slightly sunken in elliptic to oblong pits 2–2.5 mm long; **calyx** 0.5–0.75 × 2–2.5 mm, shallowly lobed, membranous, transparent when dry, **sepals** connate in basal ¾, rounded to acute apically; **petals** 3 × 1–1.5 mm, long-ovate, valvate, free nearly to base, spreading apically, acute, membranous, nearly transparent and lightly nerved when dry; **stamens** 1.5–1.75 mm high, **filaments** 1–1.5 × 0.3 mm, basally flared and connate and forming a ring around pistillode, **anthers** 0.5 mm long, elliptic to oval; **pistillode** 2 mm high, broadly columnar, tip expanded.

Pistillate in loose spirals, in bud 1.5–1.75 × 1.5 mm, conic-ovoid, yellow, slightly sunken in elliptic to oval depressions 1.5 mm long; **calyx** 0.5 × 1.5 mm, lobed, membranous, green, **sepals** imbricate in basal ½–¾, rounded to acute apically; **petals** 1.75 × 1.5–2 mm, broadly ovate, briefly connate basally, imbricate nearly to apex, erect, acute; **staminodes** 6, small; **pistil** 1 × 0.75–1 mm, conic-ovoid, **styles** short or lacking, **stigma** lobes short, recurved, not exceeding petals.

Fruits: perhaps full size but not soft ripe, 6 × 4 mm, oblong, green drying black.

Distribution: PANAMA. Chiriqui. Moist, windswept cloud forest; 1,000 m elevation or more.

Discussion: The description is from Wendland (1859) who described and named *C. microphylla* from cultivated material at the Berlin Botanic Garden that Warscewicz had introduced to European gardens from Chiriqui about the middle of the 19th century. It had disappeared from the annals of botany and horticulture and had not been recollected until recently. In 1991 we rediscovered it in moist, windswept, ridge forest on the Pacific slope in the Fortuna area of Chiriqui. *C. microphylla* is a dainty species with thin, downward-cupped, dark green pinnae markedly contracted basally. The stem is so densely white-spotted that it appears white throughout.

Staminate flowers of *C. microphylla* are distinctive in their fragile and delicate nature. Sepals, petals, and even stamens and the pistillode are membranous and nearly transparent when dry. Pistillate flowers also have a fragile and delicate nature but less so than the staminate.

Standley (1937) and others have erroneously referred material with a dwarf, stemless habit from wet forests at middle elevations of the Atlantic slope of Costa Rica near the Rio Sarapiqui to *C. microphylla.* However, this material is best referred to *C.*

→

Plate 66. A–C, Chamaedorea guntheriana. D–F, Chamaedorea lehmannii. A, B Cultivated, research collection, Los Angeles, *ex* Panama; **A.** Pistillate flowers, *Hodel 746B bis*; **B.** Staminate flowers, *Hodel 856*; **C.** Young plant, *Hodel et al. 1127,* windswept elfin cloud forest, Continental Divide, Panama. **D–F** In wet cloud forest, Baja Verapaz, Guatemala; **D, E** *Hodel & Castillo 1013*; **D.** Mature plant; **E.** Leaf sheaths arranged in a triangular fashion; **F.** Pistillate plant, *Hodel & Castillo 897* (neotype).

B
D
F

scheryi. Although *C. microphylla* and *C. scheryi* are in subgenus *Chamaedoropsis*, the decurrent pinnae, stemless habit, pinnate eophylls, and long-pedunculate, few-branched inflorescences with a short rachis that appear to arise from the ground distinguish *C. scheryi*.

51. Chamaedorea minima *Hodel*, Principes 35: 72. 1991a. Type: Cult., *D. & M. Hodel 622A* (holotype BH!). Epithet: from the Latin word meaning the smallest, in reference to the habit. **PLATE 68.**

Common Name: dwarf pumila.
Vernacular Name: none.

Habit: solitary, erect or decumbent, to 25 cm tall, often appearing stemless.

Stem: 1–2 cm diam., often creeping and buried in leaf litter, densely and prominently ringed, internodes 2–3 mm long, often with adventitious roots.

Leaves: 7–10, spreading, bifid; **sheath** 2.5 cm long, deeply split opposite petiole and clasping in a tubular manner only at base, green, minutely white-spotted; **petiole** 5–7 cm long, flat and green above, rounded and green below; **rachis** to 7 cm long, angled and green above, rounded and pale green below, minutely white-spotted; **blades** 15 × 6 cm, incised apically to ½ their length or more, lobes 4–5 cm wide at tips, acute-acuminate, coarsely toothed along margin, ± stiff, grayish dark forest green, 6–7 prominent primary nerves on each side of rachis.

Inflorescences: interfoliar, enclosed in upper rolled margins of sheath, erect-spreading; **peduncles** to 12 cm long, 3 mm wide at base and there ± flattened, 2–3 mm diam. at apex, green in flower where exposed, orange in fruit; **bracts** 6–8, prophyll 0.5 cm long, 2nd bract 1.5 cm, 3rd 2 cm, 4th 3 cm, 5th 4 cm, 6th 5 cm, 7th 3 cm, tubular, acute-acuminate, bifid, papery, brownish in flower and fruit, longitudinally striate-nerved. **Staminate** with **rachis** 2 cm long, green in flower; **rachillae** 6, these to 7 cm long, 1–1.5 mm diam., spreading or slightly drooping, green in flower.

Pistillate spicate, **rachis** or flower-bearing portion to 7 cm long, 3 mm diam. and green in flower, swollen to 4–5 mm diam. and red-orange in fruit, curved.

Flowers: Staminate in moderately dense spirals, barrel-shaped in immature bud, 1.5 × 0.75 mm, green; **calyx** low, lobed, **sepals** connate basally, broadly rounded or straight apically; **petals** valvate, free, spreading apically.

Pistillate in moderate spirals, subglobose in immature bud, 1 × 1 mm, green; **calyx** lobed, **sepals** imbricate basally; **petals** tightly imbricate apically nearly to apex, acute.

Fruits: 8–10 mm diam., ± globose, purple-black.

Distribution: COSTA RICA. Wet forest at middle elevations, exact locality unknown; perhaps extinct in the wild.

Discussion: In the late 1960s, Robert Wilson began to plant this species at his garden at San Vito in southeastern Costa Rica. Local collectors brought the plants to him but gave no specific locality other than that they grew in the mountains not too distant from San Vito. Wilson established fruiting populations of this dwarf palm in his garden (illustrated in Hodge 1981, figs. 6, 7) and frequently distributed seeds and plants to palm collectors and hobbyists, most of whom lived in the United States. Wilson called this palm "dwarf pumila," in reference to its similarity to *C. pumila*. I have seen no collections of it from the wild, so it may be extinct now. The area around San Vito from which the plants may have been originally collected has been largely deforested, adding to the possibility that the species is extinct in the wild.

Chamaedorea minima is indeed close to *C. pumila* but the larger habit and larger blades with 9–13 prominent primary nerves on each side of the rachis distinguish the latter species. In addition, the blades of *C. pumila* have an iridescent, slightly mottled bluish green color, a feature lacking in *C. minima*.

Through Wilson's efforts, *C. minima* is cultivated in Hawaii, California, Florida, and perhaps elsewhere. In California, it is susceptible to brown tipping of the leaves due to low humidity and/or high mineral content of the water.

52. Chamaedorea nubium *Standl. & Steyerm.*, Publ. Field Mus. Nat. Hist., Bot. Ser. 23: 202. 1947. Type: Guatemala, El Progreso, Sierra de las Minas, *Steyermark 43583A* (holotype F!). Epithet: is the Latin word meaning of the clouds, apparently in reference to the relatively high elevation, mountain cloud forest habitat, although the authors did not specify the application of the name. **PLATES 9(A), 69, 70.**

Common Name: none.
Vernacular Name: none.

Habit: cespitose, erect or decumbent, forming dense or loose and wide-ranging colonies to 5 × 10 m.

Stems: 8–10 mm diam., smooth, green, ringed, rooting and infrequently sprouting at nodes, uppermost internodes short, lower ones to 12 cm long, often covered by persistent leaf sheaths.

Leaves: 4–7, spreading, bifid or rarely pinnate; **sheath** 20–30 cm long, tubular, obliquely open apically and there lighter green and longitudinally striate-nerved, old sheaths drying brown, persistent; **petiole** 8–13 cm long, flattened and very slightly grooved near base and green above, rounded and pale below; **rachis** 8–15 cm long, angled and green above, rounded with pale yellowish

→

Plate 67. Chamaedorea microphylla. Staminate plant and inflorescences, *Hodel et al. 1126A*, windswept cloud forest, Chiriqui, Panama. **A.** Inflorescences; **B.** Stem and inflorescence (note densely white-spotted stem); **C.** Leaves and inflorescences.

C

band below extending onto sheath; **blades** to 45 × 30–35 cm, incised apically to ⅔ their length, green or distinctly glaucous below, lobes narrowly long-acute, cuneate-acute basally, interior margins to 20–33 cm long, exterior margins entire or slightly toothed, 12–17 prominent primary nerves on each side of rachis, these keeled and green above, light green below, 2 secondaries between each pair of primaries, tertiaries numerous, faint; or rarely **blades** partially to completely pinnate, then with **rachis** 25–33 cm long with 6–13 **pinnae** on each side, to 31 × 2 cm, linear-lanceolate, falcate, midrib and a primary nerve on each side of this, these keeled, apical pinnae slightly or much broader than others, to 7-nerved.

Inflorescences: infrafoliar, sometimes emerging through persistent sheaths, erect-spreading; **peduncles** 10–27 cm long, 2–5 mm wide at apex and there rounded, 3–5 mm wide at base and there ± flattened, erect, pale green or yellowish in flower, nodding and red-orange in fruit; **bracts** 4–5, prophyll 1.5 cm long, 2nd bract 3 cm, 3rd 6.5 cm, 4th 9.5 cm, 5th 5.5 cm and often concealed by 4th, narrowly tubular, obliquely open apically, longitudinally striate-nerved, acute-acuminate, bifid, papery and drying brown in flower, uppermost equalling or slightly exceeding peduncle; **rachises** 1–4 cm long or more, greenish or yellowish in flower, red-orange in fruit. **Staminate** with up to 15 **rachillae**, these 6–15 cm long, 1–2 mm diam., yellow, spreading and slightly drooping but not pendulous, frequently markedly undulate.

Pistillate with up to 10 **rachillae** or infrequently spicate or forked, these 8–15 cm long, 3–3.5 mm diam., erect, obtusely angled, and pale green in flower, swollen to 5 mm diam., downward-pointing, and red-orange in fruit, frequently strongly undulate when dry.

Flowers: Staminate in loose or ± moderately dense spirals but appearing as though in opposite lines, 2–2.5 mm apart, 1.5–2 × 3–3.5 mm, depressed-globose, laterally compressed, golden yellow, sunken in narrowly elliptic depressions in curves of axis; **calyx** 0.5–0.75 × 2.5–3 mm, lobed, membranous, **sepals** connate in basal ½, broadly rounded to straight apically; **petals** 2 × 2 mm, valvate, broadly ovate, connate in basal ⅓, free but incurved apically, acute, tips only slightly recurved, ± thick; **stamens** 1–1.5 mm, **filaments** very short, **anthers** 0.7–0.75 mm, bifid apically, brownish; **pistillode** 1.5 × 0.75 mm, broadly columnar, whitish.

Pistillate in loose or moderate spirals but appearing nearly distichous, 1.5–2 × 4 mm, shieldlike, pale yellow, sunken in depressions 2.5 mm long; **calyx** 1 × 4 mm, shallowly lobed, **sepals** imbricate in basal ¾, broadly rounded or straight apically; **petals** 2.5 mm long, imbricate nearly to apex and corolla opening only slightly apically, thickened in fruit, broadly triangular, drying dark brown; **staminodes** present; **pistil** shieldlike, 1.5–2 mm high, yellowish, **stigma** lobes sessile, short, exserted, bluntly pointed, dark.

Fruits: to 14 × 11 mm, ovoid-globose, black at maturity with a glaucous bloom.

Distribution: MEXICO. Chiapas. Guerrero. Oaxaca. GUATEMALA. Alta Verapaz. Baja Verapaz. El Progreso. Huehuetenango. San Marcos. Zacapa. EL SALVADOR. Chalatenango. Santa Ana. Moist or wet forest or cloud forest on the Atlantic and Pacific slopes; 1,500–2,500 m elevation.

Discussion: Standley and Steyermark (1947) described and named *C. nubium* from material collected in Guatemala. We have collected it in wet cloud forest on the Atlantic slope in Baja Verapaz, Guatemala and Chiapas, Mexico. We also collected it in oak-pine forest on the Pacific slope of Oaxaca, Mexico. *C. nubium* is an extremely handsome plant, forming large clumps several meters across of leaning, slender stems topped with large bifid leaves. In Guatemala, we found colonies 10–15 m across of loose and far-ranging stems, all of one sex. Here the plants had erect to decumbent stems that tended to root and sprout at the nodes and leaves distinctly white-glaucous below. Mexican plants had leaves that were green below.

Relatively easy to grow, *C. nubium* is not too dissimilar in habit to *C. brachypoda,* only usually larger. The generally smaller size, rhizomatous habit, and green flowers easily distinguish *C. brachypoda. C. nubium* grows more vigorously in southern California and is much superior to *C. brachypoda* there; its leaves do not yellow- and brown-tip as do those of *C. brachypoda. C. nubium* is uncommon in cultivation, until recently appearing in only a few collections in southern California. Loran Whitelock of Los Angeles introduced these in 1981 from Oaxaca, Mexico. Recently, we have collected seeds in Guatemala and distributed them as *Hodel & Castillo 899.*

Standley and Steyermark (1947, 1958) relied heavily on the undulate rachillae as a distinguishing character of *C. nubium,* a feature that I have found not to be a good diagnostic character. Rather the undulating nature is quite variable and seems to depend on the age of the rachillae when collected and dried.

53. Chamaedorea oblongata *Mart.,* Historia Naturalis Palmarum 3: 160. 1837. Type: Mexico, Veracruz, Barranca de Tioselo, *Schiede s.n.* [holotype M (photo!); isotype GOET!]. Epithet: from the Latin *oblongus* meaning oblong, in reference to the shape of the fruits or pinnae. **PLATE 71.**

Chamaedorea lunata Liebm. in Mart., Historia Naturalis Palmarum 3: 307. 1849. Type: Mexico, Veracruz, *Liebmann s.n.* (some sheets later

Plate 68. Chamaedorea minima. Cultivated. **A.** Staminate plant, *D. & M. Hodel 622A* (holotype), Jardin Botanico Robert y Catherine Wilson, Costa Rica; **B.** Two well grown plants (labeled incorrectly as *C. pumila*), *Hodel 673,* garden of F. Ketchum, California.

→

CHAMAEDOREA
PUMILA

numbered *6578, 6577*) [holotype C!; isotypes BR (photo!), FI!, GH!, NY (photo!), US!]. Epithet: from the Latin *lunatus* meaning crescent-shaped, in reference to the shape of the fruits.

Nunnezharia lunata (Liebm.) O. Kuntze, Revisio Generum Plantarum 2: 730. 1891.

Nunnezharia oblongata (Mart.) O. Kuntze, Revisio Generum Plantarum 2: 730. 1891.

Nunnezharia biloba (Hort.) O. Kuntze, Revisio Generum Plantarum 2: 731. 1891.

Nunnezharia corallocarpa (Hort.) O. Kuntze, Revisio Generum Plantarum 2: 731. 1891.

Chamaedorea fusca Standl. & Steyerm., Publ. Field Mus. Nat. Hist., Bot. Ser. 23: 201. 1947. Type: Guatemala, Izabal, *Steyermark 38123* (holotype F!). Epithet: from the Latin *fuscus* meaning brownish or very dark in color, perhaps in reference to the pistillate rachillae being blackish in the dry state.

Chamaedorea aurantiaca Hort. in H. A. Wendl., Index Palmarum 11. 1854. Not Brongn.

Chamaedorea biloba Hort. in H. A. Wendl., Index Palmarum 11. 1854.

Chamaedorea lindeniana Hort. in H. A. Wendl., Index Palmarum 13. 1854. Not H. A. Wendl.

Chamaedorea scandens Hort. in H. A. Wendl., Index Palmarum 15. 1854. Not Liebm.

Chamaedorea schiedeana Hort. in H. A. Wendl., Index Palmarum 15. 1854. Not Mart.

Morenia corallocarpa Hort. in H. A. Wendl., Index Palmarum 29. 1854.

Chamaedorea corallocarpa Hort. in Dammer, Palmenzucht und Palmenpfleg 88. 1897.

Morenia corallina Hort. in Dammer, Palmenzucht und Palmenpfleg 88. 1897. Not Karst.

Chamaedorea corallina Hort. in Anonymous, Gard. Chron. 23: 410. 1885.

Mauranthe lunata (Liebm.) O. F. Cook, Nat. Hort. Mag. 22: 83, fig. 8. 1943a. Name of no botanical standing.

Common Name: none.

Vernacular Names: caquib, xate macho—Guatemala; tepejilote, palmilla, brillosa, palmita, chate—Mexico.

Habit: solitary, erect, occasionally decumbent, 1–3 m tall.

Stem: 1–2.5 cm diam., smooth, green, ringed, internodes 4–15 cm long.

Leaves: 3–8, erect-spreading, pinnate; **sheath** 15–20 cm long, tubular, obliquely open apically, prominently costate; **petiole** 15–30 cm long, slightly grooved and green above, rounded and pale below; **rachis** 30–60 cm long, angled and green above, rounded and with a distinct yellow band below extending onto sheath; **pinnae** 5–9 on each side of rachis, middle ones 17–40 × 3.5–10 cm, variable in shape, lanceolate to rhombic-lanceolate to oblong-trapezoid, sigmoid, mostly alternate, ± leathery, thick, regularly disposed, narrowly and caudately long-acuminate or attenuate-acuminate, oblique basally, deep green, glossy above, paler below, a midrib and 2 submarginal primary nerves, these not keeled and ± obscure above, yellowish and shining below, 3 secondaries on each side of midrib, these inconspicuous as are numerous and fine tertiaries, apical pinnae shorter than middle ones and generally narrower.

Inflorescences: infrafoliar, erect-spreading, 20–70 cm long; **peduncles** 10–40 cm long, green in flower, reddish orange in fruit; **bracts** 5–7, tubular, brown in flower, thin, fibrous, acute-acuminate, bifid, longitudinally striate-nerved, uppermost equalling peduncle; **rachis** 2.5–12 cm long, green in flower, reddish orange in fruit. **Staminate** with 9–25 or more **rachillae**, these to 30 cm long, 2–3 mm diam., pendulous, slightly angled, green in flower.

Pistillate with 6–20 or more **rachillae**, these 9–16 cm long, ± stiff, ascending, spreading, 2 mm diam., green in flower, swollen and reddish orange in fruit.

Flowers: Staminate in dense spirals, 3–4 × 3–4.5 mm, ovoid to obovoid, greenish drying black, superficial or slightly sunken; **calyx** 0.5–1 × 2 mm, lobed, green, **sepals** connate in basal ½, broadly rounded to acute apically; **petals** 3–4 × 1.5–2 mm, valvate, free to nearly base and there connate briefly, erect then spreading to reflexed, acute, tips slightly recurved, thin, slightly transparent; **stamens** 1.5–2 mm long, spreading, **filaments** 0.75–1 mm long, adnate basally with petals, green, **anthers** 1 mm long, longer than filaments and briefly bifid apically, long-oblong, yellow; **pistillode** 2.5–3 mm high, columnar, slightly lobed apically, light green.

Pistillate in loose spirals, 2 × 2.5–3 mm, depressed-globose, greenish yellow drying black, scarcely sunken; **calyx** 0.5–1 × 2–2.5 mm, deeply lobed, **sepals** connate in basal ⅓, broadly rounded to acute apically; **petals** 2.5 × 2–2.5, broadly ovate, imbricate nearly to apex, connate basally briefly, thin, ± straight or acute, slightly fleshy, transparent, brown-margined, drying dark brown in fruit with abortive carpel adherent to smallest petal; **staminodes** 3, triangular; **pistil** 1.75–2 × 2 mm, depressed-globose, green, **styles** lacking, **stigma** lobes separated, sessile, angular.

Fruits: 8–14 × 6–8 mm, variable in shape, ovoid-ellipsoid or sometimes falciform or slightly lunate, attenuate at each end, or ± globose, shining black at maturity, epicarp thin, not transparent, mesocarp slightly fleshy, green, mucilaginous, aromatic, dense, fibrous; seeds 7–11 × 5–6 mm, ellipsoid.

→

Plate 69. Chamaedorea nubium. A, B Wet cloud forest, Baja Verapaz, Guatemala; **A.** Large mature clump, *Hodel & Castillo 899*; **B.** Underside of leaf showing grayish white covering, *Hodel & Castillo 884*; **C-F** Cultivated, California, *ex* Oaxaca, Mexico; **C.** Young clump, Virginia Robinson Botanical Garden, Beverly Hills; **D-F** *Hodel 663A* (staminate) *& 663B* (pistillate), garden of L. Whitelock, Los Angeles; **D.** Staminate inflorescence; **E.** Staminate flowers; **F.** Pistillate flowers.

B
D
E
F

Distribution: MEXICO. Campeche. Chiapas. Oaxaca. Quintana Roo. Tabasco. Veracruz. GUATEMALA. Alta Verapaz. Izabal. Peten. BELIZE. Belize. Cayo. HONDURAS. Atlantida. Cortes. NICARAGUA. Jinotega. Matagalpa. Moist or wet lowland forest on the Atlantic slope; to 350 m elevation; often on limestone.

Discussion: Martius (1837) described and named *C. oblongata* from material that Schiede collected in Mexico in the 1830s. Linden introduced it in 1839 from Mexico and Guatemala to European gardens and collections where it was often grown under the names *C. lunata* or *C. corallina.* Guillaumin (1923b) reported *C. lunata* as having been cultivated at the Musee de Paris in France since 1839. As the numerous horticultural names imply, *C. oblongata* was widely cultivated in European gardens and collections and this partly due to its relative ease of culture. The epithets used in horticulture, *corallina* and *corallocarpa,* are usually in reference to red fruit. However, to the best of my knowledge and after reviewing the numerous horticultural accounts and illustrations accompanying these names, I feel that these epithets refer to *C. oblongata* (because of the red fruiting rachillae) and not other red-fruited species, like *C. pinnatifrons* or species of subgenus *Morenia.*

Chamaedorea oblongata exhibits a great deal of variability in shape of leaf and of fruit over its rather wide range. Leaf shapes can be lanceolate and flat to broadly rhombic or trapezoid and cupped downward. Fruits can be nearly globose to ellipsoid or slightly lunate. *C. oblongata* is one of the most commonly cultivated species and is found in collections and gardens wherever the genus is grown. It is handsome in foliage because of the thick, leathery, lustrous, shiny leaves.

The names of three other species of similar habit, *C. neurochlamys, C. schiedeana,* and *C. concolor* (= *C. pinnatifrons*) have been erroneously applied to material of *C. oblongata* in cultivation. In fact, Krempin (1990, p. 90) illustrated *C. oblongata* but erroneously captioned the photograph as *C. concolor.* Staminate flowers with the petals free and spreading rather than connate apically and adnate to the pistillode easily distinguish *C. oblongata.* Additional differences between *C. oblongata* and these three other species are summarized below.

C. oblongata	*C. neurochlamys*	*C. schiedeana*	*C. pinnatifrons*
	pinnae/side		
5–9	7–8	10–14	2–9
	texture		
thick	medium	thin	thin
	leaf sheath		
apex green	apex white	apex green	apex white
	fruits		
black	reddish	black	red-orange

Of the four, *C. oblongata* is much more durable and is resistant to mites.

54. Chamaedorea pachecoana *Standl. & Steyerm.,* Publ. Field Mus. Nat. Hist., Bot. Ser. 23: 202. 1947. Type: Cult., *Standley 63104* (holotype F!). Epithet: honors D. M. Pacheco Herrarte of Guatemala, from whose garden the type originated. **PLATE 72.**

Common Name: none.
Vernacular Name: pacaya—Guatemala.

Habit: solitary, erect or decumbent, to 1 m tall, often flowering when appearing stemless.

Stem: 7–10 mm diam., basal portion curved and subterranean, remainder with only top exposed for up to 30 cm, densely ringed, internodes 5–10 mm long.

Leaves: 6–12 in a rosettelike crown, ascending-spreading, pinnate; **sheath** 5 cm long, long-open nearly to base, longitudinally striate-nerved, persistent; **petiole** 7–20 cm long, ± flat or grooved near base and green above, rounded and green below; **rachis** to 20 cm long, angled and green above, rounded and paler below; **pinnae** 7–11 on each side of rachis, 4.5–12 × 1.5–2 cm, broadly rhombic-lanceolate or oblong-lanceolate, sigmoid, falcate, long-acuminate, arcuate-cuneate at base, sometimes very minutely auriculate with very narrow basal attachment, thin, regularly inserted, alternate or subopposite, attached at right angles to rachis or even reflexed, especially so the middle and basal ones, terminal ones confluent and sometimes slightly longer and wider than others, apex often ± twisted, margins shallowly toothed toward tip and often verrucose toward base, all pinnae 3-nerved, secondaries slender and inconspicuous.

Inflorescences: inter- or infrequently infrafoliar, erect-spreading; **peduncles** 11–30 cm long, 2–3 mm wide at base and flattened, 1 mm diam. near apex, green in flower where exposed, orange-red in fruit; **bracts** 5–7, prophyll 1.5 cm long, others 2–12 cm, narrowly tubular, to 2 mm diam., obliquely open apically, longitudinally striated. **Staminate** with short **rachis** to 1.5 cm long, green in flower; **rachillae** 2–6, these to 12 cm long, spreading, green in flower.

Pistillate spicate or forked; **rachillae** or flower-bearing portion

→

Plate 70. Chamaedorea nubium. Large clump in clouds and mist, *D. & R. Hodel 924,* cloud forest, Chiapas, Mexico.

to 12 cm long, straight, greenish in flower, swollen and orange in fruit.

Flowers: Staminate rather densely arranged, 3 × 2 mm, ovoid to bullet-shaped, greenish yellow, slightly sunken in elliptic depressions 2.5 mm long; **calyx** 0.7–0.75 × 2.5 mm, lobed, **sepals** connate in basal ½–¾, ± acute apically; **petals** 3 × 1.5–2 mm, ovate, valvate, free nearly to base, incurved apically but not connate, obtuse; **stamens** 2–2.5 mm high, **filaments** 0.5–1 mm long, **anthers** 1–1.5 mm long, oblong-ovate, obtuse; **pistillode** 2 mm high, columnar.

Pistillate in ± lax spirals, slightly sunken; **calyx** 2 mm across, shallowly lobed, **sepals** connate and/or imbricate in basal ¾, broadly rounded apically; **petals** 2 mm long, rounded-ovate, acute, pale, lightly nerved when dry; **staminodes** present; **pistil** not seen.

Fruits: 6–8 mm diam., globose or ovoid-globose, black; abortive carpels generally adherent to corolla in fruit.

Distribution: GUATEMALA. Quetzaltenango. San Marcos. Solola. Moist or wet forest on the Pacific slope; 1,200–1,500 m elevation.

Discussion: The description is from Standley and Steyermark (1947, 1958) who stated that *C. pachecoana* is the smallest of all known palms of Guatemala and one of the most attractive. The former is certainly an arguable point since *C. tuerckheimii* is comparable in size. However, there is no doubt that *C. pachecoana* is, indeed, a most attractive plant. It is locally abundant and often grows in masses covering the ground for several square meters. Its symmetrical, rosettelike crown of numerous, ascending to spreading leaves appearing to arise from the leaf litter and bearing bright green, closely spaced, broadened pinnae is a perfect contrast to the swollen, orange, spicate inflorescences with black fruits.

Standley and Steyermark (1958) compared *C. pachecoana* in habit to *C. elegans*. The larger habit with a visible stem and fewer leaves distinguish *C. elegans* vegetatively. *C. pachecoana* is actually more closely related to *C. parvisecta* and *C. whitelockiana*. The larger habit with a well developed, above-ground stem, fewer leaves, and inflorescences with more rachillae distinguish the latter two species. Pinnae of *C. pachecoana* are occasionally auriculate at the point of attachment with the rachis, and in this regard are similar to those of *C. tenerrima*.

It is remarkable that a species as distinctive as *C. pachecoana* remained unknown for so long, especially since it was locally abundant in an area frequented by botanists in the past.

55. Chamaedorea paradoxa *H. A. Wendl.*, Bot. Zeitung Berlin 17: 29. 1859. Type: Cult., *Wendland s.n.* (holotype GOET!; isotypes K!, L!). Epithet: is a Greek word meaning unbelievable, apparently in reference to the number of petals.

Nunnezharia paradoxa (H. A. Wendl.) O. Kuntze, Revisio Generum Plantarum 2: 730. 1891.

Common Name: none.
Vernacular Name: none.

Habit: solitary, erect, 2–3 m tall.

Stem: 1.2–1.5 cm diam., dark green, minutely white-spotted, remotely ringed, internodes 2–6 cm long.

Leaves: 4–6, erect-spreading, pinnate; **sheath** 18 cm long, tubular, briefly open apically; **petiole** slightly grooved above and there calloused at base, obtusely 3-sided; **rachis** 25–35 cm long with weak green band below extending onto sheath; **pinnae** 3–5 on each side of rachis, lowermost 20–30 × 1–2.5 cm, middle ones 28–35 × 4.5–5 cm, broadly lanceolate or elongate-oblong, slightly sigmoid, long-acuminate, contracted basally, alternate, remote, erect-spreading, ± leathery, dark green, glossy, midrib and 2 primary nerves on each side of this prominent, secondaries 3 on each side of midrib, upper pinnae confluent, 5–9 cm wide with 4–5 primary nerves, 20–30 cm long on the interior margin.

Inflorescences: Staminate not seen. **Pistillate** infrafoliar, erect-spreading; **peduncle** 12–18 cm long, essentially terete; **bracts** 6–7, tubular, briefly bifid apically, uppermost exceeding peduncle; **rachis** 5–10 cm long, flexuous, angled, erect; **rachillae** 8–11, lower ones 10–13 cm long, upper 7 cm long, all angled, erect-spreading; rachis and rachillae green drying black.

Flowers: Staminate not seen. **Pistillate** sunken in oblong pits; **calyx** shallowly lobed; **petals** 3 or occasionally 6, broadly rotund-ovate, acuminate, yellow-green; **staminodes** 3 or 6, linear or semi-rotund, toothlike apically; **pistil** rotund, obtusely 3-sided, **stigma** lobes sessile, slightly recurved.

Fruits: not seen.

Distribution: MEXICO. Exact locality unknown.

Discussion: The description is from Wendland (1859) who described and named *C. paradoxa* from cultivated plants at Herrenhausen that were grown from seeds that Hartweg had collected at an unspecified locality in Mexico in the early 19th century. Wendland described the pistillate corolla as being "*triphyllum*" (3-petaled) or often "4–6-*phyllum*" (4–6-petaled). Apparently, he meant that the corolla is sometimes composed of 4–6 petals. I have examined the type and isotypes of *C. paradoxa* and, in some instances, the corolla is indeed apparently 4–6-petaled, a condition unique in *Chamaedorea*. However,

Plate 71. Chamaedorea oblongata. A. Juan Jose Castillo holds large plant, *Hodel & Castillo 876,* lowland rain forest, Izabal, Guatemala; **B.** Staminate flowers, *Hodel 768,* cultivated, garden of R. Palmer, Whittier, California; **C.** Infructescence with mature fruits, *Hodel 846,* Tikal, Guatemala; **D.** Staminate plant, cultivated, garden of R. Palmer, Whittier, California; **E.** Cultivated, Fairchild Tropical Garden, Miami, Florida.

→

B
C
E

the 4–6-petaled corolla is probably an aberation and not the normal condition for this species. In many regards, especially in the pistillate inflorescences and flowers, *C. paradoxa* seems very close to the variable *C. oblongata* and is perhaps not distinct.

According to Guillaumin (1923b), plants that were eventually to be named *C. paradoxa* were cultivated at Hooibrink de Hitzing near Vienna, Austria as early as 1809 from seeds that Hartweg had sent. If distinct, the species seems to have been lost and is not now cultivated.

56. Chamaedorea parvisecta *Burret,* Notizbl. Bot. Gart. Berlin-Dahlem 11: 742. 1933a. Type: Guatemala, Alta Verapaz, *Tuerckheim 2185* (holotype B, destroyed); Neotype (here designated): Guatemala, Baja Verapaz, *Hodel & Castillo 1102A* (BH!, isoneotype AGUAT!). Epithet: from the Latin *parvus* meaning small and *sectus* meaning division or segment, in reference to the small pinnae. **PLATE 73(A–E).**

Chamaedorea pulchra Burret, Notizbl. Bot. Gart. Berlin-Dahlem 11: 741. 1933a. Type: Guatemala, Alta Verapaz, *Tuerckheim A3* (holotype B, destroyed). Epithet: from the Latin *pulcher* meaning beautiful.

Chamaedorea digitata Standl. & Steyerm., Publ. Field Mus. Nat. Hist., Bot. Ser. 23: 200. 1947. Type: Guatemala, Huehuetenango, *Steyermark 51683* (holotype F!). Epithet: from the Latin *digitatus* referring to the manner in which appendages radiate from a common point, perhaps in reference to the pistillate rachillae.

Paranthe violaceae O. F. Cook, Nat. Hort. Mag. 22: 139. 1943b. Name of no botanical standing.

Common Name: none.
Vernacular Name: none.

Habit: solitary, slender, erect, to 3 m tall, often flowering when appearing stemless.

Stem: 5–7 mm diam., green, smooth, ringed, internodes to 10 cm long, upper part covered with persistent sheaths.

Leaves: 3–5, spreading, pinnate; **sheath** to 20 cm long, tubular, obliquely open apically and there longitudinally green-striate-nerved, persistent; **petiole** 30 cm long, grooved and green above, rounded and green below; **rachis** to 40 cm long, obtusely angled and green above, rounded and green below; **pinnae** to 10 or more on each side of rachis, larger central ones to 20 × 4.5 cm, lanceolate to broadly lanceolate, not conspicuously sigmoid, long and narrowly acuminate, strongly contracted basally, widest at middle, regularly disposed, alternate, thin-papery, dull green with a light bluish or grayish cast above, paler below, a prominent midrib, 2 ± obscure primary nerves and 2 secondaries on each side of midrib.

Inflorescences: infrafoliar, breaking through old sheaths at nodes, erect to spreading; **peduncles** 18–40 cm long; **bracts** 6–8, lowest four of these short and densely congested, less than 4 cm long, others to 15 cm long, 2.5–3.5 cm diam., tubular, brownish, longitudinally striated; **rachises** nearly lacking to 10 cm long, green in flower, red-orange in fruit. **Staminate** with 4–18 **rachillae**, these to 12 cm long, slender, drooping.

Pistillate with 3–10 **rachillae**, these 5–12 cm long, moderately slender, erect and green in flower, orange and thickened in fruit, blunt-tipped.

Flowers: Staminate in moderate or dense spirals, semiglobose in bud, at anthesis 1.75–2 × 2–2.5 mm, ± globose, cream-colored ageing to brown, slightly sunken; **calyx** 0.25–0.4 × 1.75–2 mm, deeply lobed, **sepals** connate in basal ¼ or nearly free, broadly rounded apically; **petals** 2 × 1.25–1.5 mm, long-ovate, valvate, free nearly to base, acute, faintly nerved when dry; **stamens** 1.5 mm tall, **filaments** 0.5–0.75 mm long, very slender, **anthers** 0.75 mm long, oval to oblong, basifixed; **pistillode** 1.75–2 mm tall, columnar, flared slightly basally.

Pistillate in moderate spirals, slightly sunken; calyx low, 2.5 mm across, lobed, **sepals** connate and/or imbricate basally, broadly rounded apically; **petals** 2 × 3 mm, broadly ovate, imbricate basally; **staminodes** present; **stigma** lobes short, sessile.

Fruits: 10 × 9 mm, ± globose to broadly ovoid, black, moderately fleshy, laxly disposed, seated in distinct depressions in axis.

Distribution: GUATEMALA. Alta Verapaz. Baja Verapaz. El Progreso. Huehuetenango. Zacapa. MEXICO. Chiapas. Moist or wet forest on the Atlantic slope; 1,400–2,500 m elevation.

Discussion: The description is from Burret (1933a) who described and named *C. parvisecta* from material that Tuerckheim had collected near Coban, Alta Verapaz, Guatemala in the early 1900s. Burret stated that the differences between *C. parvisecta* and *C. pulchra* were few, if any. It is unfortunate that the types of both species were destroyed and that apparently no isotypes have been found. However, the descriptions of the two species are nearly identical and, after examining numerous collections from the Coban region and Alta Verapaz that exemplify a rather large range of variability, I hold them to be the same. Since there are specimens of *C. parvisecta* in several herbaria, it is appropriate to retypify this taxon here.

Standley and Steyermark (1958) named and described *C. digitata* and I feel it represents the small

Plate 72. Chamaedorea pachecoana. In cloud forest, Quetzaltenango, Guatemala. **A.** Staminate plant, *Hodel & Castillo 908A*; **B.** Leaves and infructescence with mature fruits, *Hodel & Castillo 908B*; **C.** Fruiting plant (immature fruits), *Hodel & Castillo 900*; **D.** Fruiting plant, *Hodel & Castillo 985B*; **E.** Pistillate inflorescences and infructescences (immature fruits), *Hodel & Castillo 908B*. →

C
E

end of the range of variability of *C. parvisecta.* Also, *C. parvisecta* will flower at a very young stage when stemless and then with much reduced inflorescences. The flowering at a young stage probably accounts for the smaller dimensions of Standley's and Steyermark's description of *C. digitata.*

Chamaedorea parvisecta has leaves that become pinnate at a very early stage and is similar in this regard to *C. whitelockiana* from the Pacific slope. The spinose-tipped fruiting rachillae, erect staminate rachillae, and apically incurved nearly connate staminate petals distinguish the latter species. *C. parvisecta* is not cultivated except for a few plants in the research collection in Los Angeles.

57. Chamaedorea pittieri *L. H. Bailey,* Gentes Herb. 6: 252. 1943a. Type: Panama, Chiriqui, *Pittier 3169* (holotype US!). Epithet: honors the collector of the type and many other tropical American plants, H. H. Pittier. **PLATES 73(F), 74, 75.**

Chamaedorea hageniorum L. H. Bailey, Gentes Herb. 6: 247. 1943a. Type: Panama, Chiriqui, *von Hagen 2060* (holotype MO!; isotype BH!). Epithet: honors co-collectors of the type, Christine and Wolfgang von Hagen.

Common Name: none.
Vernacular Name: none.

Habit: solitary, erect, sometimes decumbent, to 1–2 m tall, usually flowering when appearing stemless.

Stem: 1–2 cm diam., initially creeping and subterranean or buried in leaf litter, green, densely and prominently ringed, internodes 2–3 cm long or slightly more, upper part covered with thick and durable brownish persistent leaf sheaths.

Leaves: 7–9, erect-spreading, pinnate, stiff; **sheath** 6–15 cm long, thick, durable, expanded and heavily longitudinally striated, greenish but drying dark brown and persistent; **petiole** 3–15 cm long, pale green and flattened and slightly grooved near base above and there with a raised whitish swelling, ± rounded and pale green below, stout, densely striated; **rachis** 20–40 cm long, angled and green above, rounded and pale below with a whitish band extending onto sheath; **pinnae** 5–8 on each side of rachis, to 10–20 × 2–3 cm, terminal pair ± wider, narrowly lanceolate-acuminate to narrowly oblong-lanceolate, alternate, glossy, firm, not sigmoid, gradually narrowed into straight tips, long point nearly caudate, not flat but nearly cupped downward and sometimes with revolute margins, dark forest velvety green with a ± grayish cast, a prominent midrib and 2–3 nerves on each side of this conspicuous.

Inflorescences: interfoliar but sometimes infrafoliar in fruit, long-pedunculate, erect-spreading; **peduncles** to 50 cm long, 1 cm wide at base and there flattened, 3–5 mm diam. at apex, green or pale in flower, orange in fruit where exposed; **bracts** 4, tubular, leathery, acute-acuminate, bifid, longitudinally striate-nerved, green in flower, brownish in fruit; **rachises** to 2–3 cm long, green in flower, orange in fruit. **Staminate** with 4–6 **rachillae**, these 10–20 cm long, flexuous, drooping, longitudinally striated, green in flower.

Pistillate with 2–3 **rachillae**, these 8–15 cm long, erect-ascending, striated and angled, green in flower, orange in fruit.

Flowers: Staminate in moderately dense spirals apically, more remote basally, 2.5–3 × 2.5 mm, globose-ovoid, green, slightly aromatic, sunken; **calyx** 0.5–0.75 × 1.5–3 mm, shallowly lobed, **sepals** connate and/or imbricate nearly to apex, broadly rounded apically; **petals** 2.5–4 × 2–3 mm, elliptic-oblong, valvate, free nearly to base, spreading apically, cuplike, acute, slightly recurved apically, lightly nerved; **stamens** 1.5 mm high, not equalling pistillode, **filaments** short, 0.5–0.75 mm long, whitish, **anthers** 1.5 mm long, flattened, bilobed, forming a ring around pistillode; **pistillode** 2–2.5 mm high, columnar, yellow-green, shorter than petals, slightly lobed apically.

Pistillate borne ± loosely, greenish yellow; **calyx** small, deeply lobed, **sepals** connate only briefly basally; **petals** larger, imbricate basally, spreading apically, nerved, soon spreading and chafflike; **pistil** short-oblong, obtuse.

Fruits: 10 × 6 mm, oblong, obtuse, black; perianth a small cupule with flared lobes.

Distribution: PANAMA. Chiriqui. Panama. COSTA RICA. Puntarenas. Moist forest on the Pacific slope near the Continental Divide; 1,100–1,800 m elevation.

Discussion: The description is from Bailey (1943a, 1943b) and supplemented from our living material collected near the type locality on Cerro Horqueta in Panama. *C. pittieri* is quite variable vegetatively, even among plants growing in proximity. Individuals in dense shade have longer petioles, expanded leaf blades, and longer, more remotely placed, lighter green pinnae, corresponding well with *C. hageniorum.* Nearby, no more than a few meters away, individuals in exposed situations have very short petioles, contracted leaf blades, and closely spaced, dark green pinnae that are shorter, more rounded or oblong, thicker, and almost puckered. The type of *C. pittieri* corresponds to these latter individuals from more exposed locations. There is no good, constant character to maintain these as separate species. Rather the presence of intermediate forms and

Plate 73. A–E, Chamaedorea parvisecta. F, Chamaedorea pittieri. A–E In cloud forest, Guatemala; **A.** Fruiting plant, *Hodel & Castillo 896,* Baja Verapaz (note infructescence with green immature fruits at base of plant); **B.** Plant with pinnate leaves at young age, Alta Verapaz; **C.** Few-branched infructescence (immature fruits) arises from near base of plant, *Hodel & Castillo 891,* Alta Verapaz; **D.** *Hodel & Castillo 1000,* Huehuetenango; **E.** Infructescence (immature fruits), *Hodel & Castillo 892,* Alta Verapaz. **F.** Crown of leaves, *Hodel et al. 1125,* windswept cloud forest, Chiriqui, Panama.

B
C
E
F

their similarity in flower strongly suggest that but a single species is involved here.

In the Fortuna area of Chiriqui we have made another collection that I tentatively refer to *C. pittieri*. It seems to differ from the more typical *C. pittieri* in the slender, elongated stem occasionally attaining three meters in length. When one considers the great amount of variation even between individuals within the same population, the differences between the material from Cerro Horqueta and Fortuna seem insignificant.

Vegetatively and in habit, *C. pittieri* is similar to *C. parvifolia*. Both are virtually stemless and have dark grayish green, thickened leaves. Also, they inhabit a similar type of forest. Staminate flowers with the petals connate apically and the corolla opening by lateral slits and the sigmoid pinnae drying flat and not striated distinguish *C. parvifolia*.

Other than a few plants in the research collection in Los Angeles, *C. pittieri* is not in cultivation. Of relatively easy culture, it is an attractive species due to its leafy rosette of relatively compact, stiffish, velvety green leaves.

58. Chamaedorea pochutlensis *Liebm.* in Mart., Historia Naturalis Palmarum 3: 308. 1849. Type: Mexico, Oaxaca, *Liebmann s.n.* (some sheets later numbered *6579*) (holotype C!; isotype MO!). Epithet: is named for Pochutla in Oaxaca, Mexico where the species was discovered. **PLATE 76.**

Chamaedorea karwinskyana H. A. Wendl., Allg. Gartenzeitung 21: 179. 1853c. Type: Cult., *Wendland s.n.* [holotype GOET!; isotypes C!, G (photo!), L!]. Epithet: honors Wilhelm Friedrich Karwinski von Karwin, 19th-century botanist.

Nunnezharia karwinskyana (H. A. Wendl.) O. Kuntze, Revisio Generum Plantarum 2: 730. 1891.

Nunnezharia pochutlensis (Liebm.) O. Kuntze, Revisio Generum Plantarum 2: 730. 1891.

Chamaedorea elatior Hort. in H. A. Wendl., Index Palmarum 13. 1854. Not Mart.

Chamaedorea robusta Hort.?

Common Name: none.
Vernacular Name: cannelilla—Mexico.

Habit: cespitose, erect to leaning, to 3–5 m tall or more, forming ± dense clumps 2–5 m across.

Stems: 2–3 cm diam., green, ringed, internodes 10–25 cm long, often covered with persistent sheaths.

Leaves: 3–5, erect-spreading, pinnate, to 2 m long; **sheath** to 30 cm long, tubular, obliquely open apically, longitudinally striate-nerved, green but drying pale or whitish, persistent; **petiole** to 30 cm long, ± flat and green above but toward base grooved and smooth; green and rounded below; **rachis** to 1 m long or more, angled and green above, rounded below with a pale or light green band extending onto sheath, petiole and sheath ± slightly glaucous; **pinnae** 20–33 on each side of rachis, to 40 × 2–3 cm, long-lanceolate, slightly curved or falcate or ± straight, regularly arranged, alternate or subopposite, plicate and contracted basally and there with a hard whitish swollen gland at point of attachment with rachis, acute apically, a prominent pale midrib and 2 submarginal primary nerves, secondaries and tertiaries inconspicuous.

Inflorescences: infrafoliar, erect-spreading; **peduncles** 30–40 cm long, 7–10 mm wide at base, 5 mm wide at apex, green in flower, reddish orange in fruit; **bracts** 5–7, prophyll 3 cm long, 2nd bract 6 cm, 3rd 10 cm, 4th 12 cm, 5th 15 cm, 6th 15–20 cm, 7th 8–10 cm, fibrous, dry and brownish in flower, obliquely open and bifid apically, acute-acuminate, longitudinally striate-nerved, uppermost exceeding peduncle and extending onto rachis; **rachises** 10 cm long, green in flower, reddish orange in fruit. **Staminate** with 12–25 **rachillae**, these 15–20 cm long, 2–2.5 mm diam., ± drooping, green.

Pistillate with 12–18 **rachillae**, these 15 cm long, 2 mm diam., spreading, green, ± stiff in flower, becoming ± pendulous and reddish orange in fruit.

Flowers: Staminate in ± dense spirals, 4–4.5 × 3–3.5 mm, ovoid-globose, bright yellow, strongly aromatic, slightly sunken; **calyx** 1 × 2.5–3 mm, lobed, green, **sepals** connate in basal ⅔, rounded to acute apically; **petals** 3.5–4 × 3 mm, ovate, free nearly to base, spreading, acute and slightly recurved apically, ± thick, fleshy; **stamens** 2.5 mm high, **filaments** 1–1.5 mm long, pale, **anthers** 1.5–2 mm long, not deeply 2-cleft; **pistillode** 4 mm high, columnar, equalling petals, 3-lobed apically, yellowish.

Pistillate in remote spirals, 4–5 × 3 mm, ovoid, yellow; **calyx** 1–1.5 × 3 mm, deeply lobed, green, **sepals** briefly connate and/or imbricate basally, broadly rounded apically; **petals** 3 × 3 mm, deltoid, imbricate nearly to apex, spreading, free and erect apically, acute; **pistil** 3.5 × 2.5–3 mm, ovoid-globose, green, **styles** very short or lacking, **stigma** lobes short, recurved to erect, dark, equalling petals.

Fruits: 12–13 × 8–10 mm, globose-ellipsoid, black with glaucous bloom.

Distribution: MEXICO. Colima. Durango. Guerrero. Jalisco. Michoacan. Nayarit. Oaxaca. Sinaloa. Moist forest on the Pacific slope; 50–2,000 m elevation.

Discussion: Liebmann described and named *C. pochutlensis* in Martius (1849) from material that Liebmann had collected near Pochutla in Oaxaca. It is a variable species and occurs over a rather wide range of nearly 2,000 km (1,300 miles) in western Mexico. Apparently, it is not a common plant any-

→

Plate 74. Chamaedorea pittieri. In cloud forest, Chiriqui, Panama. **A.** Staminate plant, *D. & M. Hodel 729A*; **B.** Young plant. (Note leaves and inflorescences arising from leaf litter since short stem is wholly buried.)

where and is usually found as isolated clumps on cliff sides in ravines and canyons.

Other than its original description and a few other short botanical accounts, mention of *C. pochutlensis* is lacking in the literature. On the other hand, the name *C. karwinskyana* is mentioned frequently in horticultural accounts. Wendland (1853c) described and named *C. karwinskyana* from material cultivated at Herrenhausen and originally collected in Mexico. Wendland obtained his material from Baron von Karwinski who introduced the species to Europe and who cultivated it at his garden in Munich. *C. karwinskyana* was apparently cultivated as early as the 1840s in European greenhouses and Mediterranean gardens under the name *C. elatior* (Guillaumin 1923b). Wendland (1853b) provided an explanation for the misapplication of the name when he pointed out that Martius (1837) confused *C. elatior* (which Martius himself had named in 1830) with Karwinski's species recently introduced to European gardens from Mexico. In his 1837 work, Martius illustrated this other species as *C. elatior.* Wendland (1853c) recognized this error and named the other species *C. karwinskyana.* According to Burret (1935) material referable to *C. karwinskyana* was still cultivated in European collections as *C. elatior* as recently as the 1930s.

Chamaedorea pochutlensis has been cultivated in California since the early 1900s, most of the plants there probably originating from collections made in Mexico by Edward Howard for the Doheny Estate in Los Angeles. Plants probably of the same origin were at the Vavra Estate (formerly property of the University of California at Los Angeles) in the 1960s and still exist at the Huntington Botanical Gardens in San Marino, California. The Huntington has added to its holdings of *C. pochutlensis* with recent collections from Durango and Jalisco in Mexico.

Surprisingly, most plants of *C. pochutlensis* in California are misidentified as *C. costaricana.* The two are similar in their long-pinnate leaves and clustering habit but the presence on the ligules at the apex of the leaf sheath easily distinguishes *C. costaricana,* a feature lacking in *C. pochutlensis.* Hertrich (1951) discussed and Muirhead (1961) illustrated *C. pochutlensis* at the Huntington Botanical Gardens erroneously as *Collinia elegans.*

Chamaedorea pochutlensis is relatively hardy and tolerates rather high light intensities. Old established clumps at the Huntington Botanical Gardens showed little damage at −3°C (26°F) and recovered from severe damage and defoliation at −7°C (19–20°F). Several clumps about 2–3 meters wide and tall at the Huntington are exposed to full sun yet display only slight yellowing on leaves on their south side.

A hybrid between *C. pochutlensis* and *C. graminifolia* is discussed in the chapter on hybrids. Indigenous peoples in Jalisco, Mexico use *C. pochutlensis* to treat dysentery.

59. Chamaedorea pumila *H. A. Wendl. ex Dammer,* Gard. Chron. Ser. 3, 36: 246. 1904b. Type: Costa Rica, Heredia, Rio Sarapiqui, *Wendland s.n.* [holotype K!; isotype GOET (photo!)]. Epithet: from the Latin *pumilus* meaning dwarf or short, in reference to the habit. **PLATES 2(A), 77, 78.**

Nunnezharia pumila (H. A. Wendl.) O. Kuntze, Revisio Generum Plantarum 2: 730. 1891.

Chamaedorea nana N. E. Brown, Kew Bull. 156, 1914. Type: Cult., *Brown s.n.* (holotype K!). Epithet: from the Latin *nanus* meaning dwarf, in reference to the habit.

Kinetostigma nana (N. E. Brown) Burret, Notizbl. Bot. Gart. Berlin-Dahlem 11: 318. 1932.

Common Name: none.
Vernacular Name: none.

Habit: solitary, erect to decumbent, often appearing stemless, to 50 cm tall.

Stem: 8–20 mm diam., often creeping at or near ground level, dark green, densely and prominently ringed, internodes no more than 1 cm long and usually much shorter, often covered with adventitious roots.

Leaves: 6–10, erect-spreading, bifid; **sheath** to 10 cm long, very open, splitting deeply opposite petiole and clasping in a tubular manner only near base for 1–3 cm, remainder or upper ½ of sheath not clasping stem and appearing as extension of petiole, gray-green, densely and minutely white-spotted, whitish-margined and there green longitudinally striate-nerved; **petiole** 2–5 cm long, flat, gray-green, and channeled above, margins of sheath extending upward on either side to form sides of channel, rounded and gray-green below with minute white spots; **rachis** 6–15 cm long, angled and green above, rounded and gray-green below with minute white spots; **blade** to 30 × 10–15 cm, obovate to subobovate-elliptic, incised apically ½ or more its length, lobes 10–15 cm long, thick and leathery, velvety iridescent dark gray-green, plicate, sometimes slightly mottled, acute, exterior margin toothed

→

Plate 75. Chamaedorea pittieri. In cloud forest, Chiriqui, Panama. **A.** Old plant with emergent stem and infructescences (immature fruits), *D. & M. Hodel 729B* (note long peduncles and stiff, litter-collecting leaves); **B–D.** Young plants, *Hodel et al. 1125.*

C
D

toward apex, decurrent on petiole, 9–13 prominent primary nerves on each side of rachis, these elevated above, pale below.

Inflorescences: interfoliar, erect-spreading, equalling or slightly exceeding leaves; **peduncles** 15–20 cm long, 2.5–4 mm wide, enclosed by revolute decurrent margin of leaf sheath and then emerging above petiole, green or whitish in flower, orange in fruit; **bracts** 5–6, tubular, membranous, green but rotting away before anthesis, acute, bifid, longitudinally striate-nerved. **Staminate** with **rachis** 2.5 cm long, green in flower; **rachillae** 4–10, these 8–15 cm long, 1–2 mm diam., green, recurved-drooping, simple.

Pistillate spicate or rarely furcate; **rachis** or flower-bearing portion to 8–15 cm long, recurved, ± stiff, 2 mm diam. and green in flower, becoming swollen to 3–5 mm diam. and reddish orange in fruit.

Flowers: Staminate in moderately dense spirals, 4–5 × 4 mm, globose, greenish yellow, slightly sunken; **calyx** 1 × 2.5 mm, deeply lobed, green, **sepals** connate in basal ½, rounded to acute apically; **petals** 3–4 × 3 mm, ovate, valvate, free nearly to base and there briefly connate, thick, fleshy, acute and incurved apically; **stamens** 2 mm high, **filaments** to 1 mm long, adnate basally to pistillode in a buttresslike structure, **anthers** 0.75 mm long, light yellow or pale, bilobed; **pistillode** 2.5–3.5 mm long, shorter than petals, columnar with a broad 3-angled cap, buttressed basally, light greenish yellow.

Pistillate in moderate spirals, 2.5–3 × 2–2.5 mm, conic or subglobose, greenish or greenish yellow, slightly sunken; **calyx** 1 × 2.5 mm, deeply lobed, green, **sepals** connate in basal ⅓, broadly acute apically; **petals** 3 × 3 mm, cup-shaped, imbricate nearly to apex, thick, fleshy, acute, slightly erect apically; **pistil** 1.5–2 × 2 mm, globose or subglobose, flattened apically, **stigma** lobes sessile, very short, slightly separated, whitish.

Fruits: 6–10 mm, globose, black.

Distribution: COSTA RICA. Alajuela. Cartago. Guanacaste. Heredia. San Jose. Puntarenas. Wet forest mainly on the Atlantic slope but occasionally on the Pacific slope; 400–1,500 m elevation.

Discussion: Wendland discovered *C. pumila* near San Miguel in the valley of the Rio Sarapiqui in Costa Rica. He returned to Germany and grew it in his greenhouses at Herrenhausen. Dammer (1904b) formally described and named the species after Wendland's death using the name that Wendland had suggested. In 1914, Brown described and named *C. nana* from a plant he received at Kew from the horticultural firm of Sanders & Sons Nurseries. Sanders & Sons had introduced this plant from Costa Rica and distributed it under the name *C. pumila.*

Because of its smaller habit and leaf blades and shorter petioles, Brown felt that *C. nana* was distinct from *C. pumila.* I have examined Wendland's type of *C. pumila* and Brown's type of *C. nana,* both from Kew, and can find no distinguishing differences. The flowers of the two are essentially identical. Wendland's type has leaves that are incised apically to about the middle of the leaf blade while Brown's *C. nana* is incised slightly more than half way. Also, other specimens from Costa Rica at the Missouri Botanical Garden exhibit a range of foliar variation that tends to bridge the small gap between Wendland's and Brown's specimens.

Chamaedorea pumila is similar to *C. sullivaniorum* and *C. minima*; it is somewhat intermediate between the two. *C. sullivaniorum* differs in the blade bifid apically only one-fourth to one-third its length, more nerves, staminate flowers with much thinner petals and pistillode not buttressed and grooved basally, and the pistillate flowers with the stigma lobes situated in a depression at the apex of the pistil. *C. minima* differs in its smaller, gray-green, not mottled or iridescent blades with no more than seven nerves on each side of the rachis.

All forms of *C. pumila* are very handsome plants because of their virtually stemless habit and rosette of numerous, thick, bifid leaves with a corrugated texture and elevated nerves. They would be very striking as a single specimen in a container or as a mass planting. Although cultivated to some extent in Costa Rica, the species is relatively unknown elsewhere.

60. Chamaedorea pygmaea *H. A. Wendl.,* Allg. Gartenzeitung 20: 217. 1852c. Type (syntypes): Cult., *Wendland s.n.* [C!, G (photo!), GOET!, L!] and Colombia, Socorro, *Funck & Schlim 1654* (BR!, GOET!). Epithet: from the Latin *pygmaeus* meaning dwarf, in reference to the habit. **PLATE 79(A).**

Stachyophorbe pygmaea (H. A. Wendl.) Oerst., Vidensk. Meddel. Dansk. Naturhist. Foren. Kjoebenhavn 1858: 10. 1859.

Nunnezharia pygmaea (H. A. Wendl.) O. Kuntze, Revisio Generum Plantarum 2: 731. 1891.

Chamaedorea terryorum Standl., Publ. Field Mus. Nat. Hist., Bot. Ser. 22: 326. 1940. Type: Panama, Darien, *M. & R. Terry 1452* (holotype F!; isotypes GH!, MO!). Epithet: honors the co-collectors of the type, M. E. and R. A. Terry.

Cladandra pygmaea (H. A. Wendl.) O. F. Cook, Nat. Hort. Mag. 22: 148. 1943b. Name of no botanical standing.

Common Name: none.
Vernacular Name: none.

→

Plate 76. Chamaedorea pochutlensis. Cultivated, California. **A-C.** Conservatory, Balboa Park, San Diego; **D.** Garden of P. Sullivan, Ventura.

B
D

Habit: solitary, erect or decumbent, appearing stemless, to 60 cm tall.

Stem: 1–2 cm diam., creeping at or below ground or leaf litter and attaining 30–40 cm in length, green, densely ringed, internodes to 5 mm long, often covered with persistent leaf sheaths.

Leaves: 3–8, erect-spreading, pinnate, variously pinnate or occasionally bifid; **sheath** 5–10 cm long, long-open, splitting deeply opposite petiole and tubular only near base, green with brownish margin, ageing brown and persisting, densely striated; **petiole** 5–25 cm long, slender, lightly grooved especially near base and green above, rounded and greenish below; **rachis** 12–16 cm long, angled and green above, rounded and green below; **pinnae** usually 9–12 on each side of rachis, 12–15 × 1–2.5 cm, lanceolate or elongate-lanceolate, long-attenuate, falcate, thin, lower margins briefly decurrent on rachis, upper pinnae wider and not decreasing noticeably in length toward apex of rachis, weakly toothed, a prominent midrib and one primary nerve conspicuous on each side of this above; if bifid, **blade** 20–30 × 6–15 cm, cuneate-obovate, incised apically to ⅓ its length, lobes long, thin, narrowly acuminate, narrowed basally, olive-green above in dried state, paler below, 12–14 prominent nerves on each side of rachis, 2 secondary nerves conspicuous below between each pair of primaries, tertiaries numerous and less conspicuous; or blade variously pinnate with a broad terminal pair and several basal pinnae, lower margin of lowest pinnae obscurely decurrent along petiole to sheath.

Inflorescences: inter- or infrafoliar, erect-ascending, arising from base, long-pedunculate, shorter than or about equalling leaves; **peduncles** to 20 cm long, erect; **bracts** 4–8, to 15 cm long, narrowly tubular, 3 mm wide, membranous, bifid, longitudinally striate-nerved, brown in flower; **Staminate** with **rachis** to 10 cm long, slender, straight, greenish in flower; **rachillae** 10–25, these to 6–8 cm long, longest near base and becoming progressively shorter toward apex of rachis, slender, widely spreading or ± recurved or drooping, flexuous, greenish in flower.

Pistillate spicate or forked; **rachis** or flower-bearing portion 10 cm long, curved, pale greenish in flower, orange and swollen in fruit.

Flowers: Staminate remotely arranged, green, fragrant, sessile; **calyx** 1 × 1.5 mm, shallowly lobed, **sepals** connate in basal ¾, rounded apically; **petals** 3 mm long, oblong-obovate, valvate, free nearly to base, broadly obtuse or nearly rotund to acute, fleshy, brownish in dried state; **stamens** ½–¾ as high as petals, **filaments** short, thick, **anthers** elongate.

Pistillate fairly densely arranged, green; **calyx** lobed, **sepals** bluntly rounded apically; **petals** elongated, short-pointed, imbricate basally, spreading apically; **staminodes** 6, broadly uniform; **pistil** depressed-globose, 3-lobed, **stigma** lobes sessile, recurved slightly.

Fruits: 6–8 mm long, ± oblong, black; eophyll pinnate.

Distribution: COLOMBIA. Antioquia. PANAMA. Chiriqui. Darien. COSTA RICA. Puntarenas. Wet forest and cloud forest; 200–2,000 m elevation.

Discussion: The description is from Wendland (1852c) who described and named *C. pygmaea* from cultivated plants that Linden introduced to Europe, apparently from Diablo in Socorro Province of New Granada (Colombia) prior to 1850, and is supplemented from Standley (1940). Interestingly, the word "Chiapas" is written on a syntype at Copenhagen despite the fact that Wendland stated in his 1852 article that *C. pygmaea* was collected in Colombia. However, whoever wrote "Chiapas" easily could have been confused since botanists of the time used New Granada for both Colombia and Chiapas, Mexico. *C. pygmaea* is extremely variable vegetatively with leaves being completely pinnate to bifid. The type has leaves with up to 12 pinnae on each side of the rachis. Also, *C. pygmaea* has the widest altitudinal range of any species of the genus. Although occurring mainly from 600–2,000 m elevation, a few collections from the Darien region in Panama adjacent to Colombia place it as low as 200 meters.

Standley (1940) described and named *C. terryorum* from the Darien, and the type (*M. & R. Terry 1452*) had bifid leaves. Another collection from the same population (*M. & R. Terry 1453*) has completely pinnate leaves. The latter collection fits *C. pygmaea* very well, substantiating the case that *C. terryorum* is a bifid-leaved form of *C. pygmaea.*

Chamaedorea pygmaea is close to *C. stenocarpa* but the pinnae decreasing noticeably in length toward the apex of the rachis distinguish the latter species. *C. pygmaea, C. stenocarpa, C. brachyclada, C. scheryi,* and *C. undulatifolia,* form a natural subgroup within subgenus *Chamaedoropsis* characterized by their virtually stemless habit, decurrent pinnae, long-pedunculate inflorescences ascending from or below ground level, and pinnate eophylls (Hodel and Uhl 1990b).

Although Guillaumin (1923b) reported that *C. pygmaea* was once common in European glasshouses, it is rare in cultivation today. The only living plants that we have seen are a few cultivated in a remote location of Jardin Botanico Robert y Catherine Wilson in Costa Rica. Bruce McAlpin collected these in 1974 in the upper valley of the Rio Coton, Puntarenas, Costa Rica. Like other related species from relatively high altitudes, *C. pygmaea* is difficult to cultivate and frequently brown-tips from warm temperatures, highly mineralized water, and/or low humidity.

Plate 77. Chamaedorea pumila. A. Pistillate plant, *Hodel et al. 969,* montane rain forest, Braulio Carrillo National Park, Costa Rica; **B.** Staminate plant, *Hodel 826,* cultivated, Virginia Robinson Botanical Garden, Beverly Hills, California *ex* Costa Rica.

61. Chamaedorea queroana *Hodel,* Phytologia 68: 406. 1990c. Type: Mexico, Oaxaca, *D. & R. Hodel 943B* (holotype BH!; isotype MEXU!). Epithet: honors Hermilo Quero-Rico, Mexican botanist and former Director of the Jardin Botanico, Universidad Nacional Autonoma de Mexico. **PLATE 79(B-G).**

Common Name: none.
Vernacular Name: none.

Habit: solitary, erect or decumbent, appearing stemless, to 1 m tall.

Stem: 2–3 cm diam., to 10 cm long, creeping in leaf litter, densely and prominently ringed, internodes very short, 5 mm long, covered with persistent leaf sheaths.

Leaves: 4–5, erect-spreading, regularly or variously and unevenly pinnate but occasionally bifid; **sheath** 7–14 cm long, obliquely long-open, tubular only near base, green but pale along apical margins, longitudinally striate-nerved; **petiole** 15–70 cm long, dark green, minutely but conspicuously white-spotted, flattened above and grooved near base, rounded below; **rachis** 10–40 cm long if blade pinnate, to 10 cm long if bifid, angled and green above, rounded and green below; **pinnae** to 10 on each side of rachis, middle ones longest, to 16 × 3–4 cm, lanceolate, sigmoid, acuminate, contracted basally, thin, lightly mottled green, a prominent midrib flanked by two submarginal primary nerves, 1 secondary between each of these, tertiaries numerous, faint, apical pair of pinnae with 3 primary nerves, secondaries and tertiaries numerous, faint; if bifid, **blade** to 25 × 18 cm, incised apically to more than ½ its length, exterior margins slightly toothed toward apex, 8–10 primary nerves on each side of rachis, 2 faint secondaries between each pair of primaries, tertiaries numerous, inconspicuous; or blade variously pinnate with larger terminal pair and smaller basal pinnae.

Inflorescences: infrafoliar, erect-arcuate, long-pedunculate; **peduncles** to 50 cm long, stiff, straight, 5–7 mm wide at base, 2–4 mm wide at apex, greenish in flower where exposed, bright orange in fruit; **bracts** 6–9, prophyll 1–3 cm long, 2nd bract 1.5–5 cm, 3rd 2.5–8 cm, 4th 3.5–15 cm, 5th 6–20 cm, 6th 8–12 cm, 7th 8–10 cm, 8th 7–10 cm and concealing rudimentary 9th that is 1 cm long, tubular, acute-acuminate, bifid, longitudinally striate-nerved, green to brown in flower, brown or fallen in fruit. **Staminate** with **rachis** to 10 cm long, green in flower; **rachillae** 6–13, basal ones longest, to 18 cm long, apical ones to 15 cm long, 1–1.5 mm diam., erect, pale green in flower.

Pistillate spicate or forked; **rachillae** or flower-bearing portion to 7 cm long, stiff, erect, 1–1.5 mm diam. and white to pale green in flower, 2.5–4 mm diam. and bright orange in fruit, longitudinally and finely striate-angled when dry.

Flowers: Staminate in moderate spirals, 2.5–3 × 1.5 mm, ovoid-oblong; **calyx** 0.5 × 1.5 mm, moderately lobed, thin, **sepals** connate in basal ½, rounded apically; **petals** 3 × 1.5 mm, oblong, connate in basal ⅔ and forming a distinct campanulate corolla tube, acute, free and slightly reflexed apically, white tinged with green apically; **stamens** 1–1.25 mm tall, **filaments** short or nearly lacking, to 0.3 mm long, these ± triangular, fleshy basally, **anthers** 0.8–1 mm long, slender, basifixed, flush around pistillode; **pistillode** 2–2.5 mm high, slender, grooved.

Pistillate in 6 dense spirals, 2–3 mm apart, 2.5–3 × 2.5 mm, ± ovoid, leaving superficial and rounded scars 1 mm wide; **calyx** when dry raised 0.25 mm above axis on a stalk 0.75 mm diam., explanate, 0.75-1 × 2 mm, shallowly lobed, membranous to ± fleshy, **sepals** connate and/or imbricate in basal ¾, broadly rounded apically; corolla raised 0.5 mm above calyx on central stalk when dry, **petals** 2.5–3 × 3 mm, broadly ovate, imbricate in basal ½, erect or slightly recurved apically, acute, thick, fleshy, durable, pale green tinged with dark green apically, brown above, obscurely nerved when dry; 1 **staminode** present, toothlike; **pistil** 1–1.25 × 0.75–1 mm, globose-ovoid, pale green, **styles** short, 0.25 mm long, **stigma** lobes recurved, darkened.

Fruits: 14 × 7.5 mm, ovoid, tip pointed, black; fruiting perianth persistent with petals bright shining orange, conspicuous, fleshy, to 3 mm long with a prominent midrib and lateral nerves below.

Distribution: MEXICO. Oaxaca. Wet forest on the Atlantic slope; 1,300 m elevation.

Discussion: *C. queroana* is quite distinct among Mexican members of the genus and actually appears closest to *C. pumila* from the wet Atlantic slope of Costa Rica. However, the generally much smaller habit, thick, leathery leaves, and drooping staminate rachillae distinguish *C. pumila.*

Staminate flowers of *C. queroana* are quite unusual, the connate petals with flared apexes forming a rather distinctive, campanulate corolla tube. Other than a few plants in the research collection in Los Angeles, *C. queroana* is not cultivated.

62. Chamaedorea quezalteca *Standl. & Steyerm.,* Publ. Field. Mus. Nat. Hist., Bot. Ser. 23: 204. 1947. Type: Guatemala, Quetzaltenango, *Standley 87159* (holotype F!). Epithet: from Quetzaltenango, the department in Guatemala where the type was collected. **PLATES 80, 81.**

Legnea laciniata O. F. Cook, Nat. Hort. Mag. 22: 134. 1943b ("*lacinata*"). Name of no botanical standing.

Common Name: none.
Vernacular Names: ternera, chicuilote—Guatemala.

Habit: cespitose, erect, infrequently decumbent, to 5 m tall or more, forming clumps 2–3 m across.

Stems: 1–3 cm diam., green, ringed, internodes to 20 cm long.

Plate 78. Chamaedorea pumila. A-C Cultivated, research collection, Los Angeles, *ex* Costa Rica; **A, B.** Staminate flowers and inflorescence, *Hodel 826*; **C.** Pistillate flowers, *Hodel 721*; **D.** Small fruiting plant, *Hodel & Grayum 981,* montane rain forest, Pavones, Cartago, Costa Rica; **E.** Pistillate plant, *Hodel et al. 974,* montane rain forest, Braulio Carrillo National Park, Costa Rica.

→

B
D

Leaves: 3–5, spreading, pinnate; **sheath** to 40 cm long, tubular, green, becoming dry and brownish and then persistent, durable, obliquely open apically and there a prominent fibrous ligule 5–10 cm long on each side of petiole, longitudinally striate-nerved; **petiole** 20–30 cm long, green and shallowly grooved above near base to nearly flat near apex, rounded and pale below; **rachis** to 1 m long, flat to angled and green above, rounded below with a pale band extending onto sheath; **pinnae** 15–20 on each side of rachis, to 30–45 × 3–5 cm, lanceolate, sigmoid, falcate, acuminate, regularly arranged, apical pair sometimes broader than others, a midrib and 2 primary nerves on each side of this prominent above, secondaries conspicuous, tertiaries slender, numerous, inconspicuous, pinnae with a prominent green soft pulvinus above and below.

Inflorescences: infrafoliar, emerging through or from behind old sheaths, erect although the pistillate tends to become spreading or nodding when heavily laden with fruits; **peduncles** 30–45 cm long, green in flower, orange in fruit; **bracts** 6, tubular, fibrous, obliquely open apically, brownish in flower, longitudinally striate-nerved, acute, bifid, uppermost exceeding peduncle and extending onto rachis. **Staminate** with **rachis** 10–15 cm long, green; **rachillae** 15–25, these 15–20 cm long, 2.5 mm diam., spreading to pendulous, green.

Pistillate with **rachis** shorter, to 5 cm long, green in flower, reddish orange in fruit; **rachillae** 6–10, these to 15–20 cm long, 3 mm diam., spreading, lightly longitudinally ridged, spreading and green in flower becoming drooping and red-orange in fruit, subangular and ridged when dry.

Flowers: Staminate in rather dense spirals, 2.5–4 × 2.5–4 mm, ovoid-globose, yellow ageing to orange-brown, ± aromatic, slightly sunken in elliptic depressions 2 mm long; **calyx** 0.5 × 2.5 mm, deeply lobed, yellowish or whitish, **sepals** connate and/or imbricate only briefly basally, broadly rounded to straight apically; **petals** 2–4 × 2.5–4 mm, triangular to broadly ovate, valvate, free nearly to base, spreading and erect to incurved apically, acute; **stamens** 1–1.5 mm high, ½ as high as pistillode, essentially sessile, **filaments** very short, nearly lacking, **anthers** 1–1.5 mm long, bifid apically and outwardly curved; **pistillode** 2–3.5 mm high, columnar to barrel-shaped, yellow with an angled and 3-grooved green cap.

Pistillate in remote spirals, 3 × 3.5–4 mm, depressed-globose, yellow, only slightly sunken in broadly oval pits to 1.5 mm long; **calyx** 1 × 3 mm, shallowly lobed, light green to pale, **sepals** imbricate in basal ¾, broadly rounded to straight apically; **petals** 3 × 4 mm, broadly triangular, imbricate nearly to apex, acute, shortly erect; **staminodes** short, whitish; **pistil** 2 × 2.5 mm, depressed-globose, green, **styles** lacking, lobes short, thick, pointed, recurved, sessile, ± equalling petals.

Fruits: to 10 mm diam., globose, black.

Distribution: MEXICO. Chiapas. GUATEMALA. Alta Verapaz. Baja Verapaz. Chiquimula. El Progreso. Escuintla. Huehuetenango. Quetzaltenango. San Marcos. Suchitepequez. Zacapa. EL SALVADOR. Ahuachapan. Santa Ana. HONDURAS. Atlantida. Comayagua. Morazan. Tegucigalpa. Yoro. Dense, moist or wet forest on the Pacific and Atlantic slopes; 1,000–1,500 m elevation.

Discussion: A handsome species with long-pinnate, spreading leaves and clustering, smooth, green, canelike stems, Standley and Steyermark (1947) described and named *C. quezalteca* from a pistillate collection from the lower slope of Volcan Santa Maria in Quetzaltenango, Guatemala. We found it near the type locality on the Pacific slope and at other localities throughout Guatemala. For such a little-used name, the plant is surprisingly widely cultivated in Guatemala, appearing in parks and plazas of many cities. Exceptionally fine specimens can be seen in the botanical gardens in Guatemala City. In Quetzaltenango, local residents sell the cut leaves for use in flower arrangements and bouquets. In Baja Verapaz where it is called *ternera* and *chicuilote,* the center bud is eaten, usually in a salad.

Chamaedorea quezalteca is related to and perhaps not distinct from *C. costaricana*. However, they differ in the nature of the ligules at the apex of the leaf sheath, and the difference appears to be constant over the range of the two species. In *C. quezalteca* these are fibrous and persistent, attaining a length of five to ten cm while in *C. costaricana* they are membranous with the upper portion rotting or weathering away to leave a persistent triangular base only about one to two cm high.

Chamaedorea quezalteca is rare in cultivation outside its range; the only plants recorded occur in a few collections in the greenhouse at the L. H. Bailey Hortorium at Cornell University, Ithaca, New York where they are erroneously labeled *C. pulchra* and in southern California. The plants in California were grown from seeds that Alan Bredeson of Lemon Grove collected from cultivated plants near Cerro Volcan Verde in El Salvador in 1976. Lois Rossten of Huntington Beach has a staminate plant and Richard Palmer of Whittier and Bredeson have pistillate plants. We have successfully hand-pollinated Palmer's plant with pollen from Rossten's and obtained fruits. The immature plants in the greenhouse at Cornell University were grown from seeds that

---

Plate 79. A, Chamaedorea pygmaea. B-G, Chamaedorea queroana. A. Cultivated, Jardin Botanico Robert y Catherine Wilson (*75-324*), Costa Rica, originally collected in Las Alturas de Coton by Bruce McAlpin. **B-G** In wet cloud forest, *D. & R. Hodel 943A* (staminate) *& 943B* (holotype) (pistillate), Oaxaca, Mexico; **B.** Spicate pistillate inflorescence arising from leaf litter; **C.** Large staminate plant with long peduncles, erect rachillae; **D.** Pistillate plant; **E.** Mottled leaves (note pinnate and bifid leaves); **F.** Infructescence with mature fruits arises from the leaf litter; **G.** Pistillate plant (note white inflorescences arising from base of plant).

C
E
G

the late H. E. Moore, Jr. collected in Alta Verapaz, Guatemala. We distributed seeds in 1989 of our collection from Guatemala (*Hodel & Castillo 867*). *C. quezalteca* is susceptible to infestations of mites.

63. Chamaedorea rhizomatosa *Hodel,* Phytologia 68: 401. 1990c. Type: Mexico, Oaxaca, *D. & R. Hodel 936* (holotype BH!; isotype MEXU!). Epithet: from the Latin *rhizomatosus* meaning provided with a well developed rhizome. **PLATE 82(A).**

Common Name: none.
Vernacular Name: none.

Habit: cespitose by means of long rhizomes, forming well separated loose clusters of 6–12 erect or decumbent stems in area of 4 square m, to 2.5 m tall.

Stems: 7–9 mm diam., green, ringed, internodes 2–10 cm long.

Leaves: 3–6, erect-spreading, pinnate; **sheath** to 16 cm long, tubular, green, obliquely open apically, longitudinally striate-nerved; **petiole** 10–20 cm long, flat and green above, rounded and light green below, longitudinally striate-nerved; **rachis** 25–35 cm long, angled and green above, rounded below with light yellow band extending onto sheath; **pinnae** 6–8 on each side of rachis, middle ones longest, these to 25 × 5 cm, basal ones 14 × 2 cm, lanceolate, slightly sigmoid, acuminate, contracted basally, a prominently raised midrib and two slightly less prominent submarginal primary nerves above, these yellowish below, 3–5 secondaries between midrib and each primary, tertiaries numerous, faint, apical pair of pinnae 13–17 × 5 cm with 5–7 primary nerves, 2–3 secondaries between each pair of primaries.

Inflorescences: Staminate straight, erect or spreading; **peduncle** to 15 cm long, 4 mm wide at base and there flattened, 2.5 mm diam. at apex; **bracts** 4, tubular, obliquely open apically, acute-acuminate, bifid, finely longitudinally striate-nerved, prophyll 2–3 cm long, 2nd bract 6 cm, 3rd 7 cm, 4th 5 cm and exceeding peduncle and extending onto rachis; **rachis** 5 cm long; **rachillae** 12, to 10 cm long, drooping.

Pistillate not seen.

Flowers: Staminate in moderately dense spirals, 2 mm apart, 2 × 2.5 mm in bud, globose, ± superficial but leaving elliptic scars 2.5 mm long; **calyx** 1–1.25 × 2 mm, lobed, **sepals** imbricate basally, rounded apically; **petals** 2 × 2.5 mm, broadly ovate, valvate, acute, apparently free apically; **stamens** 1.25–1.5 mm high, in a fairly tight ring around and shorter than pistillode, **filaments** 1 mm long, **anthers** 0.75–1 mm long; **pistillode** 1.75–2 mm high, ± equalling petals, columnar, flared apically.

Pistillate not seen.

Fruits: not seen.

Distribution: MEXICO. Oaxaca. Moist pine-oak forest on a steep and rocky substrate on the Pacific slope; 1,400 m elevation.

Discussion: *C. rhizomatosa* is unusual in its cespitose habit and rhizomatous stems and pinnate leaves. Only *C. brachypoda, C. stolonifera,* and *C. graminifolia* resemble *C. rhizomatosa* in their stems spreading by rhizomes or stolons. However, *C. brachypoda* and *C. stolonifera* have bifid leaves and, while *C. graminifolia* has pinnate leaves, it is amply distinct in having 20–30 narrowly linear pinnae, much stouter stems, and larger habit. Good staminate flowers of *C. rhizomatosa* are not known so its placement in subgenus *Chamaedoropsis* is somewhat tentative. We have established a few plants of *C. rhizomatosa* in the research collection in Los Angeles and hope to obtain flowers to verify its correct placement.

64. Chamaedorea rigida *H. A. Wendl. ex Dammer,* Gard. Chron., Ser. 3, 36: 246. 1904b. Type: Mexico, Oaxaca, *Galeotti 4974* (holotype BR!; isotypes GOET!, P!). Epithet: from the Latin *rigidus* meaning rigid, apparently in reference to the rigid leaf blades. **PLATES 5(C), 82(B, C).**

Nunnezharia rigida (H. A. Wendl. ex Dammer) O. Kuntze, Revisio Generum Plantarum 2: 731. 1891.

Common Name: none.
Vernacular Name: none.

Habit: solitary, erect to leaning, initially appearing stemless and flowering then but eventually to 2 m tall.

Stem: dark green, shining, smooth, prominently ringed, internodes to 4 cm long.

Leaves: 10–15, erect-spreading, bifid; **sheath** to 15 cm long, deeply split opposite petiole and tubular only in basal 3–4 cm, green, very stiff and hard, durable, nearly woody; **petiole** to 7 cm long or lacking; **rachis** 20 cm long; **blade** 35–45 × 14 cm, obovate, incised apically ½ its length, rigid, dark shining green, thickened, cuneate-angustate basally, lobes elongate-triangular, 35–45 × 13 cm, 6–10 cm wide at base, minutely toothed on apical margin, 12–15 primary nerves on each side of rachis, 2 secondaries between each pair of primaries, tertiaries numerous, faint.

Inflorescences: Staminate not seen. **Pistillate** 40–50 cm long; **peduncle** 45 cm long, 6 mm wide at base and there flattened, 2 mm diam. at apex; **bracts** 6–7, prophyll 3 cm long, 2nd bract 8 cm, 3rd 15 cm, 4th 20 cm, 5th 25–30 cm, 6th 20 cm, 7th 9 cm, tubular, longitudinally striate-nerved, acute-acuminate, bifid, uppermost exceeding peduncle; **rachis** to 3 cm long, green in flower, orange in fruit; **rachillae** 3–5, these to 12 cm long, slender, longitudinally angled, green in flower, orange in fruit, flexuous or undulate when dry.

Flowers: Staminate not seen. **Pistillate** (only fruiting perianth seen) with **calyx** 0.75 mm high, lobed; **petals** 3 × 1.5 mm, acute, faintly nerved when dry.

Fruits: 7–8 mm diam., globose, black; perianth adherent to fruit.

Plate 80. Chamaedorea quezalteca. Large, handsome clump, cultivated, University of San Carlos Botanical Garden, Guatemala. →

Distribution: MEXICO. Oaxaca. Dense, wet forest on the Atlantic slope; 1,700–1,900 m elevation.

Discussion: Dammer (1904b) described and named *C. rigida* using a name that Wendland had suggested. The description is supplemented from my own collection in Oaxaca and an examination of the type. Its placement in subgenus *Chamaedoropsis* is tentative since we are lacking staminate flowers. Vegetatively, *C. rigida* bears a resemblance to *C. stricta* but the stemless and larger habit, larger leaves with conspicuous petioles, thinner blades with parallel margins, and non-flexuous pistillate rachillae distinguish the latter species.

Chamaedorea rigida is one of the most striking members of the genus with its rigid crown of numerous, stiff, thick, dark green, bifid leaves with short or no petioles. On small plants the bifid blades seem to be arising from the leaf litter of the forest floor. The crown of stiff, ascending leaves tends to act as a litter collector, funnelling fallen leaves, twigs, and other material into the center of the plant where it can build up around the stem to 50 cm in depth. Known from only a few collections, *C. rigida* is rare in the wild and not cultivated.

65. Chamaedorea robertii *Hodel & Uhl,* Principes 34: 120. 1990c. Type: *D. & R. Hodel 737* (holotype BH!; isotype PMA!). Epithet: honors my son Robert, co-collector of the type. **PLATES 83, 84.**

Common Name: none.
Vernacular Name: none.

Habit: solitary, decumbent then shortly erect, to 1 m tall but often appearing stemless.

Stem: 2.5 cm diam., creeping at or slightly below leaf litter, green, prominently ringed, internodes 1.5 cm long.

Leaves: 5–7, erect-spreading, bifid; **sheath** to 12 cm long, splitting deeply opposite petiole, tubular only in basal ⅓, green, ragged and brown-margined apically; **petiole** 20–25 cm long, flattened and green above and slightly channeled from lower margins of blade extending downward to sheath, rounded and green below; **rachis** 20 cm long, angled and green above, rounded and green below; **blade** 40–50 × 20 cm, incised apically to nearly ½ its length, each lobe 20–25 cm long, rich green, tips 18–20 cm apart, acuminate, 12–16 raised and prominent primary nerves on each side of rachis, secondaries numerous and faint, margin conspicuously toothed.

Inflorescences: infrafoliar or sometimes interfoliar, often emerging from the leaf litter of forest floor, spicate. **Staminate** with **peduncle** to 25 cm long, 5 mm wide at base and there ± flattened, 4 mm diam. at apex, erect-ascending, pale green or yellowish where exposed in flower; **bracts** 5, prophyll 2 cm long, 2nd bract 3–4 cm, 3rd 5 cm, 4th 12 cm, 5th 15 cm, acute-acuminate, greenish at anthesis, ageing to dark brown or nearly black, tubular, especially upper ones inflated apically to 8 mm diam. for ⅔ their length, ± leathery, longitudinally striate-nerved, uppermost equalling or slightly exceeding peduncle, sometimes 5th bract very short and concealed by 4th one; **rachis** or flower-bearing portion to 10–15 cm long, 3–4 mm diam., pendulous, whitish, longitudinally ridged around each flower.

Pistillate ascending but often spreading in fruit; **peduncle** similar to that of staminate but orange in fruit; **bracts** 5, similar to those of staminate but burgundy in color when newly emerged, browning only slightly by anthesis, brownish in fruit; **rachis** to 10 cm long, ± stiff, horizontal, pale yellow or whitish in flower, 5 mm diam., orange in fruit.

Flowers: Staminate in 3 dense spirals, closely placed but not contiguous, 5 × 3.5 mm, bullet-shaped, greenish, sunken in elliptic pits 3–4 mm long; **calyx** 1–1.25 × 3.5 mm, membranous, whitish tinged with green apically, shallowly lobed, **sepals** imbricate in basal ¾, broadly rounded apically; **petals** 4 × 3 mm, valvate, free more than ½ to base, erect, acute, whitish basally; **stamens** exserted beyond petals, **filaments** 4 × 0.6 mm, columnar, clear-colored, **anthers** 1.25 mm long, bilobed, held beyond petals; **pistillode** 4 × 0.75 mm, columnar, clear-colored, 3-lobed apically.

Pistillate densely arranged, some contiguous but most not, 1–1.5 mm distant, 2.5 × 3.5 mm, subhemispherical, yellow, sunken in circular depressions 4 mm across; **calyx** 1 × 3.5 mm, pale yellowish or whitish, very shallowly and inconspicuously lobed, **sepals** imbricate nearly to apex; **petals** 2.5 × 4.5–5 mm, imbricate nearly to apex, erect, truncate and mucronate apically; **pistil** 2.5 × 3–3.5 mm, strongly depressed-globose, pale yellowish, **styles** very short or lacking, **stigma** lobes erect, pointed.

Fruits: 7 mm diam., globose, black, densely crowded.

Distribution: PANAMA. Bocas del Toro. Chiriqui. Cocle. Veraguas. COSTA RICA. Alajuela. Heredia. Guanacaste. San Jose. Dense, wet forest mainly on the Atlantic slope; 650–1,500 m elevation.

Discussion: *C. robertii* is an attractive and unusual species due to its bifid, heavily nerved leaves, stemless habit, and spicate inflorescences. The flower-bearing portion of the staminate inflorescence is pendulous and densely crowded with relatively large, white-tinged-with-green flowers. The pistillate inflorescence has densely crowded, yellow flowers and attractive, burgundy-colored bracts sheathing the peduncle.

Chamaedorea robertii is most closely related to *C. pumila* and *C. sullivaniorum.* Bluish or grayish

→

Plate 81. Chamaedorea quezalteca. A. Staminate inflorescence, *Hodel & Castillo 867,* cultivated, University of San Carlos, Guatemala; **B.** Long fibrous ligule at apex of leaf sheath is diagnostic for *C. quezalteca* (compare with that of *C. costaricana*), cultivated, research collection, Los Angeles; **C.** Fruits, cultivated, University of San Carlos Botanical Garden, Guatemala; **D.** Attractive clump, *Hodel 783,* cultivated, garden of R. Palmer, Whittier, California.

B
D

iridescent green leaves with more closely toothed margins, branched staminate inflorescences with greenish-yellow flowers, and green pistillate flowers distinguish the latter two species. Also, *C. robertii* differs from *C. pumila* in the larger, broader leaves with more nerves (12–15 rather than 10) and from *C. sullivaniorum* in the more deeply bifid blade. *C. robertii* occurs in a few collections in California, Hawaii, and Australia.

66. Chamaedorea scheryi *L. H. Bailey,* Gentes Herb. 6: 252, fig. 133. 1943a. Type: Panama, Chiriqui, *Woodson & Schery 680* (holotype MO!; isotype BH!). Epithet: honors R. W. Schery, co-collector of the type. **PLATE 85.**

Common Name: none.
Vernacular Name: none.

Habit: solitary, erect, appearing stemless, to 1–1.5 m tall.

Stem: 2–3 cm diam., creeping or erect at or below ground level or covered with leaf litter, to 15 cm tall or long, prominently ringed, nodes congested, internodes 3–5 mm long.

Leaves: 4–6, erect-spreading, pinnate; **sheath** to 15–30 cm long, long-obliquely open, splitting deeply opposite petiole and tubular only near base, green; **petiole** 30–45 cm long, much ribbed or striate laterally, flattened or slightly angled and green above, rounded and green below; **rachis** to 0.75–1 m long, slightly angled and green above, rounded and green below; **pinnae** 16–20 on each side of rachis, to 20–35 × 2.5–4 cm, linear-lanceolate to lanceolate, falcate or slightly sigmoid, narrowly acuminate, contracted basally, regularly disposed, opposite or subopposite, lower margin decurrent on rachis, glabrous, bright green, a central prominent midrib above, 2 less prominent primary nerves on either side of this, indistinct secondaries between the primaries and midrib, 5 nerves prominent below.

Inflorescences: inter- or less frequently infrafoliar, often arising from below leaf litter, erect-ascending, long-pedunculate; **peduncles** to 75 cm long, slender, erect, 1–1.5 cm wide at base and there ± flattened, 2–4 mm diam. at apex and there rounded, greenish in flower, orange in fruit; **bracts** 5–10, prophyll to 2 cm long, 2nd bract to 2.5 cm, 3rd to 5 cm, 4th to 8 cm, 5th to 11 cm, 6th and 7th to 13 cm, 8th to 10cm, 9th to 6 cm, tubular, strongly longitudinally striate-nerved, membranous, acute and bifid apically, brownish or rotting away by anthesis and in fruit, upper one equalling peduncle. **Staminate** diffuse, flowering portion to 25 cm long and broad; **rachis** 15–20 cm long, green; **rachillae** 15–30, lower ones longest, these to 10–15 cm long, uniformly spreading and diverging from rachis at nearly right angles, drooping apically, green, slender.

Pistillate few-branched; **rachis** 1–8 cm long, green in flower, orange in fruit; **rachillae** 4–9, these to 10 cm long, ± stiff, erect, parallel or slightly curving, rachillae green in flower becoming yellow-orange to orange in fruit.

Flowers: Staminate in ± dense spirals, in immature bud 2 × 1 mm, near anthesis 3 × 2.5–3 mm, ± globose, green, abruptly short-pointed, slightly sunken; **calyx** 0.5–1 × 2 mm, prominently lobed, conspicuously brown-margined, **sepals** connate in basal ⅓, slightly flared and broadly rounded apically; **petals** 2.5–3 × 1.5 mm, valvate, free nearly to base, erect-spreading, acute; **stamens** 2 mm high, leaning away from and not exceeding pistillode, **filaments** 1 mm long, pale, **anthers** 0.5–0.75 mm long, bilobed; **pistillode** 2–2.5 mm high, columnar, pale, greenish apically and there sharply 3-lobed and enlarged.

Pistillate in fairly dense spirals, 2.5 × 2–2.5 mm, ± globose, green, slightly sunken; **calyx** 0.5 × 2 mm, prominently 3-lobed, brown-margined, **sepals** free and separated basally, triangular, rounded to acute apically; **petals** 2.5 × 2.5 mm, cupped, imbricate nearly to apex, sharply acute and recurved slightly apically, thick; **staminodes** 1 mm high, acute, whitish; **pistil** 2–2.5 × 1.5–1.75 mm, inverse-conic, green; **stigma** lobes sessile, short, close, forming a pyramid-shaped structure, pale, shorter than petals.

Fruits: 5–7 mm diam., black, globose (oblong when immature); eophyll pinnate.

Distribution: PANAMA. Chiriqui. Cocle. COSTA RICA. Alajuela. Heredia. Cartago. San Jose. Wet forest mainly on the Atlantic slope but perhaps elsewhere; 800–2,000 m elevation.

Discussion: Bailey (1943a) described and named *C. scheryi* from material that Woodson and Schery had collected in Chiriqui, Panama. Apparently confined to wet forests from central Costa Rica to central Panama, it exhibits a fair amount of foliar variation over this range. Collections from the vicinity of the Rio Sarapiqui in Costa Rica differ slightly from those from Tapanti, Costa Rica in the rachis being more sharply angled above and the darker green pinnae with the lower margin more conspicuously decurrent along the rachis to the next lower pinna, and the less pronounced nerves. Material from near Boquete in Chiriqui, Panama differs from that from Costa Rica in the slightly broader pinnae with even less pronounced nerves. These differences seem insignificant when taken over the entire range.

Some material that Standley (1937) and others have referred to *C. microphylla* is most likely *C. scheryi* or *C. undulatifolia.* This confusion possibly occurred because of the tendency of *C. scheryi* to flower when very young and with leaves much reduced in size, hence the application of the name *microphylla* meaning small-leaved.

Chamaedorea scheryi is close to *C. brachyclada, C. pygmaea, C. stenocarpa,* and *C. undulatifolia,*

→

Plate 82. A, Chamaedorea rhizomatosa. B, C, Chamaedorea rigida. A. Loose clump, *D. & R. Hodel 936* (holotype), rocky slopes, pine-oak cloud forest, Oaxaca, Mexico. **B, C.** In wet cloud forest, *D. & R. Hodel 941,* Oaxaca, Mexico; **B.** Old plant with well developed stem; **C.** Young plant, stem buried in leaf litter.

B

which together form a natural subgroup within subgenus *Chamaedoropsis* (Hodel and Uhl 1990b) characterized by a stemless habit, long-pedunculate inflorescences arising from the base or leaf litter, decurrent pinnae, and pinnate eophylls.

Only a few plants of *C. scheryi* exist in cultivation in the research collection in Los Angeles and at the Jardin Botanico Robert y Catherine Wilson in Costa Rica. As is the case with the other species in this group, *C. scheryi* is somewhat difficult to maintain in a handsome and good condition in cultivation. It is susceptible to brown-tipping from low humidity and highly mineralized water and to infestations of mites and thrips.

67. Chamaedorea seifrizii *Burret,* Notizbl. Bot. Gart. Berlin-Dahlem 14: 268. 1938. Type: Mexico, Yucatan, Chichen Itza, *Seifriz s.n.* (holotype B, destroyed); Neotype (here designated): Mexico, Yucatan, Chichen Itza, *Moore 8096* (BH!). Epithet: honors the collector of the type, William Seifriz. **PLATES 9(C), 86, 87, 125(A-C, F).**

Chamaedorea erumpens H. E. Moore, Gentes Herb. 8: 232, figs. 96–97Aa. 1951. Type: Cult., *Moore 5830* (holotype BH!). Epithet: the name is a Latin word meaning breaking through, in reference to the manner in which the very short inflorescences of at least the staminate plants burst through the leaf sheaths.

Chamaedorea erumpens 'Fairchild' H. E. Moore, Gentes. Herb. 8: 233, fig. 97B, 1951. Type: Cult., *Moore 5832* (holotype BH!). Epithet: the name of the horticultural variety honors Dr. David Fairchild, well known author and botanist for the U.S.D.A. in the first half of the 20th century.

Meiota campechana O. F. Cook, Nat. Hort. Mag. 12: 138. 1943b. Name of no botanical standing.

Common Name: bamboo palm, Florida Hybrid. Vernacular Names: xate, xiat, chiat (Mayan)—Guatemala, Belize.

Habit: cespitose, erect to leaning, forming fairly dense and tight clumps of 40 or more stems to 3 × 1–2 m.

Stems: 1–2 cm diam., green, shining, white-spotted, ageing with a thin glaucous covering, conspicuously ringed, internodes 5–20 cm long.

Leaves: 4–5, erect-spreading, pinnate; **sheath** to 30 cm long, obliquely open apically, tubular, gray-green, densely and very minutely white-spotted, longitudinally striate-nerved, drying brownish, persistent; **petiole** to 10 cm long, flat and very lightly grooved near base and green or gray-green above, rounded and pale below; **rachis** 30–45 cm long, sharply angled and green above, rounded and with very faint light green band extending onto sheath; **pinnae** 5–18 on each side of rachis, median ones largest, these to 20–35 × 0.8–3 cm, basal ones smaller, these 14–20 × 0.5–1.8 cm, subapical ones 10–15 × 1–1.5 cm, apical pair 8–15 × 2.5–3 or to 9 cm wide, generally the more numerous the pinnae the narrower they are, lanceolate or long-lanceolate or linear, straight, only slightly falcate, short-acuminate or the upper merely acute, contracted basally, regularly arranged, opposite or alternate, horizontal or drooping or ascending off rachis, ± stiff or not, flat or very slightly v-shaped, concolorous, terminal ones 2–3 nerved or occasionally very broad and then to 9-nerved, remainder 1-nerved with conspicuous primary nerve and numerous closely spaced and fine and ± inconspicuous secondaries on each side of midrib, midrib prominent below, keeled, yellow.

Inflorescences: infrafoliar, bursting through base of old sheaths, erect, short, stiff. **Staminate** with **peduncle** 3.5–5.5 cm long, ascending and erect, yellow-green in flower; **bracts** 5, brown, papery, tubular, spreading apically, longitudinally striated, slightly bifid apically, lowest the shortest, upper ones largest, fourth exceeding peduncle and often concealing rudimentary 5th; **rachis** 2–4.5 cm long, erect, yellow-green in flower; **rachillae** 5–12, these 7.5–15 cm long, 2.3 mm diam., stiff, erect, mostly simple or lower ones rarely once or twice forked.

Pistillate up to 7 per stem in flower and fruit at once; **peduncle** 3.5–8.8 cm long, stiff and erect, green in flower, orange in fruit; **bracts** 5, similar to those of staminate; **rachis** 1–3 cm long, green in flower, orange in fruit; **rachillae** 4–6, these to 10 cm long, 3 mm diam., stiff, erect, green in flower, becoming swollen and orange in fruit.

Flowers: Staminate in fairly dense spirals, 3 × 3 mm, subglobose yellow, aromatic, ± sunken; **calyx** 1–1.5 × 3 mm, deeply lobed, green, **sepals** free nearly to base and there very briefly connate, sharply rounded to triangular apically; **petals** 3 × 3 mm, broadly ovate, valvate, free nearly to base and there briefly connate, spreading slightly apically but incurved, ± thickened, acute; **stamens** 2 mm high, tightly fixed around pistillode, **filaments** 1–1.5 mm long, connate basally, abruptly narrowed from a broad base, pale, **anthers** 1 mm long, brownish, dorsifixed at middle; **pistillode** 3 mm high, broadly columnar, just slightly shorter than petals, broadly 3-lobed apically, tip visible at anthesis, light yellow.

Pistillate in moderate spirals, 2–2.5 × 2.5–3 mm, depressed-globose, yellow, slightly sunken; **calyx** 1–2 × 3–3.5 mm, lobed, green with clear membranous margins, **sepals** imbricate in basal ½–⅔, broadly rounded apically; **petals** 2.5 × 2.5–3 mm, ovate, imbricate nearly to apex, cup-shaped, erect and ± acute apically, ± thick, drying brown; **staminodes** 6, minute and scalelike, bilobed; **pistil** 2 × 2–2.5 mm, subglobose, green, **stigma** lobes sessile, erect or slightly recurved, blunt.

→

Plate 83. Chamaedorea robertii. A. *Hodel et al. 975,* montane rain forest, Braulio Carrillo National Park, Costa Rica; **B.** Fruiting plant with spicate infructescence arising from base, *D. & R. M. Hodel 737* (holotype), montane rain forest, El Cope, Panama; **C-E** Cultivated, California, *ex* El Cope, Panama; **C.** Staminate flowers, *Hodel 828,* research collection, Los Angeles; **D.** Pistillate plant, *Hodel 857,* research collection, Los Angeles; **E.** Pistillate plant, Virginia Robinson Botanical Garden, Beverly Hills.

C
E

Fruits: 8 mm diam., globose, black, epicarp thin, finely reticulate, mesocarp thin, yellow, pulpy with few slender fibers adherent to thin endocarp; seeds 6 mm diam., globose.

Distribution: MEXICO. Campeche. Quintana Roo. Tabasco. Yucatan. BELIZE. Belize. Cayo. Corozal. Orange Walk. GUATEMALA. Peten. HONDURAS. Islas de la Bahia. Open or dense, moist woodland or forest; to 500 m elevation; often on limestone.

Discussion: Burret (1938) described and named *C. seifrizii* from material that William Seifriz collected near the Mayan ruins at Chichen Itza in the state of Yucatan in Mexico. Mayans cultivated it around their villages and temples as ornament and possibly for religious purposes. The Mayan names *xiat* and *chiat* mean near the edge of water, in apparent allusion to the habitat. I have observed *C. seifrizii* in Peten, Guatemala and at several localities in Belize. In each place, it occurred in low, seasonally swampy or boggy situations. In fact, local people state that it is mainly confined to *baja tierra* or low land. However, these low wet areas are only damp during the rainy season. At other times of the year they can be exceedingly dry, a situation that we found in northern Belize in May, 1991. In some instances, adjacent vegetation was actually wilting although the palms did not seem to be suffering. Along with *C. graminifolia, C. seifrizii* is certainly one of the most drought tolerant of chamaedoreas.

Although *C. seifrizii* was not named until 1938, Millspaugh (1898) and Standley (1930) encountered it at earlier dates but listed it under other names.

As interpreted here, *C. seifrizii* displays a tremendous amount of foliar variation in the wild although the flowers are essentially the same in the various types and forms. Burret's *C. seifrizii* is typified by stiffish leaves with narrow, linear, and upright pinnae. Moore's *C. erumpens* has softer leaves with broader pinnae. Moore (1951) described and named *C. erumpens* from plants cultivated at Fairchild Tropical Garden in Miami, Florida. These plants had apparently been brought in from Belize or grown from seeds that William Schipp sent and by 1950 had become quite popular in southern Florida and were used extensively as an indoor decorative as well as in the exterior landscape. Moore noted two leaf forms in *C. erumpens.* That with the pinnae regularly arranged was selected as the type. The other, with the terminal ones united and much broader, was designated as a horticultural variety honoring Dr. David Fairchild. Both types were growing together at Fairchild Tropical Garden and Moore presumed that they had come from the same lot of seeds. Moore stated that he found no differences other than in the foliage. Since both staminate and pistillate plants bore the same type of foliage and came from wild stock, the designation of a horticultural variety was probably more than a clonal selection; it did breed true from seeds.

Moore (1951) continued in his discussion and stated that *C. erumpens* was similar in floral morphology to its close relative *C. seifrizii,* and that the differences were mainly in the shape and nervation of the pinnae. *C. erumpens* had lanceolate rather than linear pinnae that were at least twice as broad as those of *C. seifrizii.* Also, *C. seifrizii* was noted as being a scrambling palm. When considered over its entire range, the shape and size of the pinnae are variable characters and of dubious merit in distinguishing between these two taxa. In fact, there is less foliar variation between Burret's typical *C. seifrizii* and Moore's typical *C. erumpens* than between the latter and its horticultural variety 'Fairchild.' In Belize, I have observed leaves of both extreme types of pinnae (*seifrizii* and *erumpens*) on the same plant. The scrambling nature of certain forms is probably not a reliable character either. Rather, it seems to be a function of age and of the amount of light; older plants in lower light seem to lean and scramble more while those in higher light are more compact, stiff, and upright. The amount of light also affects the stiffness of the leaves. Those in higher light have stiffish, somewhat v-shaped (in cross-section), upright pinnae while those in lower light have drooping, flat, softer pinnae.

Since Seifriz's holotype was destroyed and no isotypes have been found and *C. seifrizii* is common in cultivation and herbaria, it is appropriate to retypify this taxon here.

Not abundant in the wild, *C. seifrizii* is found as scattered clumps in disturbed woodland or forest. In Orange Walk, Belize, we found it growing in dis-

→

Plate 84. Chamaedorea robertii. A. Pistillate plant, *Hodel et al. 976,* montane rain forest, Braulio Carrillo National Park, Costa Rica; **B-F** Cultivated, research collection, Los Angeles, *ex* El Cope, Panama; **B.** Spicate staminate inflorescence with pendulous rachis, *Hodel 828*; **C-F** *Hodel 857*; **C.** Spicate pistillate inflorescence with straight rachis and burgundy-colored bracts; **D.** Pistillate flowers; **E.** Pistillate inflorescence; **F.** Infructescence with mature fruits.

B
D
F

turbed forest or woodland on the margins of sugarcane fields. Today, it is very widely cultivated and appears in gardens and collections in California, Hawaii, Florida, Europe, Australia, the Far East, and elsewhere. In fact, it is highly likely that more plants are in cultivation than in the wild. Often occurring naturally on limestone outcroppings in its native habitat, *C. seifrizii* is well adapted to culture in southern Florida where it has been cultivated for some time and extensive plantings now exist.

Krempin (1990, p. 91) illustrated *C. microspadix* but erroneously captioned the photograph as *C. erumpens.* Krempin (p. 92, 93) also discussed and illustrated *C. graminifolia* but the description and photograph depict *C. seifrizii.*

After *C. elegans, C. seifrizii* is the most important member of the genus commercially and is grown in large quantities as a potted specimen for indoor use or as an outdoor landscape subject in subtropical areas. About 18,200 kgs (40,000 lbs) of seeds of *C. seifrizii* are collected for commercial purposes annually, the majority originating from Mexico although a significant amount comes from plants cultivated in Florida. Recently, commercial plantations of *C. seifrizii* for seed production have been established in Hawaii, Belize, Australia, and the Orient.

The commercial industry recognizes two forms of *C. seifrizii.* One, the most popular, more or less corresponds to Burret's type with stiffish leaves and narrow, linear, upright pinnae, while the other falls into Moore's type with softer leaves and broader, flat pinnae. Extensive plantings of both forms have existed in southern Florida for years and have served as a source of breeding stock for commercial seed production. Much cross-breeding has occurred between the two and this, coupled with the natural variation within the species, has resulted in innumerable variants or breeding lines in the trade. In fact, most of the material produced commercially in Florida is called, for lack of a better term, "Florida Hybrid."

With tightly clustering, somewhat stiff and upright habit and dense foliage, *C. seifrizii* is useful as a hedge, background, or screen in either shade or sun. It is very tolerant of low light, where it takes on a softer and more open and graceful habit, often with leaning stems. It also tolerates nearly full-sun without yellowing or browning. In this situation it becomes stiffish, less graceful, and with straight erect stems; pinnae ascend off the rachis, making the blade v-shaped in cross-section. *C. seifrizii* also makes a fine potted specimen as a landscape accent. It is one of the best palms for indoor use due to its tolerance of low light and resistance to infestations of mites and is more resistant to nematodes than *C. elegans.*

Chamaedorea seifrizii is easily propagated by seeds which germinate readily within 100 days. However, some seeds may be delayed in germination or continue to germinate somewhat irregularly in cycles for two to three years after planting. The exact reason for this phenomenon is not known. A possible explanation is that the species occurs naturally in an area of definite seasonality in rainfall and on a porous substrate that is subject to fairly extensive and, at times, lengthy drying. The delayed germination or germination in cycles or spurts over several seasons may be a mechanism to ensure establishment of a sufficient number of seedlings for adequate regeneration of the species in a relatively harsh environment punctuated by dry periods. Subjecting the seeds to bottom heat of 30–32.5°C (85–90°F) may increase the speed of germination and make it more regular.

68. Chamaedorea selvae *Hodel,* Principes 35: 79. 1991a. Type: Costa Rica, Heredia, *Burger & Mata 4337* (holotype CR!; isotypes F!, US!). Epithet: from the Spanish *selva* meaning jungle (lowland tropical rainforest), in reference to the habitat.

Common Name: none.
Vernacular Name: none.

Habit: solitary (rarely cespitose? *Stevens 24461*), erect, to 2 m tall.

Stem: 2 cm diam., smooth, green ringed, internodes to 10 cm long.

Leaves: 4–5, spreading, pinnate; **sheath** green with no pale stripe extending onto rachis below; **petiole** to 35 cm long, robust; **rachis** to 66 cm long; **pinnae** 3–8 on each side of rachis, to 45 × 9 cm, broadly lanceolate, sigmoid, falcate, acuminate, thin-papery, 6–10 primary nerves prominent and elevated above, 1 secondary between each of 2 primaries, tertiaries numerous, faint, primaries paler below, end pair of pinnae sometimes very large, then each lobe to 50 × 30 cm on a rachis 40 cm long with 30

Plate 85. Chamaedorea scheryi. A, B In montane rain forest along Rio Sarapiqui, Costa Rica; **A.** Marianne Hodel holds leaves of large plant, *D. & M. Hodel 720B* (note pistillate inflorescence emerging to the right of plant from base); **B.** Plant with finely divided leaves at very young age; **C.** In wet cloud forest, *D. & M. Hodel 727,* Chiriqui, Panama; **D.** Staminate inflorescence emerging from base of plant, *Hodel & Grayum 697A,* montane rain forest, Tapanti, Costa Rica.

D

primary nerves on each side, outer margin remotely toothed toward apex.

Inflorescences: infrafoliar, attached well below the leaves; **peduncles** to 75 cm long, 7–8 mm diam., erect, ± robust; **bracts** 5–6, prophyll to 3 cm long, 2nd bract to 10 cm, 3rd to 25 cm, 4th to 40 cm, 5th to 35 cm, 6th to 30 cm, tubular, acuminate, fibrous, longitudinally striate-nerved. **Staminate** with **rachis** to 10 cm long, straight; up to 17 **rachillae,** these to 30 cm long, slender, pendulous.

Pistillate with **rachis** to 7 cm long, ± straight; up to 11 **rachillae,** these to 20 cm long, drooping in flower, pendulous and orange in fruit.

Flowers: Staminate in bud arranged in moderately dense spirals but not contiguous, 0.5–1 mm apart, 1 × 1.25–1.5 mm, ± dome-shaped, in superficial and elliptic depressions 1.25 mm long; **calyx** 0.5 × 1.5 mm, deeply lobed, thick, **sepals** connate and/or imbricate in basal ¼, rounded apically; **petals** 1 × 1.5 mm, valvate, spreading, acute; **stamens** 0.75 mm high, tightly appressed around pistillode; **pistillode** 0.75 mm high, columnar, apically lobed.

Pistillate in moderately dense spirals 1–2 mm apart, 2 × 2 mm, ± globose, leaving shallow elliptic scars 1.5 mm long; **sepals** of fruiting perianth 0.75 × 1.75 mm, imbricate basally, broadly rounded apically; **petals** of fruiting perianth 1.75 × 2.5 mm, imbricate basally, acute, petals and sepals lightly nerved on the inside; **pistil** 1.25 mm high, ovoid, **styles** short, **stigma** lobes pointed, erect, darkened.

Fruits: 7 × 5 mm, oblong, green when immature, maturing black.

Distribution: COSTA RICA. Alajuela. Heredia. Limon. San Jose. NICARAGUA. Rio San Juan. Hillsides in wet, lowland forest on the Atlantic slope; mostly below 200 m but occasionally to 700 m elevation.

Discussion: Leaves of *C. selvae* are very similar to those of *C. tepejilote,* especially in the nervation of pinnae. However, the short peduncles, contiguous staminate flowers, and prominent yellow stripe on the abaxial surface of the rachis and petiole distinguish this latter species. *C. deneversiana* is also somewhat similar vegetatively but differs in the flexuous rachises of the inflorescences, and fewer and shorter rachillae.

Perhaps *C. woodsoniana* and *C. carchensis,* all from higher altitudes, are most closely related to *C. selvae.* However, these two species are amply distinct in their narrower and more numerous pinnae (20 or more on each side of the rachis versus 3–8 for *C. selvae*). In addition, *C. woodsoniana* and *C. carchensis* have many more staminate rachillae (80–100 and 40 respectively versus 17 for *C. selvae*).

The only species of subgenus *Chamaedoropsis* that occurs in the same range as *C. selvae* is the highly variable *C. dammeriana.* However, the interfoliar inflorescences, few-branched staminate inflorescences, usually spicate or furcate pistillate inflorescences, and much smaller habit (about half the size of *C. selvae*) distinguish this latter species.

Chamaedorea selvae is rare over its range and should probably be considered endangered due to destruction of lowland forest in Limon, Heredia, and Alajuela provinces of Costa Rica and adjacent portions of Nicaragua. In fact, several of the collections are from forest remnants, indicating that suitable habitat for its growth is noncontiguous and isolated. *C. selvae* does exist in protected areas; it is well documented at La Selva as *Chamaedorea* sp. *nov. fide* Moore (Chazdon 1985, 1987, Moore and Chazdon 1985).

69. Chamaedorea skutchii *Standl. & Steyerm.,* Publ. Field Mus. Nat. Hist., Bot. Ser. 23: 206. 1947. Type: Guatemala, Quetzaltenango, *Skutch 935* (holotype F!; isotype GH!). Epithet: honors Alexander F. Skutch, collector of the type.

Common Name: none.
Vernacular Name: none.

Habit: 2.5 m tall.

Stem: 8–12 mm diam., internodes 2.5–4 cm long.

Leaves: 4–5, pinnate; **petiole** to 22 cm long or more; **rachis** 22–27 cm long, pale-backed; **pinnae** 9–11 on each side of rachis, 15–21 × 1.3–1.6 cm, subfalcate, long-attenuate, mostly alternate, 3 primary nerves slightly keeled and pale above, pale and shining below, secondaries very slender upper pair 20 cm long on upper margin, to 6 cm wide, falcate, acuminate, pale below, outer side inconspicuously toothed, 4–6-nerved.

Inflorescences: Staminate not seen. **Pistillate** infrafoliar; **peduncle** 21–35 cm long, slender, nearly 4 mm wide at apex in fruit; **rachillae** 7, these rather markedly undulate especially toward tip, orange and thickened in fruit.

Flowers: Staminate not seen. **Pistillate** borne in loose spirals, nerveless, ± longer than broad, distinctively sunken in slightly curved elliptic pits bordered on lower side by liplike margin; **calyx** 1 × 3 mm, shallowly lobed, **sepals** imbricate in basal ¼, broadly rounded apically; **petals** 1.5 mm high, imbricate, subacute to rounded, becoming brown and narrowly dark-margined in fruit, one covering abortive carpels scarcely exceeding calyx, other two twice as long as sepals; **staminodes** lacking; **pistil** not seen.

Fruits: 10 × 8 mm, obovoid-globose, black; seeds 8 mm long, globose-ovoid; abortive carpels adherent to perianth in fruit.

Plate 86. Chamaedorea seifrizii. A. Attractive clump with closely spaced stems, cultivated, Waimea Falls Park & Arboretum, Honolulu, Hawaii; **B.** Lou Thomas stands next to clump, *Hodel & Thomas 1131A,* seasonally dry woodland, Orange Walk, Belize; **C.** Fruits, cultivated, Miami, Florida; **D.** Form with broad leaflets referred to as *C. erumpens,* cultivated, garden of L. Hooper, La Habra, California.

B
D

Distribution: GUATEMALA. Quetzaltenango. Wet forest on the Pacific slope; 2,400 m elevation.

Discussion: The description is from Standley and Steyermark (1947, 1958) who described and named *C. skutchii* from pistillate material only that Skutch collected on Volcan Zunil in Quetzaltenango. Very rare and known with certainly only from the type, *C. skutchii* is close to and perhaps not distinct from *C. nubium.* Standley and Steyermark (1958) stated that *C. skutchii* differed in the imbricate pistillate sepals, shallower pits, and lack of staminodes.

70. Chamaedorea stenocarpa *Standl. & Steyerm.,* Publ. Field Mus. Nat. Hist., Bot. Ser. 23: 206. 1947. Type: Guatemala, Izabal, *Steryermark 41893* (holotype F!; isotype US!). Epithet: from the Greek *steno* meaning narrow and *carpus* meaning fruit, in reference to the elongated fruit. **PLATE 88.**

Common Name: Las Cruces elfin dwarf, Las Cruces dwarf pinnate.

Vernacular Name: none.

Habit: solitary, erect, appearing stemless, to 75 cm tall.

Stem: 1–2 cm diam., very short, creeping, 15 cm long, at or below ground or leaf litter, green, densely ringed, internodes 2–10 mm long.

Leaves: 3–5 erect-spreading, pinnate, 40–60 cm long but often flowering when much shorter; **sheath** 3–8 cm long, split deeply opposite petiole, clasping in a tubular manner only near base, narrowed upward and there not clasping at all and appearing as part of petiole, striate-nerved, green, drying brownish, persistent; **petiole** 5–25 cm long, flat or channeled toward base and green above, rounded and green below; **rachis** 14–27 cm long, angled and/or strongly verrucose and green above, pale and rounded and verrucose-margined toward apex below, narrowly margined with decurrent lower margins of pinnae; **blade** 14–32 × 10–20 cm, **pinnae** 10–20 on each side of rachis, middle ones 5–18 × 1.2–2.5 cm, apical pinnae 5–7.5 × 0.8–1.5 cm, narrowly oblong-lanceolate to lanceolate, spreading, thin, concolorous, glabrous, regularly distributed, acuminate, narrowed to ½ their width at base, basal attachment 8 mm long, slightly sigmoid, falcate, lower margin decurrent on rachis, midrib and marginal primary nerves on each side prominent above and below, these slender but prominently keeled, secondaries solitary between midrib and primaries, these very slender and inconspicuous, lower margin shallowly and often inconspicuously toothed toward apex, end pinnae 2–3-nerved, all pinnae decreasing in length noticeably in a progressive manner toward the apex of rachis.

Inflorescences: interfoliar, often emerging from below leaves at or below ground or leaf litter, erect-spreading, not exceeding leaves; **peduncles** 15–20 cm long, 1.5 mm diam. at apex, straight or slightly arcuate, erect or spreading, greenish or pale in flower, orange in fruit; **bracts** 4–5, narrowly tubular, 3 mm diam., obliquely open apically, conspicuously nerved, membranous, brownish in flower, uppermost not exceeding peduncle. **Staminate** with **rachis** to 15 cm long, green; **rachillae** 10–25, these 6–8 cm long, simple, slender, filiform, soft, ± pendulous or flexible, green.

Pistillate spicate or sometimes forked, ± stiff; **rachis** or flower-bearing portion 4–15 cm long, straight or slightly curved, green and thicker than staminate, 3 mm diam., much thicker and deep orange in fruit.

Flowers: Staminate in moderate to ± lax spirals, ovoid-globose, greenish, ± superficial; **calyx** low, lobed, **sepals** connate basally, rounded apically; **petals** briefly connate and/or imbricate basally, ovoid, free and spreading apically, acute; **stamens** ½–¾ as high as petals, **filaments** pale; **pistillode** exceeding stamens, greenish.

Pistillate in rather dense spirals, 3 × 3 mm, globose to subglobose, light green, scarcely sunken in elliptic depressions 1.8 mm long; **calyx** 1–1.5 × 2.5–3 mm, deeply lobed, light green with dark green toward margins, **sepals** connate and/or imbricate in basal ¼, rounded to acute apically; **petals** 3 × 2.5–3 mm, round-ovate to oblong, cup-shaped, imbricate nearly to apex, free and acute or obtuse and shortly erect apically, dark green toward apex; **pistil** 1.5 × 2.5 mm, depressed-globose, green, **stigma** lobes short, not equalling petals, pointed, erect, brown.

Fruits: 6–8 mm diam. or long, oblong-ellipsoid and green to yellow when immature, black and elongated to ± globose when mature, when elongated obtuse or rounded apically, obtuse basally, abortive carpels adherent to reddish orange and persistent perianth; eophyll pinnate.

Distribution: GUATEMALA. Izabal. COSTA RICA. Alajuela. Heredia. Puntarenas. PANAMA. Panama. Veraguas. Wet forest mainly on the Atlantic slope in Guatemala but the Atlantic and Pacific slopes in Costa Rica and Panama; 300–900 m elevation.

Discussion: Standley and Steyermark (1947) described and named *C. stenocarpa* from a pistillate collection only from Izabal, Guatemala. They stated that it is one of the smallest of Guatemalan palms. Although the authors made note of its distinctive oblong-ellipsoid fruits, in the mature state these become more rounded. Material from Guatemala tends to have more elongated fruits than that from Panama or Costa Rica. However, when considered over its whole range, the shape of the fruits does not seem to be a reliable diagnostic character.

Somewhat rare and locally distributed in the wild, *C. stenocarpa* is known from only a few collections from Guatemala, Costa Rica, and Panama. It inhabits wet forests at or near the Continental Divide in the latter two countries and in Guatemala is ap-

Plate 87. Chamaedorea seifrizii. A. Large clump in dense shade, *Hodel 847,* Tikal, Guatemala; **B.** Two leaf forms on same plant, cultivated, garden of L. Thomas, Belize.

→

parently restricted to Cerro San Gil in Izabal on the Atlantic slope.

Chamaedorea stenocarpa is allied to *C. pygmaea, C. scheryi, C. brachyclada,* and *C. undulatifolia.* These species form a more or less natural subgroup of *Chamaedoropsis* (Hodel and Uhl 1990b). *C. stenocarpa* is closest to *C. pygmaea* but the latter differs in the fewer pinnae that do not decrease markedly in length toward the apex of the rachis.

Robert G. Wilson of San Vito in southeastern Costa Rica was responsible for introducing *C. stenocarpa* to cultivation. In the 1960s he established fruiting populations in his garden. Hodge (1981, cover and fig. 5) handsomely illustrated one of these plants in Wilson's garden and another of Hodge's photographs of this species appeared in *Genera Palmarum* (p. 126, pl. 50D). Wilson and/or local collectors found the plants in forest remnants in the mountains above San Vito. Over the years, Wilson distributed seeds and plants as the "Las Cruces elfin dwarf" or "dwarf pinnate" *Chamaedorea* to interested collectors and hobbyists. Today, *C. stenocarpa* is cultivated in Costa Rica, Hawaii, California, Florida, Australia, and probably elsewhere.

A handsome species when well grown, *C. stenocarpa* is very effective when used as a mass planting or ground cover in a shady and humid area. Unfortunately, it and its relatives are among the most difficult of *Chamaedorea* to cultivate, at least in California. They tend to brown-tip severely due to the high mineral content of the water and low humidity. They appear somewhat stunted, bear very few leaves, and are susceptible to damage by mites and thrips. They also seem to perform better in subtropical or cooler tropical regions as opposed to hot, low elevation tropical areas. In the latter regions, the plants seem to languish and go into a slow decline eventually resulting in their demise.

71. Chamaedorea stricta *Standl. & Steyerm.,* Publ. Field Mus. Nat. Hist., Bot. Ser. 23: 207. 1947. Type: Guatemala, San Marcos, *Steyermark 37381* (holotype F!). Epithet: from the Latin *strictus,* meaning straight, in reference to the straight and erect inflorescences. **PLATES 89, 90.**

Common Name: none.
Vernacular Name: kum—Guatemala.

Habit: solitary, erect, appearing stemless, to 1.5 m tall.

Stem: 2.5–3 cm diam., curved, creeping at or below leaf litter, to 15 cm long, densely ringed, internodes to 1 cm long.

Leaves: 3–7, erect to ascending, bifid; **sheath** 8–27 cm long, long-open, attenuate apically, rather laxly striate-nerved, dark green; **petiole** 25–60 cm long, to 7 mm thick, flat and dark green above, rounded and dark green below; **rachis** 15–55 cm long, angled and dark green above, rounded and dark green below; **blade** 20–75 × 13–28 cm, rectangular, deeply incised apically, lobes long-acuminate, cuneate basally, margins ± parallel, dark nearly bluish green, ± paler below, 12–17 primary nerves on each side of rachis, these prominent below, secondaries numerous, very slender and inconspicuous.

Inflorescences: inter- or infrafoliar, arising from base of plant and often from leaf litter; **peduncles** 80–100 cm long, 2 mm wide at apex, erect, slender, stiff; **bracts** 7–10, lowest ones 7 cm long, middle ones 11–28 cm, 4 mm wide, narrowly tubular, pale green with very slender nerves, obliquely open apically. **Staminate** branched portion to 24 cm long; **rachis** 3.5 cm long, green; **rachillae** 7, these to 10 cm long, slender, spreading or slightly drooping, ± flexuous, pale green in flower.

Pistillate with **rachis** 1.5–3 cm long, green in flower, red-orange in fruit; **rachillae** 2–4, these 8–14.5 cm long, erect, straight, smooth, rounded, green in flower, rather stout and red-orange in fruit.

Flowers: Staminate borne densely, ovoid to globose, bright yellow, rather deeply sunken in narrowly elliptic depressions 2.5 mm long; **calyx** very short, 2.5 mm wide, very shallowly lobed, **sepals** connate in basal ¾; **petals** 4 mm high, broadly ovate, valvate, free nearly to base, erect, obtuse.

Pistillate in remote spirals, scarcely sunken in elliptic or oval depressions 2.5 mm long; **calyx** 3.5 mm wide, deeply lobed, **sepals** connate and/or imbricate in basal ¼–⅓; **petals** 3 × 3 mm, imbricate, nearly orbicular; **staminodes** present and prominent in fruiting corolla.

Fruits: 8 mm diam. or long, broadly rounded, subglobose; seeds nearly as large as dry fruit, globose, whitish brown.

Distribution: MEXICO. Chiapas. GUATEMALA. San Marcos. COSTA RICA. San Jose. PANAMA: Veraguas. Moist or wet forest on the Pacific slope; 850–1,900 m elevation.

Discussion: The description is from Standley and Steyermark (1947, 1958) and supplemented from our collections. Standley and Steyermark noted that of the few species with bifid leaves, *C. stricta* is outstanding for the greatly elongate, stiffly erect, radical inflorescences. It is similar to *C. rigida* but the much shorter petioles and peduncles, stiffer blade, smooth and rounded pistillate rachillae, and the plant eventually with a visible above-ground stem distinguish the latter species. *C. stricta* is a spectacular plant to see in the wild as the crown of numerous, large, bifid, dark green leaves on long petioles seems

Plate 88. Chamaedorea stenocarpa. A-D In lowland rain forest, Izabal, Guatemala; **A.** Staminate plant, *Hodel & Castillo 1022A*; **B.** Pistillate plant with infructescence, *Hodel & Castillo 1022B*; **C, D.** Young plants; **E.** Staminate plant, *Hodel 619A,* cultivated, Jardin Botanico Robert y Catherine Wilson, Costa Rica.

C
E

to arise directly from the leaf litter. The blades are not too dissimilar in color and texture from those of *C. metallica.*

Chamaedorea stricta is rare in the wild in Guatemala where it is apparently confined to Volcan de Tajumulco and Cerro de Mono in San Marcos. It has only recently been discovered in Costa Rica and Panama. Only one collection represents the species from Chiapas, Mexico. *C. stricta* does not appear to be cultivated.

72. Chamaedorea sullivaniorum *Hodel & Uhl,* Principes 34: 128. 1990c. Type: Panama, Cocle, El Valle, *D. & M. Hodel 740A* (holotype BH!; isotype PMA!). Epithet: honors Pauleen Sullivan and her late husband Joe who have supported my work in palms. **PLATES 91, 92.**

Common Name: none.
Vernacular Name: none.

Habit: solitary, decumbent, shortly erect to 25 cm tall, flowering when appearing stemless.

Stem: 1–1.5 cm diam., creeping, buried in leaf litter so as to appear stemless, light green but covered with persistent leaf sheaths, prominently ringed, internodes congested, 0.7–1 cm long.

Leaves: in rosettes of 10–15, erect-spreading, bifid; **sheath** 3–4 cm long, open, splitting deeply opposite petiole, clasping completely only at base, margins inward-rolled apically forming a tube; **petiole** 10–16 cm long, dark green, flattened above with both margins of blade decurrent and continuing down petiole to form a channel, rounded below; **rachis** to 20–25 cm long, angled and gray-green above, rounded and paler below; **blade** to 25–40 × 10–17 cm, oblong, incised apically to ⅓ its length, thick, pliable, leathery, dark velvety nearly iridescent mottled green, plicate, 15–16 prominent primary nerves above, each along ridge of a fold resulting in a corrugated effect, nerves light green or white above and below, exterior margins conspicuously toothed.

Inflorescences: interfoliar, emerging through tubelike upper rolled margins of sheath and flumelike upper surface of petiole, erect, often appearing through leaf litter of forest floor. **Staminate** with a **peduncle** to 22 cm long, 2–3 mm wide at apex, 7–9 mm wide at base; **bracts** 5, longitudinally striate-nerved, tubular, membranous, green but ± rotting away prior to flowering; **rachis** 2.5 cm long, light green; **rachillae** 4–8, these to 10 cm long, 1.5–2 mm diam., ± drooping, light green. **Pistillate** spicate or rarely forked; **peduncle** to 32 cm long, ascending to spreading, greenish or pale in flower, dull orange in fruit; **bracts** 5–6, similar to those of staminate; **rachis** or flower-bearing portion 15 to 20 cm long, 3 mm diam., strongly curved, light green in flower, thickened and orange in fruit.

Flowers: Staminate in moderately dense spirals 2 mm apart, 3.5–4 × 2.5–3 mm, ovoid, pale green to light yellow tipped with green, slightly sunken in rounded to elliptic depressions; **calyx** 0.75–1 × 2 mm, lobed, pale or light greenish yellow, **sepals** connate in basal ½, rounded apically; **petals** 3.5 × 2 mm, ovate, valvate, free nearly to base, slightly cupped inward, narrowly acute apically, thin, ± membranous; **stamens** 2 mm high, **filaments** 1–1.5 mm long, pale, **anthers** 0.5 mm long; **pistillode** 2 × 1.75–2 mm, globose basally, narrowly attenuate apically, pale to light yellowish green.

Pistillate in moderately dense spirals, 2–3 mm apart, 3 × 3–3.5 mm, ± globose, pale greenish yellow, slightly sunken in rounded to elliptic depressions; **calyx** 0.75–1 × 2.5–2.75 mm, deeply lobed, light green to yellowish green, **sepals** imbricate and/or connate in basal ⅛, rounded apically; **petals** 3.25 × 3 mm, broadly triangular, imbricate nearly to apex, acute, slightly erect apically, cupped inward, thick, fleshy; **pistil** 2 × 2 mm, depressed-globose, a depression or pit at apex, green, **stigma** lobes sessile, pointed, slightly recurved, not exceeding rim of depression, pale.

Fruits: 6–8 mm, ± globose, black.

Distribution: PANAMA. Chiriqui. Bocas del Toro. Cocle. Colon. San Blas? Veraguas. COSTA RICA. Limon. Puntarenas. San Jose. COLOMBIA. Antioquia. Dense, wet forest mainly on the Atlantic slope up to and just over the Continental Divide but also in southeastern Costa Rica on the Pacific slope; 600–1,500 m elevation on the Atlantic slope, below 400 m elevation in southeastern Costa Rica.

Discussion: *C. sullivaniorum* occurs in dense, wet forest, mainly on the Atlantic slope in western Panama, eastern Costa Rica, and northwestern Colombia. There is a report of it from San Blas in eastern Panama but this has not been verified. Where the Continental Divide is low enough, as at El Valle, Panama, *C. sullivaniorum* may occur just over it on the Pacific slope. In southeastern Costa Rica, disjunct and isolated populations occur on the Osa Peninsula and vicinity below 400 m elevation.

At the type locality in Panama, *C. sullivaniorum* is very localized. It occurs in dense, dark forest, often on the sides or bottoms of steep ravines, although we also found it in more open forest along rounded ridge tops. Before collectors decimated it near El Valle, *C. sullivaniorum* formed locally dense populations, covering the ground completely to the exclusion of nearly all other vegetation in areas about ten meters square. It grows with *C. amabilis, C. allenii, C. lucidifrons, C. serpens,* and *C. woodsoniana,* among other species. During one visit to the type locality in April, 1987, the leaf litter on the forest floor was actually dusty dry. Several months later in December, it was spongy wet.

Chamaedorea sullivaniorum is most similar to *C. pumila* but the blades incised apically for one-half

→

Plate 89. Chamaedorea stricta. Robert M. Hodel holds infructescence of large specimen, *Hodel & Castillo 995,* montane rain forest, San Marcos, Guatemala.

their length or more and with 9–13 nerves on each side of the rachis, thick staminate petals, globose pistillode not narrowly attenuate apically, and stigma lobes not sunken in a depression at the apex of the pistil distinguish the latter species. Spicate staminate inflorescences, white tinged with green staminate flowers, yellow pistillate flowers, and deeply bifid blade distinguish another similar species, *C. robertii.*

A striking plant in the wild because of its thick, leathery, heavily nerved, bifid leaves only shallowly incised apically, *C. sullivaniorum* is a handsome ornamental that exhibits a fairly wide range of foliar variation, especially in the length of the blade and petiole. In higher light, the crown is a rosette of 10–15, rigid leaves with short petioles. In extremely low light, the crown contains fewer, softer, larger leaves with longer petioles. The species is highly sought after by collectors and hobbyists who, in some instances, have completely decimated local populations (see letter to the editor by D. R. Hodel, *Principes* 32: 95, 1988). *C. sullivaniorum* is cultivated in a few gardens and collections in California, Florida, Australia, and perhaps elsewhere.

73. Chamaedorea tenerrima *Burret,* Notizbl. Bot. Gart. Berlin-Dahlem 11: 858. 1933a. Type: Guatemala, Alta Verapaz, *Lehmann 1312* (holotype B, destroyed); Neotype (here designated): Guatemala, Alta Verapaz, *Hodel & Castillo 894* (BH!; isoneotype AGUAT!). Epithet: from the Latin *tenerrimus* meaning most delicate or thin, apparently in reference to the habit. **PLATES 93-95.**

Lobia erosa O.F. Cook, Nat. Hort. Mag. 22: 148. 1943b. Name of no botanical standing.

Common Name: none.
Vernacular Name: none.

Habit: solitary, erect or decumbent, slender, to 1.5 m tall.

Stem: 3–7 mm diam., smooth, green, ringed, internodes 2–5 cm long.

Leaves: 4–5, spreading, pinnate, variously pinnate or infrequently bifid; **sheath** to 15 cm long, tubular; **petiole** 4–12 cm long, rugose, grayish; **rachis** 6–20 cm long, rugose, grayish; **blade** broadly obovate; **pinnae** 2–7 on each side of rachis, to 11 × 2.5 cm, strongly reflexed, sigmoid, lower margin produced in a prominent auricle below, outer margins minutely scarious-roughened, mucronulate and emarginate at tips of nerves and rather sharply toothed between them, a prominent midrib and a less conspicuous primary nerve on each side of this, apical pair to 15 × 8–10 cm, broadly flared, 3–8-nerved, all pinnae thin, green, concolorous; if bifid then blade incised apically to ½ its length with broadly spreading lobes, these to 19 × 14 cm wide at apex, ovate-oblong, shortly acuminate apically, rounded to subacute basally, outer margin toothed, 9–10 primary nerves on each side of rachis.

Inflorescences: infrafoliar, erect to spreading; **peduncles** 8–30 cm long, slender; **bracts** 4–6, tubular, contracted or short-acuminate, upper reaching to rachis; **rachises** to 4 cm long. **Staminate** with up to 10 **rachillae,** these slender, small, green.

Pistillate with 2–3 **rachillae,** these 3–10 cm long, very slender, recurved, downward-pointing and red in fruit.

Flowers: Staminate known only in immature state, in loose spirals, depressed-globose, superficial; **calyx** short, lobed; **petals** valvate, spreading apically; **stamens** with **anthers** entire apically; **pistillode** short, columnar.

Pistillate in ± dense spirals, 1–1.8 mm apart, 2.5–3 × 2.5–3 mm, ± globose, green; **calyx** 1 × 1.8–2 mm, moderately lobed, green, **sepals** connate and/or imbricate in basal ⅔, rounded apically; **petals** 2.5–3 × 2.5–3 mm, imbricate nearly to apex, broadly triangular, rounded to obtuse apically, only slightly erect; **staminodes** present; **pistil** 1.75 × 1.5 mm, depressed-globose-obovoid, flattened apically, **stigma** lobes short, sessile, clear-colored.

Fruits: 6–7 mm diam., globose, black, seated on superficial and widely spaced rounded cushions; seeds globose; abortive carpels apparently adherent to perianth.

Distribution: GUATEMALA. Alta Verapaz. Baja Verapaz. Wet forest on the Atlantic slope; 900–1,600 m elevation.

Discussion: The description is from Burret (1933a), Standley and Steyermark (1958), and supplemented from our collections. Unfortunately, good staminate material of *C. tenerrima* is not known. *C. tenerrima* is locally distributed in the wild, having been collected only on a few occasions in Alta Verapaz and Baja Verapaz. Although usually pinnate, plants will infrequently flower with bifid leaves. An outstanding and diagnostic feature of the pinnae are the auricles on the lower margins near the base that actually extend across the rachis on the underside. Cook (1947a, page 19) illustrated these appendages.

Since Lehmann's holotype was destroyed and isotypes have not been found and *C. tenerrima* is such a distinct plant, it is appropriate to retypify this taxon here.

Chamaedorea tenerrima easily has the most striking and unusual leaves in the genus. They are outstanding in the exceedingly broad terminal pair of pinnae, widely flaring with prominent, white, sigmoid nerves and toothed margins. Terminal pinnae are suggestive of a butterfly's large, anterior, flaring

Plate 90. Chamaedorea stricta. In montane rain forest, *Hodel & Castillo 916A* (staminate) *& 916B* (pistillate), San Marcos, Guatemala. **A.** Staminate inflorescence erect from base of plant; **B.** Leaf and staminate flowers; **C.** Immature fruits and pistillate inflorescence; **D.** Juan Jose Castillo and small plant; **E.** Plant at base of tree in deep leaf litter trapped by root buttresses.

B
D
E
BOSTON

wings. Other pinnae are much smaller, strongly contracted at the base, and sharply reflexed on the rachis. *C. tenerrima* is not cultivated except for a few plants in the research collection in Los Angeles.

74. Chamaedorea tuerckheimii (*Dammer*) *Burret,* Notizbl. Bot. Gart. Berlin-Dahlem 11: 766. 1933a. Epithet: honors the collector of the type and many other Guatemalan palms, H. B. von Tuerckheim. **PLATES 4(D), 8 (E), 96, 97.**

Malortiea tuerckheimii Dammer, Notizbl. Konigl. Bot. Gart. Berlin 4: 157. 1904a. Type: Guatemala, Alta Verapaz, *Tuerckheim 8603* (holotype B, destroyed; isotype US!).

Kinetostigma tuerckheimii (Dammer) Burret, Notizbl. Bot. Gart. Berlin-Dahlem 11: 317. 1932.

Common Name: potato-chip palm.
Vernacular Name: guonay—Mexico.

Habit: solitary, erect or briefly decumbent, 0.3–1 m tall.

Stem: 3–7 mm diam., creeping, buried in leaf litter, green, conspicuously ringed, internodes 0.6–1.2 cm long.

Leaves: 7–12, erect-ascending to spreading, bifid, forming a rosettelike crown; **sheath** 5 cm long, very open, tubular only in lower ½, closely appressed, oblique apically, margin whitish or light green and longitudinally striate-nerved; **petiole** to 5 cm long or shorter, lightly grooved and green or gray-green above, rounded and green or gray-green below, lower margins of blades lightly decurrent along petiole to sheath; **rachis** 12–20 cm long, angled and green above, rounded and gray-green or green below with a pale or whitish band extending to sheath; **blades** 12–22 × 3.5–7.5 cm, cuneate-obovate or slightly elliptic, not bifid but entire or with a very small notch at otherwise rounded or obtuse apex, closely plicate, ± stiff, cuneate basally, velvety bluish green or mottled green above, paler below, margins toothed and whitish or green, 26–30 teeth per side, these 0.75–1.5 × 0.75–1.5 mm, 13–19 primary nerves on each side of rachis, a secondary nerve between each pair of primaries.

Inflorescences: interfoliar, 10–20 cm long, erect-spreading; **peduncles** 6–10 cm long, slender, light green in flower, orange in fruit; **bracts** 4–5, slender, thin-papery, longitudinally striate-nerved, brownish in flower, uppermost largest and obliquely split on one side; **rachis** 2–9 cm long, greenish in flower, orange in fruit. **Staminate** with 7–16 **rachillae,** these to 5–6 cm long, 1.5 mm diam., spreading, whitish green.

Pistillate spicate or sometimes forked, spreading; **rachis** or flower-bearing portion 5–6 cm long, ± rounded or slightly angled, 3–4 mm thick and green in flower, thicker and orange in fruit.

Flowers: Staminate in loose spirals, 2 × 2.5–3 mm, strongly flattened-depressed-globose, whitish, slightly sunken in superficial elliptic depressions 1–1.5 mm long; **calyx** 0.75–1 × 1.5–2.5 mm, lobed, very light green to yellowish, membranous, brown-margined, **sepals** connate in basal ⅔, acute to rounded apically; **petals** 2 × 2 mm, strongly cupped, bilobed, connate basally, free apically, thick, fleshy; **stamens** 1.5 mm high, paired, one in each of 2 petal lobes, **filaments** 0.5 mm long, connate and adnate basally to pistillode, white to yellow-white, **anthers** 0.5 mm long, yellow; **pistillode** 1.5–2 × 1.5 mm high, pear-shaped, 3-lobed apically, whitish.

Pistillate in moderate spirals, 1.75–2 × 3.5 mm, strongly flattened-depressed-globose, white, sunken in elliptic to rounded depressions 1–1.5 mm long; **calyx** 0.75–1.5 × 2.5 mm, deeply lobed, green to yellowish, brown-margined, **sepals** connate in basal ⅓, broadly rounded apically; **petals** 2 × 2 mm, connate and/or imbricate basally, imbricate or valvate nearly to apex where briefly free, corolla opening by a ± triangular opening, cup-shaped, slightly erect apically, thick, fleshy; **pistil** 1.25–1.5 × 1.5 mm, ± globose, light yellowish or white, **styles** short, **stigma** lobes separated, angled, elevated, strongly recurved.

Fruits: 8.5–12 × 6–8.5 mm, ovoid to ellipsoid, black, occasionally with perianth adherent, epicarp thin, mesocarp slightly fleshy, mucilaginous, green, aromatic, endocarp membranous, nerved; seeds 7–9 × 4.5 mm, ovoid, brownish-red.

Distribution: GUATEMALA. Alta Verapaz. MEXICO. Veracruz. Wet forest on the Atlantic slope; 900–1,500 m elevation.

Discussion: *C. tuerckheimii* is one of the smallest palms known and certainly one of the most distinctive. It is exceedingly handsome because of its dwarf habit and striking leaves of unique form that are, perhaps, the smallest of any palm. Initially, only the form from Guatemala was known. Tuerckheim collected it in the mountains near Coban and introduced it to Europe in the early 1900s. The species is extremely rare or localized in the wild; there are very few collections.

When he named and described the species, Dammer (1904a) only mentioned one specimen, *Tuerckheim 8603.* Burret (1933a) mentioned an additional specimen collected several years later in 1907, *Tuerckheim 11899,* and material of this number at several herbaria (GH, L, U, US) is often identified as an isotype. However, it seems inappropriate to include the latter collection as type material since it was collected three years after Dammer described and named *C. tuerckheimii* and Dammer only mentioned *Tuerckheim 8603* in the original article.

According to Standley and Steyermark (1958), *C. tuerckheimii* frequently appeared in gardens and markets in Coban. Local peddlers, who apparently did not want to reveal its exact location, stated that it came from the *tierra caliente* or lowland forest. However, in 1990 we found it in the wild near Coban in rich, very wet, montane rain and cloud forest. Unfortunately, though, after much searching we en-

→

Plate 91. Chamaedorea sullivaniorum. In montane rain forest, El Valle, Panama. **A.** Plant with stiff rosette of short-petioled leaves, *D. & M. Hodel 740A* (holotype); **B.** Pistillate plant (immature fruits), *D. & M. Hodel 740B.*

countered only two plants, both sterile. Local collectors and plant peddlers have undoubtedly decimated wild populations over the years, and if *C. tuerckheimii* is still abundant in the Coban region, it must be so in very inaccessible areas. Today, nurseries in Coban will only infrequently offer a few plants for sale, reflecting its scarcity and/or inaccessibility in the wild.

More recently, in the 1980s and early 1990s, collectors and hobbyists have turned their attention to *C. tuerckheimii* in Veracruz, Mexico. The species in Veracruz differs slightly from the form in Guatemala in the leaves being a mottled green, somewhat broader, not as strongly plicate, and with a green instead of white margin. Like its Guatemalan counterpart, it is also rare in the wild, being very localized in the Catemaco region. Sadly however, unlike its Guatemalan counterpart, it is still more accessible and collectors and local peddlers have not hesitated to strip it from the forest in large numbers.

Chamaedorea tuerckheimii is infrequently cultivated in California, Hawaii, Florida, and perhaps elsewhere. It has earned the name "potato-chip palm" because the leaves bear a resemblance to the ruffled style potato chips. Leaves are a bright, velvety green and strongly plicate, forming a handsome, rosettelike terminal crown.

Chamaedorea tuerckheimii is susceptible to brown-tipping due to low humidity, cold temperatures, and highly mineralized water, although the Guatemalan form seems more sensitive. Also, it is very susceptible to infestations of mites. Both the Guatemalan and Mexican forms are attention-grabbers in the garden and would be effective as solitary accents or in mass plantings if the species is ever widely available. However, like many other attractive and endangered chamaedoreas, it remains to be seen if nurseries in Guatemala and Mexico will have the foresight and resources to establish cultivated mother blocks to relieve the collecting pressure on the few remaining wild plants. Cultivated mother blocks for propagative purposes and strict forest preservation will be the only way to make *C. tuerckheimii* available to palm collectors and fanciers yet ensure its survival in the wild. If not, *C. tuerckheimii* will eventually vanish from the annals of botany and horticulture, remaining only as a few dried specimens tucked away in herbaria.

Krempin (1990, p. 96) illustrated what appears to be a species of *Ptychosperma* erroneously as *C. tuerckheimii.*

75. Chamaedorea undulatifolia *Hodel & Uhl,* Principes 34: 116. 1990b. Type: Costa Rica, Monteverde, *Moore 10179* (holotype BH!). Epithet: from the Latin *undulatus* meaning undulate and *folius* meaning leaved, in reference to the leaves with the pinnae having undulate margins. **PLATE 98.**

Common Name: none.
Vernacular Name: none.

Habit: solitary, erect to briefly decumbent, appearing stemless, overall height including leaves 1–1.5 m.

Stem: short, 1–1.5 cm diam., creeping at or slightly below leaf litter, densely ringed, nodes prominent and swollen, internodes very short, to 5 mm long.

Leaves: 3–5, erect-spreading, pinnate; **sheath** to 10 cm long, splitting deeply opposite petiole, tubular only near base, green, brown-margined apically; **petiole** to 60 cm long, flat or very slightly channeled and green above, rounded and green below; **rachis** 50–75 cm long, slightly angled and green above, rounded and green below; **pinnae** 12–18 on each side of rachis, lower and central pinnae largest, these to 12.5–19 × 1.6–2.7 cm, becoming progressively smaller toward apex of rachis, lanceolate, slightly sigmoid, acuminate, alternate or subopposite, margins undulate, bright forest green, thin, lower margin decurrent along rachis, prominent midrib and 1–2 primary nerves on each side of this, these yellowish when dry, midrib with warty-roughened surface at base below.

Inflorescences: interfoliar, sometimes infrafoliar in fruit, frequently appearing to arise from ground, erect; **Staminate** to 70 cm long but often smaller; **peduncle** to 47 cm long, greenish in flower; **bracts** 6, these tubular, acute-acuminate, bifid, longitudinally striate-nerved, ± papery, green but browning in flower; **rachis** 24 cm long, greenish in flower; **rachillae** 30, lower ones longest, these to 15 cm long, those above middle shorter, to 5–11 cm long, spreading, simple or occasionally forked, green.

Pistillate to 55 cm long but often smaller; **peduncle** to 44 cm long, greenish in flower, orange in fruit; **bracts** 8, these similar to those of staminate; **rachis** very short, 0.25–2.5 cm long, green in flower, orange in fruit; **rachillae** 2–6, these 2.5–11 cm long, strongly recurved, hooklike and orange in fruit.

Flowers: Staminate green, 1.5–2 mm high; **calyx** 0.5 mm high, shallowly lobed, membranous, flaring and drying light brown with darker margins; **petals** 1.5 mm long, valvate, free, ovoid, ± acute, drying dark; **stamens** with **anthers** sessile, not deeply bifid apically; **pistillode** slightly shorter than anthers, 3-angled and ± angled-enlarged apically.

Pistillate with **calyx** flaring, shallowly lobed, similar to that of

---

Plate 92. Chamaedorea sullivaniorum. A-E Cultivated, research collection, Los Angeles, *ex* El Valle, Panama; **A.** Pistillate flowers, *Hodel 854*; **B.** Staminate flowers, *Hodel 855*; **C.** Pistillate plant, *Hodel 854*; **D.** Staminate plant, *Hodel 855*; **E.** Mottled, bifid leaf blade; **F, G** In cloud forest, Panama; **F.** *Hodel et al. 1116,* Veraguas; **G.** In cloud forest, El Valle. (Note color difference between F, G.)

C
D
F
G

staminate; **corolla** 2 mm high, **petals** valvate but separating and recurving only at tip at anthesis and corolla ± urceolate; **pistil** green.

Fruits: 8 × 6 mm, ellipsoid-globose, black; perianth persistent but not prominently nerved when dry; eophyll pinnate.

Distribution: COSTA RICA. Alajuela. Cartago. Heredia. Puntarenas. San Jose. Wet forest on the Atlantic slope or just over the Continental Divide; 800–1,700 m elevation.

Discussion: *C. undulatifolia* is related to *C. brachyclada, C. pygmaea, C. scheryi,* and *C. stenocarpa.* These form a natural subgroup within subgenus *Chamaedoropsis* (Hodel and Uhl 1990b) that is characterized by a stemless habit, long-pedunculate inflorescences, decurrent pinnae, and pinnate eophylls. *C. undulatifolia* is closest to *C. scheryi* but the non-undulate margins of the pinnae and straight pistillate rachillae distinguish the latter species.

Chazdon (1987) referred to *C. undulatifolia* as *Chamaedorea* sp. "elegantissima." Because of its stemless habit and decurrent pinnae with strikingly undulate margins, *C. undulatifolia* bears a remarkable resemblance, especially when young, to certain ferns in the genus *Polypodium.* Although the undulating margins of the pinnae are quite striking and occur even on material cultivated in Costa Rica and California, they are not readily apparent on dried herbarium material.

Chamaedorea undulatifolia, like other closely related species, is difficult to grow well and is subject to the usual yellowing and brown-tipping of pinnae. It is grown in a few collections in California and Florida.

76. Chamaedorea verecunda *Grayum & Hodel,* Principes 35: 135. 1991. Type: Panama, Chiriqui, *Hammel 6221* (holotype MO!). Epithet: from the Latin *verecundus* meaning modest or shy, in reference to the small habit, nodding flowers, and apparent scarcity. **PLATES 99, 100.**

Common Name: none.
Vernacular Name: none.

Habit: solitary, decumbent then erect, 50–75 cm tall.

Stem: 3.5–5 mm diam., creeping, to 1 m long, green, smooth, ringed, internodes 1.7–3.8 cm long, often covered with persistent sheaths.

Leaves: 8–11, spreading, bifid; **sheath** 2.2–5.3 cm long, obliquely open apically, green, prominently striate-nerved; **petiole** 1.7–4.4 cm long, flat and green above, rounded and green below, densely white-spotted; **rachis** 2.75–5.4 cm long, slightly angled and green above, rounded and green below; **blade** 9–16 × 7–10 cm, obovate, incised apically to ½ or more its length, blue-gray nearly iridescent green, margins toothed especially toward apex, 6–9 primary nerves on each side of rachis, these sharply angled on both surfaces, dark green (darker than blade).

Inflorescences: interfoliar, erect, spicate. **Staminate** with **peduncle** 13–19 cm long; **bracts** 2–3; **rachis** or flower-bearing portion to 6.5–9.7 cm long.

Pistillate with **peduncle** 14–22 cm long; **bracts** 2–3; **rachis** or flower-bearing portion to 4–5.5 cm long, orange in fruit.

Flowers: yellow, nodding, maturing in a markedly progressive manner from the apex to base of axis. **Staminate** 3.6–5.1 mm long; **calyx** 0.5–1 × 1.7–2 mm, truncate or merely notched apically, hyaline, nerveless; **petals** 2.6–3.3 × 1.8–1.9 mm, rounded to obtuse or subacute, ovate, valvate in bud, free at anthesis, narrowed basally into a tube 1.1–1.6 mm high, forming a narrowly campanulate corolla, virtually nerveless; **stamens** with **anthers** 1.2–1.5 mm long, narrowly ovate, ± narrowed apically, **filaments** very short, 0.1–0.4 mm long, adnate basally to pistillode; **pistillode** prominent, 2.3–2.8 mm high, greatly exceeding stamens, clearly differentiated and lobed apically.

Pistillate with **calyx** similar to that of the staminate, 0.4–0.5 × 1.6–1.7 mm, truncate or ± deeply undulate apically; **petals** 1.5–1.7 × 1.6–1.8 mm, broadly ovate, imbricate, obscurely 5–7-nerved, rather abruptly narrowed basally into a tube 0.4 mm long adnate to basal part of pistil, forming a broadly campanulate to suburceolate corolla; **staminodes** lacking; **pistil** (part above insertion of petals) 1.8–2.2 mm high, narrowed apically, **styles** short or lacking, **stigma** lobes separate.

Fruits: 1 cm long, black.

Distribution: PANAMA. Chiriqui. Wet forest near the Continental Divide; 1,200–1,400 m elevation.

Discussion: *C. verecunda* is known from only a few collections from western Panama. It is one of the most distinctive of the numerous, dwarf, bifid-leaved, cloud-forest *Chamaedorea* species known from Central America. Its most distinctive characters are the very small, deeply bifid, blue- and/or gray-green leaf blades; spicate inflorescences; and nodding, relatively large flowers that mature in a markedly progressive manner from the apex to the base of the axis.

Chamaedorea verecunda is closest to *C. correae* and *C. guntheriana,* both from similar habitats in central Panama. *C. correae* differs in its larger size, thicker and coarser leaf blades, and staminate inflorescence with normally 2–3 rachillae and considerably smaller flowers. *C. guntheriana* differs in its thicker, usually pinnate leaves and branched staminate inflorescences with smaller flowers that ma-

---

→

Plate 93. Chamaedorea tenerrima. In wet cloud forest, Guatemala. **A, C, D** *Hodel & Castillo 883,* Baja Verapaz; **B, E** *Hodel & Castillo 894* (neotype), Alta Verapaz; **A.** Adult leaf; **B.** Infructescences (mature fruits); **C.** Fruiting plant under log in steep ravine; **D.** Juvenile plants with bifid leaves; **E.** Juan Jose Castillo holds crown of plant.

C
E

ture in a markedly progressive manner from the base to the apex of the axis.

Other than a few plants in the research collection in Los Angeles, *C. verecunda* is not cultivated.

77. Chamaedorea volcanensis *Hodel & Castillo,* Phytologia 68: 393. 1990. Type: Guatemala, Quetzaltenango, *Hodel & Castillo 913A* (holotype BH!; isotype AGUAT!). Epithet: derived from the Spanish word for volcano, *volcan,* in reference to the habitat of this species in southwestern Guatemala. **PLATE 101.**

Common Name: none.
Vernacular Name: none.

Habit: solitary, erect, first flowering when stemless, eventually to 1.5 m tall.

Stem: 2–3 cm diam., initially creeping at or below leaf litter, then curved, to 15 cm long, rooting along its length, nodes very congested, internodes to 5 mm long, eventually above-ground stem developing with internodes to 5 cm long, erect.

Leaves: 4–5, spreading, regularly pinnate to 1.8 m long or flowering when variously and unevenly pinnate or bifid and then substantially smaller; **sheath** to 30 cm long, splitting deeply opposite petiole, obliquely long-open, tubular only in basal ½, thick, green, longitudinally striate-nerved; **petiole** to 70 cm long, flat and green above and slightly grooved near base, rounded and pale below, green and longitudinally striate-nerved laterally; **rachis** to 75 cm long, angled and green above, rounded below with a yellowish band extending onto sheath; **pinnae** 8–12 on each side of rachis, middle ones longest, to 35 × 7 cm, long-lanceolate, slightly sigmoid, falcate, acuminate, contracted basally, 5 prominent primary nerves above, apical pair usually widest, to 25 × 7.5 cm, 7–8 prominent primary nerves above, 1 secondary between each pair of primaries, basal pinnae to 25 × 4 cm, all pinnae ± thin with conspicuous primary and secondary nerves drying a distinct pale yellow below; or, if blade bifid then to 30 × 25 cm, incised apically more than ½ its length, rachis to 10 cm long, 7–9 primary nerves on each side of rachis, 2 secondaries between each pair of primaries.

Inflorescences: infrafoliar, erect from base at or below leaf litter, long-pedunculate; **peduncles** to 65 cm long, 5–7 mm wide at base and flattened, 3–5 mm diam. at apex, green where exposed in flower, orange and nodding slightly when heavily laden with fruits; **bracts** 6–7, prophyll 2.5 cm long, 2nd bract 2.5–3 cm, 3rd 4–5 cm, 4th 7–8 cm, 5th 15 cm, 6th 20 cm, 7th 20–25 cm and extending onto rachis and concealing rudimentary 8th bract, tubular, acute-acuminate, green to brown in flower, brownish in fruit, papery, longitudinally striate-nerved. **Staminate** with **rachis** 5 cm long, simple or basal portion branched with up to 5 rachillae per branch, pale green in flower; **rachillae** 2–13, these 10–15 cm long, 1 mm diam., briefly erect basally then drooping, pale green in flower, densely flowered nearly to base.

Pistillate usually few-branched with up to 5 rachillae, or spicate or furcate especially when young; **rachis** to 6 cm long, slightly longitudinally angled, greenish in flower, orange in fruit; **rachillae** or flower-bearing portion to 15 cm long, erect, rigid and whitish or very pale green in flower, slightly swollen and orange in mature fruit.

Flowers: Staminate in very dense spirals but not contiguous, 0.5–1 mm apart, in bud 1.5 × 1.5–2 mm, subglobose, older flowers 4 mm diam., ± globose, white or cream ageing to brown, only slightly sunken in superficial elliptic depressions 2.5 mm long; **calyx** well developed and prominent in bud, 1 × 2 mm, deeply but irregularly lobed, thin, membranous, brown-margined, **sepals** connate only briefly basally, rounded-acute apically; **petals** 1.5 × 2–2.5 mm, deltoid, valvate, connate basally, spreading apically to give corolla vaselike shape, fleshy but thin; **stamens** equalling or exceeding petals, **filaments** large, terete, flared basally, fleshy, those opposite sepals to 2.5 mm long, others 2.0 mm long, **anthers** 1 × 0.75–1 mm, oval, widely divergent and flattened basally, thin, attached briefly to filament distally; **pistillode** 1–1.5 mm high, columnar, flared and flattened basally, tip flat, 3-lobed.

Pistillate borne in 5 dense spirals but not contiguous, 0.5–1 mm apart, 2 mm diam., globose, whitish or cream, only slightly sunken in shallow elliptic depressions 2.5 mm long; **calyx** well developed and prominent in bud, 0.75–1 × 2–2.5 mm, lobed, thin, membranous, brown-margined, **sepals** connate in basal ½, broadly rounded apically; **petals** 2 × 1.5–2 mm, tightly imbricate nearly to apex, connate basally only briefly, straight or rounded apically; **staminodes** lacking; **pistil** 1.5 × 1.5 mm, ovoid, **stigma** lobes sessile, short, pointed, recurved.

Fruits: 12 × 8 mm, oblong or ovoid, narrowed at apical end, black.

Distribution: GUATEMALA. Quetzaltenango. San Marcos. Suchitepequez. Wet forest on the Pacific slope; 1,200–2,000 m elevation.

Discussion: *C. volcanensis* appears restricted to the vast and diverse forests of the wet, Pacific slopes of Volcan Tajumulco, Volcan Santa Maria, Volcan Zunil, and Volcan Santo Tomas in western Guatemala.

Plants will often initially flower with bifid leaves and few-branched inflorescences but will later have pinnate leaves and inflorescences with more rachillae. Without stretching the imagination too much, the staminate inflorescence with its densely flowered rachillae remind one of an erupting volcano as the rachillae are first erect then apically drooping, mimicking molten material being ejected from a crater.

Flowers of *C. volcanensis* are similar to those of *C. castillo-montii* from eastern Guatemala. However, the latter species has spicate staminate inflorescences and usually bifid leaves (or, if pinnate, then linear pinnae with a prominent midrib and less conspicuous secondary nerves). Other than a few plants in the research collection in Los Angeles, *C. volcanensis* is not cultivated.

Plate 94. Chamaedorea tenerrima. Pistillate plant (immature fruits), *Hodel & Castillo 1009,* montane rain forest, Alta Verapaz, Guatemala. →

78. Chamaedorea vulgata *Standl. & Steyerm.*, Publ. Field Mus. Nat. Hist., Bot. Ser. 23: 208. 1947. Type: Guatemala, San Marcos, *Steyermark 37735* (holotype F!). Epithet: from the Latin *vulgatus* meaning common, general, or ordinary, apparently in reference to the species initially being thought as one of the most common palms of western Guatemala. **PLATE 102.**

Common Name: none.
Vernacular Names: pacaya, cum, kip—Guatemala.

Habit: solitary, erect, 2–3 m tall.

Stem: 1–2 cm diam., green, densely and minutely white-spotted, conspicuously ringed, internodes 10 cm long or more.

Leaves: 3–5, ascending to spreading, pinnate; **sheath** to 45 cm long, long-open, distinctly white-spotted; **petiole** 50–75 cm long, 7–9 mm thick, distinctly white-spotted; **rachis** to 70 cm long, green below but distinctly white-spotted; **pinnae** 10 on each side of rachis, middle ones largest, to 50 × 9 cm, linear-lanceolate, not sigmoid, attenuately long-acuminate with nearly straight upper margin, falcate, regularly disposed, mostly alternate, thin, gradually narrowed basally to 3–3.5 cm wide on larger pinnae, green, concolorous, 5 primary nerves, keeled above, yellowish, shining and prominent below, 1–3 secondaries between primaries, shining below, not keeled above, tertiaries fine and numerous, pinnae only slightly decreasing in size toward apex and base of blade.

Inflorescences: infrafoliar, erect-spreading; **peduncles** 25–65 cm long, stout, 2 cm wide at base, 2–4 mm thick at apex in fruit; **bracts** 5–6, prophyll to 5 cm long, others to 45 cm long, 6–12 mm broad, tubular, acute, oblique apically, striate-nerved, distinctly white-spotted; **rachises** to 15 cm long, green in flower, orange in fruit. **Staminate** with 9–12 **rachillae,** these 15–25 cm long, green, pendulous.

Pistillate with 3–12 **rachillae,** these 15–27 cm long, spreading, becoming orange, thickened, and pendulous in fruit.

Flowers: Staminate in moderately dense spirals, 3 mm high, green, superficial; **calyx** 1 mm high, lobed, **sepals** connate in basal ½; **petals** valvate, free apically; **stamens** with **filaments** rather short, **anthers** elongate and generally bifid apically; **pistillode** columnar, ± thickened, 3-angled and truncate apically.

Pistillate in rather dense spirals, superficial; **calyx** 1 mm high, lobed; **petals** rather strongly imbricate, acute; **staminodes** lacking; **pistil** ovoid, **stigma** lobes sessile.

Fruits: 8–9 mm long, obovoid, black; seeds globose; abortive carpels adherent to fruit.

Distribution: GUATEMALA. Quetzaltenango. San Marcos. MEXICO. Chiapas. Moist or wet forest on the Pacific slope; 1,300–2,350 m elevation.

Discussion: The description, taken from Standley and Steyermark (1958), is more limited than that in their original work (1947) due to the inclusion of extraneous elements (probably *C. tepejilote*) in the latter. I have supplemented the description from our collections. *C. vulgata* is rather rare in the wild, having been collected on only a few occasions. The stem, sheaths, petioles, rachises, and bracts are densely covered with small but conspicuous white spots, a diagnostic feature that is apparent even on the dried type.

Because of its strongly nerved pinnae, *C. vulgata* does bear a resemblance to *C. tepejilote* but the contiguous staminate flowers, shorter peduncles, and lack of white spots densely covering the stem, leaf sheaths and petioles, and bracts distinguish the latter species. Other than a few plants in the research collection in Los Angeles, *C. vulgata* is not cultivated.

79. Chamaedorea whitelockiana *Hodel & Uhl*, Principes 34: 61. 1990a. Type: Mexico, Oaxaca, *D. & R. Hodel 935* (holotype BH!; isotype MEXU!). Epithet: honors Loran W. Whitelock who first brought this species to my attention. **PLATES 103, 104.**

Common Name: none.
Vernacular Name: none.

Habit: solitary, erect, to 4 m tall, often flowering when appearing stemless but then actually with short subterranean stem.

Stem: 1–1.3 cm diam., green, ringed, internodes to 10 cm long, often covered with persistent leaf sheaths.

Leaves: 5–7, spreading, pinnate; **sheath** to 35 cm long, tubular, obliquely open apically and there dark-margined and ragged with a brown ligule 5–8 mm long on either side of petiole, below this margin whitish and green longitudinally striate-nerved; **petiole** 10–40 cm long, finely grooved and green above, rounded and pale below; **rachis** to 35 cm long, angled and green above, rounded below with a pale band extending onto sheath; **blade** to 40 × 15 cm; **pinnae** 7–9(–13) on each side of rachis, lower middle ones longest, to 17 × 2–3.5 cm, pinnae progressively smaller toward apex of rachis, 2 basal pinnae smaller, apical pair shortest of all with 2 prominent primary nerves, lanceolate, falcate, long-acuminate, opposite or subopposite, thin-papery, apical ⅓ of each pinna drooping and curling down and under, conspicuously contracted basally and there a hard knob or callus at point of attachment, slightly iridescent gray-green above, paler below, prominent midrib and 2 ± inconspicuous primary nerves on each side of this or primaries absent, secondaries faint and inconspicuous; or when in young and stemless phase leaves finely divided, sheath 3–5 cm long, very open, tubular only near base; petiole 5–8 cm long; blade 20 × 12 cm; pinnae linear-lanceolate, straight, slightly falcate, 7 × 0.8 cm, only a midrib prominent, primary and secondary nerves very faint, inconspicuous.

Inflorescences: interfoliar but often infrafoliar in fruit, arising from behind old sheaths, erect. **Staminate** with **peduncle** 25–30

→

Plate 95. Chamaedorea tenerrima. In montane rain forest, *Hodel & Castillo 1009*, Alta Verapaz, Guatemala. **A.** Adult leaf; **B.** Crown of leaves.

cm long, 7–9 mm wide at base and there flattened, 3–4 mm wide at apex and there rounded, erect, greenish or pale in flower; **bracts** 7–8, prophyll to 3.5 cm long, 2nd bract to 6 cm, 3rd to 9 cm, 4th–7th to 10 cm, 8th to 4 cm, tubular, acute-acuminate, obliquely open apically, longitudinally striate-nerved, drying brownish and papery in flower; **rachis** to 5 cm long, slightly to strongly curved, lime-green in flower; **rachillae** 12–15, these to 10–12 cm long, 1.5–2 mm diam., erect, ± stiff, lime-green in flower.

Pistillate with **peduncle** to 40 cm long, erect, 5 mm wide at base and there flattened, 2.5 mm diam. at apex and rounded, pale or greenish in flower, orange in fruit where exposed; **bracts** similar to those of staminate; **rachis** 1–3 cm long, greenish in flower, orange in fruit; **rachillae** 6–8, these to 10 cm long, 1–2 mm diam., erect, ± stiff, slender, longitudinally ridged or angled or finely grooved, slightly spinose-tipped, greenish in flower, orange in fruit.

Flowers: Staminate in rather dense spirals, 2–3 mm apart, 3 × 4 mm, depressed-globose, yellow, slightly sunken in elliptic depressions; **calyx** 0.5 × 2.5 mm, lobed, pale yellow-green, membranous, **sepals** connate in basal ¾, rounded apically; **petals** 4 × 3 mm, broadly ovate, valvate, free nearly to base, strongly incurved apically but not connate, broadly acute; **stamens** 2–2.5 mm high, outward-leaning, **filaments** 1–1.5 mm long, pale, **anthers** 1–1.5 mm long; **pistillode** 2.25–2.75 mm high, broadly columnar, lime-green.

Pistillate in lax spirals, 4–5 mm apart, 3 × 2.5–3 mm, ovoid-globose, slightly sunken in elliptic depressions; **calyx** 1–1.5 × 2–2.5 mm, deeply lobed, light green, margins pale and membranous, **sepals** connate in basal ½–¾, broadly rounded apically; **petals** 3 × 2–2.5 mm, broadly triangular, imbricate nearly to apex, erect apically, obtuse to acute; **pistil** 2 × 2 mm, globose, flattened apically, green, **styles** short or lacking, **stigma** lobes short, pointed, dark green or brownish.

Fruits: 7–8 mm diam., globose, black.

Distribution: MEXICO. Chiapas. Oaxaca. GUATEMALA. Escuintla. Quetzaltenango. Sacatepequez. Solola. Moist to wet, pine-oak forest on the Pacific slope; 1,400–2,000 m elevation.

Discussion: Loran M. Whitelock of Los Angeles, California collected this species in 1981 in Oaxaca, Mexico. Whitelock returned to Los Angeles with seeds and plants and established them in his garden where they were first brought to my attention.

An attractive plant, *C. whitelockiana* is close to *C. parvisecta* from the Atlantic slope of Guatemala and Mexico. The pendulous staminate rachillae, distinctly smaller staminate flowers with narrowly ovate petals and erect to spreading tips, and much thicker, blunt-tipped pistillate rachillae with fairly dense flowers distinguish *C. parvisecta.*

The current description reflects the substantially larger pinnae, leaves, and stems of Guatemalan material tentatively included here, and thus differs somewhat from the original account of *C. whitelockiana* in Hodel and Uhl (1990a). Further study may reveal the Guatemalan material to represent a distinct, unnamed species. Standley and Steyermark (1958) referred the Guatemalan material to *C. aequalis* (= *C. liebmannii*), but the nerved staminate flowers with petals connate apically and the corolla opening by lateral slits distinguish *C. liebmannii.*

Chamaedorea whitelockiana exhibits a growth phase best typified as juvenile, with plants characterized by their virtually stemless habit and finely divided leaves with numerous, narrow pinnae, not unlike those of *C. elegans.* Plants are not truly stemless during this period but actually possess a short, curving, rooting subterranean stem with highly congested nodes. They begin to produce much reduced inflorescences during the stemless phase, which lasts for several years, until later forming a visible stem with normal-sized inflorescences and broader pinnae. The juvenile phase may anchor the plants securely in their substrate before an erect, aboveground stem develops.

Not widely cultivated, *C. whitelockiana* is in only a few collections in California. Plants are susceptible to infestations of thrips.

80. Chamaedorea woodsoniana *L. H. Bailey* Gentes Herb. 6: 238, fig. 131. 1943a. Type: Panama, Cocle, *Allen 1901* (holotype MO!). Epithet: honors R. L. Woodson, collector of many Panamanian plants. **PLATES 105, 106.**

Chamaedorea vistae Hodel & Uhl, Principes 34: 58. 1990a. Type: Mexico, Oaxaca, *Moore & Bunting 8913* (holotype BH!; isotype US!). Epithet: from the type locality, Vista Hermosa, an area of rugged, steep, densely forested slopes with panoramic views or vistas.

Common Name: none.
Vernacular Name: none.

Habit: solitary, erect, ± stout, 5–12 m tall.

Stem: 5–10 cm diam., green, ringed, nodes white and prominent, to 2.5 cm wide, internodes to 20 cm long, prop roots evident (at least in some cultivated material) to 60 cm up stem.

Leaves: 4–6, erect-spreading, pinnate; **sheath** 30–80 cm long, swollen, tubular, clasping, obliquely open apically, green, strongly longitudinally striate-nerved; **petiole** to 30 cm long, flattened or grooved especially toward base and green above, rounded and green below with a pale yellow or yellow-green band extending

---

Plate 96. Chamaedorea tuerckheimii. A. *Hodel 750,* cultivated, research collection, Los Angeles, *ex* Veracruz, Mexico; **B.** In montane rain forest, *Hodel & Castillo 1010,* Alta Verapaz, Guatemala.

onto rachis; **rachis** 1–1.5 m long, sharply angled and green above, rounded and pale below; petiole and underside of rachis densely covered with small, irregular pits giving living material a rough texture; **pinnae** to 36 on each side of rachis, these 30–65 × 3.5–5 cm, linear-lanceolate, ± straight or only slightly falcate, acuminate, horizontal or ascending off rachis, dull dark green, a prominent midrib and 2 primary nerves on each side of this, these conspicuous but not much raised above (only midrib prominent above in dried material), all 5 nerves strongly elevated and prominent below, secondaries numerous and faint.

Inflorescences: inter- or infrafoliar in flower, infrafoliar in fruit, erect but nodding when laden with fruits; **peduncles** to 75 cm long, ± stout, 2.5–3 cm wide at base and there flattened, 1–1.5 cm diam. at apex and there ± rounded, erect and greenish in flower, orange apically and nodding in mature fruit; **bracts** 5, prophyll to 12 cm long, 2nd bract to 30 cm, 3rd to 35 cm, 4th to 60 cm, this exceeding peduncle and concealing smaller 5th bract to 12 cm, tubular, fibrous, shredded and brownish at anthesis, often fallen in mature fruit, acute-acuminate, bifid, longitudinally striate-nerved. **Staminate** large with a rachis to 75 cm long, erect, straight, green, longitudinally ridged; **rachillae** to 100, basal ones largest, to 50 cm long, 1–2 mm diam., mostly unbranched but sometimes forked, becoming progressively shorter toward apex of rachis, drooping, green.

Pistillate with **rachis** to 40 cm long, greenish and erect in flower, orange and downward-pointing in mature fruit, basal branches ramified 1–2 times; **rachillae** to 50, these to 30 cm long, erect or ascending and dark green in flower, orange and downward-pointing in mature fruit.

Flowers: Staminate in ± dense spirals, 2.5–3 × 2.5–3.5 mm, ± globose, slightly sunken in elliptic depressions to 4 mm long; **calyx** 0.5–0.75 × 2.5–3.5 mm, prominently lobed, membranous, brown-margined, **sepals** connate and/or imbricate in basal ¼–⅓, rounded apically; **petals** 2.5–3.0 × 2.5–3 mm, broadly ovate, valvate, free nearly to base, spreading apically, acute, ± thick, margins thicker, lightly nerved on inside only; **stamens** shorter than pistillode, **filaments** to 0.3 mm long, angled, adnate basally to petals, **anthers** 1.5 × 1 mm, shorter than pistillode and tightly apressed around it, basifixed; **pistillode** 1.5–2.0 mm high, columnar-conic, nearly equalling petals, broadly flared and buttressed basally, longitudinally grooved, tip small, ± 3-lobed.

Pistillate 1.5 mm apart in 4 irregular rows, 2.5 × 2.5 mm, ± globose, orange-yellow; **calyx** 0.5–1 × 2.5 mm, prominently lobed, brown-margined, **sepals** connate and/or imbricate in basal ¼–½, rounded apically; **petals** 2–2.5 × 2–3 mm, broadly ovate, strongly imbricate nearly to apex, white-margined; **staminodes** lacking or 6, minute, toothlike, adnate to petals, white; **pistil** 2 × 1.5 mm, ovoid-globose, greenish, **styles** short or lacking, **stigma** lobes just exceeding petals, erect, pointed, reflexed, yellowish.

Fruits: to 2 × 1 cm or 1 cm diam., oblong to ± globose, golden orange ripening black.

Distribution: PANAMA. Chiriqui. Cocle. Darien. Panama. Veraguas. COSTA RICA. Alajuela. NICARAGUA. Matagalpa. HONDURAS. Atlantida. GUATEMALA. Baja Verapaz. El Pregreso. Huehuetenango. Zacapa. MEXICO. Chiapas. Oaxaca. Veracruz. COLOMBIA? Wet montane forest and cloud forest; 800–2,000 m elevation.

Discussion: Bailey (1943a) described and named *C. woodsoniana* from material Paul Allen collected near El Valle de Anton in Panama. With stems to 10 cm in diameter, it is one of the largest species of the genus. Only *C. linearis* and some forms of *C. costaricana* and *C. tepejilote* have stems as stout as those of *C. woodsoniana.*

Chamaedorea woodsoniana is a handsome and rather robust species. It is particularly striking when pistillate plants are bearing clusters of not-yet-mature, golden-orange fruits at the ends of long, arching peduncles. It differs from all other members of the genus in its combination of orange-yellow pistillate flowers, long-pedunculate inflorescences, long-pinnate leaves with pinnae prominently nerved below, and solitary habit. *C. woodsoniana* appears closest to *C. carchensis* from the Atlantic slope of Guatemala and Mexico but the black indument densely covering the petioles, fewer pinnae with fewer and less prominent primary nerves, and smaller staminate inflorescence with much fewer rachillae distinguish the latter species. Also, *C. woodsoniana* quickly develops a prominent above-ground stem while *C. carchensis* remains stemless and flowers and fruits at this stage until forming a short stem much later.

Botanists and horticulturists, me included, have confused *C. woodsoniana* with some of the larger growing forms of *Chamaedorea costaricana,* a species with cespitose stems. *C. costaricana* also differs in the ligules at the apex of the leaf sheath, shorter peduncles, rounded fruits maturing from green to black, and usually sigmoid or falcate pinnae.

The material presently cultivated as *C. woodsoniana* is actually *C. costaricana.* These plants originated from numerous introductions of seeds and young plants from the late 1960s through the 1980s from Costa Rica and Panama. Robert Wilson provided most of these from plants he had established in his garden or that he or others had collected in that part of Costa Rica or adjacent Panama. Krempin (1990, p. 97) illustrated *C. costaricana* but erroneously captioned the photograph as *C. woodsoniana.*

However, *C. woodsoniana* is frequently cultivated

Plate 97. Chamaedorea tuerckheimii. Cultivated, California. **A.** Trio of stunning plants, garden of P. Sullivan, Ventura; **B.** Staminate inflorescence, *Hodel 761,* garden of L. Hooper, La Habra.

→

but under another name, *C. vistae.* The misapplication of the name *C. woodsoniana* to a cespitose plant and naming of a species, *C. vistae,* later reduced to synonymy, is a good example of how nomenclatural problems originate. I had always assumed that *C. woodsoniana* was a cespitose species because I had encountered cespitose plants labeled as such in collections and botanical gardens and read of it in the literature. Thus, I formed a biased, preconceived idea of the habit of *C. woodsoniana.* The fact that Bailey's description and label on the type specimen of *C. woodsoniana* stated nothing of a cespitose habit did not alarm me since the habit of palms is often lacking in collection data and Bailey never saw the plant in the wild. Also, only portions of inflorescences were in the type, giving a much smaller impression of the size of these organs, and foliar material, while large, was not substantially different from that of the larger forms of *C. costaricana.*

When *C. vistae* was described and named (Hodel & Uhl 1990a), I did not consider the relationship of this solitary-stemmed species to *C. woodsoniana* since I assumed the latter was a cespitose species. I began to suspect that *C. woodsoniana* was actually solitary-stemmed upon examining additional material noted as being solitary from Costa Rica, Nicaragua, Guatemala, and the type locality in Panama that matched up very well with the type specimen of *C. woodsoniana.* I reassessed my interpretation of the habit of *C. woodsoniana* to that of solitary, leading me to suspect that *C. vistae* was not distinct.

My suspicions were confirmed when in April, 1991 we were able to collect material of a stout, solitary-stemmed *Chamaedorea* in Guatemala and at the type locality of *C. woodsoniana* in Panama that matched up well with the type of *C. woodsoniana* and yet was apparently identical to *C. vistae.* Thus, here I have the onerous task of including *C. vistae* in synonymy with *C. woodsoniana* and am reminded of what Bailey (1943a, p. 236) eloquently stated prior to naming new chamaedoreas: "Under such circumstances the adventurer in new species assumes unusual risks."

Chamaedorea woodsoniana is cultivated in gardens and collections in Hawaii, California, Florida, Australia, Costa Rica, and, perhaps, elsewhere. Krempin (1990, p. 97) illustrated *C. woodsoniana* as *Chamaedorea* sp. *9087* in the Sydney Botanical Garden. Fruiting plants occur in Hawaii at Wahiawa Botanic Garden of the Honolulu Botanic Gardens, Waimea Arboretum, and Lyon Arboretum and at Jardin Botanico Robert y Catherine Wilson near San Vito de Coto Brus in Costa Rica, and probably elsewhere. These plants have been grown from seeds collected from plants growing in the conservatory of the L. H. Bailey Hortorium at Cornell University as *BH61-1178.* The latter were grown from seeds that Moore gathered when he first collected this species in Mexico. From time to time, Moore and the Seed Bank of the International Palm Society distributed seeds produced on the plants at Cornell. Recently, Lyon Arboretum has begun to distribute seeds and seedlings of *C. woodsoniana* under the number *BH61-1178* as *C. vistae.*

Chamaedorea woodsoniana has been collected on the border of Panama and Colombia, thus I tentatively list it here as occurring in Colombia.

VI. Subgenus MORENIELLA

Chamaedorea subgenus **Moreniella.**

Key Characters: petals distinct in flowers of both sexes, thickened and persistent in fruit, staminate flowers paired or in short linear groups, petals spreading or erect, free apically; staminate inflorescences solitary at a node; fruits red.

Distribution: two species in upland forests in northeastern Mexico; to about 1,000 m elevation; often on limestone; both are commonly cultivated.

Key to the Species of Subgenus *Moreniella*

Cespitose habit; pinnae sigmoid. . . . . **81. C. microspadix**
Solitary habit; pinnae straight. . . . . **82. C. radicalis**

81. Chamaedorea microspadix *Burret,* Notizbl. Bot. Gart. Berlin-Dahlem 11: 734. 1933a. Type: Mexico, San Luis Potosi, *Pringle 3980* [(holotype B, destroyed; isotypes BM (photo!), F!, GH!, L!, M!, MICH!, MO!, NY (photo!), P!, S!, US!, VT!)]. Epithet: from the Greek *micro* meaning little or

---

→

Plate 98. Chamaedorea undulatifolia. In wet cloud forest, Monteverde, Costa Rica. **A.** Large plant, *D. & M. Hodel 695A*; **B.** Marianne Hodel holds leaves of mature plant; **C.** Young leaf. (Note distinctly undulate margins of pinnae.)

C

small and the Latin *spadix* meaning flower stalk (inflorescence), in reference to the small inflorescences. **PLATES 8(G), 9(D), 107, 108(A).**

Common Name: none.
Vernacular Name: none.

Habit: cespitose, erect to leaning, to 3 m tall, forming dense clumps several meters wide.

Stems: 1 cm diam., green, ringed, internodes 5–15 cm long.

Leaves: pinnate, spreading; **sheath** 20–30 cm long, tubular, oblique apically and there light green or whitish, green below but browning and persistent on stem with age, densely and longitudinally striate-nerved; **petiole** 15–25 cm long, strong, flat or only slightly grooved and green above, rounded, slightly paler green below especially near base; **rachis** 50 cm long, obscurely angled and green above, rounded and green below; **blade** oblong; **pinnae** 9 on each side of rachis or fewer when terminal pair is broader, to 25 × 4–5 cm, regularly arranged, alternate, lanceolate, sigmoid or falcate, velvety green above and glaucous or green below; upper pair often confluent and 2–3 times broader than others, 3 primary nerves, several secondaries and tertiaries visible above and below.

Inflorescences: infrafoliar, sometimes breaking through the old persistent sheaths, to 25 cm long; **peduncles** 10 cm long, 8 mm wide at base, 5 mm wide at apex, slender, ± flattened, green or pale where exposed in flower, green or orange where exposed in fruit; **bracts** 4–5, prophyll 3 cm long, 2nd bract 5 cm, 3rd 8 cm, 4th 7 cm, uppermost exceeding peduncle and sometimes concealing a small rudimentary bract, short-acuminate, slightly flared and bifid apically, closely or ± loosely sheathing, flattened, green becoming brown at anthesis, papery, longitudinally striate-nerved; **rachises** 2–3 cm long, green in flower, orange in fruit; **rachillae** 3–6 or slightly more, to 15 cm long, 3 mm diam., green, spreading in flower, orange and drooping in fruit.

Flowers: Staminate in clusters of 2–4 or more, 5 × 3–4 mm, obovoid-oblong, creamy-white; **calyx** 1.5 × 2.5–3 mm, deeply lobed, green, **sepals** connate in basal ¼–½, acute or broadly rounded apically; **petals** 5 × 2–3 mm, slightly concave, connate briefly basally, valvate, spreading apically, acute, ± thick and fleshy; **stamens** ⅔ as high as petals, **filaments** prominent, whitish, **anthers** shallowly 2-cleft apically and basally; **pistillode** exceeding stamens but not petals, broadly columnar, whitish.

Pistillate in moderate spirals, solitary, 5 × 3 mm, ovoid, white; **calyx** 1.5–2 × 3 mm, deeply lobed, green, relatively thick, **sepals** connate in basal ½–⅓, rounded apically; **petals** 4 × 3 mm, suborbicular, 4 × 3 mm, imbricate in basal ½–⅓, acute-acuminate, apically erect or flared outwardly; **pistil** slightly exceeding corolla, 5 × 3 mm, pear-shaped, light green, **styles** short or lacking, **stigma** lobes recurved, acute, dark.

Fruits: to 1 cm diam., ± globose, orange-red or red, fruiting perianth nerveless.

Distribution: MEXICO. Hidalgo. San Luis Potosi. Moist forest at middle elevations; usually on limestone.

Discussion: Pringle discovered *C. microspadix* in 1891 in Tamasopo Canyon, San Luis Potosi, Mexico and Burret (1933a) described and named it. It is one of the most handsome members of the genus due to its clustering habit, velvety dark green leaves, and showy clusters of bright orange-red fruits.

Chamaedorea microspadix is rather widely cultivated and appears in gardens and collections in California, Florida, Hawaii, Australia, Venezuela, Europe, and elsewhere. Seeds are handled commercially to a somewhat limited extent and plants are occasionally encountered in nurseries.

Several forms are cultivated and these vary in shape, color, and size of pinnae and the glaucous-white covering on the underside. Some forms have pinnae that are entirely green on the underside while others have pinnae that are distinctly powdery-white below.

Chamaedorea microspadix is durable, hardy, and of relatively easy culture. There appears to be some variation in tolerance of light among various forms. Some are more tolerant of higher light than others and can be grown in nearly full-sun, resulting in only slight yellowing of the leaves. Some forms appear to be more susceptible to browning and yellowing of the tips of the pinnae. Several horticultural-varietal names have been given to forms in cultivation. Among them are *C. microspadix* 'Improved,' said to be less susceptible to brown- and yellow-tipping, and *C. microspadix* 'Brentwood Select,' said to be darker green and more tolerant of higher light as well as less susceptible to brown-tipping. The late David Barry, Jr. made the latter selection and offered it at his nursery.

Surpassed by only *C. radicalis* for cold hardiness, *C. microspadix* will tolerate −2° to −4°C (25–28°F) with no appreciable leaf damage. Temperatures of −5° to −7°C (19–23°F) may damage leaves only, with plants surviving at even lower temperatures. Substantially lower temperatures have killed stems but plants survived and resprouted from the base with warmer weather. The species is also fairly resistant to nematodes.

Krempin (1990, p. 90) discussed and illustrated *C. costaricana* but the description and photograph seem to depict *C. microspadix*. Krempin (p. 91) illustrated *C. microspadix* but erroneously captioned the photograph as *C. erumpens*.

82. Chamaedorea radicalis *Mart.*, Historia Naturalis Palmarum 3: 308. 1849. Type: Mexico, Sierra

Plate 99. Chamaedorea verecunda. Pistillate plant, *Hodel et al. 1122,* wet cloud forest, Chiriqui, Panama.

Madre, *Karwinski s.n.* [holotype M!; isotypes BR (photo!), LE (photo!)]. Epithet: from the Latin *radicalis* meaning basal or arising from the root stock, in reference to the inflorescences or the apparently stemless habit. **PLATES 9(B), 108(B–F).**

Nunnezharia radicalis (Mart.) O. Kuntze, Revisio Generum Plantarum 2: 731. 1891.

Chamaedorea pringlei S. Wats., Proc. Amer. Acad. 26: 157. 1891. Type: Mexico, San Luis Potosi, Tamasopo Canyon, *Pringle 3537*? (holotype GH!) or *3737*? [C!, F!, GH!, GOET!, M!, NY (photo!), S!, US!, VT!]. Epithet: honors the collector of the type, C. G. Pringle.

Common Name: none.
Vernacular Name: none.

Habit: solitary (rarely cespitose?), erect, slender, appearing stemless or eventually to 3–4 m tall.

Stem: 2.5–3 cm diam., short, curved, subterranean and then with congested nodes, or sometimes aboveground, elongate, then green, smooth, ringed, internodes to 15 cm long.

Leaves: 4–8, erect-spreading, pinnate; **sheath** 15–25 cm long, thick, leathery, fibrous, light green, drying hard and durable, obliquely and rather long-open apically, tubular only in lower ½, light green, brown-margined, below this whitish and green-striate-nerved; **petiole** 15–30 cm long, obtusely 3-sided, grooved and light green above, rounded and light green below; **rachis** to 60–70 cm long or more, triangular, sharply angled and green above, rounded and lighter green below; **pinnae** 10–18 or more on each side of rachis, to 40 cm × 1–2.5 cm, terminal ones often wider, straight, linear-lanceolate, acuminate, slightly falcate, dark green, whitish callus at point of attachment to rachis above, alternate or subopposite, spreading, ± stiff, sometimes slightly drooping with age, ± thick, rough and raspy texture, a central midrib, 2–3 lateral primary nerves on each side above, these somewhat obscured and none prominent, midrib prominent below, secondaries and tertiaries numerous, faint.

Inflorescences: interfoliar, erect, long-pedunculate, if plant stemless often appearing to arise from base, if aboveground stem present then often infrafoliar in fruit; **peduncles** to 1 m long, equalling or exceeding leaves, to 1 cm wide, 7–8 mm thick, ± flattened, green in flower and fruit; **bracts** 6–10, prophyll to 5 cm, 2nd bract to 12 cm, 3rd to 20 cm, 4th to 26 cm, 5th, 6th, and 7th to 30 cm, 8th to 26 cm, 9th to 17 cm, uppermost exceeding peduncle and extending onto rachis and often concealing a small rudimentary bract, tubular, acuminate, obliquely long-open apically, often bifid, green at anthesis becoming brown with age, papery-fibrous, longitudinally striate-nerved. **Staminate** with **rachis** 5–15 cm long, flattened, green; **rachillae** 8–20, these 25–30 cm long, 3 mm diam., spreading, mostly simple but sometimes lower ones branched once or twice, green, minutely white-spotted.

Pistillate spicate, furcate, or with up to 10 rachillae; if branched **rachis** 4–10 cm long, erect and green in flower, spreading and green in fruit; flower-bearing portion or **rachillae** 20–25 cm long, 4–6 mm diam., obscurely angled, erect but becoming pendulous or spreading when heavily laden with fruits, green, minutely white-spotted.

Flowers: Staminate solitary distally, in ± linear groups of 2–5 basally, groups ± in spirals, 3 × 4 mm, subglobose, greenish; **calyx** 1–2 × 3 mm, deeply lobed, green becoming brown, **sepals** briefly imbricate basally, broadly rounded apically; **petals** 3 × 3 mm, broadly triangular, acute, valvate, erect, green and minutely white-spotted, **stamens** equalling or slightly exceeding petals, **filaments** stout, 2.5 × 0.75–1 mm, yellow, **anthers** 1 mm long, white; **pistillode** ± equalling stamens and petals, columnar, slightly enlarged apically, greenish.

Pistillate solitary, in ± dense spirals, 3 × 3 mm, broadly conic, green; **calyx** 1.25–2 × 2.5–3 mm, nearly equalling petals, deeply lobed, green with brown margins, **sepals** free nearly to base, rounded apically, **petals** 2–2.5 × 2.5–3 mm, ⅓–½ as high as and tight against pistil, broadly triangular, imbricate in basal ¾, acute; **pistil** 3.5 × 3 mm, subglobose or depressed-conical, green, minutely white-spotted, **stigma** lobes sessile, short, blunt, erect, dark.

Fruits: 12 × 9 mm, ellipsoid to obovoid-globose, green changing to yellow then orange and finally red with a glaucous bloom when completely ripe.

Distribution: MEXICO. Hidalgo. Nuevo Leon. San Luis Potosi. Tamaulipas. Moist, oak forest; to 1,000 m elevation; often on limestone.

Discussion: Martius (1849) described and named *C. radicalis* from material collected by Wilhelm Karwinski on the Atlantic slope of the Sierra Madre Oriental in Mexico between 21 and 22 degrees north latitude. It was cultivated as *C. pringlei* at Kew Gardens in London, England as early as 1900. Watson (1891) named and described *C. pringlei* from material that Pringle had collected. Watson cited as the type *Pringle 3527*, a specimen which I have not been able to locate. Perhaps the number is a misprint; one specimen, *3537*, that differs in just one digit is at the Gray Herbarium while other specimens, *3737*, that differ in two digits are at several herbaria (C, F, GH, GOET, M, S, US, VT).

The northernmost-occurring member of the genus, *C. radicalis* is found up to 1,000 m elevation on limestone outcroppings in oak forests. Moore (1957) reported that he observed it growing in mountain regions of eastern Mexico in the shade of oaks in Tamaulipas and Hidalgo. With the combination of high latitude and elevation, it is not surprising that *C. radicalis* is a remarkably cold-hardy palm. In fact, it is the hardiest species of the genus and can withstand −5° to −7°C (19–23°F) without displaying any leaf damage. Leaf damage may occur at temperatures below this with death of the plant

Plate 100. Chamaedorea verecunda. In wet cloud forest, *Hodel et al. 1122,* Chiriqui, Panama. **A.** Young plant with bluish sheen; **B.** Crown of leaves and spicate infructescence with immature fruits (note filiform peduncle).

probably not resulting until even lower temperatures are experienced.

Although normally solitary, there are reports of the stem being clustered (Martius 1849, Oersted 1859, Moore 1957, 1976). I have not observed this in cultivation and perhaps these reports may be attributed to several plants growing adjacent to each other or the emergent basal inflorescences which are not unlike a newly arising vegetative shoot.

A peculiar feature of this species is the "heel", the "saxophone" type of growth of the stem, that is evident on many individuals, a phenomenon shared with a few other palms (Tomlinson 1990), notably fan palms like *Sabal.* It appears to be a mechanism for secure anchorage and better survival in a rocky or otherwise harsh substrate. The "heel" appears to be a portion of the abbreviated stem, often with roots, that arises from and then submerges into the soil before arising again a few centimeters distant. The leaf bases and leaves appear to be arising directly out of the ground; in fact, basal portions of the leaf sheaths are actually below ground level. Inflorescences arise from behind these leaf sheaths and from below the ground.

With regard to the stem, there are two forms in cultivation. One is a stemless while the other has a well developed, erect, aboveground stem. The former seems to remain stemless all its life or at least for many years, and is very effective in a mass planting under trees or in shady locations. Although the stemless form is not the most elegant member of the genus, it is handsome in its own right with its stiffly ascending leaves and long-pedunculate inflorescences heavily laden with showy red fruits. Surprisingly, it is very tolerant of direct sun. Some specimens at the Huntington Botanical Gardens in San Marino, California show only slight yellowing although they occur in nearly full sun.

The form with an aboveground stem is less common in cultivation and becomes an elegant palm with age, having stems several meters in height and displaying to best advantage the arching inflorescences laden with great clusters of red fruits.

Krempin (1990, p. 95) illustrated *C. elegans* but erroneously captioned the photograph as *C. radicalis.*

A distinctive member of the genus and relatively easy to grow, *C. radicalis* is very tolerant of low light and is resistant to mites and nematodes.

VII. Subgenus MORENIOPSIS

Chamaedorea subgenus **Moreniopsis.**

Key Characters: petals distinct in flowers of both sexes, thickened and persistent in fruit, staminate flowers solitary, their petals spreading or erect and free apically; staminate inflorescences multiple at a node; fruits black.

Distribution: two species in Amazonian Bolivia, Peru, Ecuador, Colombia, and western Brazil; to about 1,000 m elevation. One species, *C. angustisecta,* is uncommonly cultivated in southern Florida.

Key to the Species of Subgenus *Moreniopsis*

Sheaths, petioles, and rachises whitish green to green and white-mottled; petals of staminate flowers bright orange throughout. . . . . **83. C. angustisecta**

Sheaths, petioles, and rachises green; petals of staminate flowers green above and orange below. . **84. C. pauciflora**

83. Chamaedorea angustisecta *Burret,* Notizbl. Bot. Gart. Berlin-Dahlem 11: 318. 1932. Type: Peru, Junin, *Killip & Smith 25091* (holotype B, destroyed; isotypes F!, NY, US!). Epithet: from the Latin *angusti* meaning narrow and *sectus* meaning cut or division, in reference to the narrow pinnae. **PLATE 109.**

Chamaedorea leonis H. E. Moore, Gentes Herb. 12: 30, figs. 1 & 2A, B. 1980. Type: Cult., *Moore 9586* (holotype BH!). Epithet: from the Latin *leo* meaning lion and in reference to Nat J. De Leon, in whose garden Moore collected the type.

Common Name: none.

Vernacular Names: siyeyi—Bolivia; palmeira, palmeirinha—Brazil.

Habit: solitary, erect, 1.5–4 m tall or more.

Plate 101. Chamaedorea volcanensis. In montane rain forest, Quetzaltenango, Guatemala. **A, D** *Hodel & Castillo 913A* (holotype); **B, C** *Hodel & Castillo 907B*; **A.** Large staminate plant; **B.** Pistillate plant, bifid leaves; **C.** Pistillate plant, pinnate leaves; **D.** Staminate inflorescence, densely flowered pendulous rachillae.

D

Stem: 2.5 cm diam., green, ringed, internodes 5 cm long.

Leaves: 5, pinnate; **sheath** 30–55 cm long, green mottled with white; **petiole** 45–100 cm long, light green mottled with white; **rachis** 1.3–1.55 m long, green or mottled with white to nearly white below; **pinnae** 30–39 on each side of rachis, apical ones 19–22 × 0.8–1.5 cm, central ones 39–48 × 2.4–3.2 cm, basal ones 27.5–30 × 0.5–1.6 cm, linear-lanceolate, slightly falcate, acute to acuminate, regularly arranged, spreading in one plane, thin, a midrib and 2 nerves on each side of this ± elevated above and prominent below.

Inflorescences: interfoliar, erect-spreading; **Staminate** 7 at a node, central one largest and opening first; **peduncle** to 80 cm long; **bracts** 5–7, prophyll 2.5 cm long, others widely separated at insertion, tubular, 2nd bract truncate, 7.5 cm long, 3rd 16.5 cm, obliquely split, 4th 22 cm, 5th 25 cm, 6th 27 cm long, 7th 24 cm; **rachis** 3.5–7.5 cm long; **rachillae** 9–16, 11–24 cm long, smooth, pendulous.

Pistillate solitary at node; **peduncle** 66–70 cm long; **bracts** 5–8, similar to those of the staminate; **rachis** 2.5–10 cm long; **rachillae** 6–16, 7–20 cm long, 4 mm diam., orange in fruit, erect to spreading.

Flowers: Staminate in loose spirals, 4 mm long, orange, intensely fragrant; **calyx** 1.5–2 mm high, deeply lobed, light green in bud, orangish with dark brown margin at anthesis, **sepals** connate basally, rounded apically; **petals** 3–4 × 2–2.4 mm, ovate, valvate, free, smooth externally, grooved internally, dark green in bud, orange at anthesis; **stamens** 2.6 mm long, shorter than pistillode, ½ as high as petals, **filaments** 2 mm long, **anthers** 1.2 mm long, briefly bifid apically and basally, lilac; **pistillode** 3.5 mm high, columnar, 3-lobed apically.

Pistillate in spirals, sessile, leaving round scars, yellow-green; **calyx** 2 mm high, adnate to receptacle for 1 mm, deeply lobed, green, **sepals** broadly rounded apically; **petals** 2.5 × 2 mm, adnate for 1 mm to receptacle, imbricate basally, valvate, erect to spreading apically, not prominently nerved; **staminodes** lacking; **pistil** depressed-globose, 3-lobed, **stigma** lobes sessile, short, erect.

Fruits: 15–18 × 8 mm, ellipsoid, black, mesocarp 0.5 mm thick, pale with long flat fibers; seeds 15 × 5 mm, ellipsoid, subfalcate.

Distribution: BOLIVIA. Beni. La Paz. BRAZIL. Acre. PERU. Amazonas. Ayacucho. Cuzco. Junin. Madre de Dios. Ucayali. Amazonian, moist or wet forest on the eastern Andean slope; to 700 m elevation.

Discussion: The description is from Burret (1932) and Moore (1980). *C. angustisecta* has been cultivated for many years in southern Florida as *C. leonis.* Moore (1980) described and named *C. leonis* from plants cultivated in Miami, Florida that were grown from seeds collected in Bolivia.

On the eastern Andean slope where *C. angustifolia* is native in Bolivia and Brazil, local inhabitants gather the inflorescences and take them indoors to perfume their houses.

Apparently only cultivated in southern Florida where it is found sparingly in a few collections and gardens, *C. angustifolia* is an attractive plant with mottled green leaf sheaths and petioles, intensely fragrant, bright orange staminate flowers, and black fruits.

84. Chamaedorea pauciflora *Mart.,* Historia Naturalis Palmarum 2: 5, t. 3, fig. 3. 1823. Type: Brazil, Rio Japura, *Martius s.n.* [holotype M (photo!); isotype GOET!]. Epithet: from the Latin *pauci* meaning few and *florus* meaning flower, perhaps in reference to the few-flowered inflorescence. **PLATE 110.**

Morenia integrifolia Trail. J. Bot. 14: 331. 1876. Type: Brazil, Rio Jurua, *Trail 1042 (157)* (holotype K; isotypes GH!, P). Epithet: from the Latin *integri* meaning entire or simple and *folius* pertaining to leaf, in reference to the simple (bifid) leaves often found on this species.

Morenia integrifolia var. *nigricans* Trail, J. Bot. 14: 331. 1876. Type: Brazil, Para, *Trail 157A–F* (holotype K; isotype GH!).

Morenia pauciflora (Mart.) Drude in Martius, Flora Brasiliensis 3: 526. 1882a.

Nunnezharia integrifolia (Trail) O. Kuntze, Revisio Generum Plantarum 2: 730. 1891.

Nunnezharia pauciflora (Mart.) O. Kuntze, Revisio Generum Plantarum 2: 730. 1891.

Chamaedorea integrifolia (Trail) Dammer, Verh. Bot. Vereins Prov. Brandenburg 48: 125. 1907a.

Chamaedorea lechleriana H. A. Wendl. ex Dammer, Gard. Chron. Ser. 3, 36: 246. 1904b. Type: Peru, St. Garan, *Lechler s.n.* (holotype K?). Epithet: honors the collector of the type, W. Lechler.

Chamaedorea amazonica Dammer, Konigl. Notizbl. Bot. Gart. Berlin 6: 263. 1915. Type: Brazil, Rio Amazonas, *Ule 5595* [holotype B, destroyed; isotypes F! (fragments), G (photo!), K, L!, MG]. Epithet: from the location where the type was collected, along the Rio Amazonas.

Common Name: none.

Vernacular Names: sangapilla—Peru; hua-co,

Plate 102. Chamaedorea vulgata. In cloud forest, *Hodel & Castillo 904,* Quetzaltenango, Guatemala. **A.** Mature plant in ravine; **B.** Conspicuously spotted petioles and sheaths characteristic of *C. vulgata*; **C.** Conspicuously spotted stem and inflorescence bracts; **D.** Infructescence (immature fruits); **E.** Pistillate plant.

B
E

suma-yuca, desodorante de indigena—Ecuador; ti aacaiba—Colombia.

Habit: solitary, erect, infrequently decumbent, 1–3 m tall, often appearing stemless.

Stem: 0.5–2 cm diam., green, prominently ringed, internodes 1–7 cm long, often with orange-brown, prickly aerial roots to 0.5 cm diam.

Leaves: 3–9, erect-spreading, 1–1.2 m long, bifid, rarely pinnate; **sheath** 15–30 cm long, tubular in basal ⅓, green, longitudinally striate-nerved, margins rough; **petiole** 5–50 cm long, 0.5 cm diam., flat to channeled and green above, rounded and pale below; **rachis** 25–60 cm long, angled and light green to yellow above, rounded below with a pale yellow stripe extending to sheath; **blade** 40–100 × 15–35 cm, ovate, bifid apically to ⅓ its length, lobes 7–15 cm wide, acute to slightly acuminate, cuneate basally, papery, glabrous, outer margin remotely toothed, pinnately ribbed with 16–23 nerves on each side of rachis, these light green to yellow below, angled and green above; or, blade with 2–10 **pinnae** on each side of rachis, 25–45 × 1.5–6.5 cm, broadly lanceolate or linear, sigmoid, subfalcate, acuminate, drying black, prominent midrib, 1–2 primaries on either side, terminal pair 4–6-nerved, secondaries few, widely spaced.

Inflorescences: interfoliar, erect, spicate. **Staminate** to 8 per node, centrally one opening first, borne on a common hypopodium, 40–70 cm long, slender, green; **peduncle** 20–42 cm long, 1.5–2 mm diam; **bracts** 3–5, prophyll 0.5–3.5 cm long, 2nd–5th bracts 2–17 cm, upper bract exceeding peduncle, tubular, bifid, membranous, pale, quickly becoming tattered and dry; **rachis** or flower-bearing portion 17–50 cm long, 1–3 mm diam., lightly grooved, green in flower.

Pistillate solitary, occasionally 2 per node; **peduncle** 25–50 cm long, 5 mm diam., enclosed in leaf sheath for ½–¾ its length; **bracts** 3–6, similar to those of the staminate; **rachis** or flower-bearing portion 20–55 cm long, 2–5 mm diam., green in flower, yellow to orange-red and pendulous in fruit.

Flowers: Staminate in loose spirals, 1.5 × 2 mm, depressed-globose, green, violet-scented, sessile; **calyx** 0.5–1.1 mm high, deeply lobed, grayish, membranous, **sepals** connate in basal ½, broadly obtuse apically; **petals** 2–2.5 × 1–1.5 mm, obicular, acute, valvate, free, deep green apically, orange basally; **stamens** 1.2–1.5 mm long, **filaments** 0.4–0.5 mm long, slender, **anthers** free, dorsifixed; **pistillode** 1.2–1.7 mm tall, umbrella-shaped.

Pistillate in loose spirals, 2–2.5 mm diam., depressed-globose, slightly sunken; **calyx** 1–1.5 mm high, deeply lobed, green, **sepals** connate in basal ½–⅓, ovate-rotund, acute apically; **petals** 1.2–2 × 1.2–2 mm, briefly imbricate basally, ovate-orbicular; **staminodes** 6, toothlike; **pistil** 1 mm high, depressed-globose, **stigma** lobes sessile, minute, slender.

Fruits: 10–15 × 5–8 mm, ellipsoid, black.

Distribution: PERU. Huanuco. Loreto. Madre de Dios. San Martin. COLOMBIA. Amazonas. BRAZIL. Acre. Amazonas. Mato Grosso. Rondonia. ECUADOR. Napo. Marona-Santiago. Pastaza. Zamora-Chinchipe. Dense, moist or wet forest in the Amazon river basin; mostly at low altitudes but to 1,100 m elevation.

Discussion: The description is from Martius (1823) who described and named *C. pauciflora* from material collected near the Rio Japura. Trail (1876), Dammer (1915), and Bergmann (1989) supplement the description. *C. pauciflora* is a handsome species with its large, bifid leaves and long, spicate inflorescences. In fruit, *C. pauciflora* is similar in habit to *C. ernesti-augustii*; however the cupped petals and solitary, branched staminate inflorescence with numerous rachillae easily distinguish the latter species. *C. pauciflora* is a common palm in some parts of its range. Kahn and Mejia (1991) reported that of 34 species of palms at a study site in Peru, *C. pauciflora* was the fifth most common overall and third most common understory species. Despite its relative abundance, *C. pauciflora* is apparently not cultivated. It is a dramatic plant, though, and would be a handsome addition to any collection.

Balslev and Barfod (1986) reported that local inhabitants in the eastern lowlands of Ecuador tie the staminate inflorescences to their bodies or bring them into their houses because of their sweet and pleasant fragrance. Some consider it a plant with magic power to be used for attracting a desired person of the opposite sex.

VIII. Subgenus STEPHANOSTACHYS

Chamaedorea subgenus **Stephanostachys** Klotzsch, Allg. Gartenzeitung 20: 363. 1852.

Chamaedorea sect. *Stachyophorbe* Liebm. ex Mart., Historia Naturalis Palmarum 3: 309. 1849.

Stachyophorbe (Liebm. ex Mart.) Liebm. ex Klotzsch, Allg. Gartenzeitung 20: 363. 1852.

Dasystachys Oerst., Videns. Meddel. Dansk. Naturhist. Foren. Kjoebenhavn 1858: 25. 1859.

Spathoscaphe Oerst., Vidensk. Meddel. Dansk. Naturhist. Foren. Kjoebenhavn 1858: 29. 1859.

Plate 103. Chamaedorea whitelockiana. A, C-E Cultivated, California; **B** In pine-oak cloud forest, Oaxaca, Mexico; **A.** Young flowering plant, *Hodel 853,* garden of L. Whitelock, Los Angeles; **B.** Young plant displaying juvenile foliage, *D. & R. Hodel 935*; **C.** Pistillate plants, Virginia Robinson Botanical Garden, Beverly Hills; **D.** Pistillate inflorescences, research collection, Los Angeles; **E.** Staminate inflorescence with erect rachillae, *Hodel 852.*

C
E

Stephanostachys (Klotzsch) Oerst., Vidensk. Meddel. Dansk. Naturhist. Foren. Kjoebenhavn 1858: 26. 1859.

Chamaedorea sect. *Stephanostachys* Burret, Notizbl. Bot. Gart. Berlin-Dahlem 11: 760. 1933a.

Key Characters: petals distinct in flowers of both sexes, thickened and persistent in fruit; staminate flowers densely crowded and contiguous in bud, their petals spreading or erect and free apically; inflorescences usually solitary, infrequently multiple at a node; fruits black or reddish.

Distribution: about ten or perhaps more species widely or locally distributed on both the Atlantic and Pacific slopes from Mexico to Colombia; to 1,500 m elevation; sometimes on limestone. With the exception of *C. nationsiana,* all species of subgenus *Stephanostachys* are presently cultivated.

Because of their unusual, distinctive inflorescences, members of subgenus *Stephanostachys* always attract attention when in flower or fruit and are favorites of palm growers everywhere. In addition to the showy staminate inflorescences with their densely packed, contiguous flowers, several species of subgenus *Stephanostachys* have pistillate inflorescences of a similar nature. Pistillate flowers are densely crowded or contiguous on a thickened spicate, furcate, or few-branched axis. Infructescences of these species are outstanding, appearing somewhat similar to an ear of corn with the fruits densely packed and often angled by mutual pressure.

Key to the Species of Subgenus *Stephanostachys*

Pistillate flowers densely crowded, contiguous on a thick spicate, forked, or few-branched axis; staminate calyx prominent, well developed in bud.
- Staminate inflorescences with 8–10 rachillae. . . . . **87. C. arenbergiana**
- Staminate inflorescences spicate or forked.
  - Staminate inflorescences multiple at a node.
    - Flowers green. . . . . **90. C. deckeriana**
    - Flowers not green. . . . . **91. C. nationsiana**
  - Staminate inflorescences solitary at a node.
    - Staminate flowers yellow; sepals connate in a cupule half as high as petals; stamens not as high as petals . . . . **85. C. allenii**
    - Staminate flowers white; sepals free nearly to base; stamens equalling or exceeding petals.
      - Staminate flowers with anthers exserted above petals; pinnae 9–11 on each side of rachis, all ± same size. . . . . **89. C. crucensis**
      - Staminate flowers with anthers not protruding beyond petals; pinnae 1–6 on each side of rachis with terminal pair as broad as or broader than others combined. . . . . **94. C. zamorae**

Pistillate flowers borne at distinct intervals or crowded in a definite spiral or vertical arrangement; staminate calyx not prominent, usually a low, membranous ring.
- Staminate and pistillate inflorescences spicate; stamens exserted beyond petals or nearly so; fruits red-orange. . . . . **92. C. oreophila**
- Staminate inflorescences branched; pistillate inflorescences branched or spicate; stamens mostly less than petals; fruits black.
  - Stems short, creeping, dichotomously branched; pinnae decurrent at base. . . . . **88. C. cataractarum**
  - Stem(s) erect, not creeping or dichotomously branched; pinnae not decurrent.
    - Inflorescences usually 2–several at a node, more rarely (especially pistillate) solitary; pistillate rachillae few, short, erect; pistillate flowers closely placed; margin of leaf sheath whitish with green nerves. . . . . **86. C. alternans**
    - Inflorescences solitary at nodes, pistillate with several long and spreading rachillae; pistillate flowers more widely spaced at distinct intervals; margin of leaf sheath green. . . . . **93. C. tepejilote**

85. Chamaedorea allenii *L. H. Bailey,* Gentes Herb. 6: 241, fig. 126. 1943a. Type: Panama, *Cocle, Allen 1909* (holotype MO!; isotype F!). Epithet: honors Paul Allen, collector of the type. **PLATES 111, 112(A).**

Common Name: none.
Vernacular Name: none.

Habit: solitary, erect or decumbent, to 2 m tall or more but sometimes flowering when stemless.

Stem: 1.5–2 cm diam., green, smooth, conspicuously ringed, internodes 3–10 cm long.

Leaves: 4–6, spreading, pinnate but sometimes flowering when bifid or variously pinnate; **sheath** 25 cm long, tubular in basal ½, white-margined apically transitioning to dark basally, longitudinally striate-nerved; **petiole** 25–35 cm long, 5–7 mm thick, lightly grooved and green above, rounded and with a light green band below; **rachis** 30–45 cm long or more, obtusely angled and green above, rounded with a light green band below extending onto sheath; **blade** 60–70 cm long, firm textured, glossy; **pinnae** 7–9 per side, 25–30 × 3–6 cm, narrowly lanceolate, slightly sigmoid or falcate, acute-acuminate, drooping at tip, ± narrowed basally, subopposite or alternate, regularly placed, 8–10 nerved, these light-colored and pronounced below, end pair of pinnae wider; if blade bifid then to 50 × 30 cm, rachis to 30 cm long, lobes to 25 cm long, 50 prominent nerves on each side of rachis.

→

Plate 104. Chamaedorea whitelockiana. In cloud forest, Quetzaltenango, Guatemala. **A.** Leaves, *Hodel & Castillo 988*; **B.** Infructescences (immature fruits), *Hodel & Castillo 988*; **C.** Staminate inflorescence, *Hodel & Castillo 911.*

B

Inflorescences: interfoliar, erect, often infrafoliar and horizontal in fruit, spicate, less often forked or with 3 rachillae. **Staminate** with **peduncle** 15–25 cm long, 5–8 mm wide at base and ± flattened, 5 mm diam. at apex and rounded, pale or white, ascending or horizontal; **bracts** 4, prophyll 2 cm long, 2nd bract 7 cm, 3rd 12 cm, 4th 15 cm, tubular basally, ± inflated apically, thin-leathery, acute-acuminate, bifid, uppermost exceeding peduncle, light green in flower; **rachis** or flower-bearing portion or **rachillae** 10–20 × 0.5–1 cm, pendulous.

Pistillate with **peduncle** 20 cm long, 8 mm wide at base and ± flattened, 6–8 mm diam. at apex and rounded, light green in flower, orange in mature fruit; **bracts** 4, similar to those of staminate but becoming tattered, brown, and ± fallen away in fruit, upper one ± equalling peduncle; **rachis** or flower-bearing portion to 10 × 1 cm, straight in flower, ± stiff, curved, swollen, and red-orange in fruit.

Flowers: Staminate in 8 dense spirals, contiguous in bud, 3.5 × 2.5 mm, ± ovoid, angled by mutual pressure, bright yellow; **calyx** prominent, 2 × 2.5 mm, scarcely lobed, whitish, **sepals** connate nearly to apex, forming a sheathing hyaline tube, ± straight apically; **petals** 3 × 2 mm, ovate, valvate, free in apical ½, spreading, rounded-acute, slightly erect, thickened, thicker distally, slightly grooved on inside; **stamens** 1.75 mm high, **filaments** thick, 1.5 × 0.5 mm, **anthers** in a close ring around pistillode, 1 × 0.5 mm; **pistillode** extending just above anthers, 2 × 0.5 mm, swollen basally.

Pistillate in 8 dense spirals, contiguous in bud, 2.5 × 3.5 mm, depressed-globose, bright yellow, sunken in shallow pits but so densely packed as to appear to be immersed in axis; **calyx** prominent, 1.5 × 3.5 mm, shallowly undulate, whitish, **sepals** connate nearly to apex, forming a tight hyaline sheath around base of flower, ± straight apically; **petals** 2.5 × 2.5 mm, broadly triangular, cupped, imbricate nearly to apex, margins rounded, straight apically but with a small acute point; **staminodes** lacking; **pistil** 2 × 2 mm, conic, fleshy, whitish, **stigma** lobes sessile, spreading or erect, prominent, barely exceeding petals.

Fruits: 7–10 mm diam., densely packed, globose but angled from mutual pressure, black, epicarp rough; seeds 5–7 mm diam., globose but slightly angled, brown.

Distribution: PANAMA. Chiriqui. Cocle. Darien. San Blas. COLOMBIA. Antioquia. Wet forest on the Pacific and Atlantic slopes; 500–1,000 m elevation.

Discussion: *C. allenii* is an attractive, single-stemmed palm with a striking, bright yellow, spicate or forked staminate inflorescence. Bailey (1943a) described and named *C. allenii* from material that Paul Allen collected near El Valle, Cocle Province, Panama. It occurs predominantly in mountain forests at middle elevations on the Pacific slope usually at or near the Continental Divide. Bailey described *C. allenii* from a pistillate collection only; fruits and staminate flowers were not seen. We were successful in establishing plants in the greenhouse in Los Angeles from the type locality. These have flowered and, fortunately, we have plants of both sexes and have successfully hand-pollinated them and obtained fruits.

Chamaedorea allenii seems to exhibit two vegetative phases. The first consists of deeply bifid leaves; later, plants develop leaves that are pinnate. Flowering and fruiting is most closely associated with the pinnate-leaved phase. However, the two phases cannot be considered strictly juvenile and adult since flowering and fruiting have been observed on some plants with bifid leaves.

Bifid-leaved forms of *C. allenii* are similar vegetatively and in fruit to *C. deckeriana.* This latter species is found mainly on the Atlantic slopes from low to middle elevations. Their ranges do not seem to overlap greatly, if at all. *C. deckeriana* is clearly distinct in the green leaf sheaths, green staminate flowers, multiple staminate inflorescences at a node, and smooth, red-orange mature fruits.

Galeano and Bernal (1987) reported *C. deckeriana* from northwestern Colombia but the accompanying description seems to refer to *C. allenii.* Chazdon (1987) reported *C. allenii* from Braulio Carrillo National Park on the Atlantic slope of Costa Rica but this species is probably *C. crucensis* or an unnamed taxon.

Rare in cultivation, *C. allenii* occurs in only a few gardens in California, Florida, and Australia. Its yellow staminate inflorescence is quite attractive but lasts for only a day or two.

86. Chamaedorea alternans *H. A. Wendl.,* Gartenflora 29: 104. 1880. Type: Cult., *Wendland s.n.* (holotype lost or destroyed); Neotype (here designated): Mexico, Veracruz, *Moore & Cetto 6260* (BH!). Epithet: is the Latin *alternans* meaning alternating, in reference to the pistillate inflorescences being solitary and then paired at alternating nodes. **PLATES 7(E), 112(B–E).**

Nunnezharia alternans (H. A. Wendl.) O. Kuntze, Revisio Generum Plantarum 2: 730. 1891.

Common Name: none.
Vernacular Name: none.

Habit: solitary, rarely cespitose? and then with few stems, erect, to 3 m or more.

Plate 105. Chamaedorea woodsoniana. Cultivated, Honolulu, Hawaii. **A, B, D, E** at Lyon Arboretum; **C** at Wahiawa Botanic Garden; **A.** Infructescence with mature fruits; **B.** Staminate inflorescence; **C.** Fruiting plant with nearly mature fruits; **D.** Group planting, mature plants; **E.** Prominent, tubular leaf sheaths and young inflorescences sheathed in bracts.

C
E

Stem: 3–4 cm diam., light green, smooth, conspicuously ringed, internodes 5–15 cm long, often with adventitious prop roots at base.

Leaves: 3–5, spreading, pinnate, to 1.75 m long; **sheath** 45 cm long, green, splitting opposite the petiole, obliquely open apically and whitish and green striate-nerved, margin brownish and ragged; **petiole** 50 cm long, ± rounded but becoming furrowed and channeled toward base and light green above, rounded and pale below; **rachis** to 0.8–1.5 m long, obtusely angled and green above, rounded below with a distinct pale band extending onto sheath; **pinnae** 6–12 on each side of rachis, lower ones largest, these 40–50 × 11–14 cm, middle ones 30–40 × 5–10 cm, elliptic-lanceolate, slightly sigmoid, falcate, acuminate, ± thin, alternate, drooping apically, well spaced, 15 cm apart, a prominent midrib and 3–4 primary nerves on each side of this, end pair wider, and 12–15-nerved.

Inflorescences: infrafoliar, erect, staminate 2–5 and pistillate 1–4 per node; **peduncles** 15–25 cm long, 1–1.5 cm wide at base, 5–8 mm wide at apex, erect, green in flower, reddish orange and nodding in fruit; **bracts** 4–6, prophyll 5 cm long, 2nd bract 12 cm, 3rd to 30 cm, 4th to 35 cm long and exceeding peduncle, sometimes a small rudimentary one concealed by largest upper bract, finely longitudinally striate-nerved, ± leathery and becoming tattered and/or deciduous in fruit, upper broad and yellow inside; **rachis** greenish in flower, reddish orange in fruit. **Staminate** with 8–15 **rachillae,** these 15–25 cm long, pendulous, greenish in flower.

Pistillate with 8–12 **rachillae,** these 10–15 cm long, ± stiff, erect and greenish white in flower, light orange and horizontal or downward-pointing in fruit.

Flowers: Staminate in dense rows, contiguous in bud, yellow; **calyx** short, ringlike; **petals** connate basally, valvate and spreading apically; **filaments** cylindric, longer than **anthers,** these ovoid; **pistillode** not exceeding stamens, briefly 3-lobed apically or sometimes lacking.

Pistillate greenish, sunken in elliptic depressions 3 mm long; **calyx** 1.5 mm high, shallowly 3-lobed; **petals** 2.5 × 3.5 mm, imbricate, acute, yellow with greenish tips.

Fruits: 2 × 1 cm, ellipsoid, black.

Distribution: MEXICO. Veracruz. Chiapas? Wet forest on the Atlantic slope; 300–700 m elevation.

Discussion: The description is from Wendland (1880) and supplemented from material in cultivation. *C. alternans* is a handsome plant with large leaves and broad, dark green pinnae. Linden apparently discovered it in Chiapas, Mexico and introduced it to European gardens in 1875. Wendland named the species from cultivated material growing at Herrenhausen that he had obtained from Linden. Apparently the type has been destroyed or lost, therefore *C. alternans* is retypified here. Although Linden stated that he collected *C. alternans* in Chiapas, I have not seen collections of it from there; however, the term Chiapas was rather loosely applied in the 19th century or perhaps there was a mix-up in labels of Linden's material.

An interesting feature of this species is the manner in which the inflorescences occur at the nodes, usually in multiples but occasionally alternating to solitary on the same individual, hence the specific epithet.

Chamaedorea alternans is close to *C. tepejilote* but the generally more numerous and broader pinnae, solitary inflorescences, and green-margined leaf sheaths distinguish the latter species. Some students of palms consider *C. alternans* to be included in *C. tepejilote.* While the latter is a highly variable species, the differences between the two are fairly constant and, at least at this time, seem to warrant their being maintained as separate species.

Several workers (Ibarra-Manriquez 1988, Oyama and Dirzo 1988, 1991, Oyama 1990, 1991, Oyama and Mendoza 1990) have discussed various ecological aspects of *C. tepejilote* in the rain forest at the Las Tuxtla Biological Station in Veracruz, Mexico. However, according to my interpretation of *C. alternans* and *C. tepejilote,* the above workers appear to be referring to *C. alternans.*

Fischer and Moore (1977) reported in detail about the development of multiple inflorescences in *C. alternans* and other species.

Chamaedorea alternans occurs sparingly in collections in California, Florida, Tahiti and perhaps elsewhere. Most plants now in cultivation originated from seeds distributed by the International Palm Society Seed Bank in the early 1980s.

87. Chamaedorea arenbergiana *H. A. Wendl.,* Index Palmarum 66. 1854. Type: Cult., *Wendland s.n.* [holotype GOET!; isotypes B (fruits, photo!), C!, G (photo!), L!]. Epithet: honors the Duke d'Arenberg-Nieppen, from whose garden the type originated. **PLATES 1(D), 8(H), 113.**

Spathoscaphe arenbergiana (H. A. Wendl.) Oerst., Vidensk. Meddel. Dansk. Naturhist. Foren. Kjoebenhavn 1858: 30. 1859.

Nunnezharia arenbergiana (H. A. Wendl.) O. Kuntze, Revisio Generum Plantarum 2: 730, 1891.

Nunnezharia latifrons (Hort.) O. Kuntze, Revisio Generum Plantarum 2: 731. 1891.

Plate 106. Chamaedorea woodsoniana. A, B In cloud forest, *Hodel & Castillo 1107A* (staminate) *& 1107B* (pistillate), Baja Verapaz, Guatemala; **A.** Infructescence (immature fruits); **B.** Staminate plant; **C, D** In montane rain forest, *Hodel et al. 1122,* El Valle, Panama; **C.** Staminate plant; **D.** Stem.

→

B
D

Chamaedorea gracilis Hort. in H. A. Wendl., Index Palmarum 13. 1854. Not Willd.
Chamaedorea latifrons Hort. in H. A. Wendl., Index Palmarum 13. 1854.
Chamaedorea oblongata Hort. in H. A. Wendl., Index Palmarum 14. 1854. Not Mart.
Chamaedorea latifolia Hort. in Nicholson, The Illustrated Dictionary of Gardening 305. 1884.
Chamaedorea densiflora Hort. in Guillaum., J. Soc. Natl. Hort. France, Ser. 4, 24: 226. 1923b.

Common Name: none.
Vernacular Names: pacaya—Honduras; chim—Guatemala.

Habit: solitary, erect, rarely decumbent, to 4 m tall, sometimes flowering when stemless.

Stem: 2–3 cm diam. green, ringed, internodes 5–15 cm long.

Leaves: 4–6, erect-spreading, pinnate, to 2–2.5 m long; **sheath** 30–45 cm long, obliquely open apically, auriculate; **petiole** 35–45 cm long, obtusely triangular, faintly channeled and green above, rounded and pale below; **rachis** 1–1.25 m long, angled and green above, rounded below with a pale band extending onto sheath; **pinnae** 8–10 on each side of rachis, largest ones 60 × 15 cm, rather elongate-oblong or oblong-lanceolate, sigmoid, ± falcate, well spaced, very long-acuminate and drooping apically, contracted basally, 5–8 prominent primary nerves keeled above, shining and yellowish below, numerous nerves of lesser orders present.

Inflorescences: infrafoliar, solitary, erect in flower, spreading or nodding in fruit; **peduncles** 15–20 cm long, ± stout, 1–1.5 cm diam., green in flower, red-orange in fruit; **bracts** 5–6, fibrous, loosely sheathing, greenish brown or brown in flower often becoming tattered, longitudinally striate-nerved, upper ones exceeding peduncle. **Staminate** with **rachis** 3 cm long, pale in flower; **rachillae** 8–10, these 12–15 cm long, 8 mm diam., pendulous, pale in flower.

Pistillate spicate or forked; **rachis** or flower-bearing portion 6–15 cm long, 1.5 cm or more diam., orange and swollen in fruit.

Flowers: Staminate in 6 dense spirals, contiguous in bud, 3 × 3.5–4 mm, depressed-globose, angled from mutual pressure, cream-colored to brownish white, ± superficial; **calyx** higher than petals and enclosing them in bud, 1.5 × 4 mm at anthesis, shallowly lobed, **sepals** connate nearly to apex, forming a sheathing tube, ± straight apically; **corolla** adnate with stamen filaments and pistillode in a very short basal stipe, **petals** 3.5–4 × 3 mm, broadly ovate, imbricate in basal ⅓–½, valvate and free apically, spreading, erect, rounded to acute, ± thick; **stamens** equalling petals, clear- or cream-colored, **filaments** prominent, distinct, **anthers** divaricate basally; **pistillode** 3.5–4 mm tall, equalling stamens and petals, columnar, apically 3-lobed, cream-colored.

Pistillate very densely crowded in perhaps as many as 20 rows and ± sunken in axis; **calyx** nearly as high as petals, shallowly lobed, membranous, **sepals** connate nearly to apex, rounded to straight apically; **petals** imbricate nearly to apex, short-pointed; **staminodes** lacking; **pistil** ovoid, 3-lobed, **stigma** lobes sessile, short, recurved.

Fruits: 12 × 19 mm, subglobose or transverse-oblong, densely packed, angled from mutual pressure, black; seeds 10–12 × 9–10 mm, oblong; abortive carpels usually separating from fruit and adhering to perianth.

Distribution: MEXICO. Chiapas. Oaxaca. Veracruz. GUATEMALA. Alta Verapaz. Baja Verapaz. Huehuetenango. San Marcos. HONDURAS. Atlantida. EL SALVADOR? NICARAGUA? COSTA RICA? PANAMA? COLOMBIA? Moist or wet forest on the Atlantic and Pacific slopes; 100–1,500 m elevation.

Discussion: Apparently introduced to European gardens prior to 1850, Wendland (1854) described and named *C. arenbergiana* from plants cultivated in the gardens of the Duke d'Arenberg-Nieppen in Belgium. Wendland was unsure of its provenance, but tentatively listed its origin as Guatemala. According to Guillaumin (1923b), it was cultivated at the Musee de Paris in France as early as 1862 as *C. densiflora.*

Chamaedorea arenbergiana occurs in moist and wet forest from Mexico through Guatemala, Honduras, and perhaps El Salvador and even further south in Central America and into northern South America. Some of the collections tentatively identified as *C. arenbergiana* which have extended its range to Costa Rica and Colombia are perhaps *C. allenii, C. crucensis,* or closely related, yet unnamed, taxa. Part of the difficulty in distinguishing between *C. arenbergiana* and related taxa (and thus establishing their true ranges) is that staminate material, which is critical for diagnosis, is infrequently found since it is fragile and deteriorates quickly. When only fruiting material is at hand, which is most often the case, diagnosis is difficult at best. I have yet to be able to establish diagnostic characters based only on foliage or fruiting material. *C. arenbergiana* is probably closest to *C. nationsiana* but the multiple, spicate staminate inflorescences distinguish this latter species.

Chamaedorea arenbergiana is an attractive ornamental with large, handsome leaves, densely flowered, spicate or forked inflorescences, and corncob-like infructescences. It always draws attention in the garden when heavily laden with densely packed, black fruits. Pinnae of *C. arenbergiana* can be up to 15 cm wide, making them among the widest in the genus. *C. arenbergiana* is cultivated in Hawaii, California, Florida, Australia, and perhaps elsewhere.

Plate 107. Chamaedorea microspadix. In full fruit, cultivated, Waimea Falls Park & Arboretum, Honolulu, Hawaii. →

88. Chamaedorea cataractarum *Mart.*, Historia Naturalis Palmarum 3: 309. 1849. Type: Mexico, Oaxaca, *Liebmann s.n.* (some sheets later numbered *10808*) (holotype C!; isotypes F!, GH!, MO!, P!, S!, U!). Epithet: from the Latin *cataracta,* meaning pertaining to cataracts and wet rocks along water courses, in reference to the habitat. **PLATES 5(A), 114.**

Stachyophorbe cataractarum (Mart.) Klotzsch, Allg. Gartenzeitung 20: 363. 1852.

Chamaedorea martiana H. A. Wendl., Allg. Gartenzeitung 21: 137. 1853b. Type: Cult., *Wendland s.n.* [holotype GOET; isotypes C!, G (photo!), L!]. Epithet: honors 19th-century German botanist Martius.

Stephanostachys martiana (H. A. Wendl.) Oerst., Vidensk. Meddel. Dansk. Naturhist. Foren. Kjoebenhavn 1858: 29. 1859.

Nunnezharia cataractarum (Mart.) O. Kuntze, Revisio Generum Plantarum 2: 730. 1891.

Nunnezharia martiana (H. A. Wendl.) O. Kuntze, Revisio Generum Plantarum 2: 730. 1891.

Nunnezharia flexuosa (Hort.) O. Kuntze, Revisio Generum Plantarum 2: 731. 1891.

Chamaedorea flexuosa Hort. in H. A. Wendl., Index Palmarum 12. 1854.

Chamaedorea lindeniana Hort. in H. A. Wendl., Index Palmarum 13. 1854. Not H. A. Wendl.

Chamaedorea atrovirens Hort. in Kerch., Les Palmiers 239. 1878. Not Mart.

Vadia jotolana O. F. Cook, Nat. Hort. Mag. 26: 12, figs. 1–3, 5–8. 1947a. Name of no botanical standing.

Vadia atrovirens (Hort.) O. F. Cook, Nat. Hort. Mag. 26: 26. 1947a. Name of no botanical standing. Not Mart.

Encheila transversa O. F. Cook, Nat. Hort. Mag. 26: 33, fig. 9. 1947a. Name of no botanical standing.

Common Names: cascade palm, cat palm.

Vernacular Name: guayita de los arroyos—Mexico.

Habit: cespitose, creeping, appearing stemless, forming dense clumps to 2 × 3 m.

Stems: 2–4 cm diam., dichotomously branched, horizontal, green, nodes prominent, congested, internodes less than 1 cm long.

Leaves: 4–5, erect-arching, pinnate, to 2 m long; **sheath** very open, tubular only at base, green; **petiole** to 30 cm long or more, channeled and green above, pale and rounded below, very flexible; **rachis** to 1 m long or more, obtusely angled and green above, rounded and pale below; **pinnae** 13–20 on each side of rachis, to 30 × 2.5 cm, linear-lanceolate, straight, briefly decurrent basally, regularly spaced, alternate, drying plicate, acute, departing rachis at an angle like a venetian blind, arching but tips ± drooping, a prominent midrib and 1 prominent primary nerve on each side of this.

Inflorescences: interfoliar or sometimes infrafoliar in fruit, erect, appearing to arise from soil or leaf litter; **peduncles** 50–70 cm long, ± rounded or slightly flattened, 7 mm wide, green in flower, orange in fruit; **bracts** 6–7, prophyll not seen, 2nd bract 10 cm long, 3rd 15 cm, 4th 18 cm, 5th 28 cm, 6th 20 cm, tubular basally, ± inflated apically, papery-fibrous, green in flower but persistent and becoming brown and tattered in fruit, longitudinally striate-nerved, uppermost exceeding peduncle and often concealing a small rudimentary one to 5 cm long. **Staminate** with **rachis** 3–8 cm long, green; **rachillae** 6–15, these 7–12 cm long, drooping.

Pistillate with **rachis** 7 cm long, rounded or angled, green in flower, red-orange in fruit; **rashillae** 10 or rarely spicate, to 15 cm long, 3–4 mm diam., erect, ± stiff, greenish in flower, orange in fruit.

Flowers: Staminate in 4 dense spirals, contiguous in bud, 2 × 2.5–3 mm, depressed-globose, angled by mutual pressure, yellowish tinged with green apically; **calyx** low, ringlike, scarcely lobed, pale, **sepals** connate nearly to apex, rounded; **petals** 2 × 2.5 mm, broadly ovate, slightly cupped, connate basally, valvate apically, rounded to straight apically; **stamens** just shorter than petals, **filaments** prominent, **anthers** 0.5 mm long; **pistillode** just shorter than or equalling stamens, columnar, briefly 3-cleft.

Pistillate close but not contiguous in bud, 2.5 × 3 mm, hemispherical or very depressed-globose, yellowish, slightly sunken in elliptic depressions 2–2.5 mm long; **calyx** 0.3–0.5 × 3 mm, ringlike, scarcely lobed, barely visible, **sepals** broadly rounded apically; **petals** 2–3 × 4 mm, broadly triangular, imbricate nearly to apex, cupped, straight to broadly rounded apically; **pistil** 2 × 3 mm, depressed-globose, light yellow, **styles** very short or lacking, **stigma** lobes short but exserted well beyond petals, pointed, recurved.

Fruits: 10 × 6–8 mm, ovoid-oblong, green changing to dark brown then black with glaucous bloom at maturity.

Distribution: MEXICO. Chiapas. Oaxaca. Tabasco. Moist or wet forest along or in streams and cataracts on the Atlantic slope; 300–1,000 m elevation; usually on limestone.

Discussion: Liebmann probably introduced *C. cataractarum* to Europe in the 1840s. He collected it during his travels in Mexico along streams and cataracts near Jocotepec in Oaxaca. Later, Martius

---

Plate 108. A, Chamaedorea microspadix. B–F, Chamaedorea radicalis. Cultivated, California. **A.** Pistillate inflorescence, Huntington Botanical Garden, San Marino. **B.** "Heel" at base of stem; **C.** Fruiting plants with well developed aboveground stem, *Hodel 820,* garden of R. Palmer, Whittier; **D.** Staminate flowers, *Hodel 793* (note arrangement in linear groups); **E.** Christina Hodel holds showy infructescences arising from nearly stemless plant, garden of R. Anderson, Sierra Madre; **F.** Infructescence with mature red fruits, *Hodel 820,* garden of R. Palmer, Whittier.

C

F

(1849) formally described and named the species. A few years later Wendland (1853b) described and named *C. martiana* from material that Linden had introduced to European gardens from Chiapas, Mexico. Wendland (1854) reported *C. martiana* growing in several gardens in Europe while Guillaumin (1923b) stated that it had been growing at the Musee de Paris since 1850 from Linden's introduction. I treated *C. martiana* as a synonym of *C. cataractarum* in an earlier paper (Hodel 1990d).

In horticulture, *C. cataractarum* has been confused with *C. atrovirens.* In 19th-century Europe, this latter name was erroneously applied to material of *C. cataractarum* and, in many instances, this is still the case today in Florida, California, Hawaii, and Australia. Krempin (1990, p. 89) discussed and illustrated *C. atrovirens* but the description and photograph depict *C. cataractarum.* On the other hand, some material grown as *C. cataractarum* is actually *C. oreophila.*

Chamaedorea cataractarum is unusual in its habitat and habit. One of the few rheophytes in the palm family, it inhabits the banks of rivers and streams of the Atlantic slope of Mexico in the states of Oaxaca, Tabasco, and Chiapas, occurring in or near water and being wholly or partially submerged during heavy rains and floods. It was originally found near waterfalls and cataracts, hence the specific epithet. In Chiapas, I observed large clumps several meters across growing along the banks of streams or small rivers. These clumps, like giant bull rushes, tended to capture and trap debris during times of high water. The debris line on the clumps of the palms was an indication of how high the water had risen.

Chamaedorea cataractarum is also unusual in its branching pattern. With age, plants tend to creep along the ground and over rocks with their horizontal stems branching in a dichotomous manner. Fisher (1974) provided a thorough and well illustrated account of this stem form. This dichotomous branching habit results in thick clumps with a sturdy network of thick, horizontal stems which grow along the ground, anchoring the clump securely. Along with flexible leaves and leaflets that tend to bend and sway with an opposing force, the creeping and securely anchored stem enables *C. cataractarum* to grow and survive in and along streams where it is occasionally inundated by swiftly moving water.

Chamaedorea cataractarum is cultivated in Hawaii, California, Florida, Venezuela, and Australia. In the United States, it is grown commercially from locally produced seeds or seeds imported from Mexico. It has gained popularity dramatically in recent years and has appeared in the indoor potted plant trade. In California, it appears to be susceptible to mites and brown tipping of leaflets due to low humidity and water with high mineral content. It grows well in Hawaii and Florida where it makes an attractive, informal border or hedge for a partially sunny location.

89. Chamaedorea crucensis Hodel, Principes 34: 166. 1990d. Type: Costa Rica, Puntarenas, San Vito, *D. & M. Hodel 706A* (holotype BH!; isotype CR!). Epithet: named for the type locality in forest remnants at Las Cruces Tropical Botanical Garden (now Jardin Botanico Robert y Catherine Wilson). **PLATE 115.**

Common Name: none.
Vernacular Name: none.

Habit: solitary, erect or rarely decumbent, 1–2 m tall.

Stem: 1.8–2 cm diam., green, smooth, ringed, 4–8 cm long, prominent brown stilt roots to 10–15 cm high basally.

Leaves: 3–5, erect-spreading, pinnate; **sheath** to 30 cm long, obliquely open apically and split opposite petiole, green but distinctly pale or yellow-margined, longitudinally striate-nerved; **petiole** to 50 cm long, slightly grooved above toward base, rounded and pale-banded below; **rachis** to 80 cm long, angled and green above, rounded below with a pale band extending onto sheath; **pinnae** 6–11 on each side of rachis, to 35 × 6.5 cm, lanceolate, sigmoid, falcate, subopposite, regularly arranged, long-acuminate, slightly drooping apically, slightly contracted basally, shining green, 7–9 prominent and equidistant primary nerves above, apical pinnae wider with more nerves.

Inflorescences: interfoliar but often infrafoliar in fruit, erect, spicate or sometimes staminate forked or rarely few-branched; **peduncles** to 30 cm long, 1–1.5 cm wide at base, 1 cm wide at apex, pale in flower, orange in fruit; **bracts** 4–6, green, deciduous. **Staminate** with rachis or flower-bearing portion or **rachillae** to 10–15 cm long, pendulous. **Pistillate** with **rachis** or flower-bearing portion to 12 cm long, straight and stiff in flower, curved, swollen, downward-pointing, and red-orange in fruit.

Flowers: densely packed, contiguous in bud, angled from mutual pressure. **Staminate** white; **calyx** prominent, 3 × 2.5–3 mm, deeply lobed, **sepals** connate in basal ⅓, straight or rounded apically, ± double along upper margins; **petals** 3.5 × 2.25 mm, valvate, spreading distally; **stamens** prominent, **filaments** 4 ×

Plate 109. Chamaedorea angustisecta. Cultivated, Fairchild Tropical Garden, Miami, Florida. **A, B.** Pistillate plants; **C.** Attractive white-green marbled and mottled sheaths and petioles.

→

C

0.75 mm, curved distally, **anthers** 1 × 0.5 mm, borne at tips of filaments and exserted above petals at anthesis, forming a stellate pattern on surface of flower; **pistillode** 2.5 × 0.5 mm, columnar, shorter than anthers.

Pistillate 2.5–3.5 × 2.5 mm, irregular, yellowish; **calyx** prominent, 1.5 × 2.5 mm, ± unlobed or only slightly notched, a ringlike sheath, **sepals** connate to apex, ± not distinct; **petals** not open, imbricate, hooded, flat spically; **staminodes** lacking; **pistil** 2 mm diam., ± globose.

Fruits: to 13 × 9 mm, ± triangular, rounded apically, densely packed, angled by mutual pressure, green when immature changing to red or orange near maturity and then black when soft-ripe; seeds to 8 × 6 mm, ellipsoid.

Distribution: COSTA RICA. Cartago. Heredia. Limon. Puntarenas. San Jose. Wet forest on the Pacific and Atlantic slopes; 1,000 m elevation.

Discussion: *C. crucensis* occurs in forest remnants adjacent to Jardin Botanico Robert y Catherine Wilson near San Vito, Puntarenas in southeastern Costa Rica. Robert Wilson established the species in his garden and over the years distributed seeds and seedlings to palm collectors and hobbyists. Moore (1971) noted it in his unpublished work on the palms of Finca Las Cruces. *C. crucensis* is cultivated in California and perhaps Florida and elsewhere.

Chamaedorea crucensis is similar to *C. allenii* but the yellow staminate flowers with the stamens (anthers) not protruding beyond the petals at anthesis distinguish the latter species. Stamens (anthers) not protruding beyond the petals at anthesis, red fruits, and the end pair of pinnae as broad as the others combined distinguish *C. zamorae,* another similar species.

Although collected with certainty only at or near the type locality, *C. crucensis* is probably more widely distributed. I refer material from adjacent provinces of Costa Rica here although staminate material is needed for confirmation.

90. Chamaedorea deckeriana (*Klotzsch*) *Hemsl.* in F. Godman & O. Salvin (eds.), Biologia Centrali-Americana, Botany 3: 404. 1885. Type: Cult., *Klotzsch s.n.* (holotype GOET!; isotypes C!, L!). Epithet: honors 19th-century German botanist G. H. Decker, from whose garden the type originated. **PLATES 116, 117(A-D).**

Stachyphorbe deckeriana Klotzsch, Allg. Gartenzeitung 20: 364. 1852.

Dasystachys deckeriana (Klotzsch) Oerst., Vidensk. Meddel. Dansk. Naturhist. Foren. Kjoebenhavn 1858: 25. 1859.

Nunnezharia deckeriana (Klotzsch) O. Kuntze, Revisio Generum Plantarum 2: 730. 1891.

Morenia deckeriana Hort. in Dammer, Palmenzucht und Palmenpflege 88. 1897.

Common Name: none.
Vernacular Name: none:

Habit: solitary, erect but sometimes decumbent, 0.3–2 m tall.

Stem: 2–3 cm diam., green, smooth, conspicuously ringed, internodes 2–5 cm long.

Leaves: 4–5, spreading, bifid, 90–125 cm long; **sheath** 15–25 cm long, obliquely long-open, tubular only in basal ⅔; **petiole** 15–25 cm long, flattened and green above with lower margins of blade faintly visible as they extend on either side to sheath, rounded and pale below; **rachis** 20–35 cm long, angled and green above, rounded and pale below; **blade** 50–70 × 25–35 cm, obovate, incised apically to ½ its length, lobes broadly lanceolate, 25–35 × 15–20 cm, cuneate basally, outer margins toothed, 20 primary nerves on each side of rachis.

Inflorescences: interfoliar, sometimes infrafoliar in fruit, spicate, erect, pendulous when heavily laden with fruits. **Staminate** 4–10 per node, middle one developing and opening first; **peduncle** 20–25 × 0.5–1 cm, rounded, green in flower; **bracts** 4–5, prophyll 3 cm long, 2nd bract 8 cm, 3rd 15 cm, 4th 15–20 cm,± loosely sheathing, becoming progressively larger distally, slightly inflated, terminal one equalling or exceeding peduncle, acute-acuminate, bifid, slightly flattened, green when young becoming brownish with age, longitudinally striate-nerved; **rachis** or flower-bearing portion 10 cm long.

Pistillate solitary; **peduncle** to 30 cm long, 1 cm wide at base, 1.5 cm wide at apex, flattened basally and apically, erect and greenish in flower, downward-curved and orange in fruit; **bracts** 5, green in flower, similar to those of staminate but fibrous and tattered in fruit; **rachis** or flower-bearing portion to 15 cm long, ± flattened, 1 cm wide, 5 mm diam., green and erect in flower, nodding, swollen to 2 cm diam., and orange in fruit.

Flowers: Staminate in 6–10 dense spirals, contiguous, 1.5–2 × 2.5–3 mm, depressed-globose, 6-angled by mutual pressure, green, emitting a distinct spicy-anise odor at anthesis, sunken; **calyx** well developed but not always prominent, often hidden by close packing of flowers, 1–1.5 × 3 mm, lobed, membranous, **sepals** connate basally, rounded apically; **petals** varying in width from 1–2 mm, usually 2 wider and 1 narrower per flower, valvate and sometimes slightly imbricate apically, erect for 1.5 mm then abruptly inflexed and tapering to flat pointed tips, tips of adjacent petals closely appressed in bud, opening as slits through which anthers are exserted at anthesis, thick, fleshy; **stamens** with **filaments** 1 × 0.75 mm, flattened, fleshy, tapering abruptly apically, **anthers** 0.75 × 0.3 mm, exserted from between petal slits on distal flat top of flower; **pistillode** 1.5 mm high, columnar, fleshy,

→

Plate 110. Chamaedorea pauciflora. In lowland Amazonian rain forest, *Hodel 1138A* (staminate) *& 1138B* (pistillate), Loreto, Peru. **A.** Spicate staminate inflorescences (note multiple arrangement at a node); **B-E** Pistillate plants; **B.** Mature plant; **C.** Spicate inflorescences and infructescences with mature fruits (inflorescences solitary at a node); **D.** Mature plant, older leaves gray with algae and lichens; **E.** Mature plant.

B
D
E

distinctly 3-lobed, rounded tip exposed in center of flower at anthesis.

Pistillate in 10–12 dense rows of up to 30 flowers per row, contiguous, 3 × 3 mm, depressed-globose, angled by close packing as those of staminate, greenish changing to greenish yellow, slightly sunken; **calyx** 1 × 2.5–3 mm, shallowly lobed, **sepals** connate nearly to apex, broadly rounded apically; **petals** 2 × 2.5 mm, subreniform, imbricate, tips inflexed to 1 mm, rounded to straight apically; **pistil** 3 × 2.5 mm, ovoid, ± 3-sided, greenish, **styles** short or lacking, **stigma** 1 mm long, exserted well beyond petals, open, erect, only very slightly if at all reflexed, pale to light yellow.

Fruits: 1–1.5 × 0.5–1 cm, ± obovoid-globose and stalked in appearance, contiguous, densely packed, flattened and/or angled by mutual pressure, green when immature becoming red-orange when soft-ripe, finally ageing blackish, epicarp smooth, mesocarp fleshy, endocarp thin; seeds 10 × 7 mm, ovoid; perianth persistent, sepals 1 mm long, petals 2 mm long.

Distribution: COSTA RICA. Alajuela. Cartago. Guanacaste. Heredia. Limon. Puntarenas. PANAMA. Bocas del Toro. Colon. San Blas. Wet forest on the Atlantic slope; to 900 m elevation.

Discussion: *C. deckeriana* is a spectacular plant with its large, bifid leaves and spicate infructescences heavily laden with densely packed, bright red-orange fruits. Warscewicz reportedly found it in Guatemala and sent seeds to Europe in 1849. Klotzsch described and named it from cultivated plants grown from these seeds. It is doubtful if its origin is Guatemala, though, since it has never been recollected there (Standley and Steyermark 1958) and has apparently never been recorded north of Costa Rica. In the 19th century, the term Guatemala was rather loosely applied to the area from that present-day state south to Costa Rica. Wendland found it in the valley of the Rio Sarapiqui in Costa Rica (Dammer 1904b) where it still occurs today. Although I reported in an earlier paper (Hodel 1990d) that a type had not been located for *C. deckeriana,* I have since examined good type material at Goettingen, Copenhagen, and Leiden.

We have successfully established plants of *C. deckeriana* in the greenhouse in Los Angeles. These were grown from seeds collected along the Rio Sucio on the Atlantic slope of Costa Rica. Fortunately, we have plants of both sexes and have sucessfully hand-pollinated them and set fruits. At anthesis, flowers of both sexes emit a fragrance best described as spicy-anise. Filling the entire greenhouse with its distinctive odor, it conjures up images of a deep, dark, primeval rain forest, just the sort of habitat in which one finds *C. deckeriana.*

See the accounts of *C. allenii* and *C. zamorae* for species which are similar to *C. deckeriana* and at times may be difficult to distinguish. Uncommon in cultivation, *C. deckeriana* occurs in a few collections in California, Florida, Hawaii, and Australia. It is also cultivated in a few gardens in Costa Rica.

91. Chamaedorea nationsiana *Hodel & Castillo,* Principes 35: 4. 1991. Type: Guatemala, Izabal, *Hodel & Castillo 1021A* (holotype BH!; isotype AGUAT!). Epithet: honors biologist and conservationist James D. Nations of Ciudad Vieja near Antigua, Guatemala who has contributed greatly to conservation and rural development efforts in that country. **PLATES 117(E), 118(A, B).**

Common Name: none.
Vernacular Name: none.

Habit: solitary, erect, to 2.5 m tall.

Stem: 2–3 cm diam., green, smooth, ringed, internodes to 10 cm long.

Leaves: 5–6, erect-spreading, pinnate; **sheath** to 30 cm long, obliquely open apically, rough- and brown-margined, longitudinally striated, green with a raised yellowish central costa; **petiole** to 75 cm long, flat or slightly channeled and green above, rounded and pale below; **rachis** to 1 m long or perhaps more, angled and green above, rounded below with a distinct yellow band extending onto sheath; **pinnae** to 11 on each side of rachis, middle ones largest, these to 53 × 9 cm, lowest pinnae to 40 × 6 cm, apical pair to 35 × 8.5 cm, lanceolate, slightly sigmoid, falcate, long-acuminate, opposite to subopposite, 8–9 prominent primary nerves above, middle one often most prominent, these pale and keeled below, 1 secondary between each pair of primaries, tertiaries numerous, faint, apical pair 10-nerved.

Inflorescences: inter- or infrafoliar in flower, infrafoliar in fruit, spicate. **Staminate** several to a node, usually 8, each with a separate prophyll and peduncular bracts but borne on a common hypopodium, center inflorescence developing first followed sequentially by others on either side; **peduncle** to 25 cm long, 5 mm wide at the base, 2–3 mm diam. at apex, ascending; **bracts** 4–5, prophyll to 5 cm long, 2nd bract to 9 cm, 3rd to 15 cm, 4th to 25 cm, 5th to 25 cm and greatly exceeding peduncle, fibrous, acute to acuminate, longitudinally striate-nerved, greenish in flower, tips brown and tattered, shredding with age; **rachis** or flower-bearing portion to 15 cm long, pendulous, densely flowered.

Pistillate solitary at a node, ascending in flower, nodding in fruit; **peduncle** to 30 cm long, 1 cm wide at base and there ± flattened, 2–5 mm diam. at apex, greenish or pale in flower, bright

---

Plate 111. Chamaedorea allenii. Mature plants, *Hodel 769A* (staminate) *& 769B* (pistillate), cultivated, research collection, Los Angeles, *ex* El Valle, Panama. **A.** Fruiting plant; **B.** Staminate inflorescence; **C.** Pistillate inflorescence; **D.** Staminate flowers; **E.** Pistillate flowers; **F.** Infructescences with mature fruits. →

B
D
F

orange and swollen in fruit; **bracts** 5, similar to those of staminate but brownish in flower becoming tattered and often fallen away in fruit; **rachis** or flower-bearing portion to 15 cm long and ± straight in flower, to 20 cm long and ± straight to strongly curved in fruit.

Flowers: Staminate in dense spirals, contiguous in bud, 2 × 1–1.5 mm, ± globose, angled by mutual pressure; **calyx** well developed, prominent, 1 × 1.5 mm, scarcely lobed, membranous, **sepals** connate to apex, straight or broadly rounded apically; **petals** 1–1.5 × 1.5 mm, triangular, valvate, spreading apically, acute; **stamens** 0.8 mm high, shorter than petals, tightly placed around but not exceeding pistillode, **filaments** short, **anthers** 0.8 mm long, oblong, ± sessile, bilobed; **pistillode** 1 mm high, columnar, flared basally and apically.

Pistillate in 5 dense spirals, contiguous, in bud 3–3.5 mm long, 2.5–3 mm wide, rhombic-shaped by mutual pressure, at anthesis 1.5–1.75 mm high, depressed-globose, ± superficial, leaving elliptic scars 2 × 0.75–1 mm; **calyx** well developed, prominent, 1.5 × 2.5–3.5 mm, scarcely lobed, ± membranous, **sepals** connate nearly to apex, straight apically; **petals** 1.5 × 2.5 mm, broadly triangular, imbricate nearly to apex, inflexed, rounded apically; **pistil** 2 mm high, ± columnar but swollen and to 1.5 mm wide basally, 0.5 mm diam. apically, terminal cap broadly flared, **stigma** lobes sessile, recurved, bifid.

Fruits: 1–1.5 cm diam., densely packed and angled from mutual pressure, black; seeds 6–9 mm diam., angled, brown.

Distribution: GUATEMALA. Izabal. Wet forest on the Atlantic slope; to 900 m elevation; on limestone.

Discussion: *C. nationsiana* is closest to *C. arenbergiana* with which it has been confused. However, the solitary, branched staminate inflorescences with up to ten pendulous rachillae distinguish *C. arenbergiana.* Fisher and Moore (1977), reporting on multiple inflorescences in palms, stated that *C. arenbergiana* had multiple staminate inflorescences. However, this is in error; Wendland's type specimen of *C. arenbergiana* and his original description clearly show this species to have solitary, branched, staminate inflorescences.

Although all the known specimens of *C. nationsiana* are from one locality in Guatemala, this highly localized distribution may be more apparent than real. Nearly all collections labeled as *C. arenbergiana* from the Atlantic slope of Guatemala and adjacent Honduras are in fruit. Some of these may actually be *C. nationsiana* since the two species are difficult to distinguish when only fruiting material is at hand.

We distributed seeds of *C. nationsiana* in December, 1989 as *C. arenbergiana* under the numbers *Hodel & Castillo 869, 873.*

92. Chamaedorea oreophila *Mart.,* Historia Naturalis Palmarum 3: 309. 1849. Type: Mexico, Oaxaca, *Liebmann s.n.* (some sheets later numbered *10810, 10811*) (holotype C!; isotypes P!, US!). Epithet: from the Greek *oreophilus* meaning mountain-loving, in reference to the habitat. **PLATES 7(D), 118(C), 119.**

Stachyophorbe montana Liebm. ex Oerst., Vidensk. Meddel. Dansk. Naturhist. Foren. Kjoebenhavn 1858: 10. 1859. (Based on *C. oreophila* Mart.) Not *C. montana* Mart. Epithet: from the Latin *montanus,* meaning pertaining to or growing in the mountains, in reference to the habit of this species.

Nunnezharia oreophila (Mart.) O. Kuntze, Revisio Generum Plantarum 2: 730. 1891.

Chamaedorea monostachys Burret, Notizbl. Bot. Gart. Berlin-Dahlem 11: 761. 1933a. Type: Mexico, Veracruz, *Seler 5183* (holotype B, destroyed). Epithet: from the Greek *mono,* meaning one, and *stachys,* meaning spike, in reference to the spicate inflorescences.

Stachyophorbe oreophila (Mart.) O. F. Cook, Nat. Hort. Mag. 22: 146. 1943b.

Stachyophorbe filipes O. F. Cook, Nat. Hort. Mag. 22: 145, fig. 15. 1943b. Name of no botanical standing.

Common Name: none.

Vernacular Names: tepejilote, tepejilote cimarron—Mexico.

Habit: solitary, erect, 1–3 m tall.

Stem: 0.6–2 cm diam., green, minutely but conspicuously white-spotted, prominently ringed, internodes 1–6 cm long.

Leaves: 5–13, erect-spreading, pinnate; **sheath** 10–25 cm long, fibrous, light green, densely white-spotted, white-margined, longitudinally green-striate-nerved, drying brown, durable, and ± persistent; **petiole** 7–35 cm long, light green with granulated white spots, slightly grooved above, rounded and pale green below; **rachis** 40–80 cm long, obtusely angled and light green above, rounded and pale green below; **pinnae** 14–25 on each side of rachis, 16–35 × 1.5–4 cm, narrowly lanceolate, falcate or slightly sigmoid, ± stiff and not drooping, acuminate apically, oblique

→

Plate 112. A, Chamaedorea allenii. B-E, Chamaedorea alternans. A. Staminate plant, spicate inflorescence, *Hodel 769A,* cultivated, research collection, Los Angeles, *ex* El Valle, Panama. **B, C, E** In lowland rain forest, *D. & R. Hodel 918A* (staminate) *& 918B* (pistillate), Catemaco, Veracruz, Mexico; **D** Cultivated, research collection, Los Angeles; **B.** Infructescences (note multiple arrangement at a node) with mature fruits; **C.** Staminate plant; **D.** Conspicuously white apex of leaf sheath with green nerves, *Hodel 751*; **E.** Staminate inflorescences (note multiple arrangement at a node).

C
E

basally, alternate or subalternate, 1.5 cm apart, dark lustrous green, prominent yellowish central midrib and 2 primary nerves on each side of this.

Inflorescences: interfoliar, erect, spicate, long-pedunculate, exceeding leaves, very slender; **peduncles** 60–140 cm long; **bracts** 6–10, upper one the largest and exceeding peduncle, to 37 cm long, slightly membranous or fibrous, acuminate, green in flower, green or brown in fruit. **Staminate** 4–8 per node; **rachis** or flower-bearing portion 10–30 cm long, 1 cm wide, pendulous, light yellow.

Pistillate solitary at a node; **rachis** or flower-bearing portion 10-30 cm long, erect or horizontal in flower, becoming pendulous in fruit.

Flowers: Staminate up to 300 per inflorescence, in dense spirals, contiguous, to 2.5–3 × 3.5 mm, ± deppressed-globose, irregularly angled by mutual pressure, slightly aromatic, yellow in bud, yellow to white at anthesis, ageing to brownish white; **calyx** 1–1.5 × 2.5–3.5 mm, lobed, membranous, green, brown-margined, **sepals** connate and/or imbricate basally rounded apically; **petals** 2–3 × 2–2.5 mm, oblong-obovate, valvate, connate basally, free apically, thin, slightly fleshy apically, slightly nerved on inside; **stamens** exceeding petals, **filaments** 2–3 mm high, cylindric, elongate, connate basally, pale yellow to greenish-orange, **anthers** 0.75–1 mm long, yellow; **pistillode** 2–2.5 mm high, slender, equalling stamens, 3-lobed apically.

Pistillate up to 200 per inflorescence in dense spirals, crowded and nearly contiguous, 1–1.5 × 3 mm, flattened-globose to conic, light yellow; **calyx** 0.5 × 3 mm, deeply lobed, pale to light green, **sepals** connate and/or imbricate briefly basally, straight or broadly rounded apically; **petals** 1.5–2.25 × 2–4 mm, broadly triangular, briefly connate basally, imbricate nearly to apex, straight apically, fleshy, thick, slightly nerved on inside; **pistil** 1.5 × 2.5–3 mm, depressed-globose, yellowish to greenish with white spots, **styles** short or lacking, **stigma** lobes separated, angled, erect, exceeding petals, sometimes slightly recurved, pale or clear-colored.

Fruits: 0.6–1.3 × 4–8 mm, ovoid-ellipsoid, greenish yellow changing to orange and then red when completely ripe, at times 2–3 adherent to one perianth, epicarp thin, slightly transparent, mesocarp thin, yellow-orange, endocarp slightly membranous, nerved; seeds 5.5–10 × 3.5–6.5 mm, ellipsoid.

Distribution: MEXICO. Veracruz. Oaxaca. Wet forest on the Atlantic slope; 1,000–1,500 m elevation; often on limestone.

Discussion: Long-pedunculate inflorescences, erect and arching well beyond the leaves, and the pendulous flower-bearing spike with either densely packed, yellow to whitish flowers or showy red fruits easily distinguish *C. oreophila.*

Martius described and named *C. oreophila* from material that Liebmann had collected in the 1840s in Oaxaca, Mexico. Liebmann (1846) had suggested the name *Stachyophorbe montana* for this material. However, Martius (1849) felt that *Stachyophorbe* was simply the same as *Chamaedorea* and, accordingly, used this latter name. Martius was precluded from using the epithet *montana* since, in the same paper, he had already used it for another Liebmann collection from Oaxaca. Martius did, though, respect Liebmann's intent for the epithet *montana* since he replaced it with *oreophila,* derived from a Greek word meaning mountain-loving. Oersted (1859) continued to maintain *Stachyophorbe* separate from *Chamaedorea.*

Later, Burret (1933a) described and named *C. monostachys,* based on collections of Seler and Galeotti in Veracruz, Mexico. In an earlier paper (Hodel 1990d), I treated *C. monostachys* as a synonym of *C. oreophila,* a possibility to which Burret, himself, alluded. Some material of *C. oreophila* in cultivation is erroneously referred to as *C. cataractarum* or, more infrequently, *C. atrovirens.* Krempin (1990, p. 93) illustrated *C. elegans* at Fairchild Tropical Garden in Miami, Florida but erroneously captioned the photograph as *C. monostachys.*

Fairly common in cultivation, *C. oreophila* appears in gardens and collections in Hawaii, California, Florida, and Australia. It is somewhat susceptible to infestation of thrips, especially on undersides of pinnae near the new, emerging leaf. Severe infestations result in defoliation and, in some instances, death of the plant.

93. Chamaedorea tepejilote *Liebm.* in Mart., Historia Naturalis Palmarum 3: 308. 1849. Type: Mexico, Oaxaca?, *Liebmann s.n.* (holotype C!; isotypes P!, US!). Epithet: is the vernacular name for the species. **PLATES 1(A), 4(C), 120, 121, 122(A), 125(G), 126.**

Chamaedorea casperiana Klotzsch, Allg. Gartenzeitung 20: 363. 1852. Type: Cult., *Klotzsch s.n.* [holotype GOET; isotypes C!, G (photo!), L!, W (photo!)]. Epithet: honors Dr. Casper, 19th-century professor in Berlin.

Stephanostachys casperiana (Klotzsch) Oerst., Vidensk. Meddel. Dansk. Naturhist. Foren. Kjoebenhavn 1858: 27. 1859.

Stephanostachys tepejilote (Liebm.) Oerst., Vidensk. Meddel. Dansk. Naturhist. Foren. Kjoebenhavn 1858: 28. 1859.

Stephanostachys wendlandiana Oerst., Vidensk.

Plate 113. Chamaedorea arenbergiana. Cultivated. **A, B** Fairchild Tropical Garden, Miami, Florida; **A.** Fruiting plant (infructescences at bottom); **B.** Spicate infructescences (immature fruits); **C.** Wide pinna, garden of B. Shannon, Hollywood, Florida; **D.** Forked pistillate inflorescence, *Hodel 689,* Huntington Botanical Garden, San Marino, California; **E.** Staminate inflorescence, *Hodel 803,* garden of P. Sullivan, California (note branched habit and solitary arrangement at a node).

C
E

Meddel. Dansk. Naturhist. Foren. Kjoebenhavn 1858: 28. 1859. Type: Cult., *Wendland s.n.* (holotype C!; isotypes GOET!, L!, M!). Epithet: honors 19th-century German horticulturist and botanist Herrmann Wendland.

Chamaedorea wendlandiana (Oerst.) Hemsl. in F. Godman & O. Salvin, Biologia Centrali-Americana, Botany 3: 407. 1885.

Nunnezharia casperiana (Klotzsch) O. Kuntze, Revisio Generum Plantarum 2: 730. 1891.

Nunnezharia tepejilote (Liebm.) O. Kuntze, Revisio Generum Plantarum 2: 731. 1891.

Nunnezharia wendlandiana (Oerst.) O. Kuntze, Revisio Generum Plantarum 2: 731. 1891.

Chamaedorea exorrhiza H. A. Wendl. ex Guillaum., Bull. Mus. Hist. Nat. (Paris) 28: 542. 1922. Type: Costa Rica, Alajuela, *Wendland s.n.* (holotype P!; isotype GOET). Epithet: from the Latin *ex,* meaning out of, and the Greek *rhiza,* meaning root, in reference to the abundant and conspicuous aerial prop roots often found emerging from the base of the stem.

Chamaedorea anomospadix Burret, Notizbl. Bot. Gart. Berlin-Dahlem 11: 763. 1933a. Type: Guatemala, Alta Verapaz, *Tuerckheim 4042* (holotype B, destroyed). Epithet: from the Latin *anomalus,* meaning abnormal, and *spadix,* meaning flower stalk, in reference to the inflorescence differing from others of the group in their small size and flowers in four rows.

Chamaedorea sphaerocarpa Burret, Notizbl. Bot. Gart. Berlin-Dahlem 11: 762. 1933a. Type: Nicaragua, Matagalpa, *Pruess 1365* (holotype B, destroyed). Epithet: from the Greek *sphaericus,* meaning spherical or globose, and *carpus* meaning pertaining to the fruit, in reference to the shape of the fruit.

Chamaedorea columbica Burret, Notizl. Bot. Gart. Berlin-Dahlem 12: 42. 1934b. Type: Colombia, Valle, *Dryander 13* (holotype B, destroyed). Epithet: named for the country where the species was discovered, Colombia.

Edanthe veraepacis O.F. Cook & C.B. Doyle, Nat. Hort. Mag. 18: 174, figs. 1–9. 1939. Name of no botanical standing.

Common Name: none.

Vernacular Names: tepejilote, pacaya, guaya, chi ib, cana verde, ixquil-quib, chimp, bojon, aula-te, chem-chem, ternero, pacaya grande—Guatemala and Mexico; palmito dulce—Costa Rica; boda, bola, nuru, cana verde—Panama.

Habit: solitary, sometimes cespitose, erect, sometimes decumbent, 2–7 m tall or more, if cespitose forming clumps to 3–4 m wide.

Stems: 2–10 cm diam., green, prominently ringed, internodes 2–15 cm long, often with more or less prominent prop roots basally.

Leaves: 3–7, spreading, pinnate; **sheath** 20–40 cm long, tubular, obliquely open apically, green, striate-nerved; **petiole** 10–30 cm long, slightly grooved and green above, rounded and pale below; **rachis** 0.5–1.4 m long, angled and green above, rounded below with a distinct yellow band extending onto sheath; **pinnae** (6–) 12–25 on each side of rachis, middle pinnae the largest, these 16–70 × 3.5–7.5 cm, broadly linear-lanceolate to long-lanceolate, sigmoid, falcate, narrowed basally, long-acuminate apically, subopposite, spreading or sometimes drooping, thin, lustrous green, 5–10 primary nerves keeled above, yellowish and shining below with 4–5 or even more secondaries interspaced, often nearly as prominent as primaries, tertiaries fine and numerous especially in large pinnae.

Inflorescences: infrafoliar, erect-spreading, solitary, 25–60 cm long; **peduncles** 6–27 cm long, stout, thick; **bracts** 4–5, to 20 cm long, lower short and truncate, upper rather prominently inflated and hoodlike, ± fibrous, drying ± woody, longitudinally striate-nerved, green becoming brownish in flower and fruit; **rachises** 1–30 cm long, green to yellow in flower, red-orange in fruit. **Staminate** with 7–50 **rachillae**, these 6–17 cm long, spreading or pendulous, yellow-green.

Pistillate with 5–20 **rachillae**, these 3–30 cm long, ± thick, spreading, straight or flexuous, ± angled, greenish yellow to green, minutely white-spotted in flower, red-orange in fruit.

Flowers: Staminate in 4–8 very dense spirals, contiguous, 2–2.5 × 3.5–5 mm, depressed-globose, irregularly shaped by mutual pressure, yellow, aromatic, seated in shallow elliptic depressions 2.5–3 mm long; **calyx** 0.5 × 3.5–5 mm, scarcely lobed, ringlike, membranous, green, partially adnate to sides of depression and similarly shaped, **sepals** connate nearly to apex, straight apically; **petals** 2–2.5 × 2.5–3.5 mm, deltoid, valvate, appearing as though connate basally due to crowding but essentially distinct, inflexed in bud, spreading apically at anthesis, ± fleshy, thick, lightly nerved on inside; **stamens** equalling or barely exceeding petals, **filaments** 1.25–1.5 mm long, green, **anthers** 0.5–0.75 mm long, ellipsoid, separated basally, entire apically, yellow; **pistillode** 0.75–1.25 mm high, shorter than or equalling stamens, slender, 3-lobed apically.

Pistillate in dense or lax spirals, 2–2.5 × 4–5 mm, conic-subglobose, greenish to yellow or whitish, slightly sunken in shallow elliptic-rounded depressions 1.5–3 mm diam.; **calyx** 0.5 × 4–5 mm, deeply or scarcely lobed, greenish, membranous, becoming undulate in fruit, **sepals** free or briefly connate and/or imbricate basally, broadly rounded apically; **petals** 2–2.5 × 4–5 mm, broadly ovate to triangular, imbricate nearly to apex, ±

Plate 114. Chamaedorea cataractarum. A, C Along river, *D. & R. Hodel 923,* Chiapas, Mexico; **B, D, E** Cultivated; **A.** Clump on rocks along river (note leaves pushed downstream by flow of water); **B.** Hedge planting, Lyon Arboretum, Honolulu, Hawaii; **C.** Along riverbank (note flood debris trapped on infructescences); **D.** Infructescences with mature fruits, *Hodel 786,* research collection, Los Angeles; **E.** Staminate (bottom) and pistillate (top) inflorescences, research collection, Los Angeles.

C
E

thick, fleshy, rounded to acute apically, usually undulate and brown-margined in fruit; **staminodes** 0–6, small, subtriangular; **pistil** 2–2.5 × 3–4.5 mm, yellow-green, globose, **styles** lacking, **stigma** lobes sessile but exserted well beyond petals, separated, recurved, angled, clear-colored.

Fruits: to 10–15(20) × 7–8 mm, ellipsoid to ovoid or nearly globose, blue-green maturing black, abortive carpels generally adherent to fruit, epicarp thin, slightly membranous, mesocarp slightly fleshy, green, aromatic, endocarp slightly membranous, fibrous; seeds 9–11 × 5–6.5 mm, ellipsoid, brownish.

Distribution: MEXICO. GUATEMALA. BELIZE. HONDURAS. EL SALVADOR. NICARAGUA. COSTA RICA. PANAMA. COLOMBIA. Moist or wet forest on the Atlantic and Pacific slopes; 0–1,600 m elevation; often on limestone.

Discussion: Liebmann apparently discovered *C. tepejilote* at Matlaluca, Veracruz, Mexico and named and described it in Martius (1849). However, I have been unable to locate any Liebmann collections of *C. tepejilote* from Veracruz. Strangely, Liebmann material identified as the holotype at Copenhagen is from Oaxaca as are the isotypes at Paris and the Smithsonian Institution.

Several collectors, including Liebmann, Linden, Warscewicz, and Wendland, sent propagative material of *C. tepejilote* to Europe, where it has been cultivated since prior to the middle of the 19th century, usually under the various names placed in synonymy here. *C. tepejilote* is widely cultivated today, appearing in gardens and collections in California, Hawaii, Florida, Australia, Europe, and elsewhere. One of the largest members of the genus, only *C. woodsoniana, C. linearis,* and the more stout and robust forms of *C. costaricana* equal or surpass *C. tepejilote* in size.

Chamaedorea tepejilote is a variable species throughout its very wide range. It occurs in moist or wet forests on a variety of substrates from southern Mexico to northern Colombia. Separate taxa have been proposed based principally on size, number of parts, and nervature of pinnae. According to Standley and Steyermark (1958), it is difficult to find constant distinguishing features and, essentially, differences are of size, not of character. They placed *C. wendlandiana* and *C. anomospadix* in synonymy with *C. tepejilote.*

In an earlier paper (Hodel 1990d) I treated *C. exorrhiza, C. sphaerocarpa,* and *C. columbica* as synonyms of *C. tepejilote.* In that paper, I maintained *C. casperiana* as separate from *C. tepejilote.* Klotzsch (1852) described and named *C. casperiana* from plants cultivated at the botanic gardens in Schonhausen near Berlin. These were grown from seeds that Warscewicz had collected in Guatemala and sent to Europe in 1849. I had not seen type material of *C. casperiana* and, based on Standley and Steyermark's (1958) assertion that the staminate calyx of *C. casperiana* was prominent and well developed, I felt the two species to be distinct. However, I have since seen good type material of *C. casperiana* at Goettingen, Copenhagen, and Leiden and the staminate calyx is variable, but not well developed and prominent. In most cases, the calyx is low, ringlike and more or less inconspicuous, like that of *C. tepejilote.* Occasionally, there will be a shriveled, membranous, toothlike sepal in the ringlike calyx, an uncommon but not unknown condition in *C. tepejilote.* When considering the highly variable nature of *C. tepejilote* though, this distinction seems insignificant.

Chamaedorea tepejilote is widely cultivated for food in southern Mexico, Guatemala, Honduras, and El Salvador. The unopened staminate inflorescences, resembling ears of corn, are sold in markets and used as a vegetable or in salads. They are called *pacaya,* a term also used for the entire plant. In Guatemala it is more intensely cultivated and exploited for market than in any other part of Central America. See the chapter on economic uses for a more extensive account of *pacaya.*

A handsome ornamental when well grown, *C. tepejilote* is a striking specimen for a background planting or as a solitary or group accent. Pistillate plants are especially attractive when bearing their heavy and relatively massive, red-orange, branched infructescences with black fruits. Although requiring some protection from the sun, *C. tepejilote* tolerates relatively high light. Its large, spreading leaves easily provide denser shade for smaller chamaedoreas requiring lower light. *C. tepejilote* is susceptible to damage from wind and infestations of mites.

Krempin (1990, p. 94) discussed and illustrated *C. neurochlamys* but the description and photograph depict *C. tepejilote.* Krempin (p. 96) also illustrated what appears to be a *Ptychosperma* but erroneously captioned the photograph as *C. tepejilote.*

→

Plate 115. Chamaedorea crucensis. In montane rain forest, Costa Rica. **A.** Staminate plant, *D. & M. Hodel 706A* (holotype), near San Vito de Coto Brus, Puntarenas (note staminate inflorescence held by hand); **B.** Fruiting plant, *Hodel & Grayum 699,* Tapanti, Cartago; **C.** Infructescences with mature fruits, *Hodel & Grayum 982,* Pavones, Cartago; **D, E.** Infructescences (immature fruits) and fruiting plant, *Hodel et al. 977,* Braulio Carrillo National Park.

B
E

94. Chamaedorea zamorae *Hodel,* Principes 34: 173. 1990d. Type: Cult., *Hodel & Bornhorst 830A* (holotype BH!). Epithet: honors Costa Rican botanist Nelson Zamora. **PLATE 122(B, C), 123.**

Common Name: none.
Vernacular Name: none.

Habit: solitary, erect, 2–3 m tall although flowering when stemless.

Stem: 2–2.5 cm diam., green, ringed, internodes 3–8 cm long.

Leaves: 3–5, erect-spreading, pinnate or rarely bifid; **sheath** 15–20 cm long, obliquely open apically and light green or nearly white and lightly striate-nerved; **petiole** 10–50 cm long, grooved and green above, rounded and pale below; **rachis** 40–60 cm long, angled and green above, rounded below with a pale band extending onto sheath; **pinnae** 3–6 on each side of rachis, all except apical pair long-lanceolate, 20–30 × 2–3 cm, weakly sigmoid, falcate, long-acuminate and drooping slightly apically, slightly contracted basally, opposite or subopposite, midrib and 4–5 prominent primary nerves on each side of this above, apical pair of pinnae very broad, as broad or broader than others combined, 12–18 cm wide and with more nerves, exterior margin coarsely toothed toward apex.

Inflorescences: interfoliar but often infrafoliar in fruit, erect to spreading; **peduncles** 15–30 cm long, erect or sometimes curved in fruit, pale or greenish in flower, red-orange in fruit; **bracts** 5–6, green, thin-papery, turning brown in fruit, uppermost opening long-oblique and exceeding peduncle. **Staminate** forked with a very short **rachis** 1–2 cm long or perhaps spicate; flower-bearing portion of **rachillae** 20–30 cm long, 1 cm wide with flowers at anthesis, pendulous.

Pistillate spicate; **rachis** or flower-bearing portion 10–15 cm long, stiff and erect in bud and in flower, becoming downward-pointing and curved in fruit.

Flowers: Staminate in 8 dense spirals, contiguous, brownish white; **calyx** prominent, 2 mm high, lobed, **sepals** connate basally, rounded to straight and broad apically, ± rectangular, 2 × 1.5 mm, thickened below across apex; **petals** 4 × 2 mm, valvate, ± fleshy, thickened along margins; **stamens** not exceeding petals, **filaments** 2.5 mm long, longer than pistillode, terete, **anthers** 1 × 0.5 mm, medifixed, at right angles to filament; **pistillode** 1.5 × 0.5 mm, columnar, slightly expanded apically and with 3 rounded lobes, ½ as high as stamens.

Pistillate in densely packed irregular rows, contiguous, 3.5 × 2.5–3.5 mm, yellow; **calyx** prominent, 2 mm high, lobed, **sepals** connate basally, widely truncate apically, 2 × 0.5 mm, thicker apically; **petals** 3 × 2.5 mm, imbricate nearly to apex, cupped with a short tip at center; **staminodes** lacking; **pistil** 2.5 × 2.25 mm, subglobose, **styles** short or lacking, **stigma** lobes closed but ± triangular.

Fruits: 10–15 × 10–12 mm, conic-globose, densely packed, contiguous, angled by mutual pressure, green, maturing red and perhaps ageing black.

Distribution: COSTA RICA. Alajuela. Cartago. Guanacaste. Heredia. Puntarenas. Wet forest on the Atlantic and Pacific slopes; to 800 m elevation or perhaps higher.

Discussion: The type of *C. zamorae* is from cultivated plants grown at Wahiawa Botanic Garden of the Honolulu Botanic Gardens in Hawaii. They were grown from seeds that Harold Koopowitz collected in 1966 near Laguna de Arenal, Costa Rica. By 1974 and up until at least 1978, they were fruiting with bifid leaves and were nearly stemless. In 1987, they were flowering and fruiting, had erect stems about 2 m tall, and, for the most part, had leaves with 3–5 narrow basal pinnae on each side of the rachis and a pair of very broadly lobed apical pinnae. *C. zamorae* is rare in cultivation, occurring only at Wahiawa Botanic Garden.

Chamaedorea zamorae is similar to *C. crucensis* and *C. allenii.* In fruit, it may be difficult to distinguish between *C. deckeriana* and bifid-leaved forms of *C. zamorae* and *C. allenii.* Some differences between these four species are summarized in the table below.

	Chamaedorea zamorae	*Chamaedorea allenii*	*Chamaedorea crucensis*	*Chamaedorea deckeriana*
pinnae/side	1–6; rarely bifid	7–9; rarely bifid	9–11	bifid
sheath margin	whitish	whitish	whitish	green
staminate inflor.	solitary	solitary	solitary	multiple
flowers	white	yellow	white	green
fruits	red > black	black	red > black	red > black
stamens	included	included	protruded	protruded

→

Plate 116. Chamaedorea deckeriana. A. Fruiting plant, *D. & M. Hodel 717,* montane rain forest, Rio Sarapiqui, Costa Rica; **B–D** *Hodel 780,* cultivated, research collection, Los Angeles; **B.** Pistillate inflorescences; **C.** Pistillate flowers; **D.** Infructescence with mature, showy red fruits.

B
D

IX. SPECIES IMPERFECTLY KNOWN

I am unable to place the following species since the types have been lost, destroyed, or are inadequate and/or the original descriptions are insufficient for diagnosis. If valid species, they are probably in subgenus *Chamaedoropsis* or subgenus *Chamaedorea.*

95. Chamaedorea atrovirens *Mart.,* Flora 35: 721. 1852. Not Hort. Type: Cult., *Martius s.n.* (holotype M?). Epithet: from the Latin *atro* meaning dark or black and *virens* meaning green, in reference to the leaves.

Nunnezharia atrovirens (Mart.) O. Kuntze, Revisio Generum Plantarum 2: 730. 1830.

Common Name: none.
Vernacular Name: none.

Habit: solitary, erect, to 2 m tall.
Stem: 1–1.5 cm diam., ringed, internodes 2–5 cm long.
Leaves: 70–90 cm long, spreading, pinnate; **sheath** 25 cm long, tubular; **petiole** 6–8 cm long, flattened or broadly grooved above, rounded below; **rachis** angled above, rounded below; **pinnae** 10–12 on each side of rachis, middle ones 20 cm long, lanceolate, uppermost shorter and broadly oblong, spreading at nearly right angles from rachis, closely spaced, acuminate, tips recurved, both sides dark green, shining under, midrib prominent and slightly raised on both sides, 8–10 clear inconspicuous primaries per pinna.
Inflorescences: Staminate infrafoliar, 50–60 cm long; **peduncle** to more than 30 cm long; **bracts** 6, prophyll 2 cm long, 2nd bract 3.5 cm, 3rd 8 cm, 4th 15 cm, 5th 30 cm, 6th 38 cm, tubular, briefly bifid apically; **rachillae** 20, simple, spreading to drooping.
Pistillate not seen.
Flowers: Staminate subglobose in bud, aromatic; **calyx** green, **sepals** transversely oblong, margins pale; **petals** ovate-oblong, obtuse, yellow, drying to olive brown or black; **stamens** shorter than petals, nearly equalling pistillode.
Pistillate not seen.
Fruits: not seen.

Distribution: MEXICO. Oaxaca. Moist or wet forest. Exact locality unknown.

Discussion: The description is from Martius (1852) who described and named *C. atrovirens* from a staminate plant in the Munich Botanical Garden that was grown from seeds Karwinski originally collected in Oaxaca without a specific locality. Martius compared it to *C. schiedeana* and *C. elatior,* stating that *C. atrovirens* was somewhat intermediate between these two species. He stated that *C. atrovirens* had many more pinnae and the petiole was more deeply grooved than that of *C. schiedeana. C. elatior* differed in the wider and more irregularly spaced, sickle-shaped or falcate, drooping pinnae. Martius' comparison with *C. elatior* is dubious, however, since there is ample evidence that he confused *C. elatior* with *C. pochutlensis.*

Martius' statement that *C. atrovirens* had many more pinnae than *C. schiedeana* is troubling since he put the number of pinnae at 20–24. It is difficult to determine if 20–24 is the total number of pinnae or number on each side of the rachis since he did not specify which case. A total of 20–24 pinnae would translate into 10–12 on each side of the rachis, about the same as *C. schiedeana.* However, if Martius meant 20–24 per side, this number of pinnae would about double that of *C. schiedeana* and be more in line with that of *C. pochutlensis.* Burret (1933a) stated that he was not able to determine the placement of *C. atrovirens* due to the meager description and material. Since the type is apparently lost and the description is inadequate, I am unable to place *C. atrovirens* even at a subgeneric level although it is probably in subgenus *Chamaedorea* or subgenus *Chamaedoropsis.* If the number of pinnae is indeed 20–24 on each side of the rachis, I suspect that *C. atrovirens* is probably not distinct from *C. pochutlensis.* If so, the fact that Martius made no mention of clustered stems may only reflect that he had a young plant that had yet to produce basal branches. The pinnae spreading at nearly right angles from the rachis, though, are more suggestive of *C. elatior.*

Chamaedorea atrovirens is unlikely to be in cultivation (if it is even a distinct species) although such an early name would have priority over many others. The name has been applied erroneously in cultivation to material of *C. cataractarum.* Krempin (1990, p. 89) discussed and illustrated *C. atrovirens* but the description and photograph depict *C. cataractarum.*

Plate 117. A–D, Chamaedorea deckeriana. E, Chamaedorea nationsiana. A–D Cultivated; **A.** Spicate staminate inflorescences (leaf sheath removed to show multiple arrangement at a node), *Hodel 797,* research collection, Los Angeles; **B.** Jerry Hunter stands with a well grown specimen in his nursery, Hilo, Hawaii; **C, D.** Staminate flowers and staminate plant, *Hodel 797,* research collection, Los Angeles. **E.** Infructescence densely packed with mature black fruits, *Hodel & Castillo 873,* lowland rain forest, Izabal, Guatemala.

→

B
D
E

96. Chamaedorea donnell-smithii Dammer, Gard. Chron., Ser. 3, 38: 43. 1905c. Type: Honduras, Rio Chamelecon, *Thieme 5537* (holotype US!). Epithet: named for American botanist John Donnell Smith.

Common Name: none.
Vernacular Name: none.

Habit: dwarf, less than 25 cm tall.
Stem: not seen.
Leaves: Pinnae 2–4 on each side of rachis, to 18 × 1.5 cm.
Inflorescences: Staminate not seen. **Pistillate** with **peduncle** to 8 cm long; **bracts** 3, prophyll to 1 cm long, 2nd bract 2.5 cm, 3rd 7 cm and exceeding peduncle; **rachillae** 4, to 10 cm long.
Flowers: Staminate not seen. **Pistillate** in ± remote spirals, 1–3 mm apart, 1–1.5 × 1.5–2 mm, subglobose, leaving shallow elliptic scars or depressions 1.5–2 mm long; **calyx** 0.25–0.4 × 1.5–2 mm, scarcely lobed, unnerved, **sepals** imbricate and/or connate nearly to apex, truncate; **petals** 1–1.5 × 2.5 mm, broadly triangular, imbricate nearly to apex and there ± truncate, ± membranous, unnerved; **pistil** 1–1.5 × 1.5–2 mm, subglobose.
Fruits: not seen.

Distribution: HONDURAS. Cortes-Santa Barbara. Forest on the Atlantic slope; 300 m elevation.

Discussion: *C. donnell-smithii* first appeared as a listed name only (Dammer 1903) and no description was provided. Later, Dammer (1905c) gave some scant information about this palm, thus validating the name. Thieme collected the species in 1888–1890 on rocky slopes along the Rio Chamelecon at about 300 m elevation near the border of the Departments of Santa Barbara and Cortes in Honduras. The species has not been listed in any of the floras of the area. The description provided here is from the meager holotype which is inadequate for placement of the species. The pistillate flowers are similar to those of *C. oblongata* and perhaps *C. donnell-smithii* is but an atypical and diminutive specimen of the former species.

→

Plate 118. A, B, Chamaedorea nationsiana. C, Chamaedorea oreophila. A, B In lowland rain forest, Izabal, Guatemala; **A.** Fruiting plant, *Hodel & Castillo 869*; **B.** Staminate inflorescences (immature), showing multiple arrangement at a node, *Hodel & Castillo 871*. **C.** Fruiting plant with mature, showy red fruits, *Hodel 765B,* cultivated, research collection, Los Angeles.

B

CHAPTER 6

HYBRIDS

Natural hybrids have not been reported and probably do not occur in *Chamaedorea.* On the other hand, several artificial hybrids have been made with species in cultivation. Unfortunately, in many of the hybrids identification of the parent plants is dubious, hence it is difficult to authenticate the species involved. The extent to which chamaedoreas will hybridize across subgeneric lines is unknown. Most reported cases of hybridization between species of different subgenera represent misidentification of parent plants. However, Richard Douglas of Walnut Creek, California has made crosses between *Chamaedorea oreophila* (subgenus *Stephanostachys*) and *C. microspadix* and *C. radicalis* (both subgenus *Moreniella*). Chamaedoreas probably will hybridize on their own in cultivation only infrequently. Most seem to be insect pollinated (Henderson 1986) and, since their natural pollinators are probably absent in cultivation, other insects must be present to perform the transfer of pollen or it must be done by hand.

Douglas' hybrids have revealed some interesting facts about inheritance in *Chamaedorea.* For example, *C. oreophila* has multiple staminate inflorescences at a node. When Douglas crossed *C. oreophila* with other species (all with solitary pistillate inflorescences), the offspring often have multiple pistillate inflorescences at a node, a rare occurrence in *Chamaedorea.* Only *C. alternans* is known to have multiple pistillate inflorescences at a node. The genes for controlling multiple inflorescences at a node seem to be dominant in many cases. When a species with cespitose habit is crossed with one of solitary habit, the offspring usually are solitary-stemmed, indicating that, in most cases, the genes controlling solitary stems are probably dominant.

Hybrids usually are not well documented and tend to confuse further an already complex situation. In addition, hybrids dilute the already small gene pool in cultivation. Most hybrids in *Chamaedorea* have not proven culturally advantageous or added significantly to the diversity of habit inherent in the existing species in the genus. Hybridizers of chamaedoreas are encouraged to document their work carefully and to prepare herbarium vouchers of each parent to avoid later confusion.

Although artificial hybrids in *Chamaedorea* were not reported until the early 1900s, Dammer (1904b) alluded to the fact that they were somewhat common in European collections by the middle 1800s. As chamaedoreas have grown in recent popularity, there has been an upsurge of hybridization. First, I present simple hybrids (those between two species) and then complex hybrids (a species crossed with a hybrid or two hybrids crossed with each other). I present them alphabetically by the given name of each hybrid. I propose new names for crosses which have none in the hopes of alleviating some of the confusion in the future. Following in parentheses are the purported parents with the staminate plant first and the pistillate one second, unless noted otherwise.

Simple Hybrids

Chamaedorea Alan Bredeson (*C. radicalis* × *C. microspadix*).

Named for Alan Bredeson of Lemon Grove, California in whose garden I first observed this cross, this hybrid is one of the best examples of a combination of two species. Made by Richard Douglas, the hybrid has the cespitose habit of *C. microspadix* but leaves and inflorescences that are exactly intermediate between the parents in shape, size, color, and texture.

Chamaedorea Amberwood (*C. microspadix* × *C. oreophila*).

Another one of Douglas' hybrids, this cross is intermediate between the parents. Staminate inflorescences are multiple at a node; pistillate plants are unknown. The hybrid is very rare and, perhaps, exists only in Douglas' collection.

→

Plate 119. Chamaedorea oreophila. A. Pistillate plant on limestone rocks, *D. & R. Hodel 953B,* near Huatusco, Veracruz, Mexico; **B–E** *Hodel 765A* (staminate) & *765B* (pistillate), cultivated, research collection, Los Angeles; **B.** Mature fruits; **C.** Staminate inflorescences (note multiple arrangement at a node); **D.** Staminate flowers; **E.** Staminate plant (note spicate inflorescences with pendulous rachis).

B
E

Chamaedorea Dick Douglas (*C. ernesti-augustii* × *C. stolonifera*).

The hybrid is named for Richard Douglas who is credited with first making it in the late 1970s. The main objective of this cross was to attain the dramatic, large, bifid leaves of *C. ernesti-augustii* and the cespitose habit of *C. stolonifera*. Unfortunately, the offspring have leaves intermediate in size between the parents and their stems are solitary [see Plate 124(A)].

Chamaedorea Douglas' Delight (*C. radicalis* × *C. oreophila*).

Still another of Douglas' hybrids, this cross displays some hybrid vigor and is intermediate in form between the parents. Stems are more robust and leaves larger than the parents. Staminate inflorescences are usually in multiples at a node, like those of *C. oreophila*, and are branched with several rachillae, as in *C. radicalis* [see Plate 124(B)]. Staminate rachillae are densely flowered and pendulous, like the spike of *C. oreophila*. Pistillate inflorescences are branched with several rachillae, as in *C. radicalis*, and, while densely flowered, the flowers are not contiguous. Pistillate inflorescences are solitary or in multiples at a node.

Chamaedorea Heidi Bornhorst (*C. tepejilote* × *C. arenbergiana*).

This hybrid is named for Heidi Bornhorst, horticulturist of Honolulu Botanic Gardens in Hawaii who assisted with collecting and identifying this taxon at their Wahiawa Botanic Garden. Perhaps an inadvertent cross, Honolulu Botanic Gardens grew it from seeds that Paul Weissich collected in 1977 from cultivated plants in Robert Wilson's garden in Costa Rica. It is not known which parent is the staminate and which is the pistillate. The hybrid is intermediate between the parents with striking infructescences having the branched habit of *C. tepejilote* but congested fruits on a thick, swollen axis as in *C. arenbergiana* [see Plate 124(C)].

Chamaedorea Horace Anderson (*C. pochutlensis* × *C. hooperiana*).

Introduced to the trade by Jack Ingwersen of Oceanside, California, this cross is named for the late Horace Anderson whose nursery in Leucadia, California specialized in chamaedoreas and other hardy palms and companion plants. Plants are vigorous and robust but intermediate in form between the parents. Impressed with the handsomeness and growth characteristics of *C. hooperiana*, Ingwersen has used it and *Chamaedorea* Horace Anderson in several other crosses although I have seen none of the offspring. He backcrossed *Chamaedorea* Horace Anderson with itself and one of its parents, *C. hooperiana*. Also, Ingwersen crossed *C. hooperiana* with *C. graminifolia*.

Chamaedorea Irving Cantor (*C. pochutlensis* × *C. graminifolia*).

Perhaps the most common hybrid *Chamaedorea* in southern California, this cross is named for the late Irving Cantor, gardener of the Vavra Estate in west Los Angeles who first made the cross in the 1960s. Cantor made the hybrid using plants at the estate when it was the property of the University of California at Los Angeles. It is not known which parent is the staminate and which is the pistillate. He reported the hybrid to be between *C. schippii* (= *C. graminifolia*) and *C. costaricana*. However, he had misidentified *C. costaricana*; the plant was actually *C. pochutlensis*.

Chamaedorea Irving Cantor has appeared in the trade to a greater extent than any other hybrid in the genus and is one of the few to display hybrid vigor. Its growth is more vigorous and the stems and leaves larger than those of the parents [see Plate 124(D, E)]. Stems are clustered, like those of both parents, and may be quite robust, to five cm in diameter or slightly more. The pinnate leaves are similar to those of *C. pochutlensis* but slightly larger. Plants grow quickly, attaining a height of three meters in about six years from seeds. Unfortunately, they are susceptible to infestations of mites which limits their use as indoor plants.

Chamaedorea Irving Cantor encompasses more variability than other hybrids in the genus. The variability may be because *C. pochutlensis* is extremely variable. Also, true or pure *C. graminifolia* is quite rare in cultivation; most material identified as such is *Chamaedorea* Irving Cantor. Although many have made this cross subsequent to Cantor, they used as parents various forms of *C. pochutlensis* and *C. gra-*

Plate 120. Chamaedorea tepejilote. Clustering form, lowland forest. **A.** Marianne Hodel and large clump, *D. & M. Hodel 844*, Tikal, Guatemala; **B.** Base of clump showing aerial roots (note machete), Stann Creek, Belize; **C.** Clump in full sun on limestone rocks, Tikal, Guatemala.

C

minifolia that may have already been crossed with *C. pochutlensis.* In many cases, hybridizers may have been inadvertently backcrossing *Chamaedorea* Irving Cantor. Krempin (1990, p. 95) illustrated *Chamaedorea* Irving Cantor but erroneously captioned the photograph as *C. schippii.*

Chamaedorea Jim Wright (*C. klotzschiana* × *C. glaucifolia*).

Named for Jim Wright of San Diego, California who first made the cross, plants are vigorous growers with leaves intermediate between the parents. Pinnae are clustered in groups along the rachis, as in *C. klotzschiana,* but are straighter and narrower.

Chamaedorea × katzeri *Loebner,* Gartenwelt 13: 159. 1909. (*C. sartorii* × *C. ernesti-augustii*).

Chamaedorea × *romana* Guillaum., J. Soc. Natl. Hort. France, Ser. 4, 24: 243. 1923b.

This cross is the first published account of a hybrid in *Chamaedorea.* According to Loebner (1909), F. Katzer reportedly made the hybrid in Pavlosk, Russia in 1899. Although originally reported to be a hybrid between *C. ernesti-augustii* and *C. concolor* (= *C. pinnatifrons*), it is doubtful if the latter species was actually the parent. From the description and illustration, it appears that it was *C. sartorii* rather than *C. concolor* that was crossed with *C. ernesti-augustii.* Marque (1923) reported the parents of this cross as *C. ernesti-augustii* and *C. sartorii.* Roster (1913) reported a cross between *C. ernesti-augustii* and *C. schiedeana* although the latter parent was probably *C. sartorii.*

Guillaumin (1923b) named a second hybrid reportedly made in Italy in 1910 between *C. ernesti-augustii* and *C. schiedeana.* However, the latter parent was also misidentified and was probably *C. sartorii.*

Many collectors have made this particular hybrid in the past. Today, *C.* × *katzeri* is not uncommon in collections and gardens. The offspring are intermediate between the two parents. Leaves are pinnate with the terminal pair very broad and with 2–4 narrow basal pinnae below these [see Plate 124(F)].

Chamaedorea U. A. Young (*C. glaucifolia* × *C. elatior*).

The cross is named for U. A. Young of Tampa, Florida who is credited with first making it in the early 1970s. Young (1974) reported this to be an accidental cross with the offspring being quite variable but intermediate between the two parents. Some of the plants tend to climb while others are self-supporting. Plants of *Chamaedorea* U. A. Young in the garden of Jim Wright have self-supporting, upright stems and leaves with pinnae somewhat similar to those of *C. elatior* but conspicuously narrower and more like those of *C. glaucifolia.* The terminal pinnae are strongly reflexed, as in *C. elatior.*

Chamaedorea Walnut Creek (*C. sartorii* × *C. stolonifera*).

Another hybrid of Richard Douglas, this cross is rare and probably does not occur outside of Douglas' collection. The hybrid is solitary-stemmed with pinnate leaves having large terminal leaflets.

Reports of several other simple hybrids have appeared in the literature but the identification of the parents is suspect. Two are listed in the 1910 *Kew Hand List of Tender Monocotyledons,* 2nd edition, pp. 84–85. They are *C.* × *kewensis* (*C. wendlandiana* × *C. lindeniana*) and the unnamed cross *C. corallina* × *C. glaucifolia.* In the case of the former, both names of the parents are synonyms, *C. wendlandiana* of *C. tepejilote* and *C. lindeniana* of *C. pinnatifrons.* Assuming that the parents were identified correctly, it seems an unlikely cross. In the case of the unnamed cross, *C. corallina* was a name applied in horticulture to *C. oblongata* and, although *C. glaucifolia* is a valid, accepted name, this too seems like an unlikely match. In both these cases, the parents are in separate subgenera.

Young (1974) reported making several crosses but at least one parent in each one was probably mis-

---

→

Plate 121. Chamaedorea tepejilote. A. Staminate inflorescences in bud, *Hodel 688,* cultivated, Huntington Botanical Garden, San Marino, California; **B.** Pistillate inflorescence, cultivated, garden of R. Palmer, Whittier, California; **C.** Staminate inflorescences, *Hodel & Grayum 703A,* along trail to river, C.A.T.I.E., Turrialba, Costa Rica; **D.** Infructescence with mature fruits, cultivated, Wahiawa Botanic Garden (*2063*), Honolulu, Hawaii; **E.** Aerial prop roots, *Hodel & Grayum 703B,* along trail to river, C.A.T.I.E., Turrialba, Costa Rica; **F.** Heidi Bornhorst stands below crown of leaves, cultivated, Waimea Falls Park & Arboretum, Honolulu, Hawaii.

B
D
F

named. One he reported was *C. tepejilote* × *C. neurochlamys.* The two species are very distinct and in separate subgenera. The latter is uncommon in cultivation, similar in habit to *C. oblongata,* and has red-orange fruits. The fact that Young stated that they were quite similar, the only difference being the lack of stilt roots in *C. neurochlamys* seems to indicate that this parent was misidentified. He also reported making a cross between the form of *C. tepejilote* with clustering stems and *C. neurochlamys.* Again, I believe that the latter parent had been misidentified.

Another cross that Young reported making was *C. oreophila* × *C. sartorii.* Both names have been misapplied in horticulture to various forms of *C. oblongata.* They are quite distinct from each other and from *C. oblongata* and all three are in different subgenera. I have seen the progeny of this cross grown from seeds supplied by Young and they are undoubtedly *C. oblongata.*

Both Barry (1967) and Young (1974) discussed the cross of *C. seifrizii* × *C. erumpens.* Since I treat the latter as a synonym of *C. seifrizii,* this hybrid is essentially a cross between two forms at two extreme ends of a highly variable species. Most material of *C. seifrizii* in the trade is actually a cross between these two extremes or forms or consists of naturally occurring intermediate forms. For lack of a better term, material of *C. seifrizii* in the trade is often called *Chamaedorea* Florida Hybrid. Pure forms of both extremes are uncommon in the trade.

Complex Hybrids

For several years Richard Douglas has been making more complex hybrids by crossing another hybrid or species with *Chamaedorea* Douglas' Delight. All are in an experimental stage and exist only in Douglas' collection.

Chamaedorea Double Delight [(*C. radicalis* × *C. oreophila*) × (*C. microspadix* × *C. radicalis*)].

The cross is a double hybrid involving *Chamaedorea* Dick Douglas and *Chamaedorea* Alan Bredeson. Pistillate offspring have multiple inflorescences at a node; staminate offspring are not known. Solitary-stemmed, the hybrid exhibits vegetative characters of the three species.

Chamaedorea Micro Delight [*C. microspadix* × (*C. radicalis* × *C. oreophila*)].

Sharing vegetative characters of all three species, the solitary-stemmed hybrid has solitary staminate and multiple pistillate inflorescences at a node, an arrangement exactly opposite that of *C. oreophila.*

Chamaedorea Oreo Delight [*C. oreophila* × (*C. radicalis* × *C. oreophila*)].

A backcross of Douglas' Delight to one of its parents, the hybrid combines vegetative characters of the three species.

Plate 122. A, Chamaedorea tepejilote. B, C, Chamaedorea zamorae. A. Cultivated, San Diego Zoo, California. **B, C** In montane rain forest, Rio Sarapiqui, Costa Rica; **B.** Pistillate plant, *D. & M. Hodel 718*; **C.** Fruiting plant (immature fruits). →

B

CHAPTER 7

ECONOMIC USES

Chamaedoreas and their products are used extensively in the horticulture and floriculture industries on a worldwide basis. In horticulture, chamaedoreas are popular interior and exterior landscape subjects for use in the ground or as potted specimens. In the floriculture trade, chamaedoreas are used to a great extent as potted decorative plants for indoor use in homes, offices, hotels, and other buildings and businesses. Also, the floricultural industry uses cut leaves of chamaedoreas as filler greenery in flower displays, arrangements, and bouquets.

Chamaedoreas even serve as a source of food to local inhabitants where the palms are indigenous. The immature, staminate inflorescence, called *pacaya,* and derived mainly from *C. tepejilote,* is cooked and either eaten alone or mixed with other foods.

Horticultural Uses

Chamaedoreas are produced on a commercial basis in the United States in Florida, California, Hawaii, and Texas. In addition to U.S. production, they are also grown in Europe, Australia, the Orient, and elsewhere. Unfortunately, little information is available on production outside the U.S. Even within the U.S. it has always been somewhat difficult to obtain an accurate count of commodity production in agriculture, and production of ornamental horticulture crops is no exception. There are several factors responsible for inaccuracy in crop reporting. Among them are a reluctance on the part of many growers to divulge information; the various and diverse marketing channels through which products can be sold and resold, even among growers; the lack of a uniform method for counting units, for example 10 separate seedlings one year may be one pot the next; and the perishable nature of many crops.

Edmonson (1989) and Hodel (1988b) reported on the commercial production of chamaedoreas in Florida [see Plate 125(C)] and California respectively, the leading areas of production in the U.S. The information presented here is based on these two sources and, although not as exact as one would want, it is a fairly accurate picture of the role of chamaedoreas in the industry.

The main species produced for use in the interior and exterior landscape in Florida and California are presented in Table 7-1 and ranked by total number of units reported grown.

Although figures were not available for Florida, Hodel (1988b) reported the wholesale value of the chamaedoreas produced in California in 1988 at \$866,876. The value and the total units produced are on the low side since growers contributed much of the information on a voluntary basis and they probably underreported the data.

The quantities and sizes of plants vary considerably. The number produced by size for the three most popular chamaedoreas in California is shown in Table 7-2. Although figures were not available for Florida, the size categories are probably similarly distributed.

Approximately 80 percent of the chamaedoreas used in horticulture in California, with the exception of *C. elegans,* were grown in Florida or Hawaii [see Plate 125(A)] and shipped to California, where they were sold immediately as finished plants or potted up and grown on and sold later as finished plants. Of this amount, about 60 percent are from Florida and 40 percent are from Hawaii. Since chamaedoreas are basically a slow-growing greenhouse crop, it is not economically feasible to grow them from seeds to finished plants in California on a commercial basis. Production costs are substantially lower in Florida and Hawaii, where the plants can be grown year-round without the use of an energy-consuming greenhouse.

However, growers in California produce *C. elegans* from seeds since it is often sold at a very young stage (one to two-leaf seedling) for use in terrariums and as small potted decorative plants. A ban on importation of palm seeds into Hawaii from Mexico because of lethal yellowing has nearly ended pro-

Plate 123. Chamaedorea zamorae. Cultivated, Wahiawa Botanic Garden (*66.915*), Honolulu, Hawaii, *ex* Laguna de Arenal, Costa Rica. **A, B.** Staminate plant and inflorescences, *Hodel & Bornhorst 830A* (holotype); **C, D** *Hodel & Bornhorst 830B*; **C.** Pistillate inflorescence; **D.** Infructescence (immature fruits).

B
D

Table 7-1. **Chamaedoreas in Commercial Horticulture in Florida and California, 1987-88.**

Species	No. of units grown Florida	California	Total
C. seifrizii (includes *C. erumpens* and *C.* Florida Hybrid)	5,738,000	25,259	5,763,259
C. elegans	3,862,000	996,094	4,858,094
C. cataractarum	29,300	5,000	34,300
C. microspadix	10,000	1,000	11,000
C. metallica	1,500	5,000	6,500
C. ernesti-augustii	3,500	500	4,000
C. costaricana	1,000	233	1,233
C. radicalis	1,100	—	1,100
Others: (<1,000 of each kind) includes *C. elatior, C. neurochlamys, C. oblongata, C. pochutlensis, C. sartorii, C. tepejilote*			6,500
		Total grown	10,685,986

duction of *C. elegans* in Hawaii. However, one large grower in Hawaii has established a plantation of *C. seifrizii* and produces about 750,000–1 million seeds annually [see Plate 125(B)]. The grower produces about 60,000 pots of *C. seifrizii* annually, with 10–15 plants per container.

Seed Sources

Chamaedorea elegans and *C. seifrizii* comprise a vast majority of the units produced and a greater part of their value. Nearly all seeds of the former and a substantial portion of the latter originate in Mexico. Florida produces a significant amount of seeds of *C. seifrizii* and is nearly self-sufficient for meeting its needs, while Belize produces a minor amount of seeds of this species [see Plate 125(F)].

The virtual lack of seed production of *C. elegans* outside of Mexico is probably due to the fact that it does not set seed readily outside its native habitat without artificial pollination. Since inexpensive seeds have been readily available from Mexico on a reliable basis, there has been no economic incentive to produce them in cultivation in Florida or elsewhere. On the other hand, *C. seifrizii* normally sets seeds readily and in copious quantities in cultivation. However, future problems, such as disease quarantines or endangered species legislation, might impact on the availability of seeds of *C. elegans* from Mexico. There is need for an understanding of the breeding mechanisms of *C. elegans* and of utilizing hand-pollination techniques to ensure an adequate seed-set.

Most of the seeds originating in Mexico are handled through two firms in southern Texas. These firms gather the seeds in Mexico, clean them, ship them to their facilities in Texas and then reclean, package, store, and distribute them throughout the United States and the world [see Plate 125(D, E)]. Several other firms in Florida, California, Hawaii, and elsewhere handle chamaedorea seeds but since all buy and sell among themselves and reship and resell, it is always difficult to attain accurate figures of the total quantities of seeds produced and their origin and destination. Europe is the largest market for seeds of *C. elegans*; in fact, Edmonson (1989) reported that one of the two largest distributors of seeds of this species exported about three-fourths of its total yearly production to the Continent. The U.S. is the largest market for seeds of *C. seifrizii.*

Table 7-3 summarizes the quantities of seeds of chamaedoreas produced on an annual basis in the commercial trade and their sources. Seeds of three species, *C. elegans, C. seifrizii,* and *C. cataractarum,*

Table 7-2. **Quantity and Sizes of the Three Most Popular Chamaedoreas Produced in California, 1987.**

		Container size					
Species	Liners	6"	8"	10"	12"	14"	>14"
C. elegans	946,000	34,551	10,293	3,745	1,055	400	50
C. seifrizii	10,000	1,235	845	9,991	2,498	980	510
C. costaricana		25	50	50		12	96

→

Plate 124. Chamaedorea Hybrids. Cultivated. **A.** *Chamaedorea* Dick Douglas, pistillate plant, garden of L. Hooper, La Habra, California; **B.** *Chamaedorea* Douglas' Delight, staminate inflorescence, garden of D. Douglas, Walnut Creek, California; **C.** *Chamaedorea* Heidi Bornhorst, *Hodel & Bornhorst 835,* pistillate inflorescences and leaf, Wahiawa Botanic Garden (*77.1344*), Honolulu, Hawaii; **D.** *Chamaedorea* Irving Cantor, Lakeside Palmetum, Oakland, California; **E.** *Chamaedorea* Irving Cantor, entrance to office building, Beverly Hills, California; **F.** *Chamaedorea* × *katzeri,* staminate plant, garden of L. Hooper, La Habra, California.

C
E
F

Table 7-3. **Quantity of Seeds Produced of Chamaedoreas in Kilograms on an Annual Basis, Source, and Number of Seeds per Kilogram (from Hodel 1988b and Edmonson 1989).**

Species	No. of kilos	Source	No. of seeds per kilo
C. elegans	100,000	100% Mexico	4,400
C. seifrizii	35,180	52% Mexico 47% Florida 1% Hawaii	3,600
C. cataractarum	12,036	94% Mexico 3% Florida 3% Hawaii	3,000
Others include: *C. costaricana, C. elatior, C. ernesti-augustii, C. metallica, C. microspadix, C. radicalis, C. tepejilote*	11,500	99% Mexico 1% Costa Rica	1,200–1,700

comprise over 90 percent of the market. Nearly all the remainder originates in Mexico. A very small portion, representing seeds of *C. costaricana,* is collected from cultivated plants near San Jose, Costa Rica. The number of seeds per kilo depends on how much the seeds have been dried.

Vosters (1975) reported that about 31,000 kgs of seeds of *C. elegans* were exported from Mexico in 1974. The 100,000 kgs exported in 1988 is a dramatic increase. About 75 percent of the seeds exported from Mexico are collected from plants in the wild, a factor that some conservationists feel may represent a threat to the survival of the species in Mexico. The remaining 25 percent are from plants cultivated for seed or foliage production. The 75:25 ratio of seeds from the wild versus cultivation is expected to be reduced as seeds from the wild become more scarce due to habitat destruction, governmental regulations restricting their export for conservation reasons, and/or extinction due to overcollecting the seeds or foliage. Currently, cost is the major factor limiting plantation-production of *Chamaedorea* seeds. Purchasing seeds from local people who gather them in the wild is much less expensive than producing seeds by cultivation in a plantation. Both Texas importers and one from California have recognized that collection of seeds from wild plants will probably be greatly restricted or prohibited for conservation reasons in the near future, and have established or plan to establish plantations for seed production in Mexico or Central America. Until governmental restrictions are in place, though, the economic impetus is to continue to look to wild plants as the source of seeds.

The number of plants of *C. elegans* that could be produced on an annual basis seems staggering. Simple arithmetic indicates that potentially 440,000,000 seedlings are possible worldwide! However, this amount is never realized since germination usually runs 60–80 percent. Also, 10–15 seedlings are usually planted together in a single container, further reducing the potential number of units produced.

As mentioned earlier, Florida is nearly self-sufficient in production of seeds of *C. seifrizii.* The species was introduced there about 50 years ago and, as a result of an unreliable and inadequate supply from Mexico, many growers established mother blocks for seed production. Since the 1970s, the seed production in Florida has reduced the pressure on wild plants in Mexico as sources of seeds. Whereas Mexico was once the sole source of seeds of *C. seifrizii,* it now has only about half the market, Florida having the other half. The five largest growers of *C. seifrizii* in Florida have 6–15 hectares (15–38 acres) each devoted to this crop. Stock plants for seed production occupy the perimeter areas of these nurseries. Growers meet their own demands for seeds and sell the excess to other growers in Florida. They feel they have greater control over quality, availability, and price and most growers have a prefer-

→

Plate 125. Economic Uses of Chamaedorea. A, B *C. seifrizii,* Kohala Nursery, Hawaii; **A.** Production of certified 10-inch pots for shipment to mainland U.S.; **B.** Plantation for seed production; **C.** Robert Bernecker, Jr. and well grown tubbed specimen of *C. seifrizii* tagged for sale in outdoor shadehouse, Bernecker's Nursery, Homestead, Florida; **D, E** Seeds of *C. elegans,* imported from Mexico, Longoria Seeds, Mission, Texas; **D.** Seeds drying before packing and shipping; **E.** Hugo Dominguez (right) inspects seeds on drying bench prior to packing; **F.** Marianne Hodel stands in seed plantation of *C. seifrizii,* Teakettle Farms, Belize; **G.** Juan Jose Castillo holds pacayas (staminate inflorescences of *C. tepejilote*), market, San Cristobal de Verapaz, Guatemala.

26
C
E
G

Table 7-4. **Country of Origin and Quantity of Leaves of *Chamaedorea* Imported into the U.S., 1986.**

Country	Quantity (in 1,000 stems)
Mexico	314,419
Guatemala	40,179
Costa Rica	4,145
Other	476
Total	359,219

ence for, and are familiar with, the cultural requirements of their particular strains. About 6,800 kgs of seeds of *C. seifrizii* can be produced annually from a plantation of four hectares (10 acres). Figures are not available for seed production of *C. elegans* since most seeds have until recently been from wild plants.

The cost of seeds is dependent on many factors and has fluctuated over the last 15 years. Depending on annual supply and demand, seeds of *C. elegans* sell anywhere from $6–20 per kilo ($3–10 per pound). Those of *C. seifrizii* and the less important species sell for $30–60 per kilo ($15–30 per pound).

No live plants of *C. elegans* or *C. seifrizii* are collected in the wild in Mexico and exported. However, about 2,273 kgs of nursery-grown, sprouted seeds of *C. elegans* are exported from Mexico and shipped to southern Texas or Florida on an annual basis. Edmonson (1989) reported that 40,000–50,000 of these are imported through Miami on a weekly basis. This is the only species of *Chamaedorea* handled in this manner.

Several other species collected in the wild or grown by hobbyists and collectors and sold to other collectors should be mentioned although they are not handled by large commercial growers. The transactions usually involve fairly limited quantities although in some instances the numbers may range to 1,000 to 3,000 plants, quantities that approach or, even in some cases, surpass numbers handled by strictly commercial growers. Trade, even in these relatively small numbers, can endanger the plants' survival in the wild since many are known only from a few, small, local populations.

Quasi-commercial species include, among others, *C. amabilis* (Costa Rica, Panama), *C. fragrans* (Peru), *C. pumila* (Costa Rica), *C. sullivaniorum* (Panama, Costa Rica), *C. tenella* (Mexico), *C. tuerckheimii* (Guatemala, Mexico), and *C. undulatifolia* (Costa Rica). Unscrupulous collectors have been known to devastate entire local populations of some of these highly ornamental species (Hodel 1988a). Most of the species occur in exclusive and narrow microclimates and have exacting and precise cultural requirements that are not easily duplicated in cultivation. They are not commercially viable species that can be cultured easily under varying growth regimes; hence, few survive the rigors of their treatment in cultivation. Prices for these range from $20–30 and up for plants in 15-cm (one-gallon) containers. Small plants are usually readily available to collectors and hobbyists in Florida and California.

Floricultural Uses

Cut leaves of several chamaedoreas are a staple item in the florist trade in California. A recent survey of the Los Angeles Wholesale Flower Market (Hodel, 1985b) revealed that leaves of about six species of *Chamaedorea* are encountered on a year-round basis. The species used include *C. elegans* (several forms), *C. oblongata, C. oreophila, C. radicalis, C. sartorii, C. tepejilote,* and several unidentified chamaedoreas.

Most cut leaves appear to originate in Mexico although some apparently come from Costa Rica, Guatemala, and Honduras. A firm in southern Texas imports a majority into the U.S. Trade names, such as "jade," "emerald," "teepee," and "commodore," are applied to the various species and forms and are used somewhat interchangeably. The trade names "commodore" and "teepee" are most likely derivatives of *Chamaedorea* and *tepejilote,* respectively, with the latter name often misapplied.

The USDA Marketing Service records the quantity of leaves of *Chamaedorea* imported into the U.S. Table 7-4 lists the quantity of leaves imported and country of origin for 1986.

Most leaves are collected from wild-growing plants although some probably originate from cultivated ones. Leaves collected from wild-growing plants may be a renewable resource if overcutting or removal of flower or fruit stalks does not harm plant populations. Heinzman and Reining (1988) reported that

→

Plate 126. Economic Uses of Chamaedorea. *Pacaya* (staminate inflorescences of *C. tepejilote*) in Guatemala. **A.** Native harvester of wild *pacaya* holds leaf and pacaya, montane rain forest, Quetzaltenango; **B.** Native seller of cultivated *pacaya* at market, San Cristobal de Verapaz (compare large size of cultivated *pacaya* with that of much smaller wild ones); **C.** Plantation of *C. tepejilote* cultivated for *pacaya* and underplanted with coffee, Alta Verapaz.

B

sustainable extraction of leaves of *C. elegans* in northern Guatemala for the floral trade was a distinct possibility but needed further investigation. Oyama and Mendoza (1990) investigated defoliation on growth, reproduction, and survival of *C. tepejilote* (probably *C. alternans*) in Mexico. They found that experimental, complete defoliation actually increased leaf production.

No dollar value of these cut leaves was available but on the Los Angeles Wholesale Flower Market a bunch of cut foliage composed of 10–12 individual leaves sold for $1–3 depending on the type, availability, and season.

Cut leaves of *C. quezalteca* are used locally in Guatemala in flower arrangements and bouquets but no information is available on quantity or price.

Dietary Uses

Chamaedoreas are of considerable economic importance throughout Central America as a source of food although their role in this seems to be diminishing steadily. Standley and Steyermark (1958) reported in great depth about their use in Guatemala. The unopened, staminate inflorescences, called *pacaya,* are enveloped in green bracts and appear somewhat like green ears of maize or corn [see Plate 125(G)]. They are sold in the markets [see Plate 126(B)] and are a popular and important vegetable comprising a substantial part of the diet of many local inhabitants. Pistillate inflorescences are not used.

Pacaya is whitish, tender, and attractive when served on the table, and of a slightly bitter but agreeable flavor. It is served raw as in a salad but more often is cooked and mixed with other dishes, such as eggs. It may be treated and used as is asparagus.

Although inflorescences of many different species of *Chamaedorea* appear in the markets and may be eaten, they differ greatly in appearance and some are shunned because of their extreme bitterness. *C. tepejilote* produces the best quality *pacaya* (the term is applied to both the plant and its edible inflorescences) in terms of flavor and size. The inflorescences of most other species are too small and are worthless for food. Plants of *C. tepejilote* from the Coban region in Alta Verapaz, Guatemala have the reputation for producing the finest quality *pacaya.* Standley and Steyermark (1958) reported that these resemble green ears of field corn when piled on the ground. Well known for their flavor and large size, they are shipped throughout Guatemala.

Chamaedorea tepejilote has been cultivated for centuries in Alta Verapaz and the most vigorous, robust, and productive plants bearing large, flavorful *pacaya* have been selected and propagated. As a result, the plants cultivated in Alta Verapaz are much larger than their wild counterparts [see Plate 126(A)]. Robust and stout, they are often taller than one-storied houses [see Plate 126(C)]. Plants are abundant in the Coban region, where they are valued more for their food than ornament. However, the plants are highly ornamental and many homes have clumps of plants adorning their yards and patios from which *pacaya* is gathered throughout the year for the table. Most *pacaya* is actually produced outside of Coban; some of the small farms in nearby Tactic and San Cristobal have veritable forests of them [see Plate 126(C) and end sheets].

In some parts of Guatemala a different dish, called *bojon,* is prepared from *Chamaedorea.* Consisting of the pith of younger stems and perhaps the leaf buds, *bojon* is either eaten raw, as in salad, or cooked in various ways. Standley and Steyermark (1958) stated that when baked, it proved to be flavorful and succulent, superior to dishes prepared from the ordinary *pacayas.* In Baja Verapaz, Guatemala, we observed that *bojon* is gathered from *C. quezalteca* and *C. woodsoniana.* The two are known locally as *chicuilote* and *ternera,* the names being used for both species.

Although on a minor scale, local inhabitants throughout its range employ various *Chamaedorea* in everyday life for a number of other uses, including medicine, shelter (thatching), and ornament. Interested readers should consult Balick and Beck (1990) for an excellent, extensive list and summary of references dealing with economic uses of palms, including *Chamaedorea.*

CHAPTER 8

CONSERVATION

It should come as no surprise that since chamaedoreas can essentially survive and reproduce only in undisturbed, primary forest, their fate is inextricably linked to that of the forest. The tropical forests are disappearing at an alarming rate. Understory palms like *Chamaedorea* are excellent indicators of the health of a tropical rain forest. Vigorous, reproducing populations of *Chamaedorea* are a good indicator that the surrounding forest is probably little disturbed, if at all.

Most chamaedoreas, like other understory palms, have two distinct disadvantages that render them particularly susceptible to the deleterious effects of deforestation. First, they require rather precise conditions for seed germination and seedling establishment. Even if chamaedoreas are left when the forest is cleared, they fail to survive let alone regenerate since moisture and shade requirements are not met in forest regrowth. Second, as mentioned earlier, most are single-stemmed and when the one growing point is destroyed or severely damaged, the palm is for all intents and purposes dead. Those with multiple stems are able to resprout but they are in a minority.

The greatest chronic threat, then, to the survival of *Chamaedorea* is destruction of forests and loss of habitat. The reasons why forest is destroyed and habitat lost are many and complex, just as are the solutions. Economic, political, and social factors all play a role. It is increasingly evident that the problem of destruction of tropical rain forests and the inherent ramifications for the future health of the planet are not simply those of the developing nations. Rather, since the developed nations are at least partly responsible because they provide the insatiable appetite for natural resources and products derived from the forest, they, too, are part of the problem and must work toward a solution. It will take a concerted, cooperative, global effort to turn the tide against destruction of tropical forests and ensure the survival of *Chamaedorea* and other forms of the tropical biota so essential to our own survival.

It needs to be emphasized that species of *Chamaedorea* must be protected. Some people would contend that maintaining species in cultivation is an adequate substitute for their protection in the wild, a theory that is somewhat shortsighted since genetic diversity of cultivated species cannot be maintained and is bound to diminish in time (Frankel 1974). Declining genetic diversity is especially acute when the original collections were made from only a few individuals of one population, natural pollinators are absent, and space limitations preclude planting the large quantities of individuals needed to maintain an adequate gene pool. Also, because of rather exacting growth requirements, many chamaedoreas cannot be cultivated successfully.

As Chazdon (1988) pointed out, genetic diversity is the evolutionary and adaptive potential within all species and is driven by the complex ecological interactions present only under natural conditions. The only way to preserve *Chamaedorea* is to preserve its habitat. Plants cultivated in botanical gardens and private collections are at best living museum pieces that do little or nothing to enhance the long-term survival of the species.

Although loss of habitat is the greatest threat to the survival of *Chamaedorea,* another more acute threat is overcollecting by acquisitive and zealous hobbyists, many with commercial interests, who do not hesitate to strip seeds and plants from the forest in great numbers. Populations of several highly ornamental species, such as *C. tuerckheimii* in Mexico and Guatemala, *C. amabilis* and *C. sullivaniorum* in Panama (Hodel 1988a), and *C. tenella* in Mexico, have declined due to overcollecting. Collecting for personal gain or profit is the least defensible of the causes leading to destruction of plants in the wild.

Collectors and cultivated collections of *Chamaedorea* can play a useful role in ensuring the survival of the genus, however. Chazdon (1988) noted that cultivated palms can be extremely valuable for educational and scientific purposes, as our use of *Chamaedorea* here in Los Angeles has demonstrated. Also, cultivated plants have touristic, esthetic, and commercial uses not to be overlooked. Once a species is in cultivation, botanical gardens can play a major role by propagating, distributing, and introducing or encouraging its wide use in the trade, lessening pressure for seed or plant collecting from wild populations. Even commercial nurseries and growers can help since they know how to propagate and distribute plants most successfully. Commercial growers in the developed nations can play a major role in the conservation effort by working with and assisting growers in the developing nations to establish in cultivation mother blocks of the horticulturally desirable species. Establishment of cultivated mother blocks is essential so that wild populations will not serve as the source of propagative material since they are unable to tolerate repeated intrusions for extraction of seeds and plants. By encouraging the use of cultivated mother blocks for propagative purposes, commercial growers can relieve the pressure on wild populations and at the same time contribute to the local economy.

Chazdon (1988) made the valid point that without major conservation efforts, palm hunting, an activity that palm enthusiasts find so rewarding, may become a thing of the past. Chazdon outlined ways in which collecting, growing, selling, and buying can be conservation-oriented. Suggestions include:

1. contributing to conservation of palm habitats through international organizations such as the Nature Conservancy International Program or the World Wildlife Fund;
2. becoming knowledgeable about your favorite palms and being an outspoken advocate of their protection;
3. visiting palms in their natural habitats and, rather than collecting seeds, taking photographs so that upon your return you can give talks and show pictures to build support for conservation;
4. having a palm auction to raise funds to contribute to conservation efforts in developing countries which often have insufficient resources for such activities.

Collectors should be careful to buy and sell only plants known to come from cultivated sources. They should refuse to patronize those individuals known for their unethical collecting habits.

Threatened Status

While it was generally recognized by the early 1970s that the tropical forests were in imminent danger, Moore (1977) first specifically discussed chamaedoreas that may be endangered or threatened. He listed as very rare and presumably endangered *C. stolonifera* from Chiapas, Mexico and *C. tuerckheimii* from Guatemala.

The threatened status of various chamaedoreas is presented in Table 8-1. In developing this information, I have consulted Johnson (1987), Dransfield *et al.* (1988), and Bernal (1989) and relied on my own observations in the field. The listings are reflective of the taxonomic and nomenclatural changes presented in this book and pertain to each species over its entire range unless otherwise noted. I have established threatened categories that more or less coincide with those in other accepted and official accounts.

Extinct species are no longer known to exist in the wild after repeated searches of their type localities and other known or likely places. **Endangered** species are unlikely to survive in the wild if causal factors continue to operate. **Vulnerable** species are likely to move into the endangered category if causal factors continue to operate. **Rare** species have small world populations that are not at present endangered or vulnerable but are at risk. **Indeterminate** species are in one of these four categories: extinct, endangered, vulnerable, or rare; information is insufficient, however, to place them precisely. **Insufficiently known** species are suspected, but not definitely known due to lack of information, to belong to any of the above categories. See the treatment of each species for further information. Species not listed in any of the above categories are probably not threatened throughout their range although they may be threatened on a local basis.

Dransfield *et al.* (1988) listed two species as not threatened, *C. integrifolia* (= *C. pauciflora*) and *C. pinnatifrons,* although they may be in a threatened category on a local basis.

Recently, some developing nations have begun to view their tropical rain forests and their components, *Chamaedorea* included, as sustainable and renewable sources of products. They have initiated studies on the ecology, management, economics, and other factors necessary to utilize a natural resource in a conservation-conscious manner without destroying that resource or significantly damaging its surrounding ecosystem. Local residents are encouraged to subscribe to this philosophy since it promises diversified, long-term, perhaps indefinite, economic possibilities, a welcomed trend away from that of viewing the forest as a nonrenewable resource, once exploited then abandoned, never regaining its species diversity and potential productivity.

Heinzman and Reining (1988) pointed out the possibility of successful management of *C. elegans* and *C. oblongata* for cut foliages in Guatemala on a sustainable and renewable basis. Their study indicates that the forest has a relatively high net value, one competitive with other land uses that convert

Table 8-1. **The Conservation Status of *Chamaedorea*.**

Endangered

C. amabilis H. A. Wendl. ex Dammer
C. brachyclada H. A. Wendl.
C. brachypoda Standl. & Steyerm.
C. correae Hodel & Uhl
C. fragrans (Ruiz & Pav.) Mart.
C. glaucifolia H. A. Wendl.
C. klotzschiana H. A. Wendl.
C. metallica O. F. Cook ex H. E. Moore
C. minima Hodel (possibly extinct)
C. pumila H. A. Wendl. ex Dammer
C. rigida H. A. Wendl. ex Dammer
C. simplex Burret
C. stolonifera H. A. Wendl. ex Hook. f.
C. stricta Standl. & Steyerm.
C. sullivaniorum Hodel & Uhl
C. tenella H. A. Wendl.
C. tenerrima Burret
C. tuerckheimii (Dammer) Burret
C. verecunda Grayum & Hodel

Vulnerable

C. adscendens (Dammer) Burret
C. cataractarum Mart.
C. geonomiformis H. A. Wendl.
C. graminifolia H. A. Wendl.
C. hooperiana Hodel
C. lucidifrons L. H. Bailey
C. microphylla H. A. Wendl.
C. microspadix Burret
C. oreophila Mart.
C. pachecoana Standl. & Steyerm.
C. pedunculata Hodel & Uhl
C. pygmaea H. A. Wendl.
C. radicalis Mart.
C. seifrizii Burret
C. selvae Hodel
C. serpens Hodel
C. smithii A. Gentry

Rare

C. alternans H. A. Wendl.
C. castillo-montii Hodel
C. chazdoniae Hodel
C. falcifera H. E. Moore
C. fractiflexa Hodel & Castillo
C. guntheriana Hodel & Uhl
C. murriensis Galeano
C. nationsiana Hodel & Castillo
C. verapazensis Hodel & Castillo

Indeterminate

C. angustisecta Burret
C. carchensis Standl. & Steyerm.
C. deneversiana Grayum & Hodel
C. foveata Hodel
C. latisecta (H. E. Moore) A. Gentry
C. lehmannii Burret
C. palmeriana Hodel & Uhl
C. parvifolia Burret
C. pittieri L. H. Bailey
C. queroana Hodel
C. rhizomatosa Hodel
C. robertii Hodel & Uhl
C. rojasiana Standl. & Steyerm.
C. sartorii Liebm.
C. scheryi L. H. Bailey
C. schiedeana Mart.
C. skutchii Standl. & Steyerm.
C. stenocarpa Standl. & Steyerm.
C. undulatifolia Hodel & Uhl
C. volcanensis Hodel & Castillo
C. vulgata Standl. & Steyerm.
C. whitelockiana Hodel & Uhl
C. zamorae Hodel

Insufficiently Known

C. allenii L. H. Bailey
C. arenbergiana H. A. Wendl.
C. crucensis Hodel
C. ernesti-augustii H. A. Wendl.
C. ferruginea H. E. Moore
C. liebmannii Mart.
C. matae Hodel
C. nubium Standl. & Steyerm.
C. paradoxa H. A. Wendl.
C. parvisecta Burret

forest. The traditional slash and burn practices may be both economically inefficient and ecologically nonsustainable. The possibility of the forest remaining intact and, with proper management, being a virtually never-ending supply of natural products and resources to supplement the economic base of the local community is attractive and would certainly reduce pressure to convert it for other purposes.

CHAPTER 9

CULTURE

With few exceptions, chamaedoreas are among the easiest palms to grow. The principles and practices of their successful culture are essentially no different from those of other palms or, for that matter, of many other tropical or subtropical plants. Adherence to basic horticultural practices and an understanding of the cultural requirements of *Chamaedorea* should, for the most part, make them easy and problem-free subjects to grow successfully.

Growers and collectors can consult many sources of information about the culture of palms. Joiner (1981) is an excellent reference on all aspects of the culture of foliage plants; nearly all his recommendations are applicable to chamaedoreas. Similarly, Burch et al. (1983) provided a guide to the commercial production of palms, including many chamaedoreas. *Principes,* the journal of the International Palm Society, has many articles containing information about the culture of palms that is applicable to chamaedoreas. Among these are discussions by Braun (1968), Doughety (1988), Read (1962), and Wagner (1982). Also, some of the more recent books on palms, including those by Blombery and Rodd (1982), Braun (1988), Hoyos and Braun (1984), Muirhead (1961), Jones (1984), Stevenson (1974), and Stewart (1981), have excellent sections on the culture of palms although some of the fertilizer recommendations are outdated.

Difficult-to-Cultivate Chamaedoreas

A few chamaedoreas are notoriously difficult to grow well and never appear as vigorous as their wild counterparts. Unfortunately, some of our choicest species fall into this category. Especially difficult to grow well are members of the pinnatifid group outlined by Hodel and Uhl (1990b), including *C. undulatifolia* and *C. stenocarpa,* that in the wild are restricted to montane rain forests with special environmental parameters. They seem to do best in a moderate or slightly cool tropical climate with little variation, conditions difficult to duplicate in cultivation. Even in places renowned for a benign climate, such as wet areas of Hawaii, the plants do not attain the same quality as in the wet montane forests of Costa Rica and Panama. In places like California, they fare even worse, tending to produce very few leaves marred by pinnae with yellow and brown tips. The "tip-burn" so characteristic of the pinnatifid group in cultivation is probably due to low atmospheric humidity and, to a certain extent, extremes of temperature.

Generally, in cultivation these plants have much reduced crowns composed of only two or three leaves, and take on a rather poor appearance. Also, they are susceptible to infestations of mites and thrips in cultivation, especially in areas of low humidity. Naturally slow-growing, they are even slower in cultivation and, more often, languish and go into a slow decline from which they seldom recover.

Growers of species in the pinnatifid group may have the best results by using a well drained medium high in organic matter and a slow-release, organic type of fertilizer, situating the plants in deep shade, and, if possible, maintaining constant temperatures between 15° and 27°C (60–80°F) and relative humidities above 80 percent. Also, periodic leaching of the root zone with distilled or rain water would be beneficial in areas where water quality is poor and/or mineral content is high.

Soils

Perhaps no other aspect plays such a wide role in the successful culture of chamaedoreas as does soil. Yet at the same time, it is one of the most misunderstood subjects and is riddled with numerous misconceptions and falsehoods.

Soils vary greatly: some are ideal garden soils; others may have physical and/or chemical limitations—too shallow, stony, alkaline, poorly drained, steep, saline, or with toxic levels of certain elements. However, chamaedoreas can be grown in nearly all soils if one is familiar with their cultural requirements and knows how to manage soil properly.

Composition of Soils

Soils can be of virtually any material as long as they provide these four basic functions for plant growth:

1. mechanical support;
2. a reservoir of water and air for plant roots;
3. a source and reservoir of plant nutrients; and

4. freedom from disease, damaging insects, and materials toxic to plant growth.

Soil is not essential for plant growth as can be seen in hydroponic (water) culture where rods and wires provide support and anchorage, and aerated water around the roots supplies oxygen. The nutrient solution in the water provides the elements essential to plant growth.

Chamaedoreas are grown in two phases:

1. the propagation and container stage, where plants are grown in containers; and
2. the landscape stage, where plants are planted out in a more or less permanent position in the landscape.

Although the objective in the first stage is to produce healthy plants for planting out, many chamaedoreas are container-grown for their entire lives.

The soils used in containers, or perhaps a better term for them is potting mixtures, have to be different than soils in the landscape because of physical limitations imposed by containers.

Potting Mixtures

Aeration is the major consideration in potting mixtures since they have an artificially high water table compared to landscape soils and do not drain as well. The container bottom acts as the water table and inhibits or reduces free-water drainage. The result is a potting mixture that is wetter in containers than the same mixture would be in the field. A way to visualize this is to imagine a sponge soaked in water lying flat and horizontally. Not much water would drain from it in this position. Now imagine the same sponge placed standing upright on end. Much more water would drain from it now, especially from the upper portion.

Another consideration is the reduced volume of soil from which the plants must obtain nutrients and water. Good potting mixtures must retain water and nutrients but at the same time be well drained and aerated. These may seem to be contradictory requirements but both can be achieved through careful selection of ingredients for potting mixtures.

Ingredients for Potting Mixtures

Little or no real soil is used in potting mixtures simply because its physical properties, as pointed out above, are inadequate for use in a container. Mixtures for containers are often referred to as artificial soils or soilless media since they contain no real or landscape soils.

Ingredients for potting mixtures must have physical and chemical properties that provide the following:

1. pore space for drainage and aeration;
2. high water-holding potential;
3. high nutrient-holding potential;
4. slightly acid pH; and
5. resistance to rapid degradation.

There is no single ingredient which meets all these requirements. For example, pure sand provides excellent aeration and drainage but its water- and nutrient-holding capacities are very limited. Peat moss easily provides these latter two requirements but lacks some of the pore space for drainage and is often too acid. By utilizing two or more components, mixtures can be devised that meet all the requirements for container-growing of chamaedoreas.

Some commonly used ingredients include peat moss, ground bark, perlite, and sand. Many other materials are also used depending on local availability and price. For example, in Hawaii volcanic cinders are often used in place of perlite to provide aeration in potting mixtures. There are as many different successful formulations of potting mixtures as there are successful growers. University researchers have devised many, and growers have added their own personal variations, but all have the same basic principles and meet the requirements outlined earlier. Baker (1957), DeWerth and Odon (1960), Mott (1967, 1971a, 1971b), Voss and Watson (1968), Boodley (1972), Poole and Waters (1972), Criley (1973), Boodley and Sheldrake (1977), and Joiner (1981) are examples of the commonly used mixtures. Keep in mind that while the ratios of the physical components are satisfactory, the fertilizer recommendations are outdated. Some of these mixtures are listed below.

University of California Soil Mix C (Baker 1957)

peat moss	1 part
fine sand	1 part

Cornell Foliage Plant Mix (Boodley 1972)

peat moss	2 parts
vermiculite	1 part
perlite	1 part

University of Florida (Poole and Waters 1973)

peat moss	2 parts
bark	2 parts
shavings	1 part

A palm collector in Portland, Oregon (Wallis 1974)

mica-peat	2 parts
perlite	2 parts
vermiculite	1 part

As one can see, there are many different variations of potting mixtures. Some are more complicated than others and require more ingredients; however, all can be used successfully. There are many commercially prepared and packaged mixes on the market that are also suitable for growing chamaedoreas although they may be more expensive than making your own. The following mixture has performed well in our research greenhouses:

peat moss	2 parts
fir bark	1 part
perlite	3 parts

To each cubic meter of mixture selected, the grower should thoroughly incorporate 4.2–6 kg (7–10 lbs/cubic yard) of dolomite and 0.3–0.9 kg (0.5–1.5 lb/cubic yard) of a complete micronutrient blend. Dolomite supplies magnesium (a nutrient needed in relatively large amounts by palms and difficult to provide by other means) and calcium as well as adjusts pH to proper levels. Incorporation of superphosphate to potting mixtures was a common recommendation and practice in the past although there is recent evidence that it may have harmful effects on growth of plants, particularly palms. Broschat and Donselman (1985) and Conover and Poole (1990) discouraged the use of superphosphate due to subsequent tie-ups of micronutrients, especially manganese. Broschat and Donselman (1985) also noted that the use of composted manures and sewage sludges can have a similar effect. Nitrogen, phosphorus, and potassium are best supplied at planting time or supplementally later, and their application is discussed in the section on nutrition.

Heat-Treating Potting Mixtures

It has been a standard horticultural practice to pasteurize (heat-treat) potting mixtures prior to their use. However, recent evidence (Broschat, pers. comm.) suggests that pasteurization of potting mixtures may not be necessary or, in some cases, even desirable. For the grower concerned with propagation of seeds and growth of rare species, though, pasteurization may be beneficial. Pasteurization will eliminate potential disease organisms that can devastate seeds and seedlings (see section on Sanitation). Pasteurization involves steam heating the soil to 80°C (175°F) for 30 minutes. Pasteurize small amounts in the home oven by placing a wooden or metal container of moist potting mixture on a rack in a large cooking pot or roasting pan. Put a layer of water in the bottom of the pan but not higher than the rack. Cover the cooking pot or pan and heat in the oven or on the range so steam is produced. Pasteurize very moist soil by wrapping it in aluminum foil and baking in the oven at 120–150°C (250–300°F) to produce active steam.

In both cases monitor soil temperature with a thermometer and maintain it at the critical level of 80°C (175°F) for 30 minutes. Put treated potting mixture into clean containers only (see section on Sanitation).

Handling Potted Plants

The culture of plants in containers is slightly different from those in the ground. Since one is dealing with restricted, finite volumes of soil, one must modify fertilizer and irrigation practices appropriately. Some of these modifications have already been alluded to and they are discussed in more detail in the sections on nutrition and irrigation.

The major operation in growing chamaedoreas in containers is repotting. Shifting of seedlings or small plants to larger containers should be done only when needed. When a firm and extensive system of roots has developed at the bottom of a container, the plant may then be repotted. In selecting a new container, choose one that is the next logical size larger. For example, plant a germinated seed in a 55-mm (2¼-inch) liner, then move it to a 10-cm (4-inch) pot, then a 15-cm (1-gallon) pot, and so on. Avoid skipping sizes, for instance, moving from a liner to a 30-cm (2-gallon) pot. Shifting plants prematurely or putting plants in too large containers can lead to overwatering and subsequent root and nutritional problems.

Landscape Soils

That chamaedoreas will grow in the rocky, limestone soils of southern Florida and the heavy clays of southern California attests to their relative tolerance of various conditions. As long as fertilizers and water are managed properly, chamaedoreas will grow in just about any soil.

Soil Texture

Soil has four basic ingredients: decomposed rock, organic matter, air, and water. These ingredients vary in abundance from soil type to soil type. Soil particles range in size from coarse, as in sand, to fine, as in silt, and to very fine, as in clay. The proportion of these particles determines the texture of the soil, whether it be sand, silt, or clay. Weight in relative terms is another way to express texture. A sandy soil is considered light, a clay one heavy.

Sandy soil is loose and well aerated; it is coarse and grainy to the touch. Water drains quickly and, as a result, it dries out rapidly. Silty soil feels smooth and slippery when wet, and like flour when dry. Clay soils feel sticky when wet and form a solid mass when dry. Silty and clay soils hold water well but are poorly aerated. Sandy loam, a mixture of all three particle sizes but with more sand, is considered an excellent soil for the cultivation of most kinds of plants. It remains crumbly, holds enough water, and is well aerated.

Soil Structure

Whereas texture refers to the relative size of the particles, structure refers to their arrangement. Sand, silt, and clay seldom occur as separate units in the soil; rather, they combine into aggregates. The size and form of these aggregates is known as the structure of the soil. Structure is an important physical characteristic of the soil. Soil structure, as in texture, affects movement of water, air, and roots.

A granular structure provides an ideal environment for plant roots. The larger pores between the granular aggregates are continuous and roots, water, and air readily penetrate them. Although water drains readily from this soil, sufficient moisture is held within the granules to supply the needs of the roots. Many chamaedoreas grow on heavy, clay soils in their native habitat. However, these soils have excellent structure and are able to withstand high rainfall without becoming water-logged and having a negative effect on the plants. Clay particles and organic matter tend to enhance granular structure in soils.

Soil pH

Soils are acid, alkaline, or neutral. The term pH expresses the amount of acidity or alkalinity. The pH scale ranges from 0–14. Acid soils are less than seven and alkaline are more than seven. Neutral is seven. The relative acidity or alkalinity of a soil is important since it determines, to a great extent, the availability of plant nutrients. Most nutrients are readily available at a pH of about six; not surprisingly, chamaedoreas grow best in soils at this pH. If the pH is not in this range, it can be modified with sulfur to increase the acidity (lower pH) or with dolomite to increase the alkalinity (raise pH).

Amendments

Soil amendments are materials used to improve the physical condition of the soil. By adding the proper amendments, one can improve the soil structure and, in turn, improve drainage, aeration, and the water- and nutrient-holding capacity of the soil. In clay soils amendments are used to help break up the soil, resulting in increased aeration, better drainage, and improved root penetration. In sandy soils they may increase the water- and nutrient-holding capacity.

Organic amendments have plant or animal origins. Some examples are compost, leaf mold, peat moss, and fir bark. As mentioned earlier, composted manures and sewage sludges are not recommended due to their tie-up of manganese. To be effective in altering the physical structure of the soil, organic matter amendments must be added in large quantities, at least 50 percent by volume. Also, in order to maintain the desired benefit, organic amendments must be replenished on a continual basis.

Another way in which organic amendments are beneficial is when used as a mulch. Schick (1980) outlined the benefits of mulches in growing palms in southern Florida. Mulches are used to inhibit weed growth, moderate soil temperatures, and keep the soil surface moist and loose. Mulches or other organic matter tend to facilitate penetration of adventitious roots that often appear at or slightly above the base of the stem of chamaedoreas and make their way toward the soil surface. Mulches will also serve as a reservoir of organic material that will enhance soil fertility and structure as they break down. They will also help reduce compaction and soil erosion due to irrigation water or rain and can help to lower soil pH.

The benefits that can be obtained from the use of organic amendments are not surprising when one visualizes how chamaedoreas grow in the wild. The forest floor is usually covered with a layer of decaying organic matter that is replenished from the forest canopy at a fairly constant rate. Even where chamaedoreas occur on limestone rock outcroppings there is a layer of organic matter. This is an ideal natural mulch and the artificial ones we develop around our cultivated plants perform the same functions.

Table 9-1. **The 16 Elements Essential to Plant Growth and their Sources.**

Source		
Air and water	Soil	
	macronutrients	
carbon hydrogen oxygen	nitrogen phosphorus sulfur	magnesium potassium calcium
	micronutrients	
	copper iron manganese boron	zinc chlorine molybdenum

Fertilizers and Nutrition

In addition to the use of organic matter in dealing with soils, the proper management of fertilizers is also important in the culture of chamaedoreas.

Researchers in Florida, including Conover and Poole (1981, 1990), Burch et al. (1983), Broschat (1984, 1990), Broschat and Meerow (1990), and Broschat and Donselman (1985), have done most of the work on fertilizers and nutritional deficiencies of ornamental palms.

Chamaedoreas, like all other plants, depend on nutrients contained in the soil to grow, flower, and produce fruits. There are 16 elements known to be required for plant growth and development. They are listed in Table 9-1.

The six macronutrients and carbon, hydrogen, and oxygen are needed in relatively large quantities by plants. Carbon, hydrogen, and oxygen are derived from the air and water. The six macronutrients come from the soil. The seven micronutrients also come from the soil and are no less essential but are needed in much smaller quantities. In fact, they can be toxic when oversupplied.

Causes of Nutrient Deficiencies

There are several factors to consider in determining the cause of a nutritional deficiency. In most situations, elements exist in sufficient quantities in the soil for proper growth and it is more likely that the deficiency is due to external factors, such as chemical interactions, root problems, temperature, and nutrient imbalances that render the element unavailable to the plant or limit the ability of the plant to take up the element. However, in some cases, macronutrients, especially nitrogen, potassium, and magnesium, may simply be present in insufficient quantities. Adherence to sound cultural practices and a complete fertilizer program are essential to avoid nutrient deficiencies and attain optimal growth of chamaedoreas. For example, if a grower neglects to add dolomite to the potting mixture because the pH is at an acceptable level, deficiencies of magnesium will result. Nitrogen, potassium, and magnesium are readily leached from sandy or coral soils, such as those in much of Florida; in these cases, a complete fertilizer program is necessary.

Chemical Interactions

Chemical interactions affect the solubility of elements and their subsequent availability to plants. Organic soils tend to fix copper. Alkaline or high pH decreases the solubility of zinc, iron, copper, and especially manganese, tying them up and rendering them unavailable to the plant. It is a common occurrence in the alkaline, coral-derived soils of southern Florida and, to a lesser extent, with some of the heavy clay soils of California. Generally, the best solution is applications of sulfur, incorporation of organic materials, and the use of mulches. Soluble forms of phosphate fertilizers can tie up iron, manganese, and zinc.

Root Problems

Root problems due to poor aeration, disease, mechanical damage, temperature extremes, and improper planting inevitably lead to nutritional disorders since damaged or low-respiring roots are unable to take up essential elements. In seedlings and young plants, root diseases caused by the fungi *Pythium, Phytophthora,* and *Rhizoctonia* damage root surfaces, directly leading to decreased uptake of nutrients. Also, damaged roots will not take up as much water, resulting in oxygen-poor and/or waterlogged soil that limits the ability of the roots to take up micronutrients, especially iron and manganese. Planting too deeply results in the same problems as oxygen-poor or waterlogged soils. Anytime root respiration is reduced, root activity decreases, leading to decreased uptake of iron and manganese and deficiencies of these elements.

Mechanical injury to roots caused by improper handling at potting, repotting, planting, or transplanting will also lead to nutrient deficiencies. Some chamaedoreas are hosts to root nematodes, especially the burrowing nematode. As the nematodes feed and damage roots, the first symptoms and, in some cases, the only symptoms that appear aboveground are those of iron and manganese deficiencies. Cool temperatures limit root activity also, resulting in decreased uptake of water and nutrients. Usually cold-induced deficiencies appear in the late winter or early spring as active growth resumes and root activity is low relative to that of the foliage. The solution is to fertilize year-round and especially in the fall before growth begins to slow down.

Nutrient Imbalances

Sometimes too much of one nutrient in relation to another can cause problems. For example, if potassium is provided in a soluble form but nitrogen is supplied by a slow-release material, more potassium than nitrogen will be leached from the soil and an imbalance between the two elements will occur. The result will be potassium deficiency. On the other hand, excessive potassium (for example, from regular applications of a potassium fertilizer but no magnesium) may lead to a deficiency in magnesium. The solution here is to treat for potassium and magnesium deficiencies simultaneously.

Insufficient Nutrients

Soils vary greatly and sometimes fail to supply in sufficient quantity all the nutrients required for the best growth of chamaedoreas. Generally, soils contain most of the elements known to be essential to plants, so one only needs to add those that are deficient in the area. For example, nitrogen is naturally low in most soils in California and Florida and additional amounts of this element are needed for optimal growth of chamaedoreas. Also, the sandy, calcareous soils of Florida are usually very deficient in potassium, magnesium, and manganese. Macronutrient deficiencies are often due to inadequate levels of the element in the soil but, as noted above, it is not so for micronutrients.

Age and Establishment

A newly planted or transplanted chamaedorea will require a substantial amount of time to establish itself in the landscape. In warmer climates, as in Florida or Hawaii, the establishment period may be as little as a year while in cooler areas, like California, it may take up to two years. During the establishment period the plant requires special care while the roots develop to the point where they have moved out of the original ball and into the surrounding soil and begin taking up nutrients. If chamaedoreas are improperly planted (see section on Planting) or poorly cared for after planting, then problems with nutrient deficiencies can arise.

Soil Type

Naturally, sandy soils are going to hold fewer nutrients than those with adequate clay or organic matter. Environmental factors, such as high rainfall and warm temperatures, will tend to exacerbate the problem. On the other hand, as noted earlier, heavy, poorly drained soils with excessive irrigation or rainfall can contribute to root problems.

Most Common Nutrient Deficiencies in Chamaedoreas

The most commonly deficient nutrients are listed below. Symptoms of deficiencies on palms may be difficult to diagnose since they are often different from those on other plants and symptoms of several deficiencies may overlap. Symptoms are the same whether the plant is container-grown or in the landscape although the causes may be different. Chase and Broschat (1991) provided an excellent reference of nutrient deficiency symptoms in palms. They included color photographs, many of *Chamaedorea.*

Nitrogen

An overall and uniform light green color and a decrease in growth characterize nitrogen deficiency. It is relatively easy to correct with applications of nitrogen fertilizers. Nitrogen deficiencies may be fairly common in the landscape, especially on light or sandy soils, but deficiencies of other elements, such as potassium, magnesium, and manganese, can be much more prevalent and severe. Nitrogen is the most commonly deficient nutrient in container-grown palms.

Potassium

Recent research has revealed the important role of potassium in a well balanced nutrition program for palms. Palms need potassium in amounts equivalent to that of nitrogen. Broschat (1990) reported the minimum potassium concentration of 1.2 percent for recently matured leaves of *C. seifrizii.* Yellowish or orange flecking, mottled or discolored pinnae, and tip and/or marginal necrosis characterize potassium deficiency on chamaedoreas. Symptoms appear on the oldest leaves first but newer leaves will be affected as the disorder becomes more severe. Affected leaves will not recover; the palm must produce a crown of new, healthy leaves.

Potassium is highly soluble and readily leached from light, sandy soils like those in southern Florida. In fact, potassium deficiency is the most widespread and severe disorder of palms in southern Florida. It is also a common disorder of palms on the young, porous, volcanic soils in some parts of Hawaii. The use of fertilizers with slow-release nitrogen but water-soluble potassium accentuates potassium deficiency on sandy soils. While the nitrogen supply will last for several months or more, heavy rain or one or two irrigations will leach the potassium completely from the soil.

Potting mixtures with their high nutrient-holding capacity and heavier landscape soils can retain potassium against leaching, hence potassium deficiency is less common in container-grown palms and in areas with heavier soil. Potassium deficiency in these cases is usually due to low potassium content of the fertilizer. Potassium deficiency is easy to correct and palms respond quickly to treatment. Applications of slow-release forms of potassium four times per year plus half as much magnesium sulfate to prevent a potassium-magnesium imbalance and resulting magnesium deficiency are usually sufficient.

Magnesium

Palms are heavy users of magnesium and deficiencies are frequently seen in palms on sandy soils. As in potassium, symptoms of magnesium deficiency appear on oldest leaves first and progress up the canopy. A broad yellow band around the margin of an otherwise green leaf characterizes magnesium deficiency. Affected leaves will not recover and the plant must replace them with new, healthy ones. Treat palms preventatively for magnesium deficiency since affected plants are slow to respond to post-symptom treatment. The use of sufficient quantities of high grade dolomite in the potting mixture and soil applications of magnesium sulfate or oxide plus slow-release potassium four times a year to palms in the landscape should prevent magnesium deficiency.

Manganese

Symptoms of manganese deficiency appear on the newest leaves first. New leaves are chlorotic, weak, reduced in size, and/or have extensive necrotic streaking. As the deficiency progresses, new leaves emerge withered, frizzled, or "scorched," finally appearing only as necrotic petiole stubs. Manganese deficiency is primarily due to the element's insolubility at high pH, making it unavailable to plants. Due to the high pH of alkaline soils, manganese is one of the most frequently deficient elements of palms in southern Florida and, to a lesser extent, in southern California. In Florida the deficiency is called "frizzletop" because of the frizzled look of the new leaves. Adjusting soil pH, applying monthly foliar sprays of 5 ml of manganese sulfate per 3.8 liters of water (1 teaspoon/gallon), or applying manganese sulfate to the soil 2–3 times a year should correct the deficiency and prevent its recurrence.

Manganese is usually not deficient on container-grown palms since the element is more soluble at the lower pHs of potting mixtures. Cool soil temperatures during winter that result in reduced root activity and uptake of micronutrients may also cause manganese deficiency, particularly in landscape soils where the element is already marginally deficient. Frizzletop will occur as air temperatures rise in spring and growth of foliage commences yet roots are still inactive in the relatively cool soil. Soil temperatures tend to lag behind those of the air, remaining cooler in the spring and warmer in the fall. The affected palm will usually grow out of the problem when soil temperatures finally rise and root activity increases.

Iron

Symptoms of iron deficiency, as in manganese, appear first on the newest leaves. Intervenal chlorosis, tip necrosis, and stunted new leaves characterize iron deficiency. Iron deficiency in palms is not due to a lack of iron in the soil, and rarely is it due to high pH, as in many other plants. Iron is nearly always present in the soil in sufficient quantities; it is poor aeration that causes deficiencies of this element. Waterlogged and poorly aerated soils, or palms planted too deeply are the primary causes of iron deficiency. Foliar sprays of 2.5 ml iron sulfate or chelate per 3.8 liters of water (½ teaspoon/gallon) may offer temporary correction; poor aeration or improper planting depth must be corrected or the problem will continue to recur. Preventative treatment by planting and repotting at the correct depth in a well drained soil or potting mixture is the only permanent solution for iron deficiency. Also, dibble, rather than surface apply, fertilizers to containers and use a potting mixture that is resistant to rapid breakdown.

Sulfur

Symptoms of sulfur deficiency are similar to those of iron. Sulfur deficiency is uncommon in palms and would probably only occur in container-grown plants when sulfate fertilizers are not used.

Deficiencies of phosphorus, calcium, copper, zinc, boron, and chlorine rarely occur if one adds dolomite and micronutrients to potting mixtures and uses a complete fertilizer.

Diagnosis and Treatment

Treatment of nutritional deficiencies is a last resort; prevention is superior to correction. Practice proper fertilization, maintain good drainage and aeration, implement sound horticultural practices such as proper planting depth and irrigation, and maintain proper pH. Macronutrient deficiencies usually can be treated by simply supplying the needed elements to the soil. As noted earlier, micronutrient deficiencies are usually traced to a cultural or environmental problem. To correct micronutrient deficiencies in most cases, the grower must alleviate the cultural or environmental cause prior to corrective application of fertilizers. Failure to do so will result in only temporary and incomplete control of the deficiency.

Remember that many micronutrient deficiencies occur because of damaged or inadequate roots. Soil applications of fertilizers may, therefore, be very slow to correct the deficiency. In these cases, it may be more advantageous to apply a foliar spray of micronutrients along with a soil application until complete recovery is achieved.

Chamaedorea Nutrition Guide

Chamaedoreas, like other palms, will suffer quickly due to insufficient or incorrect fertilization. A good fertilizer program is essential to establish a palm rapidly in containers after repotting or in the ground after planting.

There are two general categories of fertilizers. Organic fertilizers come from plant or animal sources. Most contain only moderate amounts of nutrients that are released slowly and continuously over a period of time through activity of microorganisms. Organic fertilizers improve soil structure and feed microorganisms as they decay. Examples of organic fertilizers are blood meal, bone meal, and fish meal.

Inorganic fertilizers come from chemical or mineral sources. Many are very soluble in water, making the nutrients rapidly available to plants. They are very concentrated and many people have a tendency to apply too much since only small amounts are needed. They do little to improve soil structure. Some examples of inorganic fertilizers are potassium sulfate, potassium nitrate, and ammonium sulfate.

Some inorganic fertilizers are sulfur- or resin-coated to release their nutrients continuously and evenly over extended periods of time just as organic fertilizers do through activity of microorganisms. Slow-release forms of inorganic fertilizers are formulated with specific release periods, such as 3, 6, 9, or 12 or more months at soil temperatures of 22°C (70°F). Higher soil temperatures speed up the release of nutrients and lower ones slow it down.

Slow-release forms of nutrients, either from organic sources or chemically treated inorganic forms, are best because they provide a steady stream of nutrients over a longer period rather than short bursts of fluctuating dosages. Palms seem to respond best to a steady, constant supply of nutrients, and it is no wonder since this is how they are provided in the wild from the decay of organic matter in the forest.

All packaged fertilizers show relative percentages of major nutrients in a fixed sequence. A 10-10-10 fertilizer would be 10 percent nitrogen, 10 percent phosphorus oxide, and 10 percent potassium oxide. The best ratio for palms, including chamaedoreas, is 3-1-3 (for example a 15-5-15 or 12-4-12) of nitrogen, phosphorus, and potassium. It is preferable that the fertilizer have nitrogen and potassium in the slow-release form to prevent potassium deficiency and subsequent problems related to nutrient imbalances.

Container Fertilizer Program

It is best to fertilize container-grown plants at the time of potting or repotting since this is an opportunity to place the fertilizer where it will do the most good. Dibble or place the recommended dosage of a slow-release fertilizer of 3-1-3 ratio in a little mix at the bottom of the container and then plant the palm. The fertilizer is then where the roots can come into immediate contact with it. Improved growth offsets the extra time and labor needed to place fertilizer during repotting in this manner. On the other hand, fertilizer placed on top of the mix after potting must be leached downward into the root zone, delaying the benefit of the nutrients. Nitrogen in the mix in the upper part of the pot will contribute to the premature breakdown of organic matter, leading to a decrease in aeration, and then root problems and nutritional deficiencies.

Plants in containers may need periodic, additional fertilization depending on the release period of the fertilizer. Then, of course, fertilizers must be applied as a top dressing or in liquid form. Since calcium and magnesium in the form of dolomite and micronutrients were incorporated in the potting mixture, it is usually not necessary to reapply these nutrients again. The plants will receive a new dose of these with the fresh potting mixture at each repotting.

For growers on constant liquid feed, 150 ppm of nitrogen and potassium and 50–75 ppm of magnesium are recommended. Decrease rates of fertilizers in all programs when soil temperatures drop below 17°C (65°F).

Landscape Fertilizer Program

For established plants, apply a complete, slow-release fertilizer of 3-1-3 ratio, preferably one containing magnesium and micronutrients. Broadcast at the recommended package rate and frequency, mixing into the soil thoroughly where possible and water in well. Take care not to place the fertilizer against the stem where newly emerging roots may be injured. If using soluble or rapid-release, inorganic fertilizers, a good rule-of-thumb is to apply half the recommended label rate but at twice the frequency. Soluble fertilizers need to be applied more frequently to light, sandy soils than heavy, clay ones although the total amounts are not too different over a period of time.

Salinity

Since fertilizers are essentially salts and irrigation water may contain relatively high levels of salt, salinity is an important consideration. While Koebernik (1966) showed that young plants of *C. elatior, C. seifrizii,* and *C. tepejilote* were largely unaffected or displayed only slight burn after being submerged in sea water for several hours, salinity can prove troublesome to chamaedorea growers. Use fertilizers with care and periodically leach the root zone with an extra-heavy irrigation. Avoid irrigation practices which do not wet the entire root zone at each application. "Tip-burn" and damaged roots are symptoms of high soil salinity.

Watering

Chamaedoreas grow best when soil moisture is maintained at an even, moderate level. Constantly wet soils favor the development of root-rots and other diseases. On the other hand, plants will tolerate dry periods although growth will be slowed. Fertilizer in solution in the soil will become more concentrated as the soil dries, leading to possible tip or marginal necrosis of leaves and/or roots.

The texture and structure of a soil determine the amount of available water the soil will hold at any one irrigation and also the frequency of irrigation. Although basically all soils require the same amount of water, it is the frequency of irrigation that is the key to their proper management. Heavy soils require less and light soils more frequent irrigation to maintain appropriate levels of moisture for optimal growth of chamaedoreas.

As water is lost from the soil it must be replenished. Since loss of water from the soil is dependent on many factors, including soil type, temperature, humidity, wind, light, type of plant material, and cultural practices, there is no single rule about when to water. Simply put, it is best to water when the plant needs it. Water chamaedoreas established in the ground when the soil 2–4 cm below the surface becomes dry. This may be as long as several weeks in the winter or daily in the summer. At each irrigation, try to apply enough water to wet the soil thoroughly to about 30 cm (1 foot). If the soil is well drained, it is difficult to overwater chamaedoreas.

The best way to determine how far the water has penetrated is to dig and observe the depth of its movement after an irrigation. Eventually with repeated observations one will know how long to irrigate to push water into the upper 30 cm of soil. Heavy soils on a slope will accept water at a slower rate than sandy soils on level ground. Most sprinklers apply water at too rapid a rate for most soils to absorb it, especially on a slope. Heavy soils and/or sloping ground may require running sprinklers in cycles in order to get enough moisture into the soil and still avoid run-off.

Watering chamaedoreas in containers is somewhat more straightforward. Water them when the soil is dry one cm under the surface. At each irrigation, apply enough water to wet the mixture thoroughly throughout the container and then water an additional ten percent. The additional amount ensures that there is complete movement of water throughout the container and out the bottom as well to leach accumulated salts and prevent buildup of minerals to a level that may be toxic to plants. Leaching is especially important when plants are on constant liquid fertilizer program.

However one applies water to chamaedoreas in the ground or in pots, the plants benefit from an occasional sprinkling of the foliage that helps to remove dust and other particles and insects. It is especially helpful during hot, dry weather to sprinkle the foliage frequently to raise the humidity around sensitive species.

The quality of water may play a role in growing chamaedoreas. Some areas of southern California and southern Texas, for example, have water that is relatively high in some minerals like sodium and calcium. Through evaporation and repeated application without leaching, these minerals may reach toxic levels in root zones. To counteract this mineral buildup, give chamaedoreas an extra-heavy irrigation periodically to leach the root area, another reason to have a well drained soil or potting mixture. Very sensitive and rare species, especially when young, would benefit from an occasional irrigation with distilled or rain water.

Humidity

Most chamaedoreas are rather tolerant of low humidity. In fact, those commonly grown in southern California will withstand relative humidities of 15 percent or lower for limited periods. Of course others, like those of the pinnatifid group, are more sensitive. Much of the "tip-burn" associated with these species in cultivation is attributable to low humidity and, to a lesser extent, highly mineralized water. The latter is easier to control by proper irrigation management including leaching. It is a more difficult task to control humidity but greenhouse growing, frequent sprinkling or misting, and site selection can achieve some modification.

Low humidity with cool soil temperatures and inactive roots may be especially injurious to sensitive chamaedoreas since their roots cannot begin to take up adequate amounts of water to replace the accelerated losses from the leaves. In California, winter Santa Ana winds noted for their low humidity are especially injurious to sensitive chamaedoreas.

Light

With few exceptions, chamaedoreas need shade to grow best. Not only do they need shade when young but also when mature, which should come as no surprise since we know that nearly all the species grow in the understory of dense forests. As discussed under ecology, a few species tolerate higher light but to do their best they need shade. Too high a light intensity leads to a yellowing or bleaching of leaves and if excessive, scorching or burning.

Most species do best in 1,500–3,000 foot-candles (70–80 percent shade). The more light-tolerant ones, such as *C. cataractarum, C. microspadix, C. pochutlensis, C. radicalis,* and *C. seifrizii,* can manage with 5,000–6,000 foot-candles (40–70 percent shade). Since full sun on a summer day may be 12,000–15,000 foot-candles, the need for shade is obvious. Ordinary shade cloths at 70–80 percent shade and sufficiently large trees or other plants may provide satisfactory protection. Light intensity affects fertilizer requirements; plants grown in higher light need higher amounts of fertilizer than plants of the same species growing in more shade.

Temperature

Growth of most chamaedoreas is good between 18° and 32°C (65° and 90°F). Many will tolerate higher temperatures provided watering is adjusted accordingly. Root activity is reduced when soil falls below 15°C (60°F). What is remarkable when considering the origin of chamaedoreas is their tolerance of prolonged cool weather, like that of the southern California winter. Some species will even tolerate subfreezing temperatures. Other palms and plants growing with chamaedoreas in their native habitat are usually impossible to grow outdoors in southern California. Some chamaedoreas, especially those from cool, montane rain forests, do not do well in constantly warm areas.

Cold or freezing temperatures can have a devastating effect on chamaedoreas. Many authors have documented the effect of cool or subfreezing temperatures on palms, including chamaedoreas, and offered helpful suggestions on preventing and dealing with cold- or freeze-damage to palms. Among them are Anderson (1971), Barry (1961), Dolby (1957), Hertrich (1961), Kellet (1962), Larcher and Winter (1981), Martens (1970), D. Smith (1957, 1958, 1964, 1986), and R. Smith (1973). Listing the chamaedoreas and the absolute temperatures at which they suffer damage will not gain much since many factors are involved. However, it will suffice to say that *C. radicalis* is easily the hardiest of the group. It has withstood −6°C (22°F) on several occasions with no damage and survived −13°C (10°F) with only leaf damage. *C. microspadix* is less hardy but still withstands subfreezing temperatures with no damage. Most of the other members of the genus will tolerate freezing or slightly subfreezing temperatures provided their exposure is limited and there is overhead protection against radiational heat-loss.

Cold-injury symptoms are similar to those of other stresses, such as drought, root-rots, chemical phytotoxicity, heat, and light. An accelerated rate of senescence of lower leaves, increased susceptibility to attack by insects and disease, nutritional deficiencies, chlorosis, necrosis, and slowed or retarded growth are all symptoms of cold-induced stress. These symptoms usually appear over a long period of time.

On the other hand, injury resulting from freeze-damage to tissue occurs more quickly, usually within a few hours of the onset of freezing temperatures. Water-soaking in tissues resulting from disruption of cell structure and release of cell contents is common and is followed by "burned" or "scorched" lesions, pitting, sunken areas, and discoloration. However, freeze injury to the bud may not be visible until several months following the freeze.

Growth conditions markedly determine susceptibility to cold. Plants that are actively growing tolerate much less cold than those that have mature, hardened-off foliage. Plants that are in an overall more healthy state with proper nutritional levels are less susceptible to cold damage and will recover more quickly.

Types of Frost

Two types of frost situations can occur. One is called advection frost, and it occurs when freezing air blows into the area displacing warmer air. Moderate-to-strong winds, no inversion, cloudy or clear skies, and daytime freezing temperatures usually characterize advection frosts. In some cases the temperature may remain below freezing for several successive days. Advection frosts are difficult to protect against but, fortunately, are not common in areas where chamaedoreas are grown to any great extent.

Light or no winds, temperature inversion, clear skies, and daytime temperature above freezing characterize radiation frost, the other type of situation. Radiation frosts are more common in areas where chamaedoreas are grown and, fortunately, are easier to protect against.

Radiational cooling occurs primarily at night when heat absorbed by the earth or plant surfaces during the day is reradiated into the atmosphere. Heat lost from surfaces by radiational cooling moves away from the earth's surface. The soil continues to lose heat until it is colder than the air just above it. A layer of cold air develops near the earth's surface with the warmest air 1–50 meters above (3–150 feet). The warm over cold air is called a temperature inversion because it is an inversion of normal daytime conditions where the warmer air is at the ground.

Radiational frosts do not occur on foggy, cloudy, or windy nights since these conditions tend to reduce radiational heat loss and prevent the formation or break up the inversion by mixing the lower, colder air with warmer air higher up.

During radiational freezes, plant leaves close to the ground or exposed to the open sky in a flat, horizontal position may sustain freeze-injury even though the temperature a few feet above the plant is above the freezing point. A freeze that occurred in southern California in January, 1987 illustrates the principle of radiational cooling well. During this incident, a specimen of *C. sartorii* was situated at the perimeter of the research shadehouse. Half the leaves in its crown were outside of the shadehouse and exposed to the open sky. The others were inside the shadehouse. Freeze-damage composed of "burning" and "scorching" occurred only on those leaves outside the shadehouse and then only on those portions of pinnae that were flat to the sky. Pinnae or portions of them oriented in a vertical position displayed no damage.

Methods of Cold Protection

Proper management of chamaedoreas before the winter and on a specific frost night will minimize potential damage. Since chamaedoreas are usually planted under trees or other structures for shade, this is in their favor. The same objects that provide shade will also provide protection from heat-loss due to radiation. Trees, shade cloth, or other objects overhead provide about 1–2°C (2–4°F) protection on windless nights.

A structure with a cover and sides of burlap or plastic placed around a plant only at night will offer protection also. A heat lamp or a couple of 150-watt light bulbs may add 1–2°C (2–4°F) additional protection. Do not merely drape a cover over a plant; the cold will damage foliage touching the cover.

Site selection for frost protection is important when planting chamaedoreas. Plant them in the warmest, wind-protected areas with good air drainage. Plants grown against a structure with a southern or western exposure that absorbs and retains more heat during the day than an eastern or northern exposure have an advantage. Plants exposed to prevailing cold winds are likely to be the first to suffer damage. Cold air, being heavier, tends to flow downward to the lowest spots. Some growers have reported reducing damage by setting up fans in their collections and trying to displace the cold air near the ground and mix it up with pockets of warmer air higher up.

Where excess water may be successfully drained away, two low-volume sprinklers placed on opposite sides of a plant and applying about 2 liters (½ gallon) per minute per plant to the ground beneath it may provide 1–2°C (2–4°F) protection. Sprinklers must be turned on before the temperature reaches 0°C (32°F) and should remain on until it rises above 0°C. Do not sprinkle water over the plant, as the ice buildup could cause leaf breakage. During radiational freezes with very low humidity, as is common in southern California, sprinkling will not be effective since the evaporating water may lower the temperature.

Some growers have reported using foliar applications of antitranspirants to protect plants from freeze-damage.

Planting

There are several factors to consider when planting out chamaedoreas.

Site Selection

Although most chamaedoreas are small, some, especially the large clumping types, can eventually grow to five meters tall and three meters across. One must consider their ultimate size and how they will fit into the landscape in 10–15 years. Other considerations include soil type, esthetics, and protection from the sun, wind, and frost. By knowing the soil type it is often possible to anticipate nutrient deficiencies and prevent them rather than attempt to correct them after a problem appears.

Although esthetic placement of chamaedoreas in the landscape is mainly a matter of personal taste, there are a few basic guidelines to consider. The large clumping types, like *C. costaricana* and *C. pochutlensis,* make good background plants or screens in front of which smaller types may be grown and displayed. Solitary-stemmed, medium-sized types, such as *C. oblongata* and *C. klotzschiana,* look best in groups of odd numbers unless one has many, in which case they can be massed for greatest impact. Short or dwarf species, like *C. elegans, C. metallica, C. geonomiformis,* and *C. sullivaniorum,* are very effective as a mass planting around and under medium and large growing species. The species with outstanding, dramatic leaves, such as *C. ernesti-augustii, C. amabilis,* and others with bifid leaves, can stand alone or in mass or be used in containers and positioned for maximum effect where they will always draw attention.

Preparing the Soil

It has been common practice in the past to use liberal amounts of organic matter to amend the backfill that will be placed around the plant after it has been set into the hole. Recent evidence indicates that this practice may be unnecessary and, in many cases, may be detrimental to the plant. The use of backfill amended with organic matter creates an artificial situation in which the roots may never penetrate into the surrounding poorer soil. Essentially, the sharp interface between the backfill and surrounding soil creates a giant container. The texture, structure, aeration and drainage between the two may be so different that roots may remain confined to the hole for an inordinate amount of time. Therefore, it is recommended that the same soil that is removed from the hole be used unamended as backfill. Judicious management of water and fertilizer is much more important in successful establishment of the plant than amending the backfill with organic matter.

Organic matter would be much better used at planting time by working in liberal quantities to the surface of the soil over the entire planting area and also using it to mulch around the newly planted palms. Organic matter will be more effective in improving the texture, structure, water- and nutrient-holding capacity, and aeration of the soil over the entire area than it would concentrated in a hole beneath or around the palm.

Planting Depth

The planting hole should be slightly larger than the container or rootball and of the same depth. When the palm is set in the hole, the soil line of the rootball should be equal to the grade of the surrounding soil. Chamaedoreas planted on slopes should be placed so that the top of the original rootball at the stem is even with a stick laid level across the hole. A lip must then be built on the lower side of the hole to bring the irrigation basin to its proper height on that side.

Watering

More palms are probably lost at planting and transplanting time due to improper water management than any other cause. Irrigation during the establishment period is critical. At the time of planting, a basin should be built around the palm that has a diameter slightly wider than the original rootball. Water thoroughly immediately after planting by flooding the basin. The basin can be enlarged gradually during the establishment period until it is the same diameter as the hole.

Never allow the soil to dry out completely; it is also important not to overwater. When monitoring the moisture of a newly planted palm during the establishment period, be sure to check the original rootball. Plants that were overgrown or pot-bound in containers are at great risk of drying out. Although the surrounding original soil may be moist, the rootball may have lost most of its moisture due to transpiration through the leaves.

Staking

Since chamaedoreas are small they will normally not need to be staked. However, large, overgrown plants may need staking for support until they are established. Staking should be kept to as short a time as possible; remove stakes promptly when no longer needed. Plants should be staked loosely so that they have a little give and play.

Time of Year to Plant

Chamaedoreas can be planted or transplanted at just about any time of year provided it is done with care and the rootball remains intact and undisturbed. However, they will establish more quickly and with fewer problems if planting is done during the warm months when roots are most active.

Sanitation and General Cultural Practices

Sanitation is important in order to prevent disease and pest problems or limit their damage. It is much easier to prevent diseases than to eliminate them later.

During the seed and seedling stage, chamaedoreas are very susceptible to diseases called damping-off and root-rot. Several types of damping-off may occur separately or simultaneously with seeds. Seeds may rot before germinating or seedlings may rot before emerging from the soil. Seed-rot prior to emergence is called preemergence damping-off and losses of this type are usually blamed on immature or poorly developed seeds. However, the losses are most likely due to the water molds, a collective term for the pathogenic fungi *Pythium* and *Phytophthora.*

After emergence, seedlings may develop a stem-rot at or near the soil surface and fall over. This postemergence phase is often caused by another fungi called *Rhizoctonia* although the water molds may also be active.

The water molds and another fungus, *Fusarium,* may even attack larger plants in containers and in the ground. At this stage they usually attack the roots and progress up the stem. The plant goes into a slow decline exhibiting typical stress symptoms eventually leading to death. The rate of symptom expression in the aboveground portions of plants is dependent on the rapidity of root decay.

Nature of Water Molds and Rhizoctonia

These fungal pathogens are microscopic plants which feed on the tissues of the host plant. Their potential for destroying seeds and seedlings of rare chamaedoreas cannot be overstated.

Spores or other inocula may be widely present wherever plants are or have been grown. The soil is a natural reservoir of these pathogens which easily spread in a number of ways, some of the more common being rain or splashing water; infested containers, tools, and equipment, including benches and hoses; growers' feet and hands; infected plants or seeds; and placing healthy containerized plants on the ground. In summary, anything that comes into contact with a source of inoculum can then spread the disease to healthy, uninfected plants.

Once the disease is present, there is little that can be done to control it. Most chemicals will only suppress the pathogens, not eliminate them. Repeat applications of expensive and potentially toxic materials will be necessary. It is much easier, safer, and less expensive to prevent them than to fight them later.

Pasteurize [80°C (160–180°F) for 30 minutes] media used for seed germination and potting of seedlings. Avoid handling the soil unnecessarily. Do not reuse soil that may be contaminated. Clean all reused containers and work surfaces with a solution of nine parts water to one part household bleach. Keep clean pots off the ground and segregated from uncleaned ones. Wash hands before handling plants or knocking them out of pots to check their roots. Hang the hose nozzle on a hook off the ground and never place it on the ground. Do not walk on benches that have been cleaned. Place all new acquisitions of uncertain health away from healthy plants. Better yet, carefully inspect potential acquisitions and do not accept those infected with diseases. Keep growing areas clean and neat; discard or destroy diseased material.

The imaginative chamaedorea grower can think of other ways for achieving sanitation. In summary, start with clean seeds or plants, put them in clean soil and containers, and keep them clean by practicing sanitation at every step of your operation.

Pruning or Grooming

The only reason for removing old, dead, persistent leaves from chamaedoreas is an esthetic one. No other benefit is gained from this practice. Leaves should not be torn or pulled off but rather the petiole should be cut at the leaf base. If the leaf base falls away easily, it can be removed; otherwise it is best to leave it. To remove it prematurely by pulling or tearing will wound the stem and leave entry sites for disease organisms. Also, the stem may be scarred permanently, detracting from its beauty.

Sometimes the clumping chamaedoreas have stems that have attained too great a height. These older, longer stems may be selectively removed at the base without damaging the plant or detracting from its appearance.

The serious and successful grower of chamaedoreas or any plants, for that matter, will get to know the plants on an intimate basis. Periodic inspections of both leaf surfaces, stem, and soil are essential to avoid disease, insect, and abiotic problems.

Propagation

Chamaedoreas are mainly propagated by seeds although the clumping types with multiple stems may be divided if this is done with care. A few species with multiple stems, such as *C. stolonifera* and some forms of *C. elatior,* may be air-layered and new plants obtained in this manner. Air-layering may also be used to renew old plants that have grown too tall or leggy.

Seed

The principles of seed propagation of chamaedoreas are similar to those for other palms and most other plants. It is not particularly difficult and good results can be obtained if a few factors are considered. Broschat and Donselman (1986, 1987), Braun (1988), Kiem (1958), Loomis (1958), Read (1962), Carroll (1969), Velez (1980), Wagner (1982), and Yocum (1961) all provided excellent information concerning seed-propagation of palms.

Oyama and Mendoza (1990) investigated production of inflorescences in *C. tepejilote* (probably *C. alternans*) in Mexico. Their results are interesting and may have adaptations for commercial seed producers. They found that defoliated staminate plants produced more than twice as many inflorescences as undefoliated controls while defoliated pistillate plants did not differ significantly from control plants. Although half the pistillate inflorescences of completely defoliated plants failed to develop ripe fruits, partial defoliation significantly increased fruit production. Further work is needed in this area and may provide interesting and beneficial results.

There are as many variations for germinating palm seeds as there are growers. Any method that works is suitable. It is important to use clean, fresh, mature seeds. Seeds and seedlings are susceptible to damping-off diseases. Therefore, it is better to collect seeds from the tree rather than the ground. They should be kept clean until planting and planted in clean media in clean containers. Containers of planted seeds should be kept off the ground and growers should follow the simple rules of sanitation outlined earlier at every step of their operation.

Seeds of most palms, chamaedoreas included, lose their viability rapidly. Although De Leon (1958) reported the viability of seeds of some chamaedoreas to be 4–6 weeks, delaying planting gains nothing. Seeds of chamaedoreas should be planted as soon as possible after collection. Seeds not planted immediately should be stored in plastic bags with a clean, moist medium and kept at 18–21°C (65–70°F). They may be soaked in water for 1–2 days immediately prior to planting.

It is best to collect only seeds from completely mature and soft-ripe fruits. Depending on the species, fruits at this stage are usually black, reddish, or orange to dark brown. Fruits that are full-sized but slightly under-colored may have seeds that germinate at a reduced level or not at all.

Removing the pulp from seeds of chamaedoreas prior to planting improves germination. Removal of the pulp is easily done when the fruits are mature and soft-ripe. Simply mash the fruits in a bowl or bucket and then use water to decant the pulp. Be sure and wear waterproof gloves since the juice of some fruits can be highly irritating to tender skin.

Any medium that is clean and free of diseases and well aerated yet holds moisture is suitable for planting seeds. A pasteurized medium of equal parts fine peat moss and perlite has given good results for us. Position seeds in the premoistened medium in labeled and dated plastic pots so that the seeds are at such a shallow depth that half the seed is above and the other half is below the surface. Water them gently but thoroughly again and allow them to drain. Attach a plastic bag securely over the top of the pot. Planting at a shallow depth and covering with a plastic bag ensure that the seeds will remain visible for periodic observation and run less risk of remaining too wet yet still maintain high humidity. Seeds planted in this manner need to be watered very infrequently, perhaps about once a month. Without the plastic bag, seeds need to be planted deeper to prevent drying out.

Seeds should be maintained at 22–33°C (75–90°F). Lower or higher temperatures will delay, decrease, or inhibit germination. The use of bottom heat is beneficial in germinating seeds of chamaedoreas. Electric cables and heating pads with thermostats are available that will maintain temperatures in the correct range. Other methods can be devised for supplying bottom heat. I place my seed pots in aluminum trays and, much to the chagrin of my wife, put these on our gas range in the kitchen where the pilot light maintains a steady, adequate level of heat. By varying the position and height of the trays above the range and inserting or removing layers of newspaper between the seed pots and the tray, I have been able to maintain the medium in the proper temperature range.

Many growers recommend treating their seeds with fungicides prior to planting. In most cases this is unnecessary and, in a few instances, may even be detrimental. More is gained from using clean seeds, media, and pots and following sanitation.

Seeds of most chamaedoreas will germinate in 30–120 days. The rate of germination is dependent on several factors, among them the freshness and quality of seeds, temperature, moisture levels, and natural variation within and between species. Old seeds will germinate slower or not at all. Seeds of *C. seifrizii* and *C. elegans* sometimes germinate in flushes over a period of as long as a year or more. Loomis (1958), Carroll (1969), Koebernik (1971), and Wagner (1982) listed germination times for some chamaedoreas.

Germinated seeds may be transplanted out of community seed pots into individual pots during the one- or two-leaf stage. They are more difficult to handle when smaller and their roots may be entangled if transplanting is delayed beyond the two-leaf stage. Read (1962) and Wagner (1982) gave excellent instructions for transplanting and care of newly planted, germinated palm seeds. Germinated, bareroot seedlings must be treated with care. Plant them in a clean medium in clean pots at the same level as they were in the community seed pot. Care should be taken to avoid drying of the roots during planting. Young seedlings are especially tender, and light intensity, temperature, moisture, and nutrition are critical at this time.

Artificial Pollination

Chamaedoreas do not always produce seeds readily in cultivation. The lack of natural pollinators and the dioecious nature of the species seem mainly responsible. Artificial or hand-pollination may be necessary to ensure a good seed-set. The information is provided not as an encouragement to hybridizing but rather to promote the retention of uncontaminated species in cultivation. Review Douglas (1988) and Barry (1957) for useful information about pollinating chamaedoreas.

Generally, pollen of chamaedoreas is of two types perhaps best described as dry-powdery and moist-sticky. Species having the former type are usually easy to hand-pollinate and include *C. oblongata, C. pochutlensis, C. glaucifolia, C. klotzschiana, C. geonomiformis,* and subgenus *Stephanostachys.*

Pollen of the dry-powdery type is freely liberated from the flowers and usually present in copious amounts. Hand-pollination usually involves simply shaking a staminate inflorescence or a few rachillae with flowers at anthesis over a pistillate one also with receptive flowers. Pollination should be done several times a day over a several-day period to ensure a good seed-set. Anthesis of staminate flowers with the dry-powdery type of pollen is easy to determine. A sharp tap of the inflorescence or rachillae will send forth clouds of whitish pollen.

Species having pollen of the moist-sticky type are more difficult to hand-pollinate and include *C. ernesti-augustii, C. stolonifera, C. pumila,* and *C. fragrans.* Pollen of the moist-sticky type is not freely liberated in copious quantities and usually must be gathered and mechanically transferred to receptive pistillate flowers. Anthesis of flowers with the moist-sticky type of pollen is more difficult to determine. Good indications of anthesis are a full or swollen shape, more intense petal color, dropping of some pollen when tapped, and/or fragrance.

Both types of pollen can be collected for immediate use or stored for later use. Pollen can be collected by cutting off an inflorescence or some rachillae with flowers at anthesis and placing them in a paper envelope with the corners folded to prevent pollen from sifting into these inaccessible areas. It is a good idea to hold the envelope under the inflorescence during removal so that any pollen jarred loose by the cutting will fall into it. After collection of the inflorescence or rachillae, the envelope can be tapped several times on a hard surface to dislodge the pollen from the flowers. The pollen can then be collected from the bottom of the envelope with a soft sable-hair watercolor brush and gently brushed onto receptive pistillate flowers.

Since plants may flower at different times, one may find that pollen is available before pistillate flowers are at anthesis. Pollen may be stored for several weeks and still retain a level of viability sufficient to pollinate flowers successfully. Read (1979) outlined methods for successful storage of palm pollen. Pollen should be collected and stored dry. Avoid collecting flowers during rain or heavy dew. After collecting, rachillae and attached flowers can be spread on smooth paper in a dry, warm, draft-free room for several hours to dry them as much as possible. Later, the paper can be carefully folded, and the flowers and any fallen pollen funneled into an envelope with gentle tapping. The envelope can then be tapped sharply to liberate more pollen.

Pollen and/or fragments of flowers containing pollen can then be collected and stored in gelatin capsules in an airtight container with silica gel containing a color indicator as a desiccant. Fragments of flowers should be removed when dry. Pollen must be kept dry or fungal and bacterial contamination are likely. Avoid storing more pollen or flower fragments in a capsule than can be dehydrated adequately by the desiccant. Pollen in capsules may be stored at room temperature (20°C or 70°F) but will maintain viability longer if refrigerated.

A key to successful hand-pollination of chamaedoreas is knowing when pistillate flowers are receptive to pollen. As in the staminate, good indications of anthesis in the pistillate flowers are a full or swollen shape, color changes, and/or fragrance. Unfortunately, the telltale sign of falling pollen is absent from pistillate flowers but the pistil should be swollen and the stigma lobes will appear moist or have a minute drop of a clear, honeydewlike secretion at their tips.

Pistillate flowers of many chamaedoreas have the stigma lobes exserted from the petals. It is easy to brush or shake pollen gently onto them. Other species, such as *C. sullivaniorum,* and subgenus *Eleutheropetalum* have petals which exceed and may even cover the pistil and stigma lobes. Protective petals may be removed prior to applying pollen if care is taken not to damage the stigma lobes.

A 90-degree-angle dental tool for removing petals, an illuminated magnifying glass with a flexible arm, and lots of patience are helpful when pollinating chamaedoreas. Tools and equipment used in handling pollen should be cleaned with alcohol before and after use with one species to eliminate the possibility of inadvertently making hybrids. Although there is probably little chance for hybrids of chamaedoreas to occur on their own in cultivation, it may be a prudent measure to enclose prospective seed-producing pistillate inflorescences in paper bags before and after pollination to protect them from contamination by pollen from other species. Staminate inflorescences serving as a source of pollen may be likewise protected.

Dividing Clumps

Clumping chamaedoreas with multiple stems can be propagated by dividing the clumps. Although not a method for rapid multiplication, division is used if seeds are not available or if one desires a plant of known sex. Division of clumps must be done with care. If plants are in containers it is relatively simple. Remove the plant from the container and cut or saw through the clump making sure that each new plant has viable shoots with a good root system. Replant each new plant immediately and give proper care. Usually when dividing clumps from containers the root system sustains very little damage.

Dividing a clump in the ground is slightly more complex. Select a group of well rooted stems on the perimeter of the clump. It is best first to sever these stems from the main part of the clump and then wait before excising them to allow the severed portion to establish a more extensive system of roots and adjust to any shock. The establishment period may be several months in a tropical area with rapid growth or a full year in cooler areas with slower growth, as in California.

After the establishment period, the severed portion may be carefully dug and removed, keeping the soil around the root system intact. The new plant should be potted or replanted immediately in the ground and subjected to the same judicious cultural practices as any newly planted or transplanted specimen. As noted earlier, transplanting of this sort is best performed when the roots are actively growing, usually well into the warmer months.

Air-Layering

Air-layering or marcottage can be used to renew tall or leggy plants or to reproduce clumping plants whose stems produce aerial roots. Buhler (1974), Braun (1988), and Elbert (1979) gave excellent accounts of air-layering. *C. stolonifera* is easy to propagate in this way. Select one of the longer stolons, preferably one that is curved. Place the curved portion of the stolon with a node in a pot or container or press it to the ground and cover with potting mixture. The stolon may have to be secured with a weight, such as a brick, to keep it in place. After several months a new root system should have formed where the stolon contacts the potting mixture. When well rooted, the stolon may be severed from the mother plant and can survive on its own.

Stems of chamaedoreas that produce abundant aerial roots, such as the cespitose *C. serpens, C. nubium,* and forms of *C. elatior* and the single-stemmed *C. elegans* and *C. ernesti-augustii,* can be air-layered using the same technique. Where there are aerial roots or aerial root initials, usually near a node, securely place a grapefruit-sized mass of moist sphagnum moss or potting mixture. Aluminum foil or plastic wrapped around the sphagnum moss will secure it to the stem. If potting mixture is used, cut a plastic pot in half vertically, position and secure it around the stem at the appropriate location, and fill it with the medium. When roots are formed, the stem can be cut below the air-layer and removed. The presence of aerial roots varies among species and among individuals of the same species. It appears that individuals without aerial roots or aerial root initials are difficult to air-layer.

Wild-Collected Plants

Conservation practices and regulations discourage or prohibit the use of wild-collected plants for propagative material for collectors or commercial growers. However, there may be a few instances when collecting plants from the wild is justified, such as prior to clear-cutting, burning, or bulldozing of a forest. Collected plants are usually bare-rooted for transport and remain in this state for a considerable time before replanting. In these cases, handle plants as unrooted cuttings. Select small plants so that the entire root system remains intact and is disturbed as little as possible. If the root system is cut or severely damaged, it is better to select larger plants with a well developed, erect stem. Broschat and Donselman (1990) have shown that larger plants with well developed, erect stems have better chances for survival than smaller plants when roots are cut. They discovered that root-system regeneration in palms depends on the presence of a root initiation zone that does not form until the plant has reached its maximum stem diameter and the axis elongates vertically.

Remove all leaves except the apical four to five and reduce these to half their area. Remove all soil and rinse the plants and roots thoroughly. Pack the roots in a clean, moist material. Toilet paper is usually readily available, inexpensive, and works well. After wrapping the roots and ensuring that they are moist, enclose the entire plant in a plastic bag and keep it in the shade or a dark corner of a room at moderate temperatures until time for planting. Plants prepared in such a manner can survive for at least two weeks and be reestablished successfully although it is best to plant them as soon as possible. Handled as unrooted cuttings, plant them in a very porous and well aerated medium (3 parts perlite: 1 part peat moss) and placed on bottom heat with overhead, intermittent mist.

Pest and Disease Management

Chamaedoreas are relatively problem-free but there are a few pests and diseases that are a concern to growers. The best approach to pest and disease management is correct and proper cultural and sanitation practices that prevent these problems from occurring. Be vigilant and implement a regular (weekly) pest and disease check of all your plants. As a preventive measure, clean all leaf surfaces with a forceful strong jet of water at 7–14-day intervals.

Avoid use of chemicals whenever possible, and then use them only as a last resort. Consider some of the new strategies, including use of beneficial insects, cultural practices, and less toxic pesticides, that the integrated pest management approach emphasizes.

Chemicals are potentially dangerous materials;

their proper use and handling are extremely important. Read the label carefully, understand it, and follow the instructions exactly, including all safety precautions and measures. In order to be effective, chemicals must be applied according to the label instructions. Thorough application to all plant surfaces is necessary. Follow-up, repeat applications are probably also necessary in most cases. Pesticides are available that control nearly all the pests and diseases discussed here. However, since pesticides become outdated and their labels change frequently, they are not listed here. Consult your local county extension agent or advisor for currently available pesticides and their application.

The petroleum-based carrier in many pesticides, usually listed on the label as one of the inert ingredients, can be very damaging to chamaedoreas. For this reason, and if pesticides are to be used, the wettable powder form is preferable. The potential for damage with wettable powders is much less. Remember that young plants, and especially seedlings, are more sensitive to chemicals than older plants, and special care is needed when applying pesticides to them.

Pests

Spider Mites

Spider mites are not insects or spiders although they are related to the latter. They differ from insects in having only one body segment and four pairs of legs, and they lack wings. Insects have three body segments, three pairs of legs, and may or may not have wings.

Spider mites are smaller than a pinhead, oval, soft-bodied, creamy, brown, greenish or reddish [see Plate 127(C)]. There are different types of mites but the two most common on palms are the two-spotted spider mite and the red spider mite. Adults of the former usually have two large, dark spots on each side of the body. Red spider mites are usually orange to red.

Mites use piercing-sucking mouthparts to suck juices from the leaves of chamaedoreas. Symptoms of their damage include a yellowing or bronzing of the foliage and, when closely inspected, a fine yellowish to whitish stippling or flecking. It appears almost as if the foliage has been sand-blasted with very fine particles. Most mites produce webbing on the leaf surface in which they move about and reproduce. Although mainly on the lower sides of leaves, advanced and/or severe infestations will occur on the upper side also.

Mites multiply rapidly when conditions are favorable. They increase rapidly in dusty areas during dry, warm, or hot weather. At optimal temperatures it has been estimated that a single female mite is capable of producing in one month up to 13,000,000 offspring! Fortunately not all survive. In California, mites are especially a problem in greenhouses where high humidity is not maintained. They are not usually a problem outdoors in areas with high humidity, such as Hawaii and Florida.

Chamaedoreas vary in their susceptibility to mites. Species with thin leaves seem to be more attractive to mites than those species with thick, leathery leaves.

A vigilant inspection program of leaves to monitor potential infestations before they become serious and cause damage will control mites. One can significantly reduce mite populations and avoid damage simply by a regular (weekly) spraying of both leaf surfaces with a forceful stream of plain water. This is usually all that is necessary to dislodge mites and their eggs.

Mealybugs

Mealybugs are creamy-colored or whitish, elongate-oval, slow-moving insects covered with a white, waxy coating [see Plate 127(C)]. Adult females range from about 5–8 mm ($\frac{1}{5}$ to $\frac{1}{3}$ inch) in length and often possess waxy filaments radiating from the ends or sides of their bodies.

Adult female mealybugs damage chamaedoreas by sucking juices through their piercing-sucking mouthparts. Male mealybugs live for only a few days and do not feed. The withdrawal of plant juices causes a general decline of the palm's vigor. Heavy infestations in the apical bud area can damage the newly emerging leaf, causing it to be stunted or deformed. Mealybugs tend to concentrate in large cottony-white masses that detract from the plant's appearance. Also, as they feed, they excrete a honeydewlike, sticky fluid that drips to lower plant leaves and enhances growth of sooty mold fungi. The fungi, while not pathogenic, are unsightly and, if a large enough concentration builds up, it will reduce the amount of light available to the plant for photosynthesis.

The difficulty in controlling mealybugs is due partly to their tendency to congregate in large masses, especially in protected areas on the plant, such as behind leaf bases, on lower leaf surfaces, on roots, and within bracts of the inflorescence. Also, the waxy, sticky secretions covering their bodies and egg sacs give a degree of protection from insecticides.

Spraying infested plants with a strong stream of water to dislodge and wash away mealybugs will provide some control. Handpicking or swabbing individual insects with rubbing alcohol is another method sometimes used.

Scales

Scales are related to mealybugs and damage plants in the same manner. They suck plant juices with their piercing-sucking mouthparts. Scales are small and range from 2–8 mm (1/12 to 1/3 inch) in length. There are two general categories of scales that attack chamaedoreas: soft scales and armored scales. Armored scales possess a hardened shell that is circular, elliptic, or oyster-shaped [see Plate 127(A, B)]. Soft scales are slightly larger than armored scales and lack the hardened, protective shell. Adults are white, brownish, greenish, or black in color, flattened, elliptic or hemispherical, and waxy to the touch. Unlike armored scales, soft scales excrete a honeydewlike, sticky fluid that enhances the growth of sooty mold.

Female scales lay eggs under their protective shells or coverings. Tan- or light-colored immatures, called crawlers, hatch and move about for short distances then settle and remain to feed at the same spot for extended periods. Adult female scales are essentially sedentary; it is the crawlers that are mobile.

Like mealybugs, scales may congregate in large, layered, often crusty masses on protected or exposed parts or surfaces. Scales can be controlled by scraping them off with the fingernail or a dull knife. After removing adults, crawlers can be sprayed off with a strong stream of water. Scales are difficult to control chemically because their protective shells or waxy coatings are somewhat impervious to pesticides.

Aphids

Aphids are soft-bodied insects, 2–3 mm (1/12 to 1/8 inch) in length, with a pair of long antennae and long slender legs. They vary in color from creamy to greenish, reddish, or black. Aphids damage chamaedoreas by inserting their long "beaks" and sucking plant juices. They also damage plants by secreting the honeydewlike, sticky fluid that encourages sooty mold. Aphids are usually found on young, tender plant parts like emerging or newly emerged leaves and inflorescences. Spraying with a forceful stream of water to dislodge aphids from the plants can control them rather easily.

Thrips

Thrips are minute, slender, yellowish or black, winged insects from 0.25–0.5 mm (1/50 to 1/125 inch) in length. Thrips damage chamaedoreas by rasping the leaf tissue with their mouthparts and feeding on the secreted liquids. Damage to leaves is somewhat similar to that of mites and appears as silvery, flecked areas with a salt-and-pepper appearance [see Plate 127(D)]. The presence of small, shiny black, fecal pellets deposited by feeding thrips and the absence of silken webbing distinguish thrips damage from that of spider mites.

Thrips damage can occur on any part of the plant but is usually on lower leaf surfaces or in other protected areas. *C. oreophila* seems to be particularly susceptible to thrips; infestations on basal pinnae and the newly emerging leaf in the apical bud area can kill mature plants.

Regular spraying of all leaf surfaces and the apical bud area with a strong stream of water can control thrips. Chemicals may be necessary for severe infestations.

Fungus Gnats

Fungus gnats infrequently attack chamaedoreas. The 3 mm (1/8 inch) long, winged, gray-black insects are a nuisance but essentially harmless in the adult stage. They may be seen running rapidly on potting medium, edges of containers, or plant parts. In the larval stage, though, these insects can be destructive. The beige- or whitish-colored, semi-transparent maggots are about 3 mm (1/8 inch) long. They burrow through the potting mixture, feeding on decaying organic matter, plant roots, and fungi. They are often associated with overly wet situations with visible fungal growth on the surface of the medium, hence

---

Plate 127. Pests and Diseases of Chamaedorea. A. Scale insects on stem and peduncle, *C. radicalis*; **B.** Scale insects on *C. pumila* (note white immatures); **C.** Mealybugs (left, right) and mites (tiny sandlike objects scattered throughout), underside of pinna, *C. alternans*; **D.** Thrips (black bodies) and white feeding damage, pinna, *C. oreophila*; **E.** *Gliocladium* stem rot with pink spore masses (spots) and hardened, syrupy exudate, stem and leaf sheaths, *C. elegans*.

B
E

the common name. Unfortunately, fungus gnats are nearly impossible to control without the use of chemicals. Malathion (57%) at a rate of 2.5 ml per 3.8 liters of water (½ teaspoon/gallon) used as a soil drench has given control.

Whiteflies

Adult whiteflies are small, mothlike insects, about 1.5 mm (1/16 inch) long, with a white powdery appearance and two pairs of wings. Females often lay their eggs in a circular pattern, usually on the underside of the leaf. Immatures resemble scales, and are flattened, creamy-colored, greenish, and semi-transparent. Immatures feed on plants, damaging them by sucking their juices. Whiteflies are difficult to control but fortunately they are not a common pest of chamaedoreas. Some control can be achieved by a regular and forceful spraying of the undersides of the leaves with water.

The above are the major insect and mite pests of chamaedoreas. Recently, the banana moth, a serious pest of many ornamentals in southern Florida, has been found attacking chamaedoreas there. The larvae bore into the roots, feeding in the cortex, and penetrate into the pith. The pest damages the bud in young palms, causing degeneration of the central tissues and death. In older palms leaves become bleached and posteriorly necrotic. Researchers are working on control strategies for this pest.

Snails and Slugs

Most people are familiar with snails and slugs in the garden. Snails and slugs belong to the group of animals known as mollusks that includes many of the familiar shellfish, such as mussels and clams. Similar in structure and biology, snails differ from slugs in their external spiral shell. Both glide along by means of a flat muscular organ known as a foot and leave a shiny mucus or slime trail marking their travels. Snails and slugs are active at night and on dark, cloudy, or foggy days. They are essentially inactive on bright sunny days and remain hidden in places protected from the sun and heat, such as under trash, lumber, bricks and leaves. Snails and slugs need moisture to thrive but are able to survive drought and heat by going into a dormant stage for several months or years.

Snails and slugs damage chamaedoreas by rasping, tearing, and chewing various plant parts including leaves, flowers, and fruits. Slender elongated holes in the middle of leaves or pinnae are typical of damage on chamaedoreas. Young snails and slugs can be particularly troublesome because they are difficult to find due to their small size and can hide under pots or in the drainage holes at the bottom of pots, on lower leaf surfaces, behind leaf bases, or even in the soil.

Cultural and sanitation practices, including removal of boards, stones, and other debris that serve as hiding places, will help to control snails and slugs. Consistent handpicking of snails and slugs is time consuming but effective. Water the area in the late afternoon and then, after dark, handpick and destroy or discard the pests. Snails and slugs seeking cover during the day can be trapped under boards or flower pots placed throughout the garden or yard. Boards about 30 cm square with runners along two edges to raise them three cm above the ground make excellent traps. Check traps on a daily basis and destroy or discard trapped snails and slugs.

Beer bait traps may attract snails and slugs. Place saucers or pans of beer in depressions in the ground at intervals of three meters (10 feet). Baker's yeast may be added to the beer to increase its attractiveness. The edges of the pans should be at ground level so the pests can climb in and be trapped. Check traps daily and remove and dispose of any snails and slugs.

Snails and slugs will not crawl over irritating substances, and a number of materials, like ammonium sulfate, lime, ashes, and sand, can serve as effective barriers. After good sanitation and cultural techniques have been implemented, chemical baits may be effective.

Diseases

Chamaedoreas are relatively disease-free, most problems arising from improper culture. The dangers of root-rot, especially of seedlings, due to the fungi *Rhizoctonia, Pythium, Phytophthora,* and *Fusarium* were discussed earlier in the section on sanitation. Other diseases, including burrowing and lesion nematodes, anthracnose, leaf spots, *Gliocladium* stem- and leaf-blight, and several noninfectious agents, may be troublesome on chamaedoreas.

Not surprisingly, researchers in Florida working on commercial production have performed most of the work on diseases of chamaedoreas. Chase (1987), Marlatt (1980), and Simone and Chase (1989) are excellent references on diseases of ornamental foliage plants, including chamaedoreas. Chase and Broschat (1991) is, perhaps, the best reference and contains descriptions and photos of all known diseases and physiological disorders of palms.

Nematodes

Nematodes are microscopic worms that mostly inhabit the soil and organic matter but occasionally occur aboveground. Many nematodes are not plant parasites and, in fact, are actually beneficial organisms but some are serious parasites of a number of plants, chamaedoreas included. Plant-parasitic nematodes cause many symptoms similar to those of fungal diseases and/or nutritional disorders. Esser (1967) listed 22 types of plant-parasitic nematodes isolated from several species of *Chamaedorea.* Among the 22, the burrowing nematode and lesion nematode are the most serious pests.

The burrowing nematode tunnels through roots and destroys tissues as it feeds and is a serious pest of a wide variety of plants, including a number of important agricultural crops. Root symptoms of infected chamaedoreas include blackening, disintegration, rot, and growth retardation. Red, brown, or black lesions may occur in the basal area of stems if infections are advanced. Infected plants are stunted, spindly, chlorotic, and/or necrotic. Plants are often water-stressed and nutritionally deficient due to root damage. Lesion nematodes cause symptoms similar to those of the burrowing nematode.

Chamaedorea elegans is classified as a prime host of the burrowing nematode. In areas where the burrowing nematode is present, like Florida and Hawaii, *C. elegans* and other prime hosts that are commercially produced must be grown under special conditions and certified free from this pest if they are to be shipped to other regions where the disease is not known to occur, such as California. Plants must be produced in soilless media, grown on benches above the ground, and sampled periodically for presence of the nematode. In addition, nematicides must be periodically applied to the growing area.

Other prime hosts of the burrowing nematode include the Araceae, such as philodendrons, anthuriums, and dieffenbachias, to name a few. Collectors and hobbyists often use Araceae as companion plants, thus increasing the risk of introducing burrowing nematodes into their gardens. Control of nematodes is difficult at best. For the chamaedorea collector there is no practical way to treat plants chemically that are infected with burrowing nematodes. Prevention through pest exclusion and sanitation is the only solution.

Leaf Spots

Anthracnose and leaf spots caused by the fungi *Bipolaris, Calonectria, Drechslera, Exserohilum,* and *Phytophthora* can be troublesome on chamaedoreas, especially in areas of high rainfall and humidity. They seldom occur in California and then only in greenhouses. Spots begin as tiny, nearly pinpoint, circular, chlorotic areas, usually on young leaves. If conditions for disease development are optimal, spots will enlarge, become elliptic in shape, and turn necrotic in the center with a chlorotic halo. In advanced infections, individual spots will coalesce to form larger, irregularly shaped, necrotic areas and sometimes lead to shredded pinnae in severe cases.

Growers can avoid leaf spots by keeping water on the leaves to a minimum. Washing leaves to dislodge pests and dust is best done early in the day to permit the leaves to dry out before nightfall. Several chemicals have given good control of leaf spots. Copper-containing fungicides should be avoided since the additional copper applied to the plants can be toxic and reduce their growth. Several chemicals used for insect control, including Safer's Insecticidal Soap, reduce the severity of some leaf spots.

Gliocladium Stem and Leaf Blight

Stem and leaf blight caused by the fungus *Gliocladium vermoeseni* is a serious and common problem of commercially produced chamaedoreas in Florida, especially *C. seifrizii* and *C. elegans.* In California it is not a problem except when these plants are greenhouse-grown. Several workers, including Atilano et al. (1980), Hodel (1985a), Humphrey and Keim (1984), Keim and Maire (1975), and Reynolds (1964), provided good accounts of this disease.

A leafsheath rot on the basal portions of stems characterizes *Gliocladium* stem and leaf blight. If disease conditions are optimal, dark brown necrotic areas will occur on lower leaves. Infected leaves die prematurely, resulting in a bare-looking stem. Severe and/or advanced infections often move into the hard, pithy stem tissue behind or inside the leafsheath and may kill that stem [see Plate 127(E)]. Naturally clumping, *C. seifrizii* will have reduced "breaking" or new formation of basal shoots, re-

Table 9-2. **Symptoms of Noninfectious Diseases and Some Possible Causes.**

Symptom	Possible cause
Chlorosis	
Entire plant	nitrogen deficiency; high light.
Young leaves	poor soil aeration; salinity; iron or sulfur deficiency.
Older leaves	nitrogen or magnesium deficiency; overwatering; poor soil aeration.
Leaf margins	salinity; magnesium deficiency.
Intervenal	iron or manganese deficiency.
Irregular spots	cold temperature.
Necrosis	
Water-soaked areas	cold temperature.
Tips or margins	potassium deficiency (old leaves); boron toxicity (old leaves); salinity; temperature extremes; desiccation; low humidity; root damage.
Center of leaf	sun-scorch; cold; nutrient toxicity; ethylene toxicity (poorly or unvented greenhouses).
Leaf Deformities	
Leaves abnormally small	micronutrient deficiency; salinity; high light; root-bound or -damaged plants.
Petioles elongated	low light.
Holes in leaves	mechanical injury; slugs, snails; insects.
New leaves stunted	manganese or other micronutrient deficiency; salinity; poor aeration; overwatering.
Stem Deformities	
Rot at soil line	salinity; fertilizer placed against stem; overwatering; poor aeration.
Wilting	salinity; high temperature; desiccation; low humidity; low soil moisture; root damage.
Thin, spindly, weak	low light; crowding of plants.
Root Abnormalities	
Slow development	salinity; soil temperature extremes; plant too deep; poor aeration; overwatering.
Poor or rotted roots	salinity; overwatering; poor aeration.
Stunted plants	fertilizer extremes; nutrient deficiencies; extremes of light; overwatering; poor aeration; poor roots.

sulting in a spindly plant with few stems. A dark, sappy exudate emanating from diseased areas is often associated with advanced infections. Pinkish brown spore masses are usually abundant and visible in advanced infections, especially on lower leaf sheaths.

Moist, humid conditions 20–28°C (65–85°F) favor disease development. Also, cultural practices that wound the stem, such as removal of yellowing yet not dead leaf sheaths, enhance disease spread and development. Cultural controls include providing for good air movement around plants, lowering humidity, avoiding splashing or dripping water on leaves, and, whenever possible and practical, adjusting temperature regimes to a range unfavorable to the pathogen. Chemicals may prove beneficial in suppressing the disease.

Another disease, Thielaviopsis stem rot, exhibits symptoms similar to those of *Gliocladium*. Prevention and treatment are much the same as for *Gliocladium*.

Noninfectious Diseases

Abiotic factors, including air pollution, pesticide phytotoxins, exposure to toxic materials, and deficiencies or excesses of light, water, humidity, temperature, and nutrients, are responsible for a great number of plant disorders. Most of these, including the nutritional and environmental factors, were discussed earlier. The noninfectious diseases may, in fact, exact a toll on plants exceeding that of diseases caused by living organisms.

Infectious and noninfectious diseases, acting separately or together, can cause similar symptoms on chamaedoreas. A grower should consider all possible causes and combinations when trouble-shooting. For example, a yellow, stunted seedling is probably the result of water stress and/or nutrient deficiency. Any number of disorders could be responsible, including root inactivity due to excessively cool temperatures or root loss due to excessive water, a poorly aerated medium, and/or disease.

Some symptoms of abiotic agents are presented with their possible causes in Table 9-2. Pesticide phytoxicity and damage from herbicides and other toxic materials are not included since they can cause nearly all these symptoms. Chamaedoreas are especially sensitive to petroleum-based pesticides and copper. Some fungicides and wood preservatives are sources of copper. Remember that diseases caused by living organisms and pests can cause many of these symptoms, also.

GLOSSARY

Acuminate—tapering to a point.
Acute—distinctly and sharply pointed but not drawn out.
Adnate—different organs attached to each other.
Anther—a sacklike organ at tip of the stamen and containing pollen.
Anthesis—time of shedding of the pollen (staminate) or of stigma receptivity (pistillate).
Bifid—partially divided into two equal parts; here used to refer to leaf blades that are undivided except for the 2-lobed apex; to many people a bifid leaf is a simple leaf.
Blade—expanded portion of the leaf composed of pinnae or its equivalent if bifid.
Bract—leaflike organ associated with the inflorescence.
Cespitose—clustered, in reference to multiple stems.
Calyx—collective term for the sepals, the outermost whorl of floral organs.
Clustered—with several stems.
Connate—united, organs of the same kind joined together.
Corolla—collective term for the petals, the second whorl of floral organs.
Decumbent—prostrate, lying along the ground but with the tip ascending.
Decurrent—with the lower margins of the pinnae extending along rachis to give the rachis a winged appearance.
Dioecious—with flowers of each sex borne on separate plants.
Distal—farthest from point of attachment (apical).
Elliptic—oblong with regularly rounded ends.
Endocarp—innermost layer of fruit wall.
Eophyll—first seedling leaf having a blade.
Exocarp—outermost layer of fruit wall.
Falcate—with one margin longer than another, as in a leaf; sickle-shaped.
Fibrous—composed of fibers.
Filament—stalk of the stamen bearing an anther.
Glabrous—smooth, lacking hairs or other covering.
Glaucous—covered with a grayish or greenish bloom.
Imbricate—overlapping, as shingles.
Inflorescence—structure that bears the flowers including all bracts and branches.
Infrafoliar—borne below the leaves.
Interfoliar—borne among the leaves, arising from the axil of a live leaf blade.
Internode—portion of the stem between two nodes where leaves are attached.
Intrafoliar—see infrafoliar.
Keeled—bearing or like a ridge.
Lanceolate—narrow, tapering at both ends, broader below the middle.
Leaflet—pinna or division of a leaf.
Ligule—distal or apical appendage of a leaf sheath.
Linear—narrow, several to many times longer than wide.
Mesocarp—middle layer of fruit wall.
Node—point on a stem where the leaf is or was attached, usually a ring.
Obovoid—egg-shaped, broader distally (apically).
Ovoid—egg-shaped, broader proximally (basally).
Peduncle—lower, unbranched part of the inflorescence; the stalk of the inflorescence.
Perianth—collective term for sepals and petals; used here usually in fruiting stage.
Petal—one unit of the corolla.
Petiole—stalk of a leaf.
Pinna(ae)—leaflet (leaflets) of a pinnately divided leaf.
Pinnate—divided into pinnae; refers to leaf.
Pistil—female part of the flower which receives pollen and develops into a fruit.
Pistillate—female, bearing pistil.
Pistillode—sterile or rudimentary pistil of the staminate flower.
Plicate—pleated.
Prophyll—first or lowermost bract of the inflorescence.
Proximal—closest to point of attachment (basal).
Rachilla(ae)—branch (branches) that bear flowers.
Rachis—pinnae-bearing axis of the leaf blade; or central axis where two lobes are attached of a bifid leaf blade; or main flower-bearing axis of an unbranched inflorescence; or rachillae-bearing axis of a branched inflorescence.
Rhizome—horizontal underground stem.
Rugose—densely wrinkled and having a rough texture.
Sepal—single unit of the calyx.
Sessile—lacking a stalk.
Sheath—lowermost part of the leaf clasping the stem; it is always tubular initially but splits to various degrees after maturity.
Sigmoid—S-shaped.

Solitary—single-stemmed, not cespitose.
Spicate—simple, unbranched; in reference to the inflorescence.
Stamen—male organ of a flower composed of the filament and the anther.
Staminate—male, bearing stamens.
Staminode—rudimentary nonfunctional stamen of a pistillate flower.
Stigma—terminal organ of the pistil that receives pollen.
Stolon—aboveground, prostrate, lateral stem usually rooting at the nodes.
Striated—lined.
Style—attenuate part of the pistil that ends in the stigma.
Tomentose—with tomentum.
Tomentum—covering of densely matted, short, woolly hairs.
Undulate—wavy.
Valvate—meeting exactly without overlapping.

LITERATURE CITED

Aguilar Amar, R. I. 1986. El Genero *Chamaedorea* Willd. (Palmae) en el Estado de Veracruz. Xalapa, Mexico: Universidad Veracruzana. (Thesis.)
Allen, P. H. 1956. The Rain Forests of Golfo Dulce. Stanford, California: Stanford University Press.
Anonymous. 1865. Palms as room plants. J. Hort. Pract. Gard. 33: 343.
———. 1873. Chamaedoreas. Gard. Chron. 1873: 508.
———. 1874. The indoor garden. Garden 5: 531.
———. 1876a. *Chamaedorea formosa.* Gard. Chron. 5 (New Series) (pt. 2): 724.
———. 1876b. *Chamaedorea formosa.* Forest and Pomologist 29: 246–247.
———. 1876c. The necklace-flowered *Chamaedorea.* Garden 10: 479.
———. 1877a. Neue oder empfehlenswerthe Zierpflanzen. Gartenflora 26: 53–59.
———. 1877b. *Chamaedorea formosa.* Ill. Gart. Zeitung 21: 194, t. 27.
———. 1882. Neue und empfehlenswerthe Pflanzen. Gartenflora 31: 211–218.
———. 1883. Stove palms. Garden 24: 244–245.
———. 1885. Garden palms. Gard. Chron. 23 (New Series) (pt. 2): 410.
———. 1895. Plant notes. Garden and Forest 8: 504–505, fig. 70, p. 507.
———. 1985. Know your palm. Principes 29: 188.
Anderson, L. 1971. California cold effects. Principes 15: 140.
Atilano, R. A., W. R. Llewellyn and H. M. Donselman. 1980. Control of *Gliocladium* of *Chamaedorea* palms. Proc. Florida State Hort. Soc. 93: 194–195.
Bailey, L. H. 1933. Certain palms of Panama. Gentes Herb. 3: 33–116.
———. 1935. The Standard Cyclopedia of Horticulture. New York: The Macmillan Co.
———. 1943a. New palms in Panama, and others. Gentes Herb. 6: 198–264.
———. 1943b. Palmaceae, pages 327–396. *In:* R. E. Woodson, Jr. & R. W. Schery, Flora of Panama. Ann. Missouri Bot. Gard. 30: 281–403.
Baker, K. F., (ed.) 1957. The U.C. System for Producing Healthy, Container-grown Plants. Univ. Calif. Agric. Exp. Sta. Manual 23. Berkeley, California: University of California.
Baker, R. 1980. An inventory of palms in Hawaii. Principes 24: 55–64.
Balick, M. J. and H. T. Beck (eds.). 1990. Useful Palms of the World. A Synoptic Bibliography. New York: Columbia University Press.
Balslev, H. and A. Barfod. 1986. Ecuadorian palms—an overview. Opera Bot. 92: 17–35.
——— and M. Moraes. 1989. Sinopsis de las Palmeras de Bolivia. AAU Reports 20. Risskov, Denmark: Botanical Institute Aarhus University. (In collaboration with Herbario Nacional de Bolivia.)
Barry, D., Jr. 1957. Comments on *Chamaedorea* palms. Principes 1: 91–93.
———. 1961. Notes on hardiness of palms on the French Riviera. Principes 5: 100–103.
———. 1967. Hybrids in *Chamaedorea.* Principes 11: 121–123.
Basu, S. K. 1978. A census of palms cultivated in the Indian Botanical Garden, Howrah. Principes 22: 127–135.
Bentham, G. and J. D. Hooker. 1883. Genera Plantarum. London: A. Black; W. Pamplin; Lovell Reeve & Co.; Williams & Norgate.
Bergmann, B. 1989. Slaegten *Chamaedorea* Willd. (Arecaceae) i Ecuador. Risskov, Denmark: Institute of Botany, University of Aarhus. (Thesis.)
Bernal, R. G. 1989. Endangerment of Colombian palms. Principes 33: 113–128.
Blatter, E. 1926. The Palms of British India and Ceylon. London: Oxford University Press.
Blombery, A. and T. Rodd. 1982. Palms. Sydney: Angus & Robertson.
Boodley, J. W. 1972. Soilless mixes. Horticulture 50: 38–39.
——— and R. S. Sheldrake. 1977. Cornell Peat-lite Mixes for Commercial Plant Growing. Cornell Plant Sci. Inform. Bull. 43. Ithaca, New York: Cornell University.
Braun, A. 1968. Cultivated palms of Venezuela. Principes 12: 39–103, 111–136.
———. 1983. Palmas para Interiores, Parques y Avenidas. Caracas, Venezuela: Instituto Nacional de Parques and Fundacion de Educacional Ambiental.
———. 1988. El Cultivo de las Palmas en el Tropico. Caracas, Venezuela: Tipografia Cervantes.
——— and F. Delascio Chitty. 1987. Palmas Autoctonas de Venezuela y de los Paises Adyacentes. Caracas, Venezuela: A. Braun & F. Delascio Chitty.
Breedlove, D. E. 1986. Listados Floristicos de Mexico IV. Flora of Chiapas. Mexico: Instituto de Biologia, UNAM.
Brongniart, A. 1847. Petit Capeu de palmier, Envoyee du Mexique par M. Ghiesbreght. Rev. Hort., Ser. 3, 1: 86.

Broschat, T. K. 1984. Nutrient deficiency symptoms of five species of palms grown as foliage plants. Principes 28: 6–14.
———. 1989. Potassium deficiency in south Florida ornamentals. Proc. Florida State Hort. Soc. 102: 106–108.
———. 1990. Potassium deficiency of palms in south Florida. Principes 34: 151–155.
——— and H. Donselman. 1985. Causes of palm nutrient disorders. Proc. Florida State Hort. Soc. 98: 101–102.
——— and ———. 1986. Factors affecting storage and germination of *Chrysalidocarpus lutescens* seeds. J. Amer. Soc. Hort. Sci. 111: 872–877.
——— and ———. 1987. Effects of fruit maturation, storage, presoaking, and seed cleaning on germination of three species of palm. J. Environ. Hort. 5: 6–9.
——— and ———. 1990. IBA, plant maturity, and regeneration of palm root systems. Hortscience 25: 232.
——— and A. Meerow. 1990. Palm nutrition guide. Nursery Digest 24(4): 31–34.
Buhler, T. 1974. Notes on marcottage of certain palms. Principes 18: 111–112.
Burch, D. R., R. A. Atilano and J. Reinert. 1983. Indoor Palm Production Guide for Commercial Growers. Univ. Florida Ornamental Hort. Comm. Fact Sheet OHC-8. Gainesville, Florida: University of Florida.
Burret, M. 1930. Palmae novae austroamericanae. Notizbl. Bot. Gart. Berlin-Dahlem 11: 1–19.
———. 1932. Palmae neogeae. Notizbl. Bot. Gart. Berlin-Dahlem 11: 313–327.
———. 1933a. *Chamaedorea* Willd. und verwandte Palmengattungen. Notizbl. Bot. Gart. Berlin-Dahlem 11: 724–768.
———. 1933b. Palmae neogeae IV. Notizbl. Bot. Gart. Berlin-Dahlem 11: 857–866.
———. 1934a. Palmae neogeae V. Notizbl. Bot. Gart. Berlin-Dahlem 11: 1037–1050.
———. 1934b. Palmae neogeae VI. Notizbl. Bot. Gart. Berlin-Dahlem 12: 42–44.
———. 1935. Eine kultivierte *Chamaedorea*-Art wildwachsend wiedergefunden. Notizbl. Bot. Gart. Berlin-Dahlem 12: 302–303.
———. 1936a. Die Palmengattung *Morenia* R. & P. Notizbl. Bot. Gart. Berlin-Dahlem 13: 332–339.
———. 1936b. Palmae neogeae X. Notizbl. Bot. Gart. Berlin-Dahlem 13: 339–347.
———. 1937. Plantae duqueanae. Notizbl. Bot. Gart. Berlin-Dahlem 13: 489–495.
———. 1938. Eine interessante neue *Chamaedorea*-Art aus Mexiko. Notizbl. Bot. Gart. Berlin-Dahlem 14: 268–269.
———. 1940. Palmae. *In:* F. L. E. Diels, Neue Arten aus Ecuador III. Notizbl. Bot. Gart. Berlin-Dahlem 15: 23–38.
Carroll, D. I. 1969. Letter to editor. Principes 13: 109.
Chabaud, B. 1915. Les Palmiers de la Cote d'Azure. Paris: Librairie Agricole de la Maison Rustique.
Chase, A. R. 1987. Compendium of Foliage Plant Diseases. St. Paul, Minnesota: The American Phytopathological Society.
——— and T. K. Broschat (eds.). 1991. Diseases and Disorders of Palms. St. Paul, Minnesota: The American Phytopathological Society.
Chazdon, R. 1985. The palm flora of Finca La Selva. Principes 29: 74–78.
———. 1987. The palm flora of Braulio Carrillo National Park, Costa Rica. Brenesia 28: 107–116.
———. 1988. Conservation-conscious collecting: concerns and guidelines. Principes 32: 13–17.
Chittenden, F. J. 1951. The Royal Horticultural Society Dictionary of Gardening. Oxford: Clarendon Press.
Coates-Estrada, R. and A. Estrada. 1986. Manual de Identification de Campo de los mamiferos de la Estacion de Biologica "Los Tuxtlas". Mexico: Instituto de Biologia, UNAM.
Conover, C. A. and R. T. Poole. 1981. Guide for Fertilizing Tropical Foliage Plant Crops. Univ. Florida Apopka ARC Res. Rep. RH-8H. Apopka, Florida: University of Florida.
——— and ———. 1990. Light and fertilizer recommendations for production of acclimatized potted foliage plants. Nursery Digest 24(10): 34–36, 58–59.
Cook, O. F. 1937. New household palm, *Neanthe bella.* Science 86: 120–122.
———. 1939. A second household palm, *Omanthe costaricana.* Science 90: 298–299.
———. 1943a. Household palms and related genera. Part I. Nat. Hort. Mag. 22: 83–102.
———. 1943b. Household palms and related genera. Part II. Nat. Hort. Mag. 22: 134–152, figs. 10–15.
———. 1947a. Cascade palms of southern Mexico. Nat. Hort. Mag. 26: 10–34.
———. 1947b. Climbing and creeping palms in Mexico and Guatemala, related to household palms. Nat. Hort. Mag. 26: 215–231.
——— and C. B. Doyle. 1939. The edible pacaya palm of Alta Verapaz. Nat. Hort. Mag. 18: 161–179, figs. 1–9.
——— and J. F. Joyner. 1938. A diminutive palm from Mayaland. Nat. Hort. Mag. 17: 1–12, figs. 1–8.
——— and ———. 1939. *Neanthe,* a palm for genetic study. Jour. Heredity 30: 93–103.
Cowan, C. P. 1983. Listados Floristicos de Mexico I. Flora de Tabasco. Mexico: Instituto de Biologia, UNAM.
Criley, R. A. 1973. Synthetic media for container grown plants. Misc. Pub. Univ. Hawaii Coop. Ext. Ser. 103: 31–41.
Croat, T. B. 1978. Flora of Barro Colorado Island. Stanford, California: Stanford University Press.
Croucher, J. 1872. Palms for the garden. Garden 1: 218, 283.
Dahlgren, B. E. 1959. Index of American palms. Plates. Field. Mus. Nat. Hist., Bot. Ser. 14: pl. 1–412.
Dammer, U. 1897. Palmenzucht und Palmenpfleg. Frankfurt: U. Dammer.
———. 1903. *Chamaedorea,* Vol. 6, page 54. *In:* J. Smith, Enumeratio Plantarum Guatemalensium. Baltimore, Maryland: H. N. Patterson.
———. 1904a. Eine neue Palme aus Guatemala. Notizbl. Konigl. Bot. Gart. Berlin 4: 157–158.
———. 1904b. The species of *Chamaedorea* with simple leaves. Gard. Chron., Ser. 3, 36: 202, 245–246.
———. 1905a. *Kinetostigma* Dammer—Genus novum palmarum guatemalense. Notizbl. Konigl. Bot. Gart. Berlin 4: 171–173.
———. 1905b. New or noteworthy plants—*Malortiea tuerckheimii.* Gard. Chron., Ser. 3, 38: 19.
———. 1905c. Species of *Chamaedorea* with pinnate leaves. Gard. Chron., Ser. 3, 38: 42–44.
———. 1907a. Beitrage zur Flora der Hylea nach den Sammlungen von Ules Amazonas-Expedition. Palmae. Verh. Bot. Vereins Prov. Brandenburg 48: 118–129.
———. 1907b. Die Pflanzenformationen des Amazonas-Gebietes. Bot. Jahrb. Syst. 40: 139.
———. 1915. Plantae uleanae novae. Notizbl. Konigl. Bot. Gart. Berlin 6: 261–268.

Davis, T. A. 1979. Some unusual formations in palms. Principes 23: 80–83.

De Leon, N. J. 1958. Viability of palm seeds. Principes 2: 96–98.

Decaisne, T. and C. Naudin. 1862. Manuel de l'Amateur des Jardins, III. Paris: Firmin-Didot et Cie.

De Steven, D. and F. E. Putz. 1985. Mortality rates of some rain forest palms in Panama. Principes 29: 162–165.

DeWerth, A. F. and R. E. Odon. 1960. A Standard Light Weight Growing Medium for Horticultural Specialty Crops. Texas Agric. Exp. Sta. Misc. Pub. 420. College Station, Texas: University of Texas.

Dolby, W. J. 1957. Winter protection of palms in marginal areas. Principes 1: 79–81.

Donselman, H. 1981a. Planting a Palm Tree. Univ. Florida Hort. Fact Sheet OH-46. Gainesville, Florida: University of Florida.

———. 1981b. Nutritional Deficiencies of Florida Palms. Univ. Florida Hort. Fact Sheet OH-49. Gainesville, Florida: University of Florida.

Doughty, S. 1988. Growing palms in the New Orleans area. Principes 32: 96–100.

Douglas, R. 1987. Growing chamaedoreas: six simple-leaved ones. Principes 31: 55–63.

———. 1988. Notes on pollinating the *Eleutheropetalum* group of chamaedoreas. Principes 32: 26–28.

Dransfield, J. 1978. Growth forms of rainforest palms, pages 247–268. *In:* P. B. Tomlinson and M. H. Zimmerman (eds.). Tropical Trees as Living Systems. New York: Cambridge University Press.

——— and David N. Cook. 1987. *Chamaedorea metallica.* Kew Mag. 4: 160, pl. 87.

———, O. Johnson and H. Synge. 1988. The Palms of the New World. A Conservation Census. Gland, Switzerland and Cambridge, U.K.: IUCN.

——— and N. W. Uhl. 1986. An outline of a classification of palms. Principes 30: 3–11.

Drude, O. 1881. Palmae, Vol. 3(2), pages 253–460. *In:* C. F. P. von Martius, Flora Brasiliensis. Leipzig, Munich, and Vienna: C. F. P. von Martius.

———. 1882a. Palmae, Vol. 3(2), pages 461–583. *In:* C. F. P. von Martius, Flora Brasiliensis. Leipzig, Munich, and Vienna: C. F. P. von Martius.

———. 1882b. Bemerkungen zur Nomenclature und Kulturfahigakeit der in Haage und Schmidt Pflanzen-Verzeichniss fur 1882 empfohlenen Palmen. Gart.-Zeitung 1: 178–179.

———. 1889. Palmae, Vol. 3, pages 1–93. *In:* A. Engler and K. Prantl, Die naturlichen Pflanzenfamilien II. Leipzig: Wilhelm Engelmann.

———. 1897. Palmae, Vol. 1, page 53. *In:* A. Engler and K. Prantl, Die naturlichen Pflanzenfamilien, Nachtrage. Leipzig: Wilhelm Engelmann.

Dugand, A. 1940. Palmas de Colombia. Clave diagnostica de los generos y nomina de las especies conocidas. Caldasia 1: 20–84.

———. 1961. Palms of Colombia. Principes 5: 135–144.

Edmonson, N. 1989. *Chamaedorea* Species in Cultivation [in Florida]: Threatened New World Palms in Cultivation. Report for World Wildlife Fund Project 3322. Washington, D.C.: World Wildlife Fund.

Elbert, G. A. 1979. Air layering a palm. Green Scene (Sept.): 32–34.

Endlicher, S. 1837. Genera plantarum, Palmae. Vienna: F. Beck.

Esser, R. P. 1967. Nematode pests and associates of five species of *Chamaedorea.* Principes 11: 87–91.

Essig, F. B. 1971. Observations on pollination in *Bactris.* Principes 15: 20–24.

———. 1973. Pollination in some New Guinea palms. Principes 17: 75–83.

Everett, T. H. 1981. The New York Botanical Garden Illustrated Encyclopedia of Horticulture, Vol. III. New York: Garland Publishing, Inc.

Fisher, J. B. 1974. Axillary and dichotomous branching in the palm *Chamaedorea.* Amer. J. Bot. 61: 1046–1056.

——— and H. E. Moore, Jr. 1977. Multiple inflorescences in palms (Arecaceae): their development and significance. Bot. Jahrb. Syst. 98: 573–611.

Foster R. B., B. J. Arce and T. S. Wachter. 1986. Dispersal and the sequential plant communities in Amazonia Peru floodplain, pages 357–370. *In*: A. Estrada and T. H. Fleming (eds.), Frugivores and Seed Dispersal. Dordrecht: Dr. W. Junk.

Frankel, D. H. 1974. Genetic conservation: our evolutionary responsibility. Genetics 78: 53–56.

Galeano, G. 1986. Two new species of palmae from Colombia. Brittonia 38: 60–64.

———. 1987. A new name to replace a homonym. Principes 31: 143.

———. 1991. Las Palmas de la Region de Araracuara. Estudios en la Amazonia Colombiana. Vol. 1. Bogota, Colombia: Tropenbos.

——— and R. Bernal. 1987. Palmas del Departamento de Antioquia. Region Occidental. Bogota, Colombia: Universidad Nacional de Colombia.

Gentry, A. 1986. Notes on Peruvian palms. Ann. Missouri Bot. Gard. 73: 158–165.

Gerome, J. 1911. Les *Chamaedorea* grimpants. Rev. Hort. 83: 549–571.

Glassman, S. F. 1972. A Revision of B. E. Dahlgren's Index of American Palms. Lehre, Germany: J. Cramer.

Graf, A. B. 1973. Exotica. E. Rutherford, New Jersey: Roehrs Co., Inc.

Grayum, M. H. and G. C. de Nevers. 1988. New and rare understory palms from the Peninsula de Osa, Costa Rica, and adjacent regions. Principes 32: 101–114.

——— and D. R. Hodel. 1991. Two new species of *Chamaedorea* from Panama. Principes 35: 133–138.

Guillaumin, A. 1922. Palmiers nouveaux des serres du Musee. Bull. Mus. Hist. Nat. (Paris) 28: 542–544.

———. 1923a. Les *Chamaedorea* cespiteux. Rev. Hort. 95: 406–408.

———. 1923b. Les *Chamaedorea* cultives. J. Soc. Natl. Hort. France, Ser. 4, 24: 223–244.

Habermann, W. 1902. Kreuzung von Chamaedoreen. Gartenflora 51: 70–72, pl. 9.

Hammer, N. 1985. Distribution plants, 1985. Fairchild Trop. Gard. Bull. 40: 18–24.

Hawkes, A. 1967. Some palms of Nicaragua. Principes 11: 99–104.

Heinzman, R. and C. Reining. 1988. Sustained Rural Development: Extractive Forest Reserves in the Northern Peten of Guatemala. Report to the United States Agency for International Development, World Wildlife Fund, and Yale School of Forestry and Environmental Studies. New Haven, Connecticut: Yale University School of Forestry and Environmental Studies.

Hemsley, W. B. 1885. Palmae. *In:* F. Godman and O. Salvin (eds.), Biologia Centrali-Americana, Botany 3: 400–415. London: F. Godman & O. Salvin.

———. 1904. *Chamaedorea pulchella.* Bot. Mag. 130: t. 7959.

Henderson, A. 1986. A review of pollination studies in the Palmae. Bot. Rev. 52: 221–259.

Hertrich, W. 1951. Palms and Cycads. San Marino, California: Huntington Botanical Gardens.

———. 1961. Huntington Botanical Gardens. Principes 5: 27–31.

Hesdorffer, M. 1924. Handbuch Zimmergartnerei. Berlin: M. Hesdorffer.

Heynhold, G. 1840. Nomenclator Botanicus Hortensis. Dresden and Leipzig: Arnoldischen Buchhandlung.

Heywood, V. H. 1978. Flowering Plants of the World. Oxford: Oxford University Press.

Hodel, D. R. 1985a. *Gliocladium* and *Fusarium* diseases of palms. Principes 29: 85–88.

———. 1985b. A Survey of Commodity Movement on the Los Angeles Wholesale Flower Market. Los Angeles: University of California Cooperative Extension.

———. 1988a. Letter to the editor. Principes 32: 95, 100.

———. 1988b. The Genus *Chamaedorea* in Commerce in California. Report for World Wildlife Fund Project 3322. Washington, D.C.: World Wildlife Fund.

———. 1990a. *Chamaedorea amabilis*: an ornamental species from Central America. Principes 34: 4–10.

———. 1990b. *Chamaedorea castillo-montii* (Arecaceae), a new species from Guatemala. Phytologia 68: 397–400.

———. 1990c. Three new species of *Chamaedorea* (Arecaceae) from Oaxaca, Mexico. Phytologia 68: 401–409.

———. 1990d. New species and notes on related taxa in *Chamaedorea* subgenus *Stephanostachys.* Principes 34: 160–176.

———. 1991a. New species of *Chamaedorea* from Central America. Principes 35: 72–82.

———. 1991b. The cultivated species of *Chamaedorea* with cespitose habit and pinnate leaves. Principes 35: 184–198.

——— and J. J. Castillo Mont. 1990. Two new species of *Chamaedorea* (Arecaceae) from Guatemala. Phytologia 68: 390–396.

——— and ———. 1991. Additional new species of *Chamaedorea.* Principes 35: 4–8.

——— and N. W. Uhl. 1990a. Two new species of *Chamaedorea* from Mexico. Principes 34: 58–63.

——— and ———. 1990b. A new species and synopsis of a distinctive and natural subgroup of *Chamaedorea.* Principes 34: 108–119.

——— and ———. 1990c. New species of *Chamaedorea* from Costa Rica and Panama. Principes 34: 120–133.

Hodge, W. H. 1981. Finca Las Cruces, a Costa Rican garden. Principes 25: 47–53, and cover of No. 2.

———. 1988. Cover photograph of *Chamaedorea stolonifera.* Principes 32: cover of No. 1.

Holdridge, L. R. and L. J. Poveda A. 1975. Arboles de Costa Rica, Vol. 1. San Jose, Costa Rica: Centro Cientifico Tropical.

Hooker, W. J. 1855a. *Chamaedorea ernesti-augustii* (foem.). Bot. Mag. 81: t. 4831.

———. 1855b. *Chamaedorea ernesti-augustii* (mas.). Bot. Mag. 81: t. 4837.

———. 1855c. *Chamaedorea elegans.* Bot. Mag. 81: t. 4845.

Hooker, J. D. 1865. *Morenia fragrans.* Bot. Mag. 91: t. 5492.

———. 1873. *Chamaedorea tepejilote.* Bot. Mag. 99: t. 6030.

———. 1874. *Nunnezharia geonomaeformis.* Bot. Mag. 100: t. 6088.

———. 1881. *Nunnezharia tenella.* Bot. Mag. 107: t. 6584.

———. 1884. Kew Rep. 1884: 58–59.

———. 1885. *Chamaedorea arenbergiana.* Bot. Mag. 111: t. 6838.

———. 1892. *Chamaedorea stolonifera.* Bot. Mag. 118: t. 7265.

Hooper, L. R. 1990. *Chamaedorea stolonifera.* Principes 34: 50.

Hoyos, J. and A. Braun. 1984. Palmas Tropicales Cultivadas en Venezuela. Caracas, Venezuela: Sociedad y Fundacion La Salle de Ciencias Naturales.

Humboldt, A. and A. Bonpland. 1809. Plantas Aequinoctiales. Vol. 2. Tubingen: F. Schoell & J. G. Cotta.

———, A. Bonpland and C. S. Kunth. 1816. Nova Genera et Species Plantarum. Vol. 1. Paris: Libraire grecque latine-allemande.

Humphrey, W. A. and R. Keim. 1984. Control of *Gliocladium* disease of *Chamaedorea* palms. Univ. Calif. Coop. Ext. Flower and Nursery Report Fall 1983 and Spring 1984: 5.

Ibarra-Manriquez, G. 1988. The palms of a tropical rainforest in Veracruz, Mexico. Principes 32: 147–155.

Jacquin, N. J. 1797. Plantarum Rariorum Horti Caesarei Schoenbrunnensis. Vol. 2. Vienna: C. F. Wappler.

Jahn, A., Jr. 1908. Las Palmas de la Flora Venezolana. Caracas, Venezuela: Tipografia Universal.

Johnson, D. 1987. Conservation status of wild palms in Latin America and the Caribbean. Principes 31: 96–97.

Joiner, J. N., (ed.) 1981. Foliage Plant Production. Englewood Cliffs, New Jersey: Prentice-Hall, Inc.

Jones, D. 1984. Palms in Australia. Frenchs Forest, Australia: Reed Books.

———. 1985. Palms in Colour. Frenchs Forest, Australia: Reed Books.

Kahn, F. and K. Mejia. 1991. The palm communities of two "terra firme" forests in Peruvian Amazonia. Principes 35: 22–26.

Karsten, H. 1856. Plantae columbianae. Linnaea 28: 241–281.

———. 1857. Plantae columbianae. Linnaea 28: 387–412.

———. 1866. Florae Colombiae. Vol. 2. Berlin: F. Duemmler.

Keim, R. and R. G. Maire. 1975. *Gliocladium* disease of palm. Calif. Pl. Pathl. 27: 1–2.

Keim, S. C. 1958. Propagation of palms. Principes 2: 133–138.

Kellet, J. 1969. Looking back on the Florida freeze of 1962. Principes 13: 23–35.

Kerchove de Denterghem, O. 1878. Les Palmiers. Paris: J. Rothschild.

Kimball, W. 1973. Notes on culture. Principes 17: 173.

Klotzsch, J. F. 1847. Palmae Juss., Brown, Mart., Endl., Kunth. Linnaea 20: 446–456.

———. 1852. Zwei neue Rohrpalmen aus Guatemala. Allg. Gartenzeitung 20: 352–364.

Knuth, P. O. 1904. Handbuch der Blutenbiologie. Leipzig: Wilhelm Engelmann.

Koebernik, J. 1966. Salt tolerance in young palms. Principes 10: 130–137.

———. 1971. Germination of palm seeds. Principes 15: 134–137.

Kohler, H. 1913. *Chamaedorea* ein Handelsartikel. Gartenflora 62: 197–199, pl. 24–29.

———. 1928. Aus unseren Pflanzenschatzen. Gartenwelt 32: 358–359, pl. 1–2.

Krempin, J. 1990. Palms and Cycads Around the World. Sydney: Horwitz Grahame.
Kromdijk, G. and W. R. Oudshoorn. 1967. 200 House Plants in Color. New York: McGraw-Hill Co.
Kunth, C. S. 1822. Synopsis Plantarum. Sixieme Partie: Botanique. Paris: F. G. Levrault.
———. 1841. Enumeratio Plantarum. Vol. 3. Stuttgart and Tubingen: J. G. Collae.
Kuntze, O. 1891. Revisio Generum Plantarum. Vol. 2. Leipzig: A. Felix.
Lane Publishing Co. 1988. Sunset Western Garden Book. Menlo Park, California: Lane Publishing Co.
Langlois, A. C. 1976. Supplement to Palms of the World. Gainesville, Florida: University of Florida Press.
Larcher, W. and A. Winter. 1981. Frost susceptibility of palms: experimental data and their interpretation. Principes 25: 143–152.
Le Maout, E. 1868. Traite General de Botanique. Paris: Librarie de Firmin-Didot et Cie.
Liebmann, F. M. 1846. Central Americas Palmenformer. Overs. Kongel. Danske Vidensk. Selsk. Forh. Medlemmers Arbeiter 1846: 4–10.
Linden, J. 1881. Ill. Hort. (Plantes Introduites et Mises pour la Premiere Fois dans le Commerce) 28. Bruxelles: J. Linden.
———. 1885. Catalogue General. No. 117. Bruxelles: Quartier Leopold.
Link, H. F. 1834. Ueber *Chamaedorea schiedeana* Mart. Allg. Gartenzeitung 2: 153–154.
Loebner, M. 1909. Palmen. *Chamaedorea concolor* × *ernesti augustii* (*Ch. katzeri*). Gartenwelt 13: 159–160.
Loomis, H. F. 1958. The preparation and germination of palm seeds. Principes 2: 98–102.
Macbride, J. F. 1960. Flora of Peru. Part I, No. 2. Palmae. Field Mus. Nat. Hist., Bot. Ser. 13: 321–418.
Macoboy, S. 1976. What Indoor Plant Is That? New York: Crown Publishers, Inc.
Marlatt, R. B. 1980. Noncontagious Diseases of Tropical Foliage Plants. Univ. Florida Agric. Exp. Sta. Bull. 812. Gainesville, Florida: University of Florida.
Marnier-Lapostolle, J. 1961. Palms in the Jardin Botanique "Les Cedres," France. Principes 5: 103–106.
Marque, J. P. 1923. Les *Chamaedorea sartorii* et *Ch. oblongata* aux serres du parc de la tete d'or a Lyon. Rev. Hort. 95: 446–448.
Martens, M. and H. Galeotti. 1843. Enumeratio synoptica plantarum phanerogamicarum ab Henrico Galeotti in regionibus mexicanis collectarum. Bull. Acad. Roy. Sci. Bruxelles 10: 122–124.
Martens, O. 1970. Observation of frost damage at a palm nursery. Principes 14: 66–68.
Martius, C. F. P. von. 1823. Historia Naturalis Palmarum. Vol. 2. Leipzig: T. O. Weigel.
———. 1824a. Palmarum Familia. Munich: M. Lindauer.
———. 1824b. Historia Naturalis Palmarum. Vol. 2. Leipzig: T. O. Weigel.
———. 1826. Historia Naturalis Palmarum. Vol. 2. Leipzig: T. O. Weigel.
———. 1830. Drei neue mexicanische Palmen. Linnaea 5: 204–205.
———. 1837. Historia Naturalis Palmarum. Vol. 3. Leipzig: T. O. Weigel.
———. 1838. Historia Naturalis Palmarum. Vol. 3. Leipzig: T. O. Weigel.
———. 1839. Historia Naturalis Palmarum. Vol. 3. Leipzig: T. O. Weigel.
———. 1844. Voyage dans l'Amerique Meridionale. Vol. 7, Part 3. Paris: P. Bertrand.
———. 1845. Historia Naturalis Palmarum. Vol. 3. Leipzig: T. O. Weigel.
———. 1849. Historia Naturalis Palmarum. Vol. 3. Leipzig: T. O. Weigel.
———. 1852. *Chamaedorea atrovirens* Mart., eine neue Palme aus Mexico, beschrieben. Flora 35: 721–723.
McCurrach, J. C. 1960. Palms of the World. New York: Harper & Brothers.
McGregor, L. 1987. Letter to the editor. Principes 31: 187–188.
McVaugh, R. 1987. Itinerary and gazetteer, pages 5–38. *In:* J. T. Mickel, Liebmann's Mexican Ferns (Contr. N.Y. Bot. Gard., Vol. 19). New York: New York Botanical Garden.
Meehan, T. 1877. *Chamaedorea formosa.* The Gard. Monthly & Hort. 19: 72.
Millspaugh, C. 1898. Contributions to the coastal and plain flora of Yucatan. Publ. Field Columbian Mus., Bot. Ser. 1: 355.
Minter, S. 1990. The Greatest Glass House. London: HMSO.
Moore, H. E., Jr. 1949. Other palm studies. Gentes. Herb. 8: 191–205.
———. 1951. Palmae novae variae. Gentes Herb. 8: 223–238.
———. 1957. *Chamaedorea radicalis.* Principes 1: 147.
———. 1958. *Chamaedorea falcifera.* Principes 2: 68–70.
———. 1963. An annotated checklist of cultivated palms. Principes 7: 119–182.
———. 1966. *Chamaedorea metallica*—a new species from cultivation. Principes 10: 44–50.
———. 1971. A synopsis of the palms at Finca Las Cruces. (Unpublished field key and descriptions.)
———. 1973a. A synopsis of the palms at Finca La Selva. (Unpublished guide for O.T.S. course.)
———. 1973b. The major groups of palms and their distribution. Gentes Herb. 11: 27–141.
———. 1973c. Palms in the tropical forest ecosystems of Africa and South America, pages 63–88. *In:* B. J. Meggers, E. S. Ayeusu and W. D. Duckworth (eds.), Tropical Forest Ecosystems in South America and Africa: A Comparative Review. Washington, D.C.: Smithsonian Institution Press.
———. 1975. Key to the palms at Monteverde. (Unpublished field key and descriptions.)
———. 1976. *Chamaedorea,* pages 255–257. *In:* Staff of the L. H. Bailey Hortorium, Hortus Third. New York: Macmillan Publishing Co., Inc.
———. 1977. Endangerment at the specific and generic level in palms, pages 267–282. *In:* G. T. Prance and T. S. Elias (eds.), Extinction is Forever: The Status of Threatened and Endangered Plants of the Americas. New York: New York Botanical Garden.
———. 1979. Proposal to conserve *Chamaedorea* Willd. over *Morenia* Ruiz et Pavon (Palmae). Taxon 27: 555–556.
———. 1980. Four new species of Palmae from South America. Gentes Herb. 12: 30–38.
Morley, B. 1983. Wild Flowers of the World. New York: Exeter Books (Simon & Schuster).
Mott, R. C. 1967. A potting mixture for palms. Principes 11: 52–53.
———. 1971a. Cornell foliage plant mixes. Florist's Rev. 148: 22.

———. 1971b. Potted palms for interiors. Principes 15: 121–126.

Muirhead, D. 1961. Palms. Globe, Arizona: Dale Stuart King.

Mueller, C. 1858. Palmae, pages 804–857. *In:* W. G. Walpers, Annales Botanices Systematicae. Vol. 5. Leipzig: F. Hofmeister.

Naudin, C. 1858a. *Chamaedorea ernesti-augustii* Wendland. Fl. Serres Jard. Eur. 2, Ser. 3: 122–124, t. 1357.

———. 1858b. *Chamaedorea elegans* Martius. Fl. Serres Jard. Eur. 2, Ser. 3: 133–135, t. 1362.

Nees, F. 1834. Bemerkungen uber *Chamaedorea elegans* Martius. Allg. Gartenzeitung 2: 145–146.

Nicholson, G. 1884. The Illustrated Dictionary of Gardening. London: L. Upcott Gill.

———. 1900. The 1900 Supplement to the Dictionary of Gardening. London: L. Upcott Gill.

Oersted, A. S. 1859. Palmae centroamericanae. Vidensk. Meddel. Dansk. Naturhist. Foren. Kjoebenhavn 1858: 1–54.

———. 1863. L'Amerique Centrale. Vol. 1. Copenhagen: F. S. Muhle.

Olmsted, I. 1988. Ecology and management of three palm species in Sian Ka'nan, Quintana Roo, Mexico. Report to World Wildlife Fund. Washington, D.C.: World Wildlife Fund.

Otto, F. and A. Dietrich. 1833. *Kunthia xalapensis.* Allg. Gartenzeitung 1: 1.

Oyama, K. 1990. Variation in growth and reproduction in a neotropical dioecious palm, *Chamaedorea tepejilote.* J. Ecol. 78: 648–663.

———. 1991. Seed predation by a curculionid beetle on the dioecious palm *Chamaedorea tepejilote.* Principes 35: 156–160.

——— and R. Dirzo. 1988. Biomass allocation in the doecious tropical palm *Chamaedorea tepejilote* and its life history consequences. Pl. Sp. Biol. 3: 27–33.

——— and ———. 1991. Ecological aspects of the interaction between *Chamaedorea tepejilote,* a dioecious palm, and *Calyptocephala marginipennis,* a herbivorous beetle, in a Mexican rain forest. Principes 35: 86–93.

——— and A. Mendoza. 1990. Effects of defoliation on growth, reproduction and survival of a neotropical dioecious palm *Chamaedorea tepejilote.* Biotropica 22: 119–123.

Pancoast, L. 1985. The survivors observed. Principes 29: 94–107.

Parthasarthy, M. V. 1972. Cover photograph of *Chamaedorea elatior.* Principes 16: cover of No. 1.

Poole, R. T. and W. E. Waters. 1972. Evaluation of various potting media for growth of foliage plants. Proc. Florida State Hort. Soc. 85: 395–398.

———, C. A. Conover and R. W. Henley. 1975. Parlor palm seed germination. Florist's Rev. 157: 89, 106.

Prain, O. 1912. *Chamaedorea glaucifolia.* Bot. Mag. 138: t. 8457.

———. 1913. *Morenia corallina.* Bot. Mag. 139: t. 8527.

———. 1916. *Chamaedorea nana.* Bot. Mag. 142: t. 8652.

Ramirez-Goyena, M. 1911. Flora Nicaraguensis. Vol. 2. Managua, Nicaragua: M. Ramirez-Goyena.

Read, R. W. 1962. Culture of palm seedlings after germination. Principes 6: 126–133.

———. 1979. Live storage of palm pollen. Principes 23: 33–35.

Reidel, P. 1957. Plants for Extra-Tropical Regions. Arcadia, California: California Arboretum Foundation.

Reynolds, J. E. 1964. *Gliocladium* disease of palm in Dade County, Florida. Plant Dis. Reptr. 48: 718–720.

Rodrigues, J. B. 1989. Sertum Palmarum Brasiliensium. Tomos 1 & 2. Rio de Janeiro: Jardin Botanico de Rio de Janeiro. (Reprint.)

Roster, G. 1913. Le palme coltivate in piena aria nei giardini d'Italia. Bull. Soc. Tosc. Ortic. 38: 82–90.

Ruffo, G. 1920. Le Palme di Villa Lucia. Firenze, Italia: Tip. Giuntina. Firenze.

Ruiz, H. and J. Pavon. 1794. Florae Peruvianae, et Chilensis. Prodromus. Madrid: G. Sancha.

——— and ———. 1798. Systema Vegetabilium Florae Peruvianae et Chilensis. Vol. 1. Madrid: G. Sancha.

Sastrapradja, D. S. and T. A. Davis. 1983. The Bogor Botanic Garden and its rich collection of palms. Principes 27: 18–30.

Schick, F. A. 1980. A note on mulching in south Florida. Principes 24: 91–92.

Schipper, S. D. 1926. Aus unsern Pflanzenschatzen. Gartenwelt 30: 141–142.

Schmid, R. 1970. Notes on the reproductive biology of *Asterogyne martiana* (Palmae), II: Pollination by syrphid flies. Principes 14: 39–49.

Schnizlein, A. 1849. Iconographia Familiarum Naturalium Regni Vegetabilis. Bonn: Max Cohen & Sohn.

Schultes, J. A. and J. H. Schultes. 1830. Systema Vegetabilium. Vol. 7. Stuttgart: J. G. Cottae.

Schultes, R. E. 1950. Plantae austro-americanae VI. Bot. Mus. Leafl. 14: 109–136.

Schulz, B. 1933. Bluhende Palmen im Zimmer. Gartenflora 82: 277.

Seemann, B. 1856. Popular History of Palms. London: Lovell Reeve.

Seymour, F. C. 1980. A Checklist of the Vascular Plants of Nicaragua. Phytologia Memoirs I. New York: H. N. & A. L. Moldenke.

Siebert, A. and A. Voss. 1896. Vilmorin's Blumengartnerei. Berlin: Verlagsbuchhandlung P. Parey.

Simone, G. W. and A. R. Chase. 1989. Pesticides for disease control in foliage ornamentals. Foliage Digest 12(9): 1–8.

Smith, D. 1957. Palms at high latitudes in Florida. Principes 1: 104–114.

———. 1958. Cold tolerance of the cultivated palms. Principes 2: 116–126.

———. 1964. More about cold tolerance. Principes 8: 26–39.

———. 1975. Palmologue. Principes 19: 137–146.

———. 1986. USDA hardiness zones as applied to cultivated palms, and other horticultural comments. Principes 30: 26–31.

Smith, J. D. 1905. Enumeratio Plantarum Guatemalensium. Vol. 7. Baltimore: H. N. Patterson.

Smith, R. N. 1967. Palms in south Texas. Principes 11: 94–96.

———. 1973. Letter in palmologue. Principes 17: 163–164.

Snow, D. W. and B. K. Snow. 1978. Palm fruits in the diet of the oilbird, *Steatornis caripensis.* Principes 22: 107–109.

Sousa, S. M. and E. Cabrera. 1983. Listados Floristicos de Mexico II. Flora de Quintana Roo. Mexico: Instituto de Biologia, UNAM.

Sprengel, K. 1825. Systema Vegetabilium. Vol. 2. Stuttgart: J. G. Cottae.

Spruce, R. 1869. Palmae amazonicae. J. Linn. Soc., Bot. 11: 65–83.

Standley, P. C. 1920. Trees and shrubs of Mexico. Contr. U.S. Natl. Herb. 23: 70–84.
———. 1928. Flora of the Panama Canal Zone. Contr. U.S. Natl. Herb. 27: 1–416, pl. 1–66.
———. 1930. Flora of Yucatan. Publ. Field. Columbian Mus., Bot. Ser. 3: 156–492.
———. 1931. Flora of the Lancetilla Valley, Honduras. Field Mus. Nat. Hist., Bot. Ser. 10: 109–117.
———. 1933. Flora of Barro Colorado Island, Panama. Contr. Arnold Arbor. 5: 1–168, pl. 1–21.
———. 1937. Flora of Costa Rica. Part I. Publ. Field Mus. Nat. Hist., Bot. Ser. 18: 107–128.
———. 1940. Studies of Central American plants II. Publ. Field Mus. Nat. Hist., Bot. Ser. 22: 326.
——— and S. Calderon. 1925. Lista Preliminar de las Plantas de El Salvador. San Salvador: Tipografia la Union.
——— and S. J. Record. 1936. The forests and flora of British Honduras. Field Mus. Nat. Hist., Bot. Ser. 12: 76–88.
——— and J. A. Steyermark. 1947. Studies of Central American plants VII. Publ. Field Mus. Nat. Hist., Bot. Ser. 23: 196–209.
——— and ———. 1958. Flora of Guatemala, I. Fieldiana, Bot. 24: 217–253.
Stennis, C. G. G. J. van. 1981. Rheophytes of the World. Netherlands: Sijthoff & Noordhoff, Alphen aan den Rijn.
Stevenson, G. B. 1974. Palms of South Florida. Miami: G. B. Stevenson.
Stewart, Lynette. 1981. Palms for the Home and Garden. Sydney: Angus & Robertson.
Synge, P. M. 1969. The Royal Horticultural Society Supplement to the Dictionary of Gardening. Oxford: Clarendon Press.
Tanfani, E. 1886. Sulla *Chamaedorea verschaffelti.* Bull. Soc. Tosc. Ortic. 11: 73–75, t. 3.
Thiselton-Dyer, W. T. (ed.). 1900. List of published names of plants introduced to cultivation: 1876–1896. Kew Bull. Additional Series 4: 94.
Tomlinson, P. B. 1961. Essays on the morphology of palms. V. The habit of palms. Principes 5: 83–89.
———. 1990. The Structural Biology of Palms. Oxford: Oxford University Press.
——— and H. E. Moore, Jr. 1966. Dichotomous branching in palms. Principes 10: 21–29.
Trail, J. W. H. 1876. Descriptions of new species and varieties of palms collected in the valley of the Amazon in north Brasil, in 1874. J. Bot. 14: 323–333, 353–359, t. 183.
Uhl, N. W. and J. Dransfield. 1987. Genera Palmarum. Ithaca, New York: The L. H. Bailey Hortorium and Lawrence, Kansas: The International Palm Society.
——— and H. E. Moore, Jr. 1973. The protection of pollen and ovules in palms. Principes 17: 117–149.
Van der Meersch, A. 1882. Un palmier de serre froide, *Chamaedorea elegans* Mart. Rev. Hort. Belge Etrangere 8: 17, 20, fig. 2.
Velenovsky, J. 1913. Vergleichende Morphologie der Pflanzen. IV. Supplement. Prague: Fr. Rivnac.
Velez, R. 1980. A new seed germination chamber. Principes 24: 134–135.
Verlot, B. 1869. *Chamaedorea karwinskyana.* Revue Hort. 41: 227–228, fig. 54.
Voss, R. L. and D. P. Watson. 1968. U. H. Potting Mix. Univ. Hawaii Coop. Ext. Circ. 424. Honolulu, Hawaii: University of Hawaii.
Vosters, J. 1975. Commercial use of *Chamaedorea elegans.* Principes 19: 149–150.
Wagner, R. I. 1982. Raising ornamental palms. Principes 26: 86–101.
Wait, L. H. 1961. News of the society. Principes 5: 3–7, figs. 4, 5.
Walpers, W. G. 1852. Annales Botanices Systematicae. Vol. 3. Leipzig: F. Hofmeister.
Weiss, G. 1893. Neue empfehlenswerte Pflanzen. Neubert's Deutsch. Gart.-Mag. 46: 155, 159.
Wendland, H. 1852a. Beschreibung einer neuen *Chamaedorea.* Allg. Gartenzeitung 20: 1–3.
———. 1852b. Beschreibung einer neuen *Chamaedorea.* Allg. Gartenzeitung 20: 73–74.
———. 1852c. Beschreibung einer neuen *Chamaedorea.* Allg. Gartenzeitung 20: 217–228.
———. 1853a. Uber einige Palmen Central-Americas. Allg. Gartenzeitung 21: 1–4.
———. 1853b. Beschreibung neuen chamaedoreas aus Central-America. Allg. Gartenzeitung 21: 137–140.
———. 1853c. Uber *Chamaedorea elatior* Mart., und die nahe verwandten Arten. Allg. Gartenzeitung 21: 169–172, 177–180.
———. 1854. Index Palmarum. Hannover: Libraria Hahn.
———. 1859. Beschreibung neuer Palmen. Bot. Zeitung Berlin 17: 17, 29–30, 102–103.
———. 1862. Eine neue Palme aus der Sammlung des Konigl. Berggartens zu Herrhausen. Bonplandia 10: 37–38.
———. 1880. Beitrag zur Palmenflora Amerikas. Gartenflora 29: 101–105.
Wessels Boer, J. G. 1988. Palmas indigenas de Venezuela. Pittieria 17: 1–332.
Wheelwright, N. T., W. A. Haber, K. G. Murray and C. Guindon. 1984. Tropical fruit-eating birds and their food plants: a survey of a Costa Rican lower montane forest. Biotropica 16: 173–192.
Willdenow, C. L. 1806. Species Plantarum. Vol. 4, Part 2. Berlin: G. C. Nauk.
———. 1807. De Quelques Nouveaux Palmiers de l'Amerique Meridionale. Mem. Acad. Roy. Sci. Hist. (Bruxelles) 1804: 26–42.
Wright, C. H. 1914. The genus *Morenia.* Kew Bull. 1914: 77–79.
Yocum, H. G. 1961. A method for germinating palm seeds. Principes 5: 31–32.
Young, U. A. 1974. Letter. Principes 18: 61–63.
Yuncker, T. G. 1938. A contribution to the flora of Honduras. Field Mus. Nat. Hist., Bot. Ser. 17: 316.
———. 1940. Flora of the Aguan Valley and the coastal regions near La Ceiba, Honduras. Field Mus. Nat. Hist., Bot. Ser. 9: 245–352.
Zona, S. and A. Henderson. 1989. A review of animal-mediated seed dispersal of palms. Selbyana 11: 6–21.
Zuccarini, J. G. 1834. Nachtrag. Allg. Gartenzeitung 2: 245–246.

APPENDIX A

ADDITIONAL AUTHOR CITATIONS

Additional citations of authors (other than the original one) who discussed, listed, illustrated, or otherwise treated the species are listed below. Species are arranged alphabetically by the accepted name.

Chamaedorea Willd.; Willdenow, 1807; Martius, 1824a; 1824b; 1839; Kunth, 1841; Walpers, 1852; Bentham & Hooker, 1883; Drude, 1887; 1897; Burret, 1933a; Bailey, 1933; Moore, 1973b; Dransfield & Uhl, 1986; Uhl & Dransfield, 1987. *Morenia* Ruiz & Pav.; Ruiz & Pavon, 1798; Martius, 1824a; 1837; Sprengel, 1825; Endlicher, 1837; Kunth, 1841; Wendland, 1854; Seemann, 1856; Drude, 1882a; 1889; Wright, 1914; Burret, 1933a; 1936a; Bailey, 1935; Chittenden, 1951; Dahlgren, 1959; Glassman, 1972. *Nunnezharia* Ruiz & Pav.; Ruiz & Pavon, 1798; Spruce, 1871; Kuntze, 1891. ("*Nunnezharoa*"). *Kunthia* Humb. & Bonpl.; Humboldt, Bonpland & Kunth, 1816; Kunth, 1822; Martius, 1824a; 1837; 1839, t. 142; 1849; Sprengel, 1825; Schultes, 1830; Bentham & Hooker, 1880; Drude, 1889. *Chamaedorea* sect. *Collinia* Liebm. ex Mart.; Walpers, 1852. *Stachyophorbe* (Liebm. ex Mart.) Liebm. ex Klotzsch; Liebmann, 1846, name only; Wendland, 1854; Oersted, 1859. *Chamaedorea* subgenus *Stephanostachys* Klotzsch; Wendland, 1854; Hodel, 1990d. *Chamaedorea* subgenus *Chamaedorea* (Mart.) H. A. Wendl.; Oersted, 1859; Burret, 1933a. *Chamaedorea* subgenus *Collinia* (Liebm. ex Mart.) H. A. Wendl.; Bentham & Hooker, 1880. *Chamaedorea* subgenus *Eleutheropetalum* H. A. Wendl.; Bentham & Hooker, 1880. *Collinia* (Liebm. ex Mart.) Liebm. ex Oerst.; Liebmann, 1846, name only; Burret, 1933a; McCurrach, 1960. *Chamaedorea* subgenus *Chamaedoropsis* Oerst.; Burret, 1933a, as section (in part). *Dasystachys* Oerst.; Burret, 1933a; McCurrach, 1960. *Eleutheropetalum* (H. A. Wendl.) Oerst.; Oersted, 1863, t. 3; Burret, 1933a; Langlois, 1976.

Chamaedorea adscendens (Dammer) Burret; Burret, 1934a; Standley & Record, 1936; Standley & Steyermark, 1958; Moore, 1963; 1976; Glassman, 1972; Breedlove, 1986.

Chamaedorea allenii L. H. Bailey; Bailey, 1943b; Glassman, 1972; Hodel, 1990d.

Chamaedorea alternans H. A. Wendl.; Hemsley, 1885; Standley, 1920; Guillaumin, 1923b; Burret, 1933a; Glassman, 1972; Moore, 1976; Hodel, 1990d. As *Chamaedorea tepejilote,* Ibarra-Manriquez, 1988; Oyama, 1990; 1991; Oyama & Dirzo, 1988; 1991; Oyama & Mendoza, 1990.

Chamaedorea amabilis H. A. Wendl. ex Dammer; Guillaumin, 1923b; Burret, 1933a; Standley, 1937; Chazdon, 1987; Hodel, 1990a. *Chamaedorea coclensis* L. H. Bailey; Bailey, 1943b.

Chamaedorea angustisecta Burret; Burret, 1933a; Macbride, 1960; Glassman, 1972. *Chamaedorea leonis* H. E. Moore; Hammer, 1985; Balslev & Moraes, 1989.

Chamaedorea arenbergiana H. A. Wendl.; Seemann, 1856; Mueller, 1858; Croucher, 1872; Kerchove, 1878, fig. 33; Nicholson, 1884; Anonymous, 1885; Hooker, 1885; Siebert & Voss, 1896; Dammer, 1897; Thistleton-Dyer, 1900; Ramirez-Goyena, 1911; Ruffo, 1920; Guillaumin, 1923b; Blatter, 1925; Standley, 1931; 1937; Bailey, 1935; Standley & Record, 1936; Chittenden, 1951; Standley & Steyermark, 1958; Dahlgren, 1959, pl. 92–93; McCurrach, 1960; Moore, 1963; 1976; Glassman, 1972; Stevenson, 1974; Hoyos & Braun, 1984; Jones, 1984; Hammer, 1985; Breedlove, 1986; Braun & Chitty, 1987; Hodel, 1990d. *Spathoscaphe arenbergiana* (H. A. Wendl.) Oerst.; Oersted, 1963, t. 7, fig. 29. *Chamaedorea gracilis* Hort.; Seemann, 1856. *Chamaedorea latifrons* Hort.; Seemann, 1856; Nicholson, 1884. *Chamaedorea oblongata* Hort.; Seemann, 1856. *Chamaedorea latifolia* Hort.; Anonymous, 1885.

Chamaedorea atrovirens Mart.; Wendland, 1854; Seemann, 1856; Kerchove, 1878; Burret, 1933a; Bailey, 1935; Chittenden, 1951; Moore, 1963; Glassman, 1972.

Chamaedorea brachyclada H. A. Wendl.; Guillaumin, 1923b; Burret, 1933a; Bailey, 1943b; Glassman, 1972; Grayum & de Nevers, 1988; Hodel & Uhl, 1990b.

Chamaedorea brachypoda Standl. & Steyerm.; Standley & Steyermark, 1958; Wait, 1961, fig. 4; Moore, 1963; 1976; Glassman, 1972; Graf, 1973; Braun, 1983, as *C. stolonifera*; Hoyos & Braun, 1984; Jones, 1984; Anonymous, 1985, fig. D; Braun & Chitty, 1987; Douglas, 1987; Uhl & Dransfield, 1987, illustrated p. 126; Krempin, 1990.

Chamaedorea carchensis Standl. & Steyerm.; Standley & Steyermark, 1958; Glassman, 1972.

Chamaedorea cataractarum Mart.; Walpers, 1852; Mueller, 1858; Kerchove, 1878; Standley, 1920; Burret, 1933a; Dahlgren, 1959, pl. 100–101; Moore, 1963; 1976; Glassman, 1972; Stevenson, 1974; Everett, 1981; Braun, 1983; Hoyos & Braun, 1984; Jones, 1984; Breedlove, 1986; Braun & Chitty, 1987; Krempin, 1990; Krempin, 1990, as *C. atrovirens*; Minter, 1990; Hodel, 1990d; 1991b. *Stachyophorbe cataractarum* (Mart.) Klotzsch; Liebmann, 1846, name only; Oersted, 1859. *Chamaedorea martiana* H. A. Wendl.; Wendland, 1854; Seemann, 1856; Decaisne & Naudin, 1862; Croucher, 1872; Kerchove, 1878; Nicholson, 1884; Anonymous, 1885; Dammer, 1897; Velenovsky, 1913; Standley, 1920; Guillaumin, 1923b; Burret, 1933a; Chittenden, 1951; Dahlgren, 1959, pl. 127; Moore, 1963; Glassman, 1972; Everett, 1981; Breedlove, 1986. *Chamaedorea flexuosa* Hort.; Seemann, 1856. *Chamaedorea lindeniana* Hort.; Seemann, 1856. *Vadia jotolana* O. F. Cook; Cook, 1947b.

Chamaedorea costaricana Oerst.; Kerchove, 1878; Burret, 1933a; Standley, 1937; Allen, 1956, as *C. woodsoniana*; Dahlgren, 1959, pl. 104; Moore, 1963; 1976; Glassman, 1972; Graf, 1973; Holdridge & Poveda, 1975; Blombery & Rodd, 1982; Braun, 1983; Jones, 1984; Braun & Chitty, 1987; Lane, 1988; Krempin, 1990, as *C. woodsoniana*; Hodel, 1991b. *Chamaedorea biolleyi* Guillaum.; Guillaumin, 1923a; 1923b. *Chamaedorea linearia* L. H. Bailey; Bailey, 1943b; Galeano & Bernal, 1987. *Chamaedorea seibertii* L. H. Bailey; Bailey, 1943b. *Omanthe costaricana* O. F. Cook; Cook, 1943a; Hertrich, 1951.

Chamaedorea dammeriana Burret; Standley, 1937; Glassman, 1972. *Chamaedorea wedeliana* L. H. Bailey; Bailey, 1943b; Glassman, 1972.

Chamaedorea deckeriana (Klotzsch) Hemsl.; Croucher, 1872, name only; Anonymous, 1885; Dammer, 1904b; Guillaumin, 1923b; Standley & Steyermark, 1958; Chazdon, 1985; 1987;

Hodel, 1990d. *Stachyphorbe deckeriana* Klotzsch; Wendland, 1854. *Dasystachys deckeriana* (Klotzsch) Oerst.; 1863, t. 6; Burret, 1933a; Dahlgren, 1959, pl. 178.

Chamaedorea donnell-smithii Dammer; Dammer, 1903, name only; Glassman, 1972.

Chamaedorea elatior Mart.; Heynhold, 1840; Wendland, 1853c; 1854; Oersted, 1859; Croucher, 1872; Kerchove, 1878; Nicholson, 1884; Anonymous, 1885; Knuth, 1904; Gerome, 1911; Roster, 1913; Standley, 1920; Guillaumin, 1923a; 1923b; Burret, 1933a; Bailey, 1935; Chittenden, 1951; Standley & Steyermark, 1958; Dahlgren, 1959, pl. 106–107; McCurrach, 1960; Moore, 1963; 1976; Glassman, 1972; Parthasarathy, 1972; Graf, 1973, as *C. martiana*; Stevenson, 1974; Everett, 1981; Blomberry & Rodd, 1982; Jones, 1984; Aguilar Amar, 1986; Breedlove, 1986; Braun & Chitty, 1987; Ibarra-Manriquez, 1988; Krempin, 1990; Minter, 1990; Hodel, 1991b. *Chamaedorea scandens* Liebm.; Liebmann, 1846, name only; Walpers, 1852; Wendland, 1854; Seemann, 1856; Mueller, 1858; Decaisne & Naudin, 1862; Kerchove, 1878; Nicholson, 1884; Anonymous, 1885; Guillaumin, 1923b; Burret, 1933a; Dahlgren, 1959, pl. 152–154; Breedlove, 1986. *Chamaedorea affinis* Liebm.; Liebmann, 1846, name only; Walpers, 1852; Mueller, 1858; Oersted, 1859; Kerchove, 1878; Standley, 1920; Burret, 1933a; Dahlgren, 1959, pl. 90–91; Glassman, 1972. *Chamaedorea montana* Liebm.; Liebmann, 1846, name only; Walpers, 1853; Wendland, 1854; Mueller, 1858; Oersted, 1859; Kerchove, 1878; Standley, 1920; Burret, 1933a; Dahlgren, 1959, pl. 129–131; Glassman, 1972. *Chamaedorea resinifera* H. A. Wendl.; Wendland, 1854; Seemann, 1856; Oersted, 1859; Guillaumin, 1923b; Burret, 1933a; Dahlgren, 1959, pl. 151. *Chamaedorea desmoncoides* H. A. Wendl.; Wendland, 1854; Seemann, 1856; Oersted, 1859; Kerchove, 1878; Nicholson, 1884; Anonymous, 1885; Dammer, 1897; Roster, 1913; Ruffo, 1920; Burret, 1933a; Bailey, 1935; Chittenden, 1951; Reidel, 1957; Dahlgren, 1959, pl. 105. *Chamaedorea elatior* var. *desmoncoides* H. A. Wendl. ex Dammer; Guillaumin, 1923b. *Chamaedorea scandens* var. *bambusoides* H. A. Wendl. ex Dammer; Guillaumin, 1923b. *Chamaedorea bambusoides* Gerome; Ruffo, 1920; Guillaumin, 1923a; 1923b; Bailey, 1935; Reidel, 1957. *Chamaedorea bambusoides* var. *graminifolia* Gerome; Guillaumin, 1923b. *Chamaedorea bambusoides* var. *juncea* Gerome; Guillaumin, 1923b. *Chamaedorea regia* Hort.; Seemann, 1856. *Chamaedorea repens* Hort.; Seemann, 1856. *Chamaedorea robusta* Hort.; Seemann, 1856. *Anothea scandens* O. F. Cook; Cook, 1947b.

Chamaedorea elegans Mart.; Martius, 1837; 1845, t. R, fig. 3; 1849, t. Z(9), fig. 4, t. Z(17), fig. 7; Nees, 1834; Zuccarini, 1834; Heynhold, 1840; Kunth, 1841; Walpers, 1852; Wendland, 1854; Seemann, 1856; Mueller, 1858; Croucher, 1872; Decaisne & Naudin, 1862; Anonymous, 1873; 1876a; 1877b; 1882; 1885; Kerchove, 1878; Drude, 1882b; Van der Meersch, 1882; Nicholson, 1884; 1900, as *C. graminifolia*; Anonymous, 1895, as *C. glaucifolia*; Seibert & Voss, 1896; Dammer, 1897; Kohler, 1913; 1928; Roster, 1913; Chabaud, 1915; Ruffo, 1920; Standley, 1920; Guillaumin, 1923b; Bailey, 1935; Hertrich, 1951, as *C. graminifolia*; Reidel, 1957; Standley & Steyermark, 1958; McCurrach, 1960, as *C. graminifolia*; Moore, 1963; 1976; Kromdijk, 1967; Glassman, 1972; Graf, 1973; Stevenson, 1974; Everett, 1981; Stewart, 1981; Blombery & Rodd, 1982; Braun, 1983; Cowan, 1983; Hoyos & Braun, 1984; Jones, 1984; Aguilar Amar, 1986; Breedlove, 1986; Braun & Chitty, 1987; Lane, 1988; Krempin, 1990; Krempin, 1990, as *C. monostachys* and p. 95 illustrated as *C. radicalis*; Minter, 1990. *Chamaedorea elegans* var. *angustifolia* M. Martens & Galeotti; Walpers, 1849; Wendland, 1854; Kerchove, 1878. *Chamaedorea humilis* Liebm. ex Mart.; Walpers, 1852; Wendland, 1854; Mueller, 1858; Oersted, 1859; Standley, 1920; Burret, 1933a; Dahlgren, 1959, pl. 117–119; Breedlove, 1986. *Chamaedorea deppeana* Klotzsch; Wendland, 1854. *Chamaedorea helleriana* Klotzsch; Wendland, 1854; Nicholson, 1884. *Collinia elegans* (Mart.) Oerst.; Burret, 1933a; Dahlgren, 1959, pl. 165–168; McCurrach, 1960; Synge, 1969; Macoboy, 1976. *Collinia humilis* (Liebm. ex Mart.) Oerst.; Liebmann, 1846, name only; Oersted, 1859. *Chamaedorea pulchella* Linden ex Hemsl.; Linden, 1885, name only; Thistleton-Dyer, 1900; Guillaumin, 1923b; Chittenden, 1951; Synge, 1969. *Kunthia deppii* Hort. in Zucc.; Martius, 1849; Klotzsch, 1852. *Chamaedorea lindeniana* Hort.; Seemann, 1856. *Chamaedorea martiana* Hort.; Seemann, 1856. *Chamaedorea elegantissima* Hort.; Anonymous, 1885; Dammer, 1897; Roster, 1913. *Neanthe bella* O. F. Cook; Cook & Joyner, 1938; 1939; Cook, 1943a. *Neanthe elegans* (Mart.) O. F. Cook; Cook & Joyner, 1939; Cook, 1943a; Chittenden, 1951. *Chamaedorea neesiana* O. F. Cook; Cook, 1943a.

Chamaedorea ernesti-augustii H. A. Wendl.; Wendland, 1854; Hooker, 1855a; 1855b; Seemann, 1856; Mueller, 1858; Naudin, 1858a; Decaisne & Naudin, 1862; Anonymous, 1865; 1873; 1885; Croucher, 1872; Kerchove, 1878; Nicholson, 1884; 1885; Weiss, 1893; Dammer, 1897; 1904b; Smith, 1905; Knuth, 1904; Kohler, 1913, pl. 28 illustrated as *C. concolor*; Roster, 1913; Chabaud, 1915; Ruffo, 1920; Standley, 1920; Guillaumin, 1923; Hesdorffer, 1924; Bailey, 1935; Standley & Record, 1936, as *C. geonomiformis*; Hertrich, 1951; Standley & Steyermark, 1958; McCurrach, 1960, p. 43 illustrated as *C. geonomiformis*; Moore, 1963; 1976; Glassman, 1972; Graf, 1973; Stevenson, 1974; Everett, 1981; Blombery & Rodd, 1982; Braun, 1983; Hoyos & Braun, 1984; Jones, 1984; Aguilar Amar, 1986; Breedlove, 1986; Braun & Chitty, 1987; Douglas, 1987; Ibarra-Manriquez, 1988; Lane, 1988; Krempin, 1990; Krempin, 1990, p. 92 illustrated as *C. geonomiformis*; Minter, 1990. *Eleutheropetalum ernesti-augustii* (H. A. Wendl.) Oerst.; Oersted, 1863, t. 3, figs. 7–30; Burret, 1933a; Chittenden, 1951; Dahlgren, 1959, pl. 205–206; Langlois, 1976. *Chamaedorea latifrons* Hort.; Seemann, 1856. *Chamaedorea simplicifrons* Hort.; Wendland, 1854; Seemann, 1856; Nicholson, 1884; Chabaud, 1915; Chittenden, 1951. *Hyospathe elegans* Hort.; Seemann, 1856; Burret, 1933a.

Chamaedorea falcifera H. E. Moore; Moore, 1976; Standley & Steyermark, 1958; Glassman, 1972; Minter, 1990.

Chamaedorea ferruginea H. E. Moore; Glassman, 1972.

Chamaedorea fragrans (Ruiz & Pav.) Mart.; Martius, 1849; Sprengel, 1825; Heynhold, 1840; Kunth, 1841; Wendland, 1854; Mueller, 1858; Decaisne & Naudin, 1862; Kerchove, 1878; Drude, 1882a; Anonymous, 1885; Burret, 1933a; Dahlgren, 1959, pl. 111; Macbride, 1960; McCurrach, 1960; Moore, 1963; 1976; Glassman, 1972; Braun & Chitty, 1987; Douglas, 1987; Balslev & Moraes, 1989; Minter, 1990. *Nunnezharia fragrans* Ruiz & Pav.; Oerst., 1859; 1863, t. figs. 18–20; Spruce, 1871; Kuntze, 1891. *Chamaedorea pavoniana* H. A. Wendl. ex Dammer; Guillaumin, 1923b; Burret, 1933a; Macbride, 1960. *Chamaedorea ruizii* H. A. Wendl. ex Dammer; Anonymous, 1885; Dammer, 1897; Guillaumin, 1923b; Burret, 1933a. *Chamaedorea verschaffeltii* Hort.; Tanfani, 1886. *Chamaedorea gratissima* Hort.; Thistleton-Dyer, 1900; Guillaumin, 1923b.

Chamaedorea geonomiformis H. A. Wendl.; Wendland, 1854; Seemann, 1856; Oersted, 1859; 1863, t. 5, figs. 1–17; Croucher, 1872; Anonymous, 1876c; 1885; 1883; Kerchove, 1878; Nicholson, 1884; Dammer, 1897; 1904b; Guillaumin, 1923b; Hesdorffer, 1924; Standley, 1931; Burret, 1933a; Bailey, 1935; Yuncker, 1938; Chittenden, 1951; Reidel, 1957; Standley & Steyermark, 1958, fig. 41F; Dahlgren, 1959, pl. 113; Moore, 1963; 1976; Glassman, 1972; Cowan, 1983; Jones, 1984; Breedlove, 1986; Douglas, 1987; Galeano & Bernal, 1987; Lane, 1988. *Nunnezharia geonomiformis* (H. A. Wendl.) Hook. f.; Kuntze, 1891. *Chamaedorea fenestrata* Hort.; Seemann, 1856. *Chamaedorea humilis* Hort.; Seemann, 1856; Burret, 1933a. *Geonoma fenestrata* Hort.; Seemann, 1856; Dammer, 1904b. *Geonoma humilis* Hort.; Burret, 1933a.

Chamaedorea glaucifolia H. A. Wendl.; Mueller, 1858; Oersted, 1859; Decaisne & Naudin, 1862; Croucher, 1872; Kerchove, 1878; Nicholson, 1884; Anonymous, 1885; 1895 as *C. elegans*; Dammer, 1897; Prain, 1912; Guillaumin, 1923; Burret, 1933; Bailey, 1935 as *C. elegans*; Chittenden, 1951; Reidel, 1957; Dahlgren, 1959, pl. 114; McCurrach, 1960; Muirhead, 1961 as *C. graminifolia*; Moore, 1963; 1967; Synge, 1969; Glassman, 1972; Stevenson, 1974; Blombery & Rodd, 1982; Braun, 1983; Hoyos & Braun, 1984; Jones, 1984; Braun & Chitty, 1987; Lane, 1988; Krempin, 1990; Minter, 1990. *Chamaedorea crucifolia* Hort.; Hooker, 1884; Dammer, 1897.

Chamaedorea graminifolia H. A. Wendl.; Seemann, 1856; Mueller, 1858; Oersted, 1859; Croucher, 1872; Kerchove, 1878, t. 13; Anonymous, 1883; 1885; Nicholson, 1884; Dammer, 1897; Roster, 1913; Ruffo, 1920; Guillaumin, 1923b; Standley, 1920; Standley & Calderon, 1925; Standley, 1931; Burret, 1933a; 1934; Standley & Record, 1936; Yuncker, 1938; 1940; Chittenden, 1951; Dahlgren, 1959, pl. 115; Moore, 1963; 1976; Hawkes, 1967; Glassman, 1972; Seymour, 1980; Everett, 1981; Hodel, 1991b. *Chamaedorea schippii* Burret; Standley & Record, 1936; Standley & Steyermark, 1958; Moore, 1963; 1976; Glassman, 1972; Breedlove, 1986; Krempin, 1990.

Chamaedorea klotzschiana H. A. Wendl.; Seemann, 1856; Mueller, 1858; Oersted, 1859; Kerchove, 1878; Standley, 1920; Guillaumin, 1923b; Burret, 1933a; Dahlgren, 1959, pl. 121–122; Moore, 1963; 1976; Glassman, 1972; Graf, 1973; Stevenson, 1974; Blombery & Rodd, 1982; Jones, 1984; Aguilar Amar, 1986; Braun & Chitty, 1987; Lane, 1988; Krempin, 1990; Minter, 1990.

Chamaedorea latisecta (H. E. Moore) A. Gentry; Galeano & Bernal, 1987. *Morenia latisecta* H. E. Moore; Glassman, 1972.

Chamaedorea lehmannii Burret; Standley & Steyermark, 1958; Glassman, 1972.

Chamaedorea liebmannii Mart.; Walpers, 1853; Wendland, 1854; Mueller, 1858; Kerchove, 1878; Standley, 1920; Burret, 1933a; Dahlgren, 1959, pl. 124–125; Glassman, 1972. *Collinia elatior* Liebm. ex Oerst.; Liebmann, 1846, name only. *Chamaedorea lepidota* H. A. Wendl.; Wendland, 1854; Seemann, 1856; Oersted, 1859; Kerchove, 1878; Guillaumin, 1923b; Burret, 1933a; Dahlgren, 1959, pl. 126; Glassman, 1972; Moore, 1976; Aguilar Amar, 1986. *Chamaedorea velutina* Hort.; Seemann, 1856. *Chamaedorea aequalis* Standl. & Steyerm.; Standley & Steyermark, 1958; Glassman, 1972; Breedlove, 1986.

Chamaedorea linearis (Ruiz & Pav.) Mart.; Martius, 1849; Sprengel, 1825; Knuth, 1841; Wendland, 1854; Mueller, 1858; Kerchove, 1878. *Morenia linearis* (Ruiz & Pav.) Burret; Burret, 1936a; Dahlgren, 1959, pl. 329; Macbride, 1960; Glassman, 1972. *Morenia fragrans* Ruiz & Pav.; Ruiz & Pavon, 1798; Sprengel, 1825; Martius, 1837; 1844, t. 3, fig. 1, t. 16; 1849; Knuth, 1841; Seemann, 1856; Kerchove, 1878; Drude, 1882a; Nicholson, 1884, as *C. fragrans*; Wright, 1914; Guillaumin, 1923b, as *C. fragrans*; Bailey, 1935; Burret, 1936a; Chittenden, 1951, as *C. fragrans*; Macbride, 1960; Glassman, 1972; Heywood, 1978, p. 302, fig. 10, as *C. fragrans*. *Kunthia montana* Humb. & Bonpl.; Humboldt, Bonpland & Kunth, 1816; Kunth, 1822; Sprengel, 1825; Martius, 1826, t. 101, figs. 6 and 7; 1837; 1839, t. 142; 1849; Schultes, 1830, Drude, 1882a. *Morenia poeppigiana* Mart.; 1839, t. 140–141; 1849; Knuth, 1841; Seemann, 1856; Mueller 1858; Hooker, 1865; Spruce, 1871; Kerchove, 1878; Drude, 1882a; Wright, 1914; Burret, 1936a; Dahlgren, 1959, pl. 330–331; Macbride, 1960; Glassmann, 1972. *Morenia corallina* Karst.; Karsten, 1866, t. 177; Mueller, 1858; Prain, 1913; 1914; Bailey, 1935; Burret, 1936a; Dugand, 1940; Chittenden, 1951; Glassman, 1972. *Morenia lindeniana* H. A. Wendl.; Croucher, 1872 as *C. lindeniana*; Drude, 1882a; Wright, 1914; Burret, 1936a; Dugand, 1940; Chittenden, 1951; Glassman, 1972. *Morenia montana* (Humb. & Bonpl.) Burret; Burret, 1936a; Dugand, 1940; Glassman; 1972. *Morenia caudata* Burret; Glassman, 1972; Balslev & Barfod, 1986. *Morenia macrocarpa* Burret; Macbride, 1960; Glassman, 1972. *Morenia robusta* Burret; Dugand, 1940; 1961, as *C. robusta* although no formal transfer was made; Glassman, 1972. *Chamaedorea polyclada* Burret; Glassman, 1972; Balslev & Barfod, 1986. *Morenia microspadix* Burret; Glassman, 1972; Balslev & Barfod, 1986. *Chamaedorea poeppigiana* (Mart.) A. Gentry; Balslev & Moraes, 1989; Bergmann, 1989. *Chamaedorea formosa* Hort.; Anonymous, 1876a; 1876b; 1877a; 1877b; 1885; Nicholson, 1884; 1900; Thistleton-Dyer, 1900; Guillaumin, 1923b; Bailey, 1935.

Chamaedorea lucidifrons L. H. Bailey; Bailey, 1943b; Glassman, 1972.

Chamaedorea macrospadix Oerst.; Kerchove, 1878; Nicholson, 1884; Burret, 1933a; Standley, 1937; Glassman, 1972.

Chamaedorea metallica O. F. Cook ex H. E. Moore; Glassman, 1972; Graf, 1973; Stevenson, 1974; Langlois, 1976, as *Eleutheropetalum metallica* (no formal transfer made); Moore, 1976; Heywood, 1978, as *C. tenella*; Everett, 1981; Blombery & Rodd, 1982; Braun, 1983; Hoyos & Braun, 1984; Jones, 1984; Aguilar Amar, 1986; Braun & Chitty, 1987; Douglas, 1987; Dransfield & Cook, 1987; Uhl & Dransfield, 1987, illustrated p. 93; Lane, 1988, as *C. tenella*; Krempin, 1990; Minter, 1990.

Chamaedorea microphylla H. A. Wendl.; Croucher, 1872; Kerchove, 1878; Nicholson, 1884; Anonymous, 1885; Dammer, 1897; Guillaumin, 1923b; Burret, 1933a; Standley, 1937; Bailey, 1943b; Dahlgren, 1959, pl. 128; Glassman, 1972.

Chamaedorea microspadix Burret; Wait, 1961, fig. 5; Moore, 1963; 1976; Glassman, 1972; Stevenson, 1974; Stewart, 1981; Blombery & Rodd, 1982; Braun, 1983; Hoyos & Braun, 1984; Jones, 1984; 1985; Braun & Chitty, 1987; Lane, 1988; Krempin, 1990; Krempin, 1990, as *C. costaricana* and p. 91 illustrated as *C. erumpens*; Minter, 1990; Hodel, 1991b.

Chamaedorea murriensis Galeano; *Chamaedorea macroloba* Galeano; Galeano & Bernal, 1987.

Chamaedorea neurochlamys Burret; Standley & Steyermark, 1958; Moore, 1963; 1976; Glassman, 1972; Cowan, 1983; Sousa & Cabrera, 1983; Breedlove, 1986.

Chamaedorea nubium Standl. & Steyerm.; Standley & Steyermark, 1958; Glassman, 1972; Breedlove, 1986.

Chamaedorea oblongata Mart.; Martius, 1845, t. Z(4); 1849; Heynhold, 1840; Kunth, 1841; Mueller, 1858; Kerchove, 1878;

Nicholson, 1884; Anonymous, 1885; Dammer, 1897, illustrated as *Morenia corallina*; Ruffo, 1920; Guillaumin, 1923b; Marque, 1923; Kohler, 1928, pl. 1, as *C. concolor*; Burret, 1933a; Chittenden, 1951; Reidel, 1957; Standley & Steyermark, 1958, fig. 41G; Dahlgren, 1959, pl. 132–135; Moore, 1963; 1976; Glassman, 1972; Stevenson, 1974; Seymour, 1980; Braun, 1983; Morley, 1983; Hoyos & Braun, 1984; Jones, 1984; Aguilar Amar, 1986; Breedlove, 1986; Braun & Chitty, 1987; Ibarra-Manriquez, 1988; Krempin, 1990; Krempin, 1990, p. 90 illustrated as *C. concolor*; Minter, 1990. *Chamaedorea lunata* Liebm.; Liebmann, 1846, name only; Walpers, 1852; Wendland, 1854; Seemann, 1856; Mueller, 1858; Oersted, 1859; Croucher, 1872; Nicholson, 1884; Dammer, 1903; Standley, 1920; Guillaumin, 1923b. *Chamaedorea aurantiaca* Hort.; Seemann, 1856. *Chamaedorea biloba* Hort.; Seemann, 1856. *Chamaedorea lindeniana* Hort.; Seemann, 1856. *Chamaedorea scandens* Hort.; Seemann, 1856. *Chamaedorea schiedeana* Hort.; Seemann, 1856. *Chamaedorea corallina* Hort.; Kohler, 1913; Roster, 1913; Ruffo, 1920.

Chamaedorea oreophila Mart.; Walpers, 1852; Kerchove, 1878; Burret, 1933a; Dahlgren, 1959, pl. 136–138; Moore, 1963; 1976; Glassman, 1972; Graf, 1973, as *C. cataractarum*; Morley, 1983; Aguilar Amar, 1986; Lane, 1988, as *C. cataractarum*; Hodel, 1990d. *Stachyophorbe montana* Liebm. ex Oerst.; Liebmann, 1846, name only; Oersted, 1863, t. fig. 17. *Chamaedorea monostachys* Burret; Moore, 1963; Glassman, 1972; Aguilar Amar, 1986. *Stachyophorbe filipes* O. F. Cook; Cook, 1947a.

Chamaedorea pachecoana Standl. & Steyerm.; Standley & Steyermark, 1958; Glassman, 1972.

Chamaedorea paradoxa H. A. Wendl.; Kerchove, 1878; Guillaumin, 1923b; Burret, 1933a; Dahlgren, 1959, pl. 140; Glassman, 1972.

Chamaedorea parvifolia Burret; Standley, 1937; Glassman, 1972.

Chamaedorea parvisecta Burret; Standley & Steyermark, 1958; Glassman, 1972. *Chamaedorea pulchra* Burret; Standley & Steyermark, 1958; Glassman, 1972. *Chamaedorea digitata* Standl. & Steyerm.; Standley & Steyermark, 1958; Glassman, 1972.

Chamaedorea pauciflora Mart.; Martius, 1849; Kunth, 1841; Wendland, 1854; Mueller, 1858; Kerchove, 1878; Jahn, 1908; Burret, 1933a; Dugand, 1940; Dahlgren, 1959, pl. 141; Macbride, 1960; Dugand, 1961; Glassman, 1972; Galeano, 1991. *Morenia integrifolia* Trail; Kerchove, 1878; Drude, 1882a; Wright, 1914. *Morenia integrifolia* var. *nigricans* Trail; Drude, 1882a; Wright, 1914. *Chamaedorea integrifolia* (Trail); Dammer, 1907b; Burret, 1933a; Schultes, 1950; Macbride, 1960; Dugand, 1961; Glassman, 1972; Balslev & Barfod, 1986; Bergmann, 1989; Rodrigues, 1989; Kahn & Mejia, 1991.

Chamaedorea pinnatifrons (Jacq.) Oerst.; Kerchove, 1878; Burret, 1933a; Dahlgren, 1959, pl. 142–143; Braun, 1968; Glassman, 1972; Braun, 1983; Hoyos & Braun, 1984; Jones, 1984; Balslev & Barfod, 1986; Braun & Chitty, 1987; Wessels Boer, 1988; Bergmann, 1989; Krempin, 1990; Galeano, 1991. *Chamaedorea gracilis* Willd.; Willdenow, 1805, name only; Martius, 1842b; 1837; 1849; Sprengel, 1825; Kunth, 1841; Klotzsch, 1847; Wendland, 1854; Mueller, 1858; Decaisne & Naudin, 1862; Drude, 1882a; Jahn, 1908; Guillaumin, 1923b. *Chamaedorea concolor* Mart.; Martius, 1849; Kunth, 1841; Mueller, 1858; Oersted, 1859; Decaisne & Naudin, 1862; Kerchove, 1878; Anonymous, 1885; Dammer, 1897; Nicholson, 1900; Ruffo, 1920; Guillaumin, 1923b; Burret, 1933a; Dugand, 1940; Chittenden, 1951; Dahlgren, 1959, pl. 103; Moore, 1963; Glassman, 1972; Aguilar Amar, 1986; Breedlove, 1986; Ibarra-Manriquez, 1988; Krempin, 1990; Minter, 1990. *Chamaedorea lanceolata* (Ruiz & Pav.) Kunth; Martius, 1844, t. 6, fig. 3; 1849, t. 16A; Walpers, 1852; Mueller, 1858; Oersted, 1859; Kerchove, 1878; Drude, 1881, t. 125; 1882a; 1887; Dammer, 1907; Burret, 1933a; Schultes, 1950; Dahlgren, 1959, pl. 123; Macbride, 1960; Dugand, 1961; Glassman, 1972; Wessels Boer, 1988; Balslev & Moraes, 1989. *Chamaedorea conocarpa* Mart.; Martius, 1849; Walpers, 1849; Wendland, 1854; Mueller, 1858; Kerchove, 1878; Burret, 1933a; Glassman, 1972; Balslev & Moraes, 1989. *Chamaedorea lindeniana* H. A. Wendl.; Wendland, 1854; Heynhold, 1840, name only; Seemann, 1856; Oersted, 1859; Decaisne & Naudin, 1862; Croucher, 1872; Kerchove, 1878; Nicholson, 1884; Anonymous, 1885; Standley, 1920; Guillaumin, 1923b; Yuncker, 1938; 1940; Chittenden, 1951 as *C. karwinskyana*; Breedlove, 1986. *Chamaedorea bartlingiana* H. A. Wendl.; Seemann, 1856; Mueller, 1858; Oersted, 1859; Kerchove, 1878; Guillaumin, 1923; Burret, 1933; Dugand, 1940; Dahlgren, 1959, pl. 94; Glassman, 1972; Wessels Boer, 1988. *Chamaedorea brevifrons* H. A. Wendl.; Seemann, 1856; Oersted, 1859; Kerchove, 1878; Nicholson, 1884; Guillaumin, 1923b; Burret, 1933a; Dugand, 1940; Dahlgren, 1959, pl. 97–98; Glassman, 1972; Galleano & Bernal, 1987. *Chamaedorea flavovirens* H. A. Wendl.; Seemann, 1856; Mueller, 1858; Oersted, 1859; Kerchove, 1878; Guillaumin, 1923b; Burret, 1933a; Bailey, 1943a; Dahlgren, 1959, pl. 109–110. *Chamaedorea bracteata* H. A. Wendl.; Kerchove, 1878; Guillaumin, 1923b; Burret, 1933a; Dahlgren, 1959, pl. 96; Glassman, 1972. *Chamaedorea bifurcata* Oerst.; Kerchove, 1878; Burret, 1933a; Standley, 1937; Dahlgren, 1959, pl. 95; Glassman, 1972. *Chamaedorea membranacea* Oerst.; Kerchove, 1878; Ramirez-Goyena, 1911; Burret, 1933; Hawkes, 1967; Glassman, 1972. *Chamaedorea pacaya* Oerst.; Kerchove, 1878; Standley, 1931; 1937; Burret, 1933a; Bailey, 1943a, fig. 256; 1943b; Dahlgren, 1959, pl. 139; McCurrach, 1960; Moore, 1963; 1976; Glassman, 1972. *Nunnezharia geonomoides* Spruce; Kuntze, 1891. *Chamaedorea geonomoides* (Spruce) Drude; Burret, 1933a; Macbride, 1960; Glassman, 1972. *Chamaedorea boliviensis* Dammer; Burret, 1933a; Macbride, 1960; Glassman, 1972. *Chamaedorea depauperata* Dammer; Burret, 1933a; Macbride, 1960; Glassman, 1972. *Chamaedorea minor* Burret; Burret, 1933a; Glassman, 1972. *Chamaedorea concinna* Burret; Dugand, 1940; 1961; Glassman, 1972. *Chamaedorea dryanderae* Burret; Dugand, 1940; 1961; Glassman, 1972; Galeano & Bernal, 1987. *Chamaedorea heilbornii* Burret; Glassman, 1972; Balslev & Barfod, 1986. *Chamaedorea herrerae* Burret; MacBride, 1960; Glassman, 1972. *Chamaedorea holmgrenii* Burret; Glassman, 1972; Balslev & Barfod, 1986. *Chamaedorea hoppii* Burret; Glassman, 1972; Balslev & Barfod, 1986. *Chamaedorea kalbreyeriana* H. A. Wendl. ex Burret; Dugand, 1940; Glassman, 1972. *Chamaedorea macroloba* Burret; Glassman, 1972; Balslev & Moraes, 1989. *Chamaedorea micrantha* Burret; Standley & Steyermark, 1958; Glassman, 1972. *Chamaedorea rhombea* Burret; Standley, 1937; Glassman, 1972. *Chamaedorea aguilariana* Standl. & Steyerm.; Standley & Steyermark, 1958; Glassman, 1972; Breedlove, 1986. *Chamaedorea lindeniana* Hort.; Seemann, 1856.

Chamaedorea pittieri L. H. Bailey; Bailey, 1943b; Glassman, 1972. *Chamaedorea hageniorum* L. H. Bailey; Bailey, 1943b; Glassman, 1972.

Chamaedorea pochutlensis Liebm.; Liebmann, 1846, name only;

Walpers, 1852; Mueller, 1858; Oersted, 1859; Kerchove, 1878; Standley, 1920; Burret, 1933a; Dahlgren, 1959, pl. 144–146; Glassman, 1972; Moore, 1976; Breedlove, 1986; Krempin, 1990; Hodel, 1991b. *Chamaedorea karwinskyana* H. A. Wendl.; Wendland, 1854; Seemann, 1856; Oersted, 1858; Decaisne & Naudin, 1862; Anonymous, 1874; 1883; Nicholson, 1900, as synonym of *C. elatior*; Roster, 1913; Standley, 1920; Guillaumin, 1923a; 1923b; Burret, 1933a; 1935; Schulz, 1933; Dahlgren, 1959, pl. 120; Glassman, 1972; Moore, 1976. *Chamaedorea elatior ex parte* Mart., Zuccarini, 1834, p. 246; Martius, 1837, p. 157, t. 126, t. R, figs. 1 and 2 s, t. Z2, fig. 12, Z4, figs. 6–10, Z6, figs. 4–33; Knuth, 1841, t. 3, p. 171; Guillaumin, 1923b; Kohler, 1913, pl. 26; *Chamaedorea elatior* Hort.; Seemann, 1856.

Chamaedorea pumila H. A. Wendl. ex Dammer; Guillaumin, 1923b; Burret, 1933a; Standley, 1937; Dahlgren, 1959, pl. 147; Moore, 1963; Glassman, 1972; Chazdon, 1987. *Chamaedorea nana* N. E. Brown; Prain, 1916; Guillaumin, 1923b; Burret, 1933a; Standley, 1937; Chittenden, 1951; Glassman, 1972.

Chamaedorea pygmaea H. A. Wendl.; Wendland, 1854; Seemann, 1856; Croucher, 1872; Kerchove, 1878; Anonymous, 1885; Dammer, 1897; Standley, 1920; Guillaumin, 1923b; Burret, 1933a; Dugand, 1940; Bailey, 1943b; Dahlgren, 1959, pl. 148; Glassman, 1972; Hodel & Uhl, 1990b. *Stachyophorbe pygmaea* (H. A. Wendl.) Oerst.; Oersted, 1863, t. 4, figs. 1–16. *Chamaedorea terryorum* Standl.; Bailey, 1943b.

Chamaedorea quezalteca Standl. & Steyerm.; Standley & Steyermark, 1958; Glassman, 1972; Hodel, 1991b. *Legnea lacinata* O. F. Cook; Cook, 1943a, fig. 7.

Chamaedorea radicalis Mart.; Walpers, 1852; Wendland, 1854; Seemann, 1856; Mueller, 1858; Oersted, 1859; Kerchove, 1878; Standley, 1920; Burret, 1933a; Moore, 1957; Dahlgren, 1959, pl. 149–150; McCurrach, 1960; Moore, 1963; 1976; Glassman, 1972; Graf, 1973; Stevenson, 1974; Braun, 1983; Hoyos & Braun, 1984; Jones, 1984; Braun & Chitty, 1987; Lane, 1988; Minter, 1990. *Chamaedorea pringlei* S. Wats.; Guillaumin, 1923b; Burret, 1933a; Bailey, 1935; Chittenden, 1951; Hertrich, 1951.

Chamaedorea rigida H. A. Wendl. ex Dammer; Burret, 1933a; Glassman, 1972.

Chamaedorea rojasiana Standl. & Steyerm.; Standley & Steyermark, 1958; Glassman, 1972; Breedlove, 1986.

Chamaedorea sartorii Liebm; Liebmann, 1846, name only; Walpers, 1852; Wendland, 1854; Hooker, 1855c, as *C. elegans*; Seemann, 1856; Mueller, 1858; Naudin, 1858b, as *C. elegans*; Decaisne & Naudin, 1862; La Maout, 1868, as *C. latifolia*; Croucher, 1872, as *C. elegans*; Kerchove, 1878; Nicholson, 1884, fig. 417; 1900; Anonymous, 1885; Dammer, 1897; Habermann, 1902, pl. 9, as *C. concolor*; Roster, 1913; Ruffo, 1920; Standley, 1920; Guillaumin, 1923b; Marque, 1923; Schipper, 1926, as *C. concolor*; Bailey, 1935; Hertrich, 1951; Dahlgren, 1959, pl. 102, as *C. concolor*; 1959, pl. 112, as *C. geonomiformis*; Reidel, 1957; Moore, 1963; 1976; Glassman, 1972; Aguilar Amar, 1986. *Chamaedorea aurantiaca* Brongn.; Decaisne & Naudin, 1862; Guillaumin, 1923b. *Eleutheropetalum sartorii* (Liebm.) Oerst.; Burret, 1933a; Chittenden, 1951; Dahlgren, 1959, pl. 207–210; Langlois, 1976. *Chamaedorea hartwegii* Hort.; Dammer, 1897; Guillaumin, 1923b. *Chamaedorea mexicana* Hort.; Wendland, 1854; Seemann, 1856; Nicholson, 1884; Dammer, 1897; Ruffo, 1920. *Chamaedorea wobstiana* Hort.; Thistleton-Dyer, 1900; Guillaumin, 1923b.

Chamaedorea schiedeana Mart.; Martius, 1837; 1839, t. 138; 1845, t. R, figs. 4–5, t. V, fig. 1, t. Z, fig. 3 (1–9), t. Z2, fig. 13, t. Z5, fig. 6; 1849, t. Z9, fig. 2 (1–41), t. Z17, fig. 8, t. Z23, figs. 1–2; Link, 1834; Zuccarini, 1834; Heynhold, 1840; Knuth, 1841; Schnizlein, 1849; Wendland, 1854; Seemann, 1856; Mueller, 1858; Oersted, 1859; Decaisne & Naudin, 1862; Kerchove, 1878; Roster, 1920; Standley, 1920; Guillaumin, 1923b; Standley & Calderon, 1925; Burret, 1933a; Dahlgren, 1959, pl. 155–157; Moore, 1963; 1976; Glassman, 1972; Cowan, 1983; Aguilar Amar, 1986. *Chamaedorea gracilis* Hort.; Seemann, 1856. *Chamaedorea speciosa* Hort.; Seemann, 1856.

Chamaedorea scheryi L. H. Bailey; Bailey, 1943b; Glassman, 1972; Chazdon, 1987, as *C. aff. graminifolia*; Hodel & Uhl, 1990b.

Chamaedorea seifrizii Burret; Standley, 1930, as *C. graminifolia*; Dahlgren, 1959, pl. 116, as *C. graminifolia*; McCurrach, 1960; Moore, 1963; 1976; Glassman, 1972; Graf, 1973; Stevenson, 1974; Everett, 1981; Stewart, 1981; Blombery & Rodd, 1982; Braun, 1983; Sousa & Cabrera, 1983; Hoyos & Braun, 1984; Jones, 1984; Braun & Chitty, 1987; Krempin, 1990; Krempin, 1990, p. 93 illustrated as *C. graminifolia*; Hodel, 1991b. *Chamaedorea erumpens* H. E. Moore; Standley & Steyermark, 1958; McCurrach, 1960; Moore, 1963; 1976; Glassman, 1972; Graf, 1973; Stevenson, 1974; Macoboy, 1976; Everett, 1981; Stewart, 1981; Braun, 1983; Sousa & Cabrera, 1983; Hoyos & Braun, 1984; Jones, 1984; Braun & Chitty, 1987; Lane, 1988; Minter, 1990.

Chamaedorea simplex Burret; Standley & Steyermark, 1958; Glassman, 1972; Breedlove, 1986.

Chamaedorea skutchii Standl. & Steyerm.; Standley & Steyermark, 1958; Glassman, 1972.

Chamaedorea stenocarpa Standl. & Steyerm.; Standley & Steyermark, 1958; Glassman, 1972; Hodge, 1981, fig. 5; Uhl & Dransfield, 1987, p. 126, illustrated as *Chamaedorea* sp.; Hodel & Uhl, 1990b.

Chamaedorea stolonifera H. A. Wendl. ex Hook. f.; Anonymous, 1885, name only; Dammer, 1897; 1904b; Nicholson, 1900; Thistleton-Dyer, 1990; Standley, 1920; Guillaumin, 1923b; Burret, 1933a; Hertrich, 1951; Moore, 1963; 1976; Glassman, 1972; Langlois, 1976, as *Eleutheropetalum stolonifera* (no formal transfer made); Everett, 1981; Douglas, 1987; Hodge, 1988; Hooper, 1990; Krempin, 1990; Minter, 1990.

Chamaedorea stricta Standl. & Steyerm.; Standley & Steyermark, 1958; Glassman, 1972.

Chamaedorea tenella H. A. Wendl.; Nicholson, 1884; Anonymous, 1885; Dammer, 1897; 1904b; Standley, 1920; Guillaumin, 1923b; Burret, 1933a; Chittenden, 1951; Reidel, 1957; Moore, 1963; 1966, fig. 3; 1976; Glassman, 1972; Jones, 1984; Breedlove, 1986; Grayum & de Nevers, 1988. *Nunnezharia tenella* (H. A. Wendl.) Hook. f.; Kuntze, 1891.

Chamaedorea tenerrima Burret; Standley & Steyermark, 1958; Glassman, 1972; Breedlove, 1986. *Lobia erosa* Cook; Cook, 1947a.

Chamaedorea tepejilote Liebm.; Liebmann, 1846, name only; Walpers, 1852; Mueller, 1858; Hooker, 1873; Kerchove, 1878; Nicholson, 1884; Standley, 1920; Guillaumin, 1923b; Burret, 1933a; Bailey, 1935; Standley & Steyermark, 1958; Dahlgren, 1959, pl. 158–160; McCurrach, 1960; Moore, 1963; 1976; Synge, 1969; Glassman, 1972; Graf, 1973, as *C. pacaya*; Stevenson, 1974; Seymour, 1980; Everett, 1981; Blombery & Rodd, 1982; Cowan, 1983; Hoyos & Braun, 1984; Jones, 1984; Aguilar Amar, 1986; Breedlove, 1986; Braun & Chitty, 1987; Chazdon, 1987; Lane, 1988; Hodel, 1990d; 1991b; Krempin, 1990;

Krempin, 1990, as *C. neurochlamys*; Minter, 1990. *Chamaedorea casperiana* Klotzsch; Wendland, 1854; Seemann, 1856; Mueller, 1858; Croucher, 1872; Kerchove, 1878; Anonymous, 1885; Dammer, 1897; Guillaumin, 1923b; Burret, 1933a; Standley & Steyermark, 1958; Dahlgren, 1959, pl. 99; Glassman, 1972; Hodel, 1990d. *Stephanostachys casperiana* (Klotzsch) Oerst.; Oersted, 1963, t. 7, figs. 1–11. *Stephanostachys tepejilote* (Liebm.) Oerst.; Oersted, 1863, t. 7, figs. 26–28. *Stephanostachys wendlandiana* Oerst.; Oersted, 1863, t. 7, figs. 12–25. *Chamaedorea wendlandiana* (Oerst.) Hemsl.; Croucher, 1872, name only; Kerchove, 1878; Anonymous, 1885; Nicholson, 1884; Dammer, 1897; 1903; Standley, 1920; 1928; 1933; Guillaumin, 1923b; Standley & Calderon, 1925; Bailey, 1943a, figs. 62–63; 1943b; Burret, 1933a; Chittenden, 1951; Allen, 1956; Dahlgren, 1959, pl. 162–163; Croat, 1978; Seymour, 1980; Krempin, 1990 ('*wendlandi*'). *Chamaedorea exorrhiza* H. A. Wendl. ex Guillaum.; Dammer, 1905b, name only; Guillaumin, 1923b; Burret, 1933a; Standley, 1937; Dahlgren, 1959, pl. 108; Moore, 1973a; 1975; Chazdon, 1985; 1987. *Chamaedorea sphaerocarpa* Burret; Hawkes, 1967; Glassman, 1972. *Chamaedorea columbica* Burret; Burret, 1937; Dugand, 1940; 1961; Glassman, 1972. *Edanthe veraepacis* O. F. Cook & C. B. Doyle; Chittenden, 1951.

Chamaedorea tuerckheimii (Dammer) Burret; Standley & Steyermark, 1958; Aguilar Amar, 1986. *Malortiea tuerckheimii* Dammer; Dammer, 1905b; Smith, 1905.

Chamaedorea vulgata Standl. & Steyerm.; Standley & Steyermark, 1958; Glassman, 1972.

Chamaedorea warscewiczii H. A. Wendl.; Kerchove, 1878; Nicholson, 1884; Dammer, 1903; Guillaumin, 1923b; Burret, 1933a; Standley, 1937; Standley & Steyermark, 1958; Dahlgren, 1959, pl. 161; Glassman, 1972; Moore, 1973; Chazdon, 1985; 1987. *Chamaedorea latipinna* L. H. Bailey; Bailey, 1943b; Glassman, 1972.

Chamaedorea woodsoniana L. H. Bailey; Bailey, 1943b; Krempin, 1990, as *Chamaedorea* sp. *9087* (at Sydney Botanical Garden).

APPENDIX B

SELECTED SPECIMENS EXAMINED

Species are listed alphabetically and following each are samples of the more than 5,000 specimens that I have examined in various herbaria. An attempt was made to select specimens that are representative of the morphological and geographical ranges within a species. Types are not cited here since they are listed under the treatment of each species.

Chamaedorea adscendens. BELIZE. Toledo: *Schipp 1223* (BM, F, GH, MICH, MO, S). GUATEMALA. Alta Verapaz: *Steyermark 44473, 45607* (F). Peten: *Hodel 1101A, 1101B* (AGUAT, BH).

Chamaedorea allenii. PANAMA. Chirirqui: *Churchill 5280* (MO). Cocle: *Hodel 742* (BH, PMA); *Gentry 3633* (MO). Darien: *Duke 5042* (GH, MO); *Garwood 2742A* (F); *McDonagh 449* (BM). San Blas: *Sagden 623* (MO).

Chamaedorea alternans. MEXICO. Veracruz: *Cedillo 102* (F, GH); *Hodel 918* (BH, MEXU); *Iltis 27251* (WIS); *Moore 6365, 8929* (BH).

Chamaedorea amabilis. COLOMBIA. Choco: *Forero 3085* (MO). COSTA RICA. Alajuela: *Burger 11920* (F); *Hodel 716A, 716B* (BH, CR); *Moore 6568, 6630, 6742* (BH); *Liesner 14663, 14881* (MO); *Smith 1659* (F, GH, MO). Cartago: *Lent 1420* (F). San Jose: *Gomez 20871* (MO). PANAMA. Cocle: *Hodel 739, 741A, 741B* (BH, PMA); *Knapp 4914* (MO). Colon: *de Nevers 4060* (MO).

Chamaedorea angustisecta. BOLIVIA. La Paz: *Rusby 534, 1073* (GH). BRAZIL. Acre: *Pinard 848* (NY). PERU. Amazonas: *Hutchison 3890* (BH). Ayacucho: *Killip 22864* (F). Junin: *Gentry 1982* (F). Madre de Dios: *Kahn 2146* (BH). Ucayali: *Gentry 58542* (F). Cuzco: *Dudley 11508* (NA).

Chamaedorea arenbergiana. COLOMBIA. Antioquia: *Kalbreyer 1894* (GOET). COSTA RICA. Without specific locality: *Koschny s.n.* (M). GUATEMALA. Alta Verapaz: *Hodel 1011* (AGUAT, BH); *Standley 91939* (F). Baja Verapaz: *Hodel 1015, 1104* (AGUAT, BH). Huehuetenango: *Hodel 1005* (AGUAT, BH); *Steyermark 48846* (F). San Marcos: *Hodel 996* (AGUAT, BH); *Steyermark 37362* (F). HONDURAS. Atlantida: *Standley 3351, 54156, 54157* (F, GH). MEXICO. Chiapas: *Breedlove 33109* (MEXU); *Croat 46653* (MO); *Hodel 929A, 929B* (BH, MEXU). Oaxaca/Veracruz: *Hammel 15566* (MO). PANAMA. San Blas: *Pittier 4276* (F).

Chamaedorea brachyclada. COSTA RICA. San Jose: *Chavarria 107* (CR). Puntarenas: *Grayum 3351, 3352, 3365, 9280* (MO); *Hodel 705A, 705B* (BH, CR); *Maas 1387* (F, U). PANAMA. Chiriqui: *Croat 15904* (MO).

Chamaedorea brachypoda. GUATEMALA. Izabal: *Hodel 881* (AGUAT, BH); *Moore 8221* (BH); *Standley 23915, 24483* (GH); *38749* (F).

Chamaedorea carchensis. GUATEMALA. Alta Verapaz: *Hodel 888A, 888B* (AGUAT, BH); *Molina 12136* (F). Huehuetenango: *Hodel 1002* (AGUAT, BH). MEXICO. Chiapas: *Breedlove 21780* (CAS); *Dressler 1502* (GH).

Chamaedorea castillo-montii. GUATEMALA. Izabal: *Hodel 868B, 1016, 1019* (AGUAT, BH). Alta Verapaz: *Hodel 1007* (AGUAT, BH).

Chamaedorea cataractarum. MEXICO. Chiapas: *Cook 127* (US); *Hodel 923* (BH, MEXU); *Matuda 3650* (F, GH, MICH, US). Oaxaca; *Galeotti 4989* (P), *4989D* (BH, L, GOET); *Mexia 9132* (GH, S, U). Tabasco: *Schulz s.n.* (BH).

Chamaedorea chazdoniae. COSTA RICA. Heredia: *Chazdon 160, 198, 200* (CR); *Grayum 6773* (MO). Limon: *Ocampo s.n.* (CR).

Chamaedorea correae. PANAMA. Cocle: *Antonio 3037* (MO); *Hodel 1113A, 1113B* (BH, PMA); *Lewis 1175* (BH, MO). Colon: *Churchill* 5547 (MO). Panama: Croat *26068* (MO); *Knapp 5916* (MO). Veraguas: *Hodel 1118* (BH, PMA); *Mori 6715, 6775* (MO).

Chamaedorea costaricana. COSTA RICA. Alajuela: *Croat 43601* (MO). Cartago: *Hodel 698A, 698B, 978A, 978B* (BH, CR); *Wendland 16, 24* (GOET). Guanacaste: *Garwood 680* (BM, F); *Grayum 6214* (MO). Heredia: *Burger 7687* (F). Puntarenas: *Gomez 21564, 21565, 21566* (MO); *Hammel 13942* (MO); *Hodel 693A, 693B* (BH, CR); *Moore 6549, 9436, 9437, 10151* (BH). San Jose: *Grayum 5138* (MO). HONDURAS. El Paraiso: *Pastor 102* (UNAH). Ocotepeque: *Nelson 7984* (UNAH). NICARAGUA. Chontales: *Bunting 1194* (F). Granada: *Baker 195* (CAS, GH, RSA). Jinotega: *Bunting 966* (F). Madruz: *Williams 20275* (F). Managua: *Standley 8384* (F). Matagalpa *Hawkes 2156* (C); *Williams 27673, 29201* (F). PANAMA. Bocas del Toro: *Kirkbride 604* (MO). Chiriqui: *Allen 3521* (BH, MO); *Hammel 1528* (BH, MO); *Hodel 736A, 736B* (BH, PMA); *Wagner 476* (GOET), *s.n.* (M); *Woodson 502* (MO), *948* (BH, MO). Cocle: *Hodel 747* (BH, PMA). Herrera: *Hammel 4308* (MO). Los Santos: *Lewis 2927* (BH). Veraguas: *de Nevers 6793* (MO).

Chamaedorea crucensis. COSTA RICA. Cartago: *Chacon 1427, 1492* (MO); *Hodel 699, 982* (BH, CR); *Lent 1080, 1258, 2213* (F); *Moore 6689* (BH). Heredia: *Gomez 20093, 20095* (MO). Limon: *Standley 37181* (F). Puntarenas: *Hammer 64* (FTG); *Hodel 706B* (BH, CR); *Maas 1501* (F, U); *Moore 10505* (BH). San Jose: *Hodel 977* (BH, CR).

Chamaedorea dammeriana. COSTA RICA. Alajuela: *Burger 11736, 11745* (F); *Croat 68174* (MO); *Liesner 5173* (MO); *Moore 6583* (BH). Guanacaste: *Garwood 715* (BM, F); *Grayum 4980, 4981, 4982, 4983, 4984* (MO); *Moore 6725* (BH). Heredia: *Skutch 3462* (GH, K, S), *3464* (GH, K, MO, S); *Wendland s.n.* (GOET, L). Limon: *Hodel 954A, 954B, 956* (BH, CR); *Grayum 4397, 5797* (MO); *Pittier 8645* (BH), *8646* (BH, BR). Puntarenas: *Grayum 9314.* Without specific locality, *Pittier 7633* (BR). PANAMA. Colon: *de Nevers 6228, 6241* (CAS). Panama: *Knapp 1377* (CAS).

Chamaedorea deckeriana. COSTA RICA. Alajuela: *Hodel 717, 718* (BH, CR); *Lent 2781, 2892* (F); *de Nevers 7804, 7809, 7811* (MO). Guanacaste: *Grayum 5011* (MO). Heredia: *Maas 1261* (U). Limon: *Grayum 3537* (MO); *Jimenez 2130, 2808* (F). Puntarenas: *Grayum* 9252 (CR). Without specific locality, *Wendland 3* (GOET). PANAMA. Colon: *de Nevers 4062, 6877* (MO). San Blas: *de Nevers 7913* (CAS).

Chamaedorea deneversiana. COSTA RICA. Guanacaste: *Grayum 5000* (CR, MO). ECUADOR. Carchi: *Madison 7062* (BH). Cotopaxi: *Sparre 17178* (S). Pichincha: *Sparre 15188* (S). PANAMA. San Blas: *de Nevers 5553B, 6292* (MO). Veraguas: *Hodel 1114* (BH, PMA); *Mori 5352* (MO).

Chamaedorea elatior. GUATEMALA. Alta Verapaz: *Cook 123, 149, 167* (F). Huehuetenango: *Steyermark 49513* (GH). Sacatepequez: *Standley 63157* (F). San Marcos: *Hodel 914* (AGUAT, BH); *Steyermark 37101, 37539* (F). MEXICO. Chiapas: *Breedlove 21860* (RSA), *56621* MEXU). Oaxaca: *Galeotti 4972* (GH, GOET); *Vera 3796* (BH, MICH). Puebla: *Vera 3325* (BH, MICH). Veracruz: *Hodel 921, 949* (BH, MEXU); *Purpus 2301* (F), *15409, 15785* (MICH); *Vasquez 324, 326* (F); *Vera 3013* (MICH). CULTIVATED. California: research collection, Los Angeles, *Hodel 805* (BH), *ex* Veracruz, Mexico.

Chamaedorea elegans. GUATEMALA. Alta Verapaz: *Muenscher 12048* (F); *Steyermark 45178, 45428* (F). Peten: *Bartlett 12166* (F, MICH), *12181, 12607* (MICH); *Hodel 845, 849* (BH); *Ortiz 213, 308* (F). MEXICO. Chiapas: *Breedlove 15678* (F, MICH); *Linden s.n.* (GOET); *Martinez 6946* (F, MEXU). Hidalgo: *Croat 66009* (CAS). Oaxaca: *Galeotti 4989* (P), *4989C* (GH); *Linden s.n.* (P); *Moore 8911* (BH). Puebla: *Vera 3341, 3667* (MICH). San Luis Potosi: *Kenoyer 1036* (F), *4003* (GH, MICH); *Rzedowski 9946, 10108* (MICH). Tabasco: *Matuda 3275* (F, GH, MEXU, MICH). Veracruz: *Hodel 946* (BH, MEXU); *Liebmann s.n.* (*6584*) (C, GH, MICH, MO); *Nee 22785, 25048* (F); *Purpus 2100* (L, P), *14101* (MICH). CULTIVATED. Belize: Cayo, *Hodel 1134* (BH), *ex* Water Hole, Cayo, Belize.

Chamaedorea ernesti-augustii. BELIZE: Cayo: *Hodel 840* (BH); *Lundell 6182, 6207* (MICH); *Gentle 2396, 2397* (MICH). Stann Creek: *Gentle 2945, 3511* (MICH); *Gentry 7980* (F). Toledo: *Boutin 5007, 5051, 5178, 5188* (HNT); *Croat 24306* (F); *Gentle 4270, 5104* (MEXU). GUATEMALA. Alta Verapaz: *Williams 42094, 42103, 42106* (F). Huehuetenango: *Steyermark 49279* (F). Izabal: *Hodel 879, 1018* (AGUAT, BH). Peten: *Bartlett 12630* (MICH); *Hodel 848* (BH). HONDURAS. Atlantida: *Standley 52690* (GH). Comayagua: *Edwards 588, 590* (GH); *Yuncker 6135* (F, GH, S, U). Cortes: *Molina 3561* (F). Yoro: *von Hagen 1035* (F). MEXICO. Chiapas: *Breedlove 15695* (MICH), *21007, 30853* (MEXU); *Hodel 931* (BH, MEXU); *Perino 3097* (F). Oaxaca: *Moore 6356* (BH); *Williams 9584* (F). Tabasco: *Matuda 3443* (F, GH, MEXU, MICH). Veracruz: *Hodel 920A, 920B* (BH, MEXU); *Nee 22784, 29892, 29977, 30005* (F).

Chamaedorea falcifera. GUATEMALA. Izabal: *Jones 3245* (F, MICH), *3250* (F); *Steyermark 38157, 39191, 39651, 39984* (F). Alta Verapaz: *Watson 195* (GH).

Chamaedorea foveata. MEXICO. Chiapas: *Breedlove 33677, 31354* (CAS). Oaxaca: *Hodel 937, 939B* (BH, MEXU); *McVaugh 22358* (MICH).

Chamaedorea fractiflexa. GUATEMALA. Quetzaltenango: *Hodel 905, 912B, 986* (AGUAT, BH). San Marcos: *Williams 26845* (F). MEXICO. Chiapas: *Breedlove 41648* (CAS).

Chamaedorea fragrans. PERU. Cuzco: *Davis 1318* (F). Huanuco: *Moore 8550* (BH); *Schunke 9477* (F), *9882* (F, U). Junin: *Iltis 257* (WIS). San Martin: *Klug 4367* (F, GH, S, U); *Williams 5719, 5759* (F).

Chamaedorea geonomiformis. BELIZE. Toledo: *Boutin 5005, 5006, 5070, 5086, 5087, 5189* (HNT). GUATEMALA. Alta Verapaz: *Watson 255* (GH), *307* (F, GH, S). Izabal: *Harmon 1911* (F, MO); *Hodel 872, 1023A, 1023B* (AGUAT, BH). Peten: *Hodel 1100A, 1100B* (AGUAT, BH). HONDURAS. Atlantida: *Standley 52727, 53197, 53878, 55631, 56558* (F, GH); *Yuncker 4995* (F, GH, MICH). MEXICO. Without specific locality: *Botteri 134* (GH).

Chamaedorea glaucifolia. MEXICO. Chiapas: *Bauml 543* (HNT); *Carlson 1517* (F); *Hodel 928* (BH, MEXU); *Laughlin 766* (BH, F); *Miranda 6146* (MEXU); *Moore 6318* (MEXU).

Chamaedorea graminifolia. BELIZE. Cayo: *Hodel 1128A, 1128B, 1129A, 1129B, 1132A, 1132B* (BH); *Keim 536* (BH); *Lundell 6853* (MICH). Toledo: *Davidse 32139* (CAS). COSTA RICA. Alajuela: *Koschny s.n.* (GOET); *Cook 78* (US). Limon: *Standley 49006* (US). GUATEMALA. Alta Verapaz: *Steyermark 45612* (F); *Cook 80* (US). CULTIVATED. USA. California: Doheny Estate, Los Angeles, *Bailey 9125* (BH), *ex* Tehuantepec, Mexico.

Chamaedorea guntheriana. PANAMA. Panama: *Dwyer 8090*

(MO); *Foster 1887* (MO); *Hodel 1127* (BH, PMA); *Mori 2378, 2732, 3794, 6094* (MO); *Witherspoon 8496* (MO).

Chamaedorea hooperiana. MEXICO. Veracruz: *Dressler 91* (GH); *Hodel 922* (BH, MEXU).

Chamaedorea klotzschiana. MEXICO. Veracruz: *Castillo 555* (F); *Rees 1643* (F); *Vazquez 784* (GH); *Vovides 136* (WIS).

Chamaedorea lehmannii. GUATEMALA. Baja Verapaz: *Hodel 1013, 1105* (AGUAT, BH). El Progreso: *Steyermark 43804, 43805, 43587* (F). Zacapa: *Steyermark 29890, 42554, 42868, 42885* (F).

Chamaedorea liebmannii. GUATEMALA. Huehuetenango: *Hodel 1004A, 1004B* (AGUAT, BH). MEXICO. Chiapas: *Hodel 926* (BH, MEXU); *Matuda 18085* (F); *Thorne 40279* (RSA). Oaxaca: *Hodel 942, 944* (BH, MEXU); *Mexia 9278* (F, GH, S, U), *9279* (F, GH); *Moore 8912* (BH). Puebla: *Harrell 441* (MEXU). Veracruz: *Beaman 47209* (F); *Calzada 220* (F), *272* (CAS, F, GH); *Galeotti 4971* (GOET, P), *4975* (F); *Nee 24967* (F).

Chamaedorea linearis. BOLIVIA. La Paz: *Weddell 4222* (F). COLOMBIA. Antioquia: *Kalbreyer 1377A, 1891, 1899, 1902* (GOET); *Moore 10221* (BH). Magdalena: *Kirkbride 2284* (BH). Santa Marta: *Smith 2479* (F, GH, VT). Santander: *Killip 19226* (GH). Santo Domingo: *Kalbreyer 1700* (GOET). Valle: *Cuatrecasas 21173, 22731* (F). Without specific locality: *Funck 782* (GOET). ECUADOR. Carchi: *Madison 7194* (BH). Esmeraldas: *Sparre 15321, 15494* (S). Guayas: *Dodson 1658* (WIS). Los Rios: *Essig 340, 341* (USF). Napo-Pastaza: *Balslev 10346* (F); *Asplund 19412* (S); *Essig 361* (USF). Pichincha: *Sparre 14770* (S). Tungurahua: *Asplund 18645* (S). Without a specific locality: *Eggers 14311* (M, US), *15094* (C); *Wiggins 11081* (CAS). PERU. Amazonas: *Knapp 6580* (F). Cuzco: *Vargas 15629* (BH). Huanuco: *Mexia 8300* (F, GH, S, U). Junin: *Iltis 256* (WIS); *Killip 24672, 25820* (F). San Martin: *Melin 163* (S); *Schunke 3424* (F); *Williams 7335* (F). Tumbes: *Simpson 391* (F). VENEZUELA. Lara: *Steyermark 110158* (U, WIS).

Chamaedorea lucidifrons. COLOMBIA. Without specific locality: *Kalbreyer 1569* (GOET). PANAMA. Cocle: *Hodel 743, 1110* (BH, PMA).

Chamaedorea macrospadix. COSTA RICA. Alajuela: *Molina 17322* (F). Cartago: *Hodel 701A, 701B* (BH, CR); *Lent 300* (F). Limon: *Hodel 957* (BH, CR). Puntarenas: *Burger 4501, 4784* (F); *Gomez 19653, 19658* (MO); *Grayum 9144* (MO); *Hodel 692A, 692B, 711, 712A, 712B, 713A, 713B* (BH, CR); *Moore 6533, 6550, 6552, 9439, 9445* (BH). San Jose: *Croat 43416* (MO); *Skutch 2437* (S).

Chamaedorea matae. COSTA RICA. Puntarenas: *Burger 8905* (F, MO); *Essig 700811-1* (BH); *Grayum 5946* (MO); *Hodel 715* (BH, CR); *Liesner 2841* (CR, MO), *3259* (CR); *Moore 6527, 6539* (BH); *Raven 21561* (F, RSA). PANAMA. Herrera: *Hammel 4225* (MO).

Chamaedorea metallica. MEXICO. Veracruz: *Chazaro 176* (F); *Hodel 950* (BH, MEXU).

Chamaedorea microphylla. PANAMA. Chiriqui: *Antonio 2856* (MO); *Hammel 2118* (BH, MO); *Hodel 1126A, 1126B* (BH, MO, PMA).

Chamaedorea microspadix. MEXICO. Hidalgo: *Moore 2962, 5079* (BH); *Rodriguez 5683* (CAS). San Luis Potosi: *Bartholomew 3457* (CAS); *Edwards 908* (F); *Kenoyer 1036* (F).

Chamaedorea murriensis. COLOMBIA. Antioquia: *Kalbreyer 1373, 1829, 1878, 1890* (GOET). PANAMA. Darien: *Folsom 4296* (MO); *Gentry 28690* (MO); *Hartman 4590, 4801, 8529* (MO); *Mori 5452* (MO).

Chamaedorea nationsiana. GUATEMALA. Izabal: *Hodel 869, 871, 873, 880, 1017, 1021B* (AGUAT, BH); *Steyermark 41863* (F).

Chamaedorea neurochlamys. BELIZE. Cayo: *Gentle 2494* (MICH); *Gentry 7734* (F); *Hodel 842A, 842B* (BH). Stann Creek: *Schipp 296* (F, GH, MICH, S). Toledo: *Boutin 5000, 5125, 5172* (HNT); *Peck 616* (GH). GUATEMALA. Alta Verapaz: *Hodel 1006* (AGUAT, BH); *Watson 134, 404* (GH). Huehuetenango: *Steyermark 49295* (F). Izabal: *Steyermark 38755, 39184* (F). Peten: *Steyermark 45368* (F). HONDURAS. Atlantida: *Standley 52907, 55388, 56744* (F, GH); *Tellez 8794* (MEXU); *Yuncker 8517* (F). Comayagua: *Molina 5844, 10822* (F). Cortes: *Zuniga 525* (UNAH); *Barkley 40791* (GH). La Paz: *Molina 13871* (F). Ocotepeque: *Blackmore 3868* (UNAH). MEXICO. Campeche: *Shepherd 113* (WIS). Chiapas: *Carlson 2101* (F); *Martinez 6979* (MEXU). Quintana Roo: *Cabrero 5575* (MEXU).

Chamaedorea nubium. EL SALVADOR. Chalatenango: *Tucker 1099* (F, MICH). Santa Ana: *Molina 16899* (F). GUATEMALA. Alta Verapaz: *Forther 2096* (M). Baja Verapaz: *Hodel 884, 899, 1014* (AGUAT, BH). El Progreso: *Forther 2333* (M). Huehuetenango: *Steyermark 48612* (F). San Marcos: *Croat 40942* (BH, MO); *Williams 26844* (F). Zacapa: *Steyermark 42384, 42892* (F). MEXICO. Chiapas: *Bartholomew 55642* (CAS), *Breedlove 15244* (F, MICH), *24929* (MEXU), *33619, 55692* (CAS); *Croat 47664* (MO); *Heath 753, 963* (MEXU); *Hodel 924* (BH, MEXU); *Laughlin 418* (MEXU, WIS); *Matuda 15458, 1928* (F); *Thorne 46414* (RSA). Guerrero: *McVaugh 2, 115* (MICH); *Hinton 14057* (MICH, P, RSA); *Reveal 4272* (MICH). Oaxaca: *Hodel 938* (BH, MEXU).

Chamaedorea oblongata. BELIZE. Belize: *Bartlett 11345* (MICH); *Gentle 1679* (MICH). Cayo: *Dwyer 10908* (F); *Gentle 2466* (MICH); *Hodel 841* (BH); *Lundell 6361* (MICH). GUATEMALA. Alta Verapaz: *Williams 40503, 42093* (F); *Watson 55* (GH). Izabal: *Hodel 876* (AGUAT, BH); *Steyermark 38123* (F). Peten: *Bartlett 12146* (MICH); *Hodel 846* (BH); *Lundell 2878* (MICH); *Molina 15819* (F); *Ortiz 1427* (F). HONDURAS. Atlantida: *Bangham 230* (GH). Cortes: *Molina 3474* (F). MEXICO. Campeche: *Lundell 1235* (F, WIS). Chiapas: *Breedlove 71299* (CAS); *Martinez 6948* (F); *Miranda 5920* (MEXU). Oaxaca: *Martinez 127* (GH); *Vera 2511* (MICH). Quintana Roo: *Sanders 9828, 9945* (UCR); *Tellez 3410* (MEXU). Tabasco: *Matuda 3232* (F, GH, MEXU, MICH). Veracruz: *Castillo 2002, 2379* (F); *Hodel 951* (BH, MEXU); *Nee 29737, 29891* (F); *Purpus 16196* (C); *Rosas 972* (GH, U). NICARAGUA. Jinotega: *Standley 10460* (F). Matagalpa: *Molina 20455, 31207* (F); *Williams 23843, 27888* (F).

Chamaedorea oreophila. MEXICO. Veracruz: *Avendano 399* (F); *Bourgeau 3260* (P); *Castillo 1365* (F); *Hernandez 1597* (F); *Hodel 953A, 953B* (BH, MEXU).

Chamaedorea pachecoana. GUATEMALA. Quetzaltenango: *Hodel 900, 908, 985A, 985B* (AGUAT, BH); *Standley 87009, 87011* (F); *Steyermark 33333* (F). San Marcos: *Hodel 989A, 989B, 993* (AGUAT, BH); *Steyermark 37386* (F). Solola: *Steyermark 47383* (F).

Chamaedorea palmeriana. COSTA RICA. Alajuela: *Chacon 1204* (MO); *Gomez 23328* (MO). Cartago: *Gomez 18739, 19248, 19264* (MO). Heredia: *Chazdon 181, 193, 194* (CR). Limon: *Gomez 23514, 23548* (MO). San Jose: *Chacon 1704, 1731* (CR, MO); *Grayum 6121, 6122* (MO); *Hodel 970, 972* (BH, CR). PANAMA. Chiriqui: *Churchill 5538, 5539, 6115, 6116, 6123, 6124* (MO); *Hampshire 945* (BM); *Hodel 726B, 1119A, 1119B* (BH, PMA); *Knapp 5077, 5535, 5604* (MO). Veraguas: *Mori 7667* (MO).

Chamaedorea parvifolia. COSTA RICA. Alajuela: *Brenes 18272,*

18275 (CR), *19121* (F); *Lent 1970* (F). Cartago: *Liesner 14614* (CR, MO); *Stork 2864* (GH). Puntarenas: *Burger 9576* (CR, F); *Haber 2399* (CR); *Hammel 13873* (CR). San Jose: *Chavarria 93, 94, 97* (CR); *Grayum 5139* (CR, MO); *Hodel 965B* (BH, CR).

Chamaedorea parvisecta. GUATEMALA. Alta Verapaz: *Cook 121, 141, 142, 143, 144, 145, 177* (US); *Hodel 891, 892* (AGUAT, BH); *Standley 91426* (F); *Williams 40730* (F). Baja Verapaz: *Hodel 896, 1102B* (AGUAT, BH); *Williams 43304, 43310* (F). El Progreso: *Steyermark 43493* (F). Huehuetenango: *Hodel 1000, 1001* (AGUAT, BH). MEXICO. Chiapas: *Breedlove 33618, 55696, 67896* (CAS); *Croat 46631* (MO).

Chamaedorea pauciflora. BRAZIL. Acre: *Forero 6387* (BH); *Prance 7684* (BH). Mato Grosso: *Krukoff 1685* (F). COLOMBIA. Amazonas: *Schultes 6676, 13697, 13699* (BH). ECUADOR. Napo-Pastaza: *Asplund 9166, 9427, 9444* (S); *Shemluck 307* (F). PERU. Huanuco: *Schunke 1439, 1645, 1695, 2323* (F). Loreto: *Croat 18477* (BH); *Hodel 1138A, 1138B* (BH, USM); *Kahn 2019* (BH); *Killip 28345* (F); *Williams 2920, 3005* (F).

Chamaedorea pedunculata. COSTA RICA. Puntarenas: *Burger 4461* (F); *Hodel 708B* (BH, CR); *Moore 10507* (BH).

Chamaedorea pinnatifrons. BOLIVIA. La Paz: *Krukoff 10481* (F, GH, S, U). Santa Cruz: *Krapovickas 31624* (F). Sara: *Steinback 5424* (GH), *5524* (F). Yapacani: *Kuntze s.n.* (F). BRAZIL. Amazonas: *Trail 1037, 1038* (GH). COLOMBIA. Antioquia: *Kalbreyer 1476, 1893* (GOET). Caldas: *Dryander 48* (F); Killip 8788 (GH). Caqueta: *von Sneidern s.n.* (S). Cauca: *Pennell 7644* (GH). Magdalena: *Grant 10727* (WIS). Narino: *von Sneidern 4653* (S). Santa Marta: *Smith 2477* (F, GH). Santander: *Killip 17040, 18996, 19452, 20413* (GH). Tolima: *von Sneidern s.n.* (S). Valle: *Cuatrecasas 18238, 22159* (F); *Maas 1834* (U). COSTA RICA. Alajuela: *Burger 9842, 10730* (F); *Dryer 79, 188* (F); *Utley 3915, 4029* (F). Cartago: *Hodel 700, 979, 980* (BH, CR); *Moore 6690* (BH). Heredia: *Burger 6372* (CR, F); *Hodel 723* (BH, CR); *Pittier 374* (BR); *Utley 2064* (F). Limon: *Grayum 4397* (MO); *Hodel 955* (BH, CR). Puntarenas: *Hodel 694A, 694B, 704, 707* (BH, CR); *Raven 20897* (F); *Tonduz 3491* (BH). San Jose: *Grayum 5137* (CR); *Hodel 963, 964, 966, 967, 973* (BH, CR). ECUADOR. Azuay: *Steyermark 52778* (F). Morona-Santiago: *Sparre 19021* (S). Napo-Pastaza: *Asplund 19090* (S); *Harling 3593* (S). Otavalo: *Heinrichs 735* (F, M). Pichincha: *Sparre 14679* (S). Santiago-Zamora: *Steyermark 54725* (F). Tena: *Heinrichs 300* (M). Zamora: *Dodson 837* (WIS). EL SALVADOR. Ahuachapan: *Standley 20165* (GH). Sosonate: *Tucker 1357* (F, MICH). GUATEMALA. Alta Verapaz: *Hodel 886, 889* (AGUAT, BH); *Standley 89817, 90093, 90494, 90842* (F); *Williams 40655* (F). Chimaltenango: *Standley 62282* (F). Escuintla: *Standley 64764, 64814* (F). Huehuetenango: *Steyermark 48787* (F). Quetzaltenango: *Hodel 902, 909* (AGUAT, BH); *Standley 86890* (F); *Skutch 928* (GH). San Marcos: *Hodel 990* (AGUAT, BH). Santa Rosa: *Standley 60727, 78512* (F). Solola: *Forther 2504* (M). Suchitepequez: *Steyermark 35278* (F). HONDURAS. Comayagua: *Yuncker 6086* (F, GH, MICH, U, S). Cortes: *Nelson 7988* (UNAH). La Paz: *Molina 14074* (F). Olancho: *Croat 64360* (CAS). MEXICO. Chiapas: *Hodel 932* (BH, MEXU); *Matuda 18034* (F, MEXU), *30178, 30179* (MO). Oaxaca: *Galeotti 4986* (F, GOET, L, P); *Vera 3412* (MICH). Puebla: *Vera 3324* (MICH). Veracruz: *Castillo 350* (F); *Hodel 917* (BH, MEXU); *Liebmann s.n.* (C); *Nee 22786, 24686, 25025, 26100* (F); *Purpus 16207* (C, GH). PANAMA. Bocas del Toro: *Kennedy 1239* (F). Chiriqui: *Hodel 728, 730, 733, 734, 735* (BH, PMA); *Woodson 270, 681* (MO), *702* (GH, MO). Colon: *Hammel 3118* (MO). Darien: *Gentry 13687, 14031* (MO). Panama: *de Nevers 8987* (CAS); *Dressler 4889* (BH); *Mori 4186* (WIS), *5138* (MO). San Blas: *de Nevers 3690, 6154, 6206* (MO). Veraguas: *Hodel 1117A, 1117B* (BH, PMA); *Mori 4805* (MEXU). PERU. Ayacucho: *Killip 22760* (F). Cuzco: *Herrera 2017* (F). Huanuco: *Asplund 13273* (S); *Moore 8357* (BH); *Plowman 5923* (BH); *Schunke 2302, 2831* (F). Junin: *Iltis 298* (WIS); *Killip 24929, 25532* (F). Loreto: *Gentry 27511* (F); *Jaramillo 4290* (F); *Williams 3411, 3419* (F). Madre de Dios: *Foster 2714, 3431, 5417* (F). San Martin: *Gentry 25734* (BH); *Schunke 3136* (BH), *4768, 6141, 8537* (F), *6038* (F, RSA), *6136* (BH). Without specific locality: *Macbride 4813* (F). VENEZUELA. Aragua: *Bunting 4694* (WIS). Caracas: *Vogl 157* (F). Merida: *Steyermark 55783, 56112, 56623* (F); *de Bruijn 1283* (S, U). Tachira: *Steyermark 96754* (U). Zulia: *Steyermark 105553* (U).

Chamaedorea pittieri. COSTA RICA. Puntarenas: *Pittier 3452* (BR), *3455* (BH). PANAMA. Chiriqui: *Croat 33050, 33432* (MO); *Davidson 133* (MO), *188* (F); *Folsom 1755* (PMA); *Hodel 729A, 729B, 1125A, 1125B* (BH, PMA); *Mori 5980* (MO); *Pittier 3168* (US). Panama: *Dwyer 8495* (F, SCZ); *Peralta 28* (PMA).

Chamaedorea pochutlensis. MEXICO. Colima: *Sanders 8146* (RSA, UCR), *10241* (UCR). Durango: *Kimnach 1689, 1781* (HNT). Guerrero: *Moore 6202* (BH). Jalisco: *Boutin 3113* (HNT); *McVaugh 10222, 20444, 23385* (MICH); *Mexia 1573* (F, GH). Michoacan: *Hinton 15984* (RSA); *Rzedowski s.n.* (MICH). Nayarit: *Boutin 2090* (HNT); *McVaugh 15265* (MICH), *16517* (BH, MICH). Oaxaca: *Hodel 934A, 934B, 940* (BH, MEXU). Sinaloa: *Gentry 5605* (GH, MICH), *7281* (MICH, RSA); *Rose 1793* (GH).

Chamaedorea pumila. COSTA RICA. Alajuela: *Brenes 6112, 13500* (F), *18271* (CR); *Chacon 1780, 1789, 1893* (CR); *Herrera 172* (CAS); *Liesner 14736* (MO). Cartago: *Croat 56789, 56818* (MO); *Hodel 981* (BH, CR). Guanacaste: *Standley 45632* (F). Heredia: *Chazdon 190, 191* (CR); *Grayum 7762* (MO). San Jose: *Hodel 969, 974* (BH, CR).

Chamaedorea pygmaea. COLOMBIA. Antioquia: *Kalbreyer 1895* (GOET). COSTA RICA. Puntarenas: *Davidse 28488* (CR, MO); *Gomez 21683* (CR, MO); *Maas 1488, 1489* (BH, U). PANAMA. Chiriqui: *Croat 48854* (MO); *Woodson 585* (MO). Darien: *de Nevers 8268* (MO); *Mori 5388* (MO); *Terry 1451* (F, GH) *1453* (F, GH, MO).

Chamaedorea queroana. MEXICO. Oaxaca: *Croat 48049* (MO); *Hodel 943A* (BH, MEXU); *Moore 8914* (BH).

Chamaedorea quezalteca. EL SALVADOR. Ahuachapan: *Standley 20137* (F, GH). Santa Ana: *Carlson 680* (F). GUATEMALA. Alta Verapaz: *Hodel 895* (AGUAT, BH); *Moore 8224* (BH). Baja Verapaz: *Hodel 1103* (AGUAT, BH). Chiquimula: *Standley 31576* (F), El Progreso: *Steyermark 43016* (F). Escuintla: *Standley 64860* (F). Huehuetenango: *Hodel 998* (AGUAT, BH). Quetzaltenango: *Hodel 901* (AGUAT, BH); *Skutch 947* (GH); *Standley 65423, 68331, 68401, 86912* (F). San Marcos: *Hodel 997* (AGUAT, BH); *Williams 26216, 26218* (F). Suchitepequez: *Steyermark 46610, 46673* (F). Zacapa: *Steyermark 29867* (F). HONDURAS. Atlantida: *Standley 59342* (F). Comayagua: *Yuncker 6025* (F, GH, MICH, S, U), *6183* (GH), *6184* (F, GH, MICH, U). Morazan: *Glassman 2037* (GH); *Moore 6754* (BH). Tegucigalpa: *von Hagen 1283* (F). Yoro: *von Hagen 1026* (F). MEXICO. Chiapas: *Breedlove 16095, 21785, 31337, 67428, 69095* (CAS); *Matuda 1929* (F, GH, MEXU, MICH).

Chamaedorea radicalis. MEXICO. Hidalgo: *Moore 5035* (BH, MICH). Nuevo Leon: *Meyer 2590* (BH). San Luis Potosi: *Four-*

nier 1119, 1163 (P); *McVaugh 10438* (MICH); *Pringle 3737* (C, F, GH, GOET, M, S, US, VT). Tamaulipas: *Dressler 2022* (GH); *Moore 6156* (BH); *Reznicek 8430* (WIS); *Stanford 2080* (CAS, U).

Chamaedorea rigida. MEXICO. Oaxaca: *Hodel 941* (BH, MEXU); *Rzedowski 33787* (WIS).

Chamaedorea robertii. COSTA RICA. Alajuela: *Chacon 2200* (CR); *Molina 17314* (F). Guanacaste: *Haber 5874* (MO). Heredia: *Chazdon 196* (CR). San Jose: *Hodel 971, 975, 976* (BH, CR). PANAMA. Chiriqui: *Churchill 5263, 5264* (MO); *Hodel 1120A, 1120B* (BH, PMA); *Mori 7814* (MO); *de Nevers 6854* (MO). Cocle: *Folsom 3243* (MO); *Hammel 2421* (MO); *Maas 2742* (U). Veraguas: *de Nevers 6765* (MO); *Hodel 1115* (BH, PMA).

Chamaedorea rojasiana. GUATEMALA. Quetzaltenango: *Hodel 903, 906, 910, 984A, 984B, 987* (AGUAT, BH); *Standley 68308* (F); *Steyermark 33541, 37363* (F). San Marcos: *Hodel 915, 991* (BH, AGUAT). Solola: *Steyermark 47355* (F). Suchitepequez: *Steyermark 35248* (F). MEXICO. Chiapas: *Breedlove 15291* (F); *Croat 65182* (CAS); *Hodel 925, 927* (BH, MEXU); *Purpus 7421* (F, GH, MO); *Thorne 40278* (RSA).

Chamaedorea sartorii. HONDURAS. Atlantida: *Standley 52654, 56755, 56767, 56894* (F, GH); *Yuncker 4585* (F, GH, MICH). Olancho: *Croat 64387, 64463* (CAS). MEXICO. Oaxaca: *Liebmann s.n.* (C). Puebla: *Vera 3695, 3696* (BH, MICH). Veracruz: *Castillo 2000* (F); *Galeotti 4975* (F), *4984A* (GOET); *Hodel 947* (BH, MEXU); *Purpus 16209* (C, F, GH); *Ross 664, 1115* (FI).

Chamaedorea scheryi. COSTA RICA. Alajuela: *Barringer 2199* (F); *Hodel 720A, 720B* (BH, CR). Cartago: *Baker 209* (CR, F); *Croat 68254* (MO); *Hodel 697A, 697B* (BH, CR). Heredia: *Burger 11150* (CR, F); *Chazdon 179, 180* (CR); *Grayum 7359, 7360* (CR, MO). San Jose: *Chazdon 220* (CR); *Grayum 6124* (MO). PANAMA. Chiriqui: *Churchill 6193, 6194* (MO); *Hodel 727A, 727B, 1121* (BH, PMA). Cocle: *Folsom 6445* (MO); *de Nevers 6382* (MO); *Hammel 3552* (MO); *Hodel 738* (BH, PMA).

Chamaedorea schiedeana. MEXICO. Oaxaca: *Liebmann s.n.* (C). Puebla: *Asplund 824* (S); *Moore 6313* (BH); *Vera 3339* (MICH). Veracruz: *Avendano 398, 470* (F); *Bourgeau 2190* (P); *Hodel 952A, 952B* (BH, MEXU); *Nee 26366, 26417* (F); *Nevling 2418* (F, GH).

Chamaedorea seifrizii. BELIZE. Belize: *O'Neill 8334* (WIS). Corozal: *Davidse 32751, 32605* (CAS); *Gentle 824* (MICH, WIS). Orange Walk: *Hodel 1131A, 1131B* (BH). GUATEMALA. Peten: *Bartlett 12446* (MICH); *Hodel 847, 850* (BH). HONDURAS. Islas de la Bahia: *Jordahn s.n.* (F); *Molina 20715* (F). MEXICO. Campeche: *Cabrera 6990* (MEXU); *Reznicek 222* (MICH); *Shepherd 111, 209* (WIS). Qintana Roo: *Cabrera 1165* (MEXU); *Tellez 145, 3142* (MEXU), *3557* (CAS). Tabasco: *Matuda 3153* (F, GH, MEXU, MICH). Yucatan: *Gaumer 420* (F), *23215, 24083* (F, GH); *Lundell 7318, 7529* (MICH); *Ortega 349* (RSA, UCR). CULTIVATED. Belize: Cayo, *Hodel. 1133A, 1133B* (BH), *ex* Cayo, Belize.

Chamaedorea selvae. COSTA RICA. Alajuela: *Donnell-Smith 6847* (K). Heredia: *Grayum 9625, 9627* (CR); *Hammer 112* (FTG); *Moore 10122, 10500* (BH). Limon: *Davidson 6979* (RSA); *Liesner 15449* (CR, WIS); *Stevens 24461* (CR). San Jose: *Chacon 1625* (CR); *Zamora 446* (CR). NICARAGUA. Rio San Juan: *Bunting 873* (F).

Chamaedorea serpens. PANAMA. Cocle: *Hodel 1111* (BH, PMA). Panama: *Fairchild s.n.* (BH); *Porter 4956* (MEXU).

Chamaedorea simplex. COLOMBIA. Antioquia: *Kalbreyer 1667* (GOET). GUATEMALA. Alta Verapaz: *Hodel 887A, 887B, 1008A, 1008B* (AGUAT, BH); *Standley 71602, 71638, 89900, 90134* (F); *Tuerckheim 1078, 11961* (GH). Baja Verapaz: *Tuerckheim 1702* (P). Huehuetenango: *Hodel 1003A, 1003B* (AGUAT, BH). MEXICO. Chiapas: *Breedlove 68693, 68694* (CAS).

Chamaedorea stenocarpa. COSTA RICA. Alajuela: *Brenes 4924* (F), *6111* (CR); *Smith 1993* (GH). Heredia: *Roig 12* (F, CR). Puntarenas: *Hobbs 583CR101* (CR); *Wilson 66-33* (BH). GUATEMALA Izabal: *Hodel 870, 1022A, 1022B* (AGUAT, BH). PANAMA. Panama: *Kennedy 687* (F). Veraguas: *Antonio 4040* (MO); *Croat 27694* (MO); *Maas 1656* (U); *Mori 3113* (MO), *3843* (MO, WIS); *Dressler 5021* (BH).

Chamaedorea stolonifera. MEXICO. Chiapas: *Hodel 933A, 933B* (BH, MEXU); *Miranda 5712, 6145* (MEXU); *Moore 6323* (BH).

Chamaedorea stricta. COSTA RICA. San Jose: *Chavarria 109* (CR); *Zamora 1428* (CR, MO). GUATEMALA. San Marcos: *Hodel 916A, 916B, 995* (AGUAT, BH); *Steyermark 37252* (F). MEXICO. Chiapas: *Matuda 15344* (F). PANAMA. Veraguas: *Antonio 1861* (MO); *Hamilton 4020* (MO); *Hammel 4615* (MO).

Chamaedorea sullivaniorum. COLOMBIA. Antioquia: *Kalbreyer 1874* (GOET). COSTA RICA. Limon: *Gomez 23486* (CR). Puntarenas: *Grayum 4050, 4051* (MO). San Jose: *Burger 10122* (CR, F). PANAMA. Chiriqui: *Croat 60214* (MO); *Hammel 14665* (CAS). Cocle: *de Nevers 6747* (BH, CAS); *Hodel 740B, 1108A, 1108B* (BH, PMA). Colon: *Hammel 3121* (MO). Veraguas: *Antonio 2968* (MO); *Hammel 4722* (MO); *Hodel 1116* (BH, PMA); *Maas 5029* (U).

Chamaedorea tenella. COSTA RICA. Puntarenas: *Croat 59766A, 59766B* (MO); *Grayum 7358* (MO); *Hodel 714* (BH, CR); *Matae 501* (CR); *Moore 6526* (BH). MEXICO. Chiapas: *Breedlove 29748* (MEXU); *Hernandez 1288* (BH); *Hodel 930* (BH, MEXU). Veracruz: *Calzada 1803* (F); *Hodel 919* (BH, MEXU).

Chamaedorea tenerrima. GUATEMALA. Alta Verapaz: *Hodel 1009* (AGUAT, BH); *Standley 90728* (F); *Moore 8225* (BH). Baja Verapaz: *Croat 41214* (MO); *Hodel 883, 1012* (AGUAT, BH).

Chamaedorea tepejilote. BELIZE. Cayo: *Dwyer 11441* (F); *Hodel 843* (BH); *Lundell 6285* (MICH). Stann Creek: *Schipp 544* (F, GH, MICH); *Gentle 2770* MICH). Toledo: *Boutin 5003, 5004* (HNT). COLOMBIA. Cauca: *Kalbreyer 1512* (GOET). Choco: *Gentry 28702* (BH); *Leon 1947* (BH). Valle: *Cuatrecasas 14991, 15154* (F). COSTA RICA. Alajuela: *Dryer 184, 693* (F); *Smith 6846, 6847* (GH); *Utley 3240* (F). Cartago: *Hodel 703A, 703B* (BH, CR); *Lent 107, 305, 4070* (F). Guanacaste: *Barringer 4065* (F); *Burger 9116* (F). Heredia: *Burger 5835* (F); *Smith 487* (WIS). *Sperry 763* (F). Limon: *Baker 114* (F, U); *Rossbach 3871* (GH). Puntarenas: *Hodel 696* (BH, CR); *Utley 1175* (F). San Jose: *Poveda 1099* (F). EL SALVADOR. San Salvador: *Standley 19388* (GH). GUATEMALA. Alta Verapaz: *Steyermark 52135* (F). Izabal: *Hodel 1020* (AGUAT, BH); *Standley 24484* (GH). Peten: *Bartlett 12186, 12218* (MICH); *Hodel 844* (BH); *Lundell 2082, 4335* (MICH). Quetzaltenango: *Rodriguez 1628* (P); *Steyermark 33830, 33835, 33836* (F). Sacatepequez: *Smith 2134* (GH). San Marcos: *Standley 68750* (F). Santa Rosa: *Smith 4301* (GH); *Standley 79065* (F). HONDURAS. Atlantida: *Yuncker 8516* (F, GH, MICH, S); *Zuniga 448* (UNAH). Cortes: *Zuniga 843* (UNAH). Ocotepeque: *Nelson 1665* (UNAH). MEXICO. Chiapas: *Breedlove 9985* (F, MICH), *57967* (CAS); *Dressler 1565* (GH, MEXU); *Hansen 1624* (MEXU, MICH, WIS); *Matuda 17416, 17534* (F, MEXU); *Miranda 6598, 6975* (MEXU). Oaxaca: *Bauml 468* (HNT); *Galeotti 4978* (GOET), *s.n.* (K). Tabasco: *Matuda 3466* (F, MEXU,

MICH). Veracruz: *Dorantes 2533, 3035* (F); *Hodel 948* (BH, MEXU); *Rosas 898* (GH, U); *Vera 3898* (MICH). PANAMA. Chiriqui: *McPherson 7781* (CAS); *Wagner 479* (GOET, M). Cocle: *Bartlett 16680* (MICH). Colon: *Folsom 3867* (MEXU). Darien: *Gentry 13552* (MO), *28588* (PMA), *28702* (MO). Panama: *Bartlett 16741, 16786* (MICH); *Correa 2987* (PMA); *Folsom 7845* (F); *Hodel 724* (BH, PMA); *Luteyn 1546* (F). Veraguas: *Mori 4876* (MEXU).

Chamaedorea tuerckheimii. GUATEMALA. Alta Verapaz: *Hodel 1010* (AGUAT, BH); *Tuerckheim 11899* (GH, L, U, US). MEXICO: Veracruz: *Cedillo 171* (F).

Chamaedorea undulatifolia. COSTA RICA. Alajuela: *Lent 3007* (F); *Liesner 14886* (CR, MO), *15476* (MO); *Smith 1005* (F, GH), *20, 666, 1329* (F). Cartago: *Grayum 3691 3692* (MO); *Lent 2214* (F). Heredia: *Grayum 7345* (CR); *Valerio 1590* (F). Puntarenas: *Almeda 5389* (CAS); *Dryer 149, 150B, 185, 273* (CR, F); *Hodel 695A, 695B* (BH, CR); *Moore 10180* (BH). San Jose: *Chazdon 225, 236* (CR), *235* (BH).

Chamaedorea verapazensis. GUATEMALA. Alta Verapaz: *Cook 199* (US); *Hodel 890, 893* (AGUAT, BH). Baja Verapaz: *Hodel 885, 898B, 1106* (AGUAT, BH); *Williams 43311* (F). El Progreso: *Steyermark 43590* (F). Huehuetenango: *Hodel 999* (AGUAT, BH). Zacapa: *Steyermark 29932, 42555* (F).

Chamaedorea verecunda. PANAMA. Chiriqui: *de Nevers 8807* (CAS); *Hammel 2124, 2254* (MO); *Hodel 1122* (BH, PMA); *Kirkbride 942* (MO).

Chamaedorea volcanensis. GUATEMALA. Quetzaltenango: *Hodel 907, 913B* (AGUAT, BH); *Standley 86696* (F); *Steyermark 33467, 35247* (F). San Marcos: *Hodel 992, 994* (AGUAT, BH). Suchitepequez: *Steyermark 35247* (F).

Chamaedorea vulgata. GUATEMALA. Quetzaltenango: *Hodel 904* (AGUAT, BH); *Standley 67344* (F). MEXICO. Chiapas: *Breedlove 31683* (CAS).

Chamaedorea warscewiczii. COSTA RICA. Alajuela: *Burger 11206* (F). Cartago: *Hodel 702A, 702B* (BH, CR); *Wendland 32* (GOET). Guanacaste: *Garwood 733* (BM); *Schatz 1099* (WIS). Heredia: *Hodel 722* (BH, CR). Limon: *Tonduz 12925* (BM, M, P). San Jose: *Hodel 968* (BH, CR). PANAMA. Cocle: *Hodel 744* (BH, PMA). Colon: *Folsom 3739* (MEXU). Darien: *Folsom 6352* (BH, MEXU); *Gentry 28606* (BH, PMA). Panama: *Alston 8913* (BM); *Bartlett 16932* (MICH); *Croat 1975* (PMA); *Hodel 725A, 725B* (BH, PMA); *Kennedy 667* (F); *Mendez 152* (PMA); *Mori 4960* (MEXU, MO), *7690* (PMA); *Porter 4129* (BH). Veraguas: *Croat 23101* (BH, MEXU, PMA); *Mori 4805* (BH).

Chamaedorea whitelockiana. GUATEMALA. Escuintla: *Standley 61298* (F). Quetzaltenango: *Hodel 911, 988A, 988B* (AGUAT, BH); *Skutch 926, 1535* (GH). Sacatepequez: *Harmon 2367* (MO). Solola: *Steyermark 47411* (F). MEXICO. Chiapas: *Matuda 18281* (MEXU).

Chamaedorea woodsoniana. COSTA RICA. Alajuela: *Molina 17333* (GH). GUATEMALA. Baja Verapaz: *Hodel 1107A, 1107B* (AGUAT, BH). El Progreso: *Forther 2334* (M). Huehuetenango: *Steyermark 48737* (F). Zacapa: *Steyermark 42875* (F). HONDURAS. Atlantida: *Standley 56742* (F, GH, US); *Zuniga 442, 1452* (UNAH). MEXICO. Chiapas: *Breedlove 33069* (CAS). Oaxaca: *Hodel 945* (BH, MEXU). Veracruz: *Beaman 6520* (BH); *Diggs 2697* (BH, F); *Gomez-Pompa 5475* (F); *Nee 25030* (BH, F); *Nevling 2514* (F); *Sousa 3499* (CAS). NICARAGUA. Matagalpa: *Williams 23948, 29268* (F). PANAMA. Chiriqui: *Churchill 5396, 6199* (MO), *6167* (MEXU); *Croat 66743, 66746* (MO); *de Nevers 6802* (MO); *Hampshire 10* (BM, PMA); *Knapp 2098* (MO). Cocle: *Croat 13470* (MO); *de Nevers 6329* (MO); *Hodel 1112* (BH, PMA). Darien: *Gentry 7037* (MO); *Hammel 1321* (MO). Panama: *Croat 12128, 22828, 25246* (MO); *Fairchild s.n.* (BH). Veraguas: *Allen 4377* (BH); *Hamilton 3966* (MO).

Chamaedorea zamorae. COSTA RICA. Alajuela: *Hodel 719* (BH, CR). Cartago: *Moore 6741* (BH). Heredia: *Moore 6563* (BH). Puntarenas: *Burger 5451, 5471* (F); *Essig 700808-1, 700814-1, 700814-2* (BH); *Moore 6551* (BH); *Raven 21625* (F); *Utley 1063* (F).

APPENDIX C

AUTHOR'S COLLECTIONS

A list of Hodel *Chamaedorea* collection numbers and determinations is provided since seeds of many of my collections have been distributed by number only. "Cult." preceding the country of origin signifies that the collection was made from cultivated material; otherwise all collections were made in the wild. Original herbarium vouchers of all collections were deposited in the Bailey Hortorium (BH) and, in most cases, duplicates of wild collected material were deposited in the country of collection: Mexico (MEXU), Guatemala (AGUAT), Costa Rica (CR), Panama (PMA), and Peru (USM).

Number	Species	Origin
619A	stenocarpa	Cult., Costa Rica
619B	stenocarpa	Cult., Costa Rica
622A	minima	Cult., Costa Rica
622B	minima	Cult., Costa Rica
624	robertii	Cult., Costa Rica
626A	sullivaniorum	Cult., Costa Rica
626B	sullivaniorum	Cult., Costa Rica
663A	nubium	Cult., USA, California
663B	nubium	Cult., USA, California
664	sullivaniorum	Cult., USA, California
667	fragrans	Cult., USA, California
668	oblongata	Cult., USA, California
671A	neurochlamys	Cult., USA, California
671B	neurochlamys	Cult., USA, California
673	minima	Cult., USA, California
674	brachyclada	Cult., USA, California
676	pinnatifrons	Cult., USA, California
678	pinnatifrons	Cult., USA, California
679	tepejilote	Cult., USA, California

682A	geonomiformis	Cult., USA, California
682B	geonomiformis	Cult., USA, California
683	microspadix	Cult., USA, California
684	brachypoda	Cult., USA, California
685	klotzschiana	Cult., USA, California
686	elegans	Cult., USA, California
687	whitelockiana	Cult., USA, California
688	tepejilote	Cult., USA, California
689	arenbergiana	Cult., USA, California
690	hooperiana	Cult., USA, California
691	pinnatifrons	Cult., USA, California
692A	macrospadix	Costa Rica
692B	macrospadix	Costa Rica
693A	costaricana	Costa Rica
693B	costaricana	Costa Rica
694A	pinnatifrons	Costa Rica
694B	pinnatifrons	Costa Rica
695A	undulatifolia	Costa Rica
695B	undulatifolia	Costa Rica
696	tepejilote	Costa Rica
697A	scheryi	Costa Rica
697B	scheryi	Costa Rica
698A	costaricana	Costa Rica
698B	costaricana	Costa Rica
699	crucensis	Costa Rica
700	pinnatifrons	Costa Rica
701A	macrospadix	Costa Rica
701B	macrospadix	Costa Rica
702A	warscewiczii	Costa Rica
702B	warscewiczii	Costa Rica
703A	tepejilote	Costa Rica
703B	tepejilote	Costa Rica
704	pinnatifrons	Costa Rica
705A	brachyclada	Costa Rica
705B	brachyclada	Costa Rica
706A	crucensis	Costa Rica
706B	crucensis	Costa Rica
707	pinnatifrons	Costa Rica
708A	pedunculata	Costa Rica
708B	pedunculata	Costa Rica
709A	costaricana	Cult., Costa Rica
709B	costaricana	Cult., Costa Rica
710A	costaricana	Cult., Costa Rica
710B	costaricana	Cult., Costa Rica
711	macrospadix	Costa Rica
712A	macrospadix	Costa Rica
712B	macrospadix	Costa Rica
713A	macrospadix	Costa Rica
713B	macrospadix	Costa Rica
714	tenella	Costa Rica
715	matae	Costa Rica
716A	amabilis	Costa Rica
716B	amabilis	Costa Rica
717	deckeriana	Costa Rica
718	deckeriana	Costa Rica
719	zamorae	Costa Rica
720A	scheryi	Costa Rica
720B	scheryi	Costa Rica
722	warscewiczii	Costa Rica
723	pinnatifrons	Costa Rica
724	tepejilote	Panama
725A	warscewiczii	Panama
725B	warscewiczii	Panama
726A	palmeriana	Panama
726B	palmeriana	Panama
727A	scheryi	Panama
727B	scheryi	Panama
728	pinnatifrons	Panama
729A	pittieri	Panama
729B	pittieri	Panama
730	pinnatifrons	Panama
731A	costaricana	Panama
731B	costaricana	Panama
732	costaricana	Panama
733	pinnatifrons	Panama
734	pinnatifrons	Panama
735	pinnatifrons	Panama
736A	costaricana	Panama
736B	costaricana	Panama
737	robertii	Panama
738	scheryi	Panama
739	amabilis	Panama
740A	sullivaniorum	Panama
740B	sullivaniorum	Panama
741A	amabilis	Panama
741B	amabilis	Panama
742	allenii	Panama
743	lucidifrons	Panama
744	warscewiczii	Panama
745	serpens	Panama
746	guntheriana	Panama
746Abis	guntheriana	Cult., USA, California
746Bbis	guntheriana	Cult., USA, California
747	costaricana	Panama
748	costaricana	Cult., USA, California
749	oblongata	Cult., USA, California
750	tuerckheimii	Cult., USA, California
751	alternans	Cult., USA, California
752	pochutlensis	Cult., USA, California
753	pochutlensis	Cult., USA, California
754	pochutlensis	Cult., USA, California
755A	ernesti-augustii	Cult., USA, California
755B	ernesti-augustii	Cult., USA, California
756A	sartorii	Cult., USA, California
756B	sartorii	Cult., USA, California
757	stolonifera	Cult., USA, California
758	schiedeana	Cult., USA, California
759A	microspadix	Cult., USA, California
759B	microspadix	Cult., USA, California
760	schiedeana	Cult., USA, California
761	tuerckheimii	Cult., USA, California
762	costaricana	Cult., USA, California
763A	schiedeana	Cult., USA, California
763B	schiedeana	Cult., USA, California
764	graminifolia	Cult., USA, California
765A	oreophila	Cult., USA, California
765B	oreophila	Cult., USA, California
766	tepejilote	Cult., USA, California
767	elegans	Cult., USA, California
768	oblongata	Cult., USA, California
769A	allenii	Cult., USA, California
769B	allenii	Cult., USA, California
770	neurochlamys	Cult., USA, California
771	stolonifera	Cult., USA, California

772	hooperiana	Cult., USA, California
773	pochutlensis	Cult., USA, California
774	costaricana	Cult., USA, California
775	neurochlamys	Cult., USA, California
776	crucensis	Cult., USA, California
777A	costaricana	Cult., USA, California
777B	costaricana	Cult., USA, California
778	neurochlamys	Cult., USA, California
779	pochutlensis	Cult., USA, California
780	deckeriana	Cult., USA, California
781	schiedeana	Cult., USA, California
782	oblongata	Cult., USA, California
783	quezalteca	Cult., USA, California
784	graminifolia	Cult., USA, California
785	aff. carchensis	Cult., USA, California
786A	cataractarum	Cult., USA, California
786B	cataractarum	Cult., USA, California
787	elegans	Cult., USA, California
788	pochutlensis	Cult., USA, California
789	pochutlensis	Cult., USA, California
790A	schiedeana	Cult., USA, California
790B	schiedeana	Cult., USA, California
791	schiedeana	Cult., USA, California
792A	geonomiformis	Cult., USA, California
792B	geonomiformis	Cult., USA, California
793	radicalis	Cult., USA, California
794A	seifrizii	Cult., USA, California
794B	seifrizii	Cult., USA, California
795	seifrizii	Cult., USA, California
796	glaucifolia	Cult., USA, California
797	deckeriana	Cult., USA, California
798A	oblongata	Cult., USA, California
798B	oblongata	Cult., USA, California
799A	metallica	Cult., USA, California
799B	metallica	Cult., USA, California
800	quezalteca	Cult., USA, California
801	tepejilote	Cult., USA, California
802A	klotzschiana	Cult., USA, California
802B	klotzschiana	Cult., USA, California
803	arenbergiana	Cult., USA, California
804	hooperiana	Cult., USA, California
805	elatior	Cult., USA, California
806	tepejilote	Cult., USA, California
807	glaucifolia	Cult., USA, California
808	metallica	Cult., USA, California
809	radicalis	Cult., USA, California
810	pinnatifrons	Cult., USA, California
811A	elatior	Cult., USA, California
811B	elatior	Cult., USA, California
812	pochutlensis	Cult., USA, California
813	pochutlensis	Cult., USA, California
814	quezalteca	Cult., USA, California
815	pinnatifrons	Cult., USA, California
816	pinnatifrons	Cult., USA, California
817	metallica	Cult., USA, California
818	fragrans	Cult., USA, California
819	glaucifolia	Cult., USA, California
820	radicalis	Cult., USA, California
821	seifrizii	Cult., USA, California
822	fragrans	Cult., USA, California
823	geonomiformis	Cult., USA, California
824A	stenocarpa	Cult., USA, California
824B	stenocarpa	Cult., USA, California
825	amabilis	Cult., USA, California
826	pumila	Cult., USA, California
827	brachypoda	Cult., USA, California
828	robertii	Cult., USA, California
830A	zamorae	Cult., USA, Hawaii
830B	zamorae	Cult., USA, Hawaii
831	linearis	Cult., USA, Hawaii
832	warscewiczii	Cult., USA, Hawaii
833	stenocarpa	Cult., USA, Hawaii
834	arenbergiana	Cult., USA, Hawaii
835	Heidi Bornhorst	Cult., USA, Hawaii
840	ernesti-augustii	Belize
841	oblongata	Belize
842A	neurochlamys	Belize
842B	neurochlamys	Belize
843	tepejilote	Belize
844	tepejilote	Guatemala
845	elegans	Guatemala
846	oblongata	Guatemala
847	seifrizii	Guatemala
848	ernesti-augustii	Guatemala
849	elegans	Guatemala
850	seifrizii	Guatemala
851	tuerckheimii	Cult., USA, California
852	whitelockiana	Cult., USA, California
853	whitelockiana	Cult., USA, California
854	sullivaniorum	Cult., USA, California
855	sullivaniorum	Cult., USA, California
856	guntheriana	Cult., USA, California
857	robertii	Cult., USA, California
858	tenella	Cult., USA, California
859	tuerckheimii	Cult., USA, California
860	tuerckheimii	Cult., USA, California
861	brachypoda	Cult., USA, California
862	elatior	Cult., USA, California
863	tenella	Cult., USA, California
864	stolonifera	Cult., USA, California
865	brachyclada	Cult., USA, California
866	fragrans	Cult., USA, California
867	quezalteca	Cult., Guatemala
868A	castillo-montii	Guatemala
868B	castillo-montii	Guatemala
869	nationsiana	Guatemala
870	stenocarpa	Guatemala
871	nationsiana	Guatemala
872	geonomiformis	Guatemala
873	nationsiana	Guatemala
874	geonomiformis	Guatemala
875	geonomiformis	Guatemala
876	oblongata	Guatemala
877	geonomiformis	Guatemala
878	nationsiana	Guatemala
879	ernesti-augustii	Guatemala
880	nationsiana	Guatemala
881	brachypoda	Guatemala
882	oblongata	Guatemala
883	tenerrima	Guatemala
884	nubium	Guatemala
885	verapazensis	Guatemala
886	pinnatifrons	Guatemala
887A	simplex	Guatemala

887B	simplex	Guatemala
888A	carchensis	Guatemala
888B	carchensis	Guatemala
889	pinnatifrons	Guatemala
890	verapazensis	Guatemala
891	parvisecta	Guatemala
892	parvisecta	Guatemala
893	verapazensis	Guatemala
894	tenerrima	Guatemala
895	quezalteca	Guatemala
896	parvisecta	Guatemala
897	lehmannii	Guatemala
898A	verapazensis	Guatemala
898B	verapazensis	Guatemala
899	nubium	Guatemala
900	pachecoana	Guatemala
901	quezalteca	Guatemala
902	pinnatifrons	Guatemala
903	rojasiana	Guatemala
904	vulgata	Guatemala
905	fractiflexa	Guatemala
906	rojasiana	Guatemala
907	volcanensis	Guatemala
908A	pachecoana	Guatemala
908B	pachecoana	Guatemala
909	pinnatifrons	Guatemala
910	rojasiana	Guatemala
911	whitelockiana	Guatemala
912A	fractiflexa	Guatemala
912B	fractiflexa	Guatemala
913A	volcanensis	Guatemala
913B	volcanensis	Guatemala
914	elatior	Guatemala
915	rojasiana	Guatemala
916A	stricta	Guatemala
916B	stricta	Guatemala
917	pinnatifrons	Mexico
918A	alternans	Mexico
918B	alternans	Mexico
919	tenella	Mexico
920A	ernesti-augustii	Mexico
920B	ernesti-augustii	Mexico
921	elatior	Mexico
922	hooperiana	Mexico
923	cataractarum	Mexico
924	nubium	Mexico
925	rojasiana	Mexico
926	leibmannii	Mexico
927	rojasiana	Mexico
928	glaucifolia	Mexico
929A	arenbergiana	Mexico
929B	arenbergiana	Mexico
930	tenella	Mexico
931	ernesti-augustii	Mexico
932	pinnatifrons	Mexico
933A	stolonifera	Mexico
933B	stolonifera	Mexico
934A	pochutlensis	Mexico
934B	pochutlensis	Mexico
935	whitelockiana	Mexico
936	rhizomatosa	Mexico
937	foveata	Mexico
938	nubium	Mexico
939A	foveata	Mexico
939B	foveata	Mexico
940	pochutlensis	Mexico
941	rigida	Mexico
942	liebmannii	Mexico
943A	queroana	Mexico
943B	queroana	Mexico
944	liebmannii	Mexico
945	woodsoniana	Mexico
946	elegans	Mexico
947	sartorii	Mexico
948	tepejilote	Mexico
949	elatior	Mexico
950	metallica	Mexico
951	oblongata	Mexico
952A	schiedeana	Mexico
952B	schiedeana	Mexico
953A	oreophila	Mexico
953B	oreophila	Mexico
954A	dammeriana	Costa Rica
954B	dammeriana	Costa Rica
955	pinnatifrons	Costa Rica
956	dammeriana	Costa Rica
957	macrospadix	Costa Rica
963	pinnatifrons	Costa Rica
964	pinnatifrons	Costa Rica
965A	parvifolia	Costa Rica
965B	parvifolia	Costa Rica
966	pinnatifrons	Costa Rica
967	pinnatifrons	Costa Rica
968	warscewiczii	Costa Rica
969	pumila	Costa Rica
970	palmeriana	Costa Rica
971	robertii	Costa Rica
972	palmeriana	Costa Rica
973	pinnatifrons	Costa Rica
974	pumila	Costa Rica
975	robertii	Costa Rica
976	robertii	Costa Rica
977	crucensis	Costa Rica
978A	costaricana	Costa Rica
978B	costaricana	Costa Rica
979	pinnatifrons	Costa Rica
980	pinnatifrons	Costa Rica
981	pumila	Costa Rica
982	crucensis	Costa Rica
983	pinnatifrons	Cult., USA, California
984A	rojasiana	Guatemala
984B	rojasiana	Guatemala
985A	pachecoana	Guatemala
985B	pachecoana	Guatemala
986	fractiflexa	Guatemala
987	rojasiana	Guatemala
988A	whitelockiana	Guatemala
988B	whitelockiana	Guatemala
989A	pachecoana	Guatemala
989B	pachecoana	Guatemala
990	pinnatifrons	Guatemala
991	rojasiana	Guatemala
992	volcanensis	Guatemala
993	pachecoana	Guatemala

994	volcanensis	Guatemala	1104	arenbergiana	Guatemala
995	stricta	Guatemala	1105	lehmannii	Guatemala
996	arenbergiana	Guatemala	1106	verapazensis	Guatemala
997	quezalteca	Guatemala	1107A	woodsoniana	Guatemala
998	quezalteca	Guatemala	1107B	woodsoniana	Guatemala
999	verapazensis	Guatemala	1108A	sullivaniorum	Panama
1000	parvisecta	Guatemala	1108B	sullivaniorum	Panama
1001	parvisecta	Guatemala	1109A	amabilis	Panama
1002	carchensis	Guatemala	1109B	amabilis	Panama
1003A	simplex	Guatemala	1110	lucidifrons	Panama
1003B	simplex	Guatemala	1111	serpens	Panama
1004A	liebmannii	Guatemala	1112	woodsoniana	Panama
1004B	liebmannii	Guatemala	1113A	correae	Panama
1005	arenbergiana	Guatemala	1113B	correae	Panama
1006	neurochlamys	Guatemala	1114	deneversiana	Panama
1007	castillo-montii	Guatemala	1115	robertii	Panama
1008A	simplex	Guatemala	1116	sullivaniorum	Panama
1008B	simplex	Guatemala	1117A	pinnatifrons	Panama
1009	tenerrima	Guatemala	1117B	pinnatifrons	Panama
1010	tuerckheimii	Guatemala	1118	correae	Panama
1011	arenbergiana	Guatemala	1119A	palmeriana	Panama
1012	tenerrima	Guatemala	1119B	palmeriana	Panama
1013	lehmannii	Guatemala	1120A	robertii	Panama
1014	nubium	Guatemala	1120B	robertii	Panama
1015	arenbergiana	Guatemala	1121	scheryi	Panama
1016	castillo-montii	Guatemala	1122	verecunda	Panama
1017	nationsiana	Guatemala	1125A	pittieri	Panama
1018	ernesti-augustii	Guatemala	1125B	pittieri	Panama
1019	castillo-montii	Guatemala	1126A	microphylla	Panama
1020	tepejilote	Guatemala	1126B	microphylla	Panama
1021A	nationsiana	Guatemala	1127	guntheriana	Panama
1021B	nationsiana	Guatemala	1128A	graminifolia	Belize
1022A	stenocarpa	Guatemala	1128B	graminifolia	Belize
1022B	stenocarpa	Guatemala	1129A	graminifolia	Belize
1023A	geonomiformis	Guatemala	1129B	graminifolia	Belize
1023B	geonomiformis	Guatemala	1131A	seifrizii	Belize
1100A	geonomiformis	Guatemala	1131B	seifrizii	Belize
1100B	geonomiformis	Guatemala	1132A	graminifolia	Belize
1101A	adscendens	Guatemala	1132B	graminifolia	Belize
1101B	adscendens	Guatemala	1133A	seifrizii	Cult., Belize
1102A	parvisecta	Guatemala	1133B	seifrizii	Cult., Belize
1102B	parvisecta	Guatemala	1134	elegans	Cult., Belize
1103A	quezalteca	Guatemala	1138A	pauciflora	Peru
1103B	quezalteca	Guatemala	1138B	pauciflora	Peru

APPENDIX D

GEOGRAPHICAL LISTING OF SPECIES OF *CHAMAEDOREA* BY COUNTRY

Mexico

alternans
arenbergiana
atrovirens
carchensis
cataractarum
elatior
elegans
ernesti-augustii
ferruginea
foveata
fractiflexa
geonomiformis?
glaucifolia
graminifolia
hooperiana
klotzschiana
liebmannii
metallica
microspadix
neurochlamys
nubium
oblongata
oreophila
paradoxa
parvisecta
pinnatifrons
pochutlensis

queroana
quezalteca
radicalis
rhizomatosa
rigida
rojasiana
sartorii
schiedeana
seifrizii
simplex
stolonifera
stricta
tenella
tepejilote
tuerckheimii
vulgata
whitelockiana
woodsoniana

Belize
adscendens
elegans
ernesti-augustii
geonomiformis
graminifolia
neurochlamys
oblongata
seifrizii
tepejilote

Guatemala
adscendens
arenbergiana
brachypoda
carchensis
castillo-montii
elatior
elegans
ernesti-augustii
falcifera
fractiflexa
geonomiformis
graminifolia
lehmannii
liebmannii
nationsiana
neurochlamys
nubium
oblongata
pachecoana
parvisecta
pinnatifrons
quezalteca
rojasiana
seifrizii
simplex
skutchii
stenocarpa
stricta
tenerrima
tepejilote
tuerckheimii
verapazensis
volcanensis
vulgata
whitelockiana
woodsoniana

Honduras
arenbergiana
brachypoda
costaricana
donnell-smithii
ernesti-augustii
geonomiformis
neurochlamys
oblongata
nubium
pinnatifrons
quezalteca
sartorii
seifrizii
tepejilote
woodsoniana

El Salvador
arenbergiana?
nubium
pinnatifrons
quezalteca
tepejilote

Nicaragua
arenbergiana?
costaricana
oblongata
pinnatifrons
selvae
tepejilote
woodsoniana

Costa Rica
amabilis
arenbergiana?
brachyclada
chazdoniae
costaricana
crucensis
dammeriana
deckeriana
deneversiana?
graminifolia?
macrospadix
matae
minima
palmeriana
parvifolia
pedunculata
pinnatifrons
pittieri
pumila
pygmaea
robertii
scheryi
selvae
stenocarpa
stricta
sullivaniorum
tenella
tepejilote
undulatifolia
warscewiczii
woodsoniana
zamorae

Panama
allenii
amabilis
arenbergiana?
brachyclada
correae
costaricana
dammeriana
deckeriana
deneversiana
guntheriana
lucidifrons
matae
microphylla
palmeriana
pinnatifrons
pittieri
pygmaea
robertii
scheryi
serpens
stenocarpa
stricta
sullivaniorum
tepejilote
verecunda
warscewiczii
woodsoniana

Colombia
allenii
amabilis
arenbergiana?
costaricana?
geonomiformis?
latisecta
linearis
lucidifrons?
murriensis
pauciflora
pinnatifrons
pygmaea
simplex?
sullivaniorum
tepejilote
woodsoniana?

Venezuela
linearis
pinnatifrons

Peru
angustisecta
fragrans
linearis
pauciflora
pinnatifrons
smithii

Ecuador
deneversiana
linearis
pauciflora
pinnatifrons

Brazil
angustisecta
pauciflora
pinnatifrons

Bolivia
angustisecta
linearis
pinnatifrons

APPENDIX E

VEGETATIVE KEY TO THE CULTIVATED SPECIES OF *CHAMAEDOREA*

Leaves bifid or entire but not pinnate.
  Solitary habit.
    Leaf blades entire. .... **74. C. tuerckheimii**
    Leaf blades bifid.
      Stem creeping, short, nearly lacking.
        Leaf blades with 7 or fewer primary nerves on each side of rachis. .... **51. C. minima**
        Leaf blades with 9 or more primary nerves on each side of rachis.
          Leaf blades with 9–13 primary nerves on each side of rachis. .... **59. C. pumila**
          Leaf blades with 15–16 primary nerves on each side of rachis. .... **72. C. sullivaniorum**
      Stem elongate, visible.
        Leaves with metallic-blue sheen. .... **3. C. metallica**
        Leaves without metallic-blue sheen.
          Leaf blades obscurely nerved.
            Leaf blades less than 15 cm long, bifid apically to ¼ their length. .... **32. C. tenella**
            Leaf blades more than 15 cm long, bifid apically ⅓–½ their length. .... **13. C. geonomiformis**
          Leaf blades conspicuously nerved.
            Leaf blades with margins distinctly serrate, bifid apically to less than ⅓ their length. .... **9. C. amabilis**
            Leaf blades with margins not distinctly serrate, bifid apically ⅓ or more their length.
              Leaf blades thick, 12–18 primary nerves on each side of rachis. .... **2. C. ernesti-augustii**
              Leaf blades ± thin, less than 10 or 20 or more primary nerves on each side of rachis.
                Leaf blades with 20 or more primary nerves on each side of rachis. .... **90. C. deckeriana**
                Leaf blades with 10 primary nerves on each side of rachis. .... **27. C. pinnatifrons**
  Cespitose habit.
    Stems spreading and multiplying by stolons or rhizomes.
      Stems with stolons. .... **5. C. stolonifera**
      Stems with rhizomes. .... **37. C. brachypoda**
    Stems spreading and multiplying by lateral, basal shoots, not by stolons or rhizomes.
      Leaf blades bifid apically to more than ½ their length, lobes narrowly divergent. .... **47. C. fragrans**
      Leaf blades bifid apically to no more than ½ their length, lobes broadly divergent. .... **52. C. nubium**
Leaves pinnate.
  Cespitose habit.
    Stems very short, creeping, dichotomously branched. .... **88. C. cataractarum**
    Stems elongate, not creeping or dichotomously branched.
      Stems climbing, vining; terminal pinnae reflexed, hooklike. .... **10. C. elatior**
      Stems not climbing, vining; terminal pinnae not reflexed, hooklike.
        Leaf sheaths with ligules at apex.
          Ligules long-lanceolate, to 5 cm or more long, brown, fibrous. .... **62. C. quezalteca**
          Ligules triangular, short, less than 3 cm long, green with brown margin. .... **42. C. costaricana**
        Leaf sheaths lacking ligules at apex.
          Pinnae with several prominent primary nerves, these keeled above, drying yellow. .... **93. C. tepejilote**
          Pinnae with a dominant midrib, other nerves much less conspicuous, not keeled above and drying yellow.
            Pinnae 16 or fewer on each side of rachis.
              Pinnae straight. .... **67. C. seifrizii**
              Pinnae sigmoid. .... **81. C. microspadix**
            Pinnae 20 or more on each side of rachis.
              Pinnae narrowly linear, gray-green, drying strongly plicate; stems rhizomatous in open clusters. .... **15. C. graminifolia**
              Pinnae lanceolate or long-lanceolate, green, not drying strongly plicate; multiple stems in dense clusters.
                Leaf sheaths drying brownish, thick, durable, nearly woody, persistent throughout stem; new shoots emerging from tops of sheaths; pinnae thick. .... **16. C. hooperiana**
                Leaf sheaths drying pale or whitish, not thick, durable or woody, persistent only near tops of stems; new shoots emerging from base of bare stems; pinnae not thick. .... **58. C. pochutlensis**
  Solitary habit.
    Plants appearing stemless.
      Pinnae not decurrent on rachis.
        Pinnae lanceolate, green, smooth textured, with dominant midrib. .... **20. C. macrospadix**
        Pinnae linear, dark green, rough textured, lacking a dominant midrib. .... **82. C. radicalis**

Pinnae decurrent on rachis.
Leaves with 20 or more linear-lanceolate, ± straight pinnae on each side of rachis. .... **36. C. brachyclada**
Leaves with fewer than 20, lanceolate, ± sigmoid pinnae on each side of rachis.
Small plants; leaves less than 70 cm long, margins of pinnae not undulate; pistillate rachillae spicate or furcate, straight. .... **70. C. stenocarpa**
Larger plants; leaves 90 cm long or more, margins of pinnae undulate; pistillate rachillae usually 3–5, strongly recurved and hooklike. .... **75. C. undulatifolia**
Plants with visible elongate stems.
Stem climbing, vining; terminal pinnae reflexed, hooklike. .... **10. C. elatior**
Stem not climbing or vining; terminal pinnae not reflexed, hooklike.
Leaves with metallic-blue sheen. .... **3. C. metallica**
Leaves green without metallic-blue sheen.
Leaf sheaths and petioles green- and white-mottled. .... **83. C. angustisecta**
Leaf sheaths and petioles not green- and white-mottled.
Pinnae clustered in groups along rachis.
Pinnae long-linear, gray-green. .... **14. C. glaucifolia**
Pinnae lanceolate, green. .... **17. C. klotzschiana**
Pinnae not clustered in groups along rachis.
Leaf sheaths conspicuously whitish and green-striate-nerved apically.
Pinnae with 3 prominent nerves above. .... **23. C. neurochlamys**
Pinnae with 7–8 prominent nerves above. .... **86. C. alternans**
Leaf sheaths ± green throughout or if whitish apically then only briefly so.
Leaf sheaths long-open, tubular only near base.
Leaf sheaths thick, durable, densely and prominently striated; pinnae thick, lacking a dominant midrib. .... **82. C. radicalis**
Leaf sheaths not thickened, durable, densely and prominently striated; pinnae with dominant midrib.
Pinnae thin. .... **1. C. elegans**
Pinnae thick. .... **20. C. macrospadix**
Leaf sheaths ± tubular throughout.
Pinnae ± straight or only falcate.
Terminal pinnae as broad or broader than all others combined. .... **94. C. zamorae**
Terminal pinnae same size or only slightly larger than other individual pinnae.
Pinnae with nerves angled and prominently raised above, ± smooth below. .... **7. C. linearis**
Pinnae with nerves smooth above, angled and prominently raised below.
Pinnae with 5 prominent nerves; petiole rough; stem and sheath green. .... **80. C. woodsoniana**
Pinnae with a dominant midrib, other nerves not prominent; petiole smooth; stem and sheath conspicuously white-spotted. .... **92. C. oreophila**
Pinnae conspicuously sigmoid.
Pinnae with dominant midrib, other nerves not prominent.
Pinnae thick, leathery. .... **53. C. oblongata**
Pinnae thin-papery. .... **79. C. whitelockiana**
Pinnae lacking dominant midrib, primary nerves ± equally prominent.
Pinnae 12 or more on each side of rachis. .... **93. C. tepejilote**
Pinnae 11 or fewer on each side of rachis.
Pinnae with 3 primary nerves above. .... **4. C. sartorii**
Pinnae with 5 or more primary nerves above.
Nerves elevated below. .... **34. C. warscewiczii**
Nerves not elevated below.
Nerves keeled above.
Leaf sheaths green throughout; generally large plants with pinnae to 60 × 15 cm. **87. C. arenbergiana**
Leaf sheaths pale green or whitish apically; generally small plants with pinnae to 35 × 6.5 cm.
Nerves green above. .... **89. C. crucensis**
Nerves pale above. .... **85. C. allenii**
Nerves not keeled above.
Pinnae 10–14 on each side of rachis. .... **29. C. schiedeana**
Pinnae 2–8 on each side of rachis. .... **27. C. pinnatifrons**

APPENDIX F

INDEX OF NAMES

Names in bold are accepted ones as treated here. Those in italics are synonyms. Number following accepted name at its entry is the page where that species is treated.

Nunnezharia pygmaea (H. A. Wendl.) O. Kuntze = **Chamaedorea pygmaea** H. A. Wendl.
Nunnezharia radicalis (Mart.) O. Kuntze = **Chamaedorea radicalis** Mart.
Nunnezharia regia (Hort.) O. Kuntze = **Chamaedorea elatior** Mart.
Nunnezharia repens (Hort.) O. Kuntze = **Chamaedorea elatior** Mart.
Nunnezharia resinifera (H. A. Wendl.) O. Kuntze = **Chamaedorea elatior** Mart.
Nunnezharia rigida (H. A. Wendl. ex Dammer) O. Kuntze = **Chamaedorea rigida** H. A. Wendl. ex Dammer
Nunnezharia sartorii (Liebm.) O. Kuntze = **Chamaedorea sartorii** Liebm.
Nunnezharia schiedeana (Mart.) O. Kuntze = **Chamaedorea schiedeana** Mart.
Nunnezharia tenella (H. A. Wendl.) Hook. f. = **Chamaedorea tenella** H. A. Wendl.
Nunnezharia tepejilote (Liebm.) O. Kuntze = **Chamaedorea tepejilote** Liebm.
Nunnezharia warscewiczii (H. A. Wendl.) O. Kuntze = **Chamaedorea warscewiczii** H. A. Wendl.
Nunnezharia wendlandiana (Oerst.) O. Kuntze = **Chamaedorea tepejilote** Liebm.
Omanthe costaricana O. F. Cook = **Chamaedorea costaricana** Oerst.
Paranthe violaceae O. F. Cook = **Chamaedorea parvisecta** Burret
Platythea graminea O. F. Cook = **Chamaedorea elatior** Mart.
Spathoscaphe Oerst. = **Chamaedorea** Willd.
Spathoscaphe arenbergiana (H. A. Wendl.) Oerst. = **Chamaedorea arenbergiana** H. A. Wendl.
Stachyophorbe (Liebm. ex Mart.) Liebm. ex Klotzsch = **Chamaedorea** Willd.
Stachyophorbe cataractarum (Mart.) Klotzsch = **Chamaedorea cataractarum** Mart.
Stachyophorbe deckeriana Klotzsch = **Chamaedorea deckeriana** (Klotzsch) Hemsl.
Stachyophorbe filipes O. F. Cook = **Chamaedorea oreophila** Mart.
Stachyophorbe montana Liebm. ex Oerst. = **Chamaedorea oreophila** Mart.
Stachyophorbe oreophila (Mart.) O. F. Cook = **Chamaedorea oreophila** Mart.
Stachyophorbe pygmaea (H. A. Wendl.) Oerst. = **Chamaedorea pygmaea** H. A. Wendl.
Stephanostachys (Klotzsch) Oerst. = **Chamaedorea** Willd.
Stephanostachys casperiana (Klotzsch) Oerst. = **Chamaedorea tepejilote** Liebm.
Stephanostachys martiana (H. A. Wendl.) Oerst. = **Chamaedorea cataractarum** Mart.
Stephanostachys tepejilote (Liebm.) Oerst. = **Chamaedorea tepejilote** Liebm.
Stephanostachys wendlandiana Oerst. = **Chamaedorea tepejilote** Liebm.
Tuerckheimia adscendens Dammer ex J. D. Smith = **Chamaedorea adscendens** (Dammer) Burret
Vadia atrovirens Hort. ex O. F. Cook = **Chamaedorea cataractarum** Mart.
Vadia jotolana O. F. Cook = **Chamaedorea cataractarum** Mart.

EXCLUDED NAMES

The following names apply to species in other genera.

Chamaedorea falcaria L. H. Bailey, Gentes Herb. 6: 254. 1943a. = **Hyospathe lehmannii** Burret
Chamaedorea fenestrata Hort. in H. A. Wendl., Index Palmarum 28. 1854. = **Reinhardtia gracilis** (H. A. Wendl.) Drude ex Dammer
Chamaedorea fibrosa H. A. Wendl., Index Palmarum 57. 1854. = **Synechanthus fibrosus** (H. A. Wendl.) H. A. Wendl.
Chamaedorea ghiesbreghtii Hort. in Kerch., Les Palmiers 239. 1878. = name of uncertain application probably used for material of **Calyptrogyne ghiesbreghtiana** (Linden ex H. A. Wendl.) H. A. Wendl.
Collinia fibrosa (H. A. Wendl.) Oerst., Vidensk. Meddel. Dansk. Naturhist. Foren. Kjoebenhavn 5. 1859. = **Synechanthus fibrosus** (H. A. Wendl.) H. A. Wendl.
Morenia chonta Philippi = **Juania australis** (Mart.) Drude ex Hook. f.